PRAISE FOR
PRINCIPLES AND PRACTICES OF INTERCONNECTION NETWORKS

The scholarship of this book is unparalleled in its area. This text is for interconnection networks what Hennessy and Patterson's text is for computer architecture — an authoritative, one-stop source that clearly and methodically explains the more significant concepts. Treatment of the material both in breadth and in depth is very well done . . . a must read and a slam dunk! — Timothy Mark Pinkston, University of Southern California

[This book is] the most comprehensive and coherent work on modern interconnection networks. As leaders in the field, Dally and Towles capitalize on their vast experience as researchers and engineers to present both the theory behind such networks and the practice of building them. This book is a necessity for anyone studying, analyzing, or designing interconnection networks. — Stephen W. Keckler, The University of Texas at Austin

This book will serve as excellent teaching material, an invaluable research reference, and a very handy supplement for system designers. In addition to documenting and clearly presenting the key research findings, the book's incisive practical treatment is unique. By presenting how actual design constraints impact each facet of interconnection network design, the book deftly ties theoretical findings of the past decades to real systems design. This perspective is critically needed in engineering education. — Li-Shiuan Peh, Princeton University

Principles and Practices of Interconnection Networks is a triple threat: comprehensive, well written and authoritative. The need for this book has grown with the increasing impact of interconnects on computer system performance and cost. It will be a great tool for students and teachers alike, and will clearly help practicing engineers build better networks. — Steve Scott, Cray, Inc.

Dally and Towles use their combined three decades of experience to create a book that elucidates the theory and practice of computer interconnection networks. On one hand, they derive fundamentals and enumerate design alternatives. On the other, they present numerous case studies and are not afraid to give their experienced opinions on current choices and future trends. This book is a "must buy" for those interested in or designing interconnection networks. — Mark Hill, University of Wisconsin, Madison

This book will instantly become a canonical reference in the field of interconnection networks. Professor Dally's pioneering research dramatically and permanently changed this field by introducing rigorous evaluation techniques and creative solutions to the challenge of high-performance computer system communication. This well-organized textbook will benefit both students and experienced practitioners. The presentation and exercises are a result of years of classroom experience in creating this material. All in all, this is a must-have source of information. — Craig Stunkel, IBM

Principles and Practices of Interconnection Networks

Principles and Practices of Interconnection Networks

William James Dally

Brian Towles

AMSTERDAM • BOSTON • HEIDELBERG • LONDON
NEW YORK • OXFORD • PARIS • SAN DIEGO
SAN FRANCISCO • SINGAPORE • SYDNEY • TOKYO

Morgan Kaufmann is an imprint of Elsevier

Publishing Director:	*Diane D. Cerra*
Senior Editor:	*Denise E. M. Penrose*
Publishing Services Manager:	*Simon Crump*
Project Manager:	*Marcy Barnes-Henrie*
Editorial Coordinator:	*Alyson Day*
Editorial Assistant:	*Summer Block*
Cover Design:	*Hannus Design Associates*
Cover Image:	*Frank Stella*, Takht-i-Sulayan-I (1967)
Text Design:	*Rebecca Evans & Associates*
Composition:	*Integra Software Services Pvt., Ltd.*
Copyeditor:	*Catherine Albano*
Proofreader:	*Deborah Prato*
Indexer:	*Sharon Hilgenberg*
Interior printer	*The Maple-Vail Book Manufacturing Group*
Cover printer	*Phoenix Color Corp.*

Morgan Kaufmann Publishers is an imprint of Elsevier
500 Sansome Street, Suite 400, San Francisco, CA 94111

This book is printed on acid-free paper.

Figure 3.10 © 2003 Silicon Graphics, Inc. Used by permission. All rights reserved.

Figure 3.13 courtesy of the Association for Computing Machinery (ACM), from James Laudon and Daniel Lenoski, "The SGI Origin: a ccNUMA highly scalable server," Proceedings of the International Symposium on Computer Architecture (ISCA), pp. 241-251, 1997. (ISBN: 0897919017) Figure 10.

Figure 10.7 from Thinking Machines Corp.

Figure 11.5 courtesy of Ray Mains, Ray Mains Photography, http://www.mauigateway.com/~raymains/.

Designations used by companies to distinguish their products are often claimed as trademarks or registered trademarks. In all instances in which Morgan Kaufmann Publishers is aware of a claim, the product names appear in initial capital or all capital letters. Readers, however, should contact the appropriate companies for more complete information regarding trademarks and registration.

Permissions may be sought directly from Elsevier's Science & Technology Rights Department in Oxford, UK: phone: (+44) 1865 843830, fax: (+44) 1865 853333, e-mail: permissions@elsevier.com.uk. You may also complete your request on-line via the Elsevier homepage (http://elsevier.com) by selecting "Customer Support" and then "Obtaining Permissions."

Library of Congress Cataloging-in-Publication Data

Dally, William J.
    Principles and practices of interconnection networks / William
    Dally, Brian Towles.
        p. cm.
    Includes bibliographical references and index.
    ISBN 0-12-200751-4 (alk. paper)
       1. Computer networks-Design and construction.
    2. Multiprocessors. I. Towles, Brian. II. Title.

    TK5105.5.D3272003
    004.6'5–dc22

ISBN: 0-12-200751-4   2003058915

For information on all Morgan Kaufmann publications,
visit our Web Site at *www.mkp.com*

Printed and bound by CPI Group (UK) Ltd, Croydon, CR0 4YY

Transferred to Digital Print 2011

Contents

Acknowledgments

We are deeply indebted to a large number of people who have contributed to the creation of this book. Timothy Pinkston at USC and Li-Shiuan Peh at Princeton were the first brave souls (other than the authors) to teach courses using drafts of this text. Their comments have greatly improved the quality of the finished book. Mitchell Gusat, Mark Hill, Li-Shiuan Peh, Timothy Pinkston, and Craig Stunkel carefully reviewed drafts of this manuscript and provided invaluable comments that led to numerous improvements.

Many people (mostly designers of the original networks) contributed information to the case studies and verfied their accuracy. Randy Rettberg provided information on the BBN Butterfly and Monarch. Charles Leiserson and Bradley Kuszmaul filled in the details of the Thinking Machines CM-5 network. Craig Stunkel and Bulent Abali provided information on the IBM SP1 and SP2. Information on the Alpha 21364 was provided by Shubu Mukherjee. Steve Scott provided information on the Cray T3E. Greg Thorson provided the pictures of the T3E.

Much of the development of this material has been influenced by the students and staff that have worked with us on interconnection network research projects at Stanford and MIT, including Andrew Chien, Scott Wills, Peter Nuth, Larry Dennison, Mike Noakes, Andrew Chang, Hiromichi Aoki, Rich Lethin, Whay Lee, Li-Shiuan Peh, Jin Namkoong, Arjun Singh, and Amit Gupta.

This material has been developed over the years teaching courses on interconnection networks: 6.845 at MIT and EE482B at Stanford. The students in these classes helped us hone our understanding and presentation of the material. Past TAs for EE482B Li-Shiuan Peh and Kelly Shaw deserve particular thanks.

We have learned much from discussions with colleagues over the years, including Jose Duato (Valencia), Timothy Pinkston (USC), Sudha Yalamanchili (Georgia Tech), Anant Agarwal (MIT), Tom Knight (MIT), Gill Pratt (MIT), Steve Ward (MIT), Chuck Seitz (Myricom), and Shubu Mukherjee (Intel). Our practical understanding of interconnection networks has benefited from industrial collaborations with Justin Rattner (Intel), Dave Dunning (Intel), Steve Oberlin (Cray), Greg Thorson (Cray), Steve Scott (Cray), Burton Smith (Cray), Phil Carvey (BBN and Avici), Larry Dennison (Avici), Allen King (Avici), Derek Chiou (Avici), Gopalkrishna Ramamurthy (Velio), and Ephrem Wu (Velio).

Denise Penrose, Summer Block, and Alyson Day have helped us throughout the project.

We also thank both Catherine Albano and Deborah Prato for careful editing, and our production manager, Marcy Barnes-Henrie, who shepherded the book through the sometimes difficult passage from manuscript through finished product.

Finally, our families: Sharon, Jenny, Katie, and Liza Dally and Herman and Dana Towles offered tremendous support and made significant sacrifices so we could have time to devote to writing.

Preface

Digital electronic systems of all types are rapidly becoming *commmunication limited*. Movement of data, not arithmetic or control logic, is the factor limiting cost, performance, size, and power in these systems. At the same time, buses, long the mainstay of system interconnect, are unable to keep up with increasing performance requirements.

Interconnection networks offer an attractive solution to this communication crisis and are becoming pervasive in digital systems. A well-designed interconnection network makes efficient use of scarce communication resources — providing high-bandwidth, low-latency communication between clients with a minimum of cost and energy.

Historically used only in high-end supercomputers and telecom switches, interconnection networks are now found in digital systems of all sizes and all types. They are used in systems ranging from large supercomputers to small embedded systems-on-a-chip (SoC) and in applications including inter-processor communication, processor-memory interconnect, input/output and storage switches, router fabrics, and to replace dedicated wiring.

Indeed, as system complexity and integration continues to increase, many designers are finding it more efficient to route packets, not wires. Using an interconnection network rather than dedicated wiring allows scarce bandwidth to be shared so it can be used efficiently with a high duty factor. In contrast, dedicated wiring is idle much of the time. Using a network also enforces regular, structured use of communication resources, making systems easier to design, debug, and optimize.

The basic principles of interconnection networks are relatively simple and it is easy to design an interconnection network that efficiently meets all of the requirements of a given application. Unfortunately, if the basic principles are not understood it is also easy to design an interconnection network that works poorly if at all. Experienced engineers have designed networks that have deadlocked, that have performance bottlenecks due to a poor topology choice or routing algorithm, and that realize only a tiny fraction of their peak performance because of poor flow control. These mistakes would have been easy to avoid if the designers had understood a few simple principles.

This book draws on the experience of the authors in designing interconnection networks over a period of more than twenty years. We have designed tens of networks that today form the backbone of high-performance computers (both message-passing and shared-memory), Internet routers, telecom circuit switches, and I/O interconnect. These systems have been designed around a variety of topologies including crossbars, tori, Clos networks, and butterflies. We developed wormhole routing and virtual-channel flow control. In designing these systems and developing these methods we learned many lessons about what works and what doesn't. In this book, we share with you, the reader, the benefit of this experience in the form of a set of simple principles for interconnection network design based on topology, routing, flow control, and router architecture.

Organization

The book starts with two introductory chapters and is then divided into five parts that deal with topology, routing, flow control, router architecture, and performance. A graphical outline of the book showing dependences between sections and chapters is shown in Figure 1. We start in Chapter 1 by describing what interconnection networks are, how they are used, the performance requirements of their different applications, and how design choices of topology, routing, and flow control are made to satisfy these requirements. To make these concepts concrete and to motivate the remainder of the book, Chapter 2 describes a simple interconnection network in detail: from the topology down to the Verilog for each router. The detail of this example demystifies the abstract topics of routing and flow control, and the performance issues with this simple network motivate the more sophisticated methods and design approaches described in the remainder of the book.

The first step in designing an interconnection network is to select a topology that meets the throughput, latency, and cost requirements of the application given a set of packaging constraints. Chapters 3 through 7 explore the topology design space. We start in Chapter 3 by developing topology metrics. A topology's bisection bandwidth and diameter bound its achievable throughput and latency, respectively, and its path diversity determines both performance under adversarial traffic and fault tolerance. Topology is constrained by the available packaging technology and cost requirements with both module pin limitations and system wire bisection governing achievable channel width. In Chapters 4 through 6, we address the performance metrics and packaging constraints of several common topologies: butterflies, tori, and non-blocking networks. Our discussion of topology ends at Chapter 7 with coverage of concentration and topology *slicing*, methods used to handle bursty traffic and to map topologies to packaging modules.

Once a topology is selected, a routing algorithm determines how much of the bisection bandwidth can be converted to system throughput and how closely latency approaches the diameter limit. Chapters 4 through 11 describe the routing problem and a range of solutions. A good routing algorithm load-balances traffic across the channels of a topology to handle adversarial traffic patterns while simultaneously exploiting the locality of benign traffic patterns. We introduce the problem in Chapter 8 by considering routing on a ring network and show that the naive *greedy* algorithm gives poor performance on adversarial traffic. We go on to describe oblivious

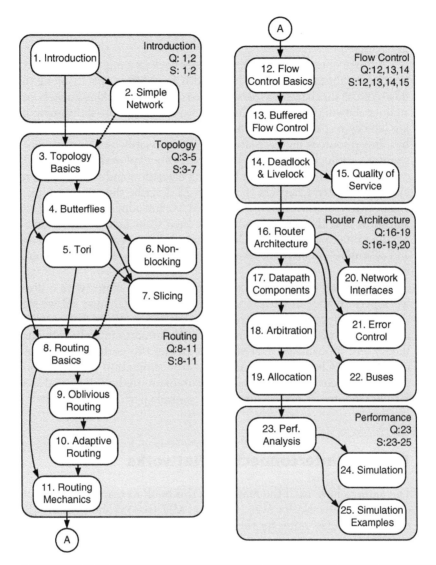

Figure 1   Outline of this book showing dependencies between chapters. Major sections are denoted as shaded areas. Chapters that should be covered in any course on the subject are placed along the left side of the shaded areas. Optional chapters are placed to the right. Dependences are indicated by arrows. A solid arrow implies that the chapter at the tail of the arrow must be understood to understand the chapter at the head of the arrow. A dotted arrow indicates that it is helpful, but not required, to understand the chapter at the tail of the arrow before the chapter at the head. The notation in each shaded area recommends which chapters to cover in a quarter course (Q) and a semester course (S).

routing algorithms in Chapter 9 and adaptive routing algorithms in Chapter 10. The routing portion of the book then concludes with a discussion of routing mechanics in Chapter 11.

A flow-control mechanism sequences packets along the path from source to destination by allocating channel bandwidth and buffer capacity along the way. A good flow-control mechanism avoids idling resources or blocking packets on resource constraints, allowing it to realize a large fraction of the potential throughput and minimizing latency respectively. A bad flow-control mechanism may squander throughput by idling resources, increase latency by unnecessarily blocking packets, and may even result in deadlock or livelock. These topics are explored in Chapters 12 through 15.

The policies embedded in a routing algorithm and flow-control method are realized in a router. Chapters 16 through 22 describe the microarchitecture of routers and network interfaces. In these chapters, we introduce the building blocks of routers and show how they are composed. We then show how a router can be pipelined to handle a flit or packet each cycle. Special attention is given to problems of *arbitration* and *allocation* in Chapters 18 and 19 because these functions are critical to router performance.

To bring all of these topics together, the book closes with a discussion of network performance in Chapters 23 through 25. In Chapter 23 we start by defining the basic performance measures and point out a number of common pitfalls that can result in misleading measurements. We go on to introduce the use of queueing theory and probablistic analysis in predicting the performance of interconnection networks. In Chapter 24 we describe how simulation is used to predict network performance covering workloads, measurement methodology, and simulator design. Finally, Chapter 25 gives a number of example performance results.

Teaching Interconnection Networks

The authors have used the material in this book to teach graduate courses on interconnection networks for over 10 years at MIT (6.845) and Stanford (EE482b). Over the years the class notes for these courses have evolved and been refined. The result is this book.

A one quarter or one semester course on interconnection networks can follow the outline of this book, as indicated in Figure 1. An individual instructor can add or delete the optional chapters (shown to the right side of the shaded area) to tailor the course to their own needs.

One schedule for a one-quarter course using this book is shown in Table 1. Each lecture corresponds roughly to one chapter of the book. A semester course can start with this same basic outline and add additional material from the optional chapters.

In teaching a graduate interconnections network course using this book, we typically assign a research or design project (in addition to assigning selected exercises from each chapter). A typical project involves designing an interconnection network (or a component of a network) given a set of constraints, and comparing the performance of alternative designs. The design project brings the course material together

Table 1 One schedule for a ten-week quarter course on interconnection networks. Each chapter covered corresponds roughly to one lecture. In week 3, Chapter 6 through Section 6.3.1 is covered.

Week	Topic	Chapters
1	Introduction	1, 2
2	Topology	3, 4
3	Topology	5, (6)
4	Routing	8, 9
5	Routing	10, 11
6	Flow Control	12, 13, 14
7	Router Architecture	16, 17
8	Arbitration & Allocation	18, 19
9	Performance	23
10	Review	

for students. They see the interplay of the different aspects of interconnection network design and get to apply the principles they have learned first hand.

Teaching materials for a one quarter course using this book (Stanford EE482b) are available on-line at http://cva.stanford.edu/ee482b. This page also includes example projects and student papers from the last several offerings of this course.

About the Authors

Bill Dally received his B.S. in electrical engineering from Virginia Polytechnic Institute, an M.S. in electrical engineering from Stanford University, and a Ph.D. in computer science from Caltech. Bill and his group have developed system architecture, network architecture, signaling, routing, and synchronization technology that can be found in most large parallel computers today. While at Bell Telephone Laboratories, Bill contributed to the design of the BELLMAC32 microprocessor and designed the MARS hardware accelerator. At Caltech he designed the MOSSIM Simulation Engine and the Torus Routing Chip, which pioneered wormhole routing and virtual-channel flow control. While a Professor of Electrical Engineering and Computer Science at the Massachusetts Institute of Technology, his group built the J-Machine and the M-Machine, experimental parallel computer systems that pioneered the separation of mechanisms from programming models and demonstrated very low overhead synchronization and communication mechanisms. Bill is currently a professor of electrical engineering and computer science at Stanford University. His group at Stanford has developed the Imagine processor, which introduced the concepts of stream processing and partitioned register organizations. Bill has worked with Cray Research and Intel to incorporate many of these innovations in commercial parallel computers. He has also worked with Avici Systems to incorporate this technology into Internet routers, and co-founded Velio Communications to commercialize high-speed signaling technology. He is a fellow of the IEEE, a fellow of the ACM, and has received numerous honors including the ACM Maurice Wilkes award. He currently leads projects on high-speed signaling, computer architecture, and network architecture. He has published more than 150 papers in these areas and is an author of the textbook *Digital Systems Engineering* (Cambridge University Press, 1998).

Brian Towles received a B.CmpE in computer engineering from the Georgia Institute of Technology in 1999 and an M.S. in electrical engineering from Stanford University in 2002. He is currently working toward a Ph.D. in electrical engineering at Stanford University. His research interests include interconnection networks, network algorithms, and parallel computer architecture.

CHAPTER 1

Introduction to Interconnection Networks

Digital systems are pervasive in modern society. Digital computers are used for tasks ranging from simulating physical systems to managing large databases to preparing documents. Digital communication systems relay telephone calls, video signals, and Internet data. Audio and video entertainment is increasingly being delivered and processed in digital form. Finally, almost all products from automobiles to home appliances are digitally controlled.

A digital system is composed of three basic building blocks: logic, memory, and communication. Logic transforms and combines data — for example, by performing arithmetic operations or making decisions. Memory stores data for later retrieval, moving it in time. Communication moves data from one location to another. This book deals with the communication component of digital systems. Specifically, it explores *interconnection networks* that are used to transport data between the subsystems of a digital system.

The performance of most digital systems today is limited by their communication or interconnection, not by their logic or memory. In a high-end system, most of the power is used to drive wires and most of the clock cycle is spent on wire delay, not gate delay. As technology improves, memories and processors become small, fast, and inexpensive. The speed of light, however, remains unchanged. The pin density and wiring density that govern interconnections between system components are scaling at a slower rate than the components themselves. Also, the frequency of communication between components is lagging far beyond the clock rates of modern processors. These factors combine to make interconnection the key factor in the success of future digital systems.

As designers strive to make more efficient use of scarce interconnection bandwidth, interconnection networks are emerging as a nearly universal solution to the system-level communication problems for modern digital systems. Originally developed for the demanding communication requirements of multicomputers, interconnection networks are beginning to replace buses as the standard system-level interconnection. They are also replacing dedicated wiring in special-purpose systems as designers discover that routing packets is both faster and more economical than routing wires.

1.1 Three Questions About Interconnection Networks

Before going any further, we will answer some basic questions about interconnection networks: What is an interconnection network? Where do you find them? Why are they important?

What is an interconnection network? As illustrated in Figure 1.1, an interconnection network is a programmable system that transports data between terminals. The figure shows six terminals, T1 through T6, connected to a network. When terminal T3 wishes to communicate some data with terminal T5, it sends a *message* containing the data into the network and the network delivers the message to T5. The network is programmable in the sense that it makes different connections at different points in time. The network in the figure may deliver a message from T3 to T5 in one cycle and then use the same resources to deliver a message from T3 to T1 in the next cycle. The network is a system because it is composed of many components: buffers, channels, switches, and controls that work together to deliver data.

Networks meeting this broad definition occur at many scales. On-chip networks may deliver data between memory arrays, registers, and arithmetic units within a single processor. Board-level and system-level networks tie processors to memories or input ports to output ports. Finally, local-area and wide-area networks connect disparate systems together within an enterprise or across the globe. In this book, we restrict our attention to the smaller scales: from chip-level to system level. Many excellent texts already exist addressing the larger-scale networks. However, the issues at the system level and below, where channels are short and the data rates very

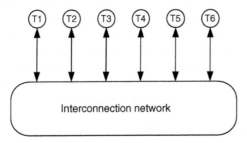

Figure 1.1 Functional view of an interconnection network. *Terminals* (labeled T1 through T6) are connected to the network using *channels*. The arrowheads on each end of the channel indicate it is *bidirectional*, supporting movement of data both into and out of the interconnection network.

high, are fundamentally different than at the large scales and demand different solutions

Where do you find interconnection networks? They are used in almost all digital systems that are large enough to have two components to connect. The most common applications of interconnection networks are in computer systems and communication switches. In computer systems, they connect processors to memories and input/output (I/O) devices to I/O controllers. They connect input ports to output ports in communication switches and network routers. They also connect sensors and actuators to processors in control systems. Anywhere that bits are transported between two components of a system, an interconnection network is likely to be found.

As recently as the late 1980s, most of these applications were served by a very simple interconnection network: the multi-drop bus. If this book had been written then, it would probably be a book on bus design. We devote Chapter 22 to buses, as they are still important in many applications. Today, however, all high-performance interconnections are performed by point-to-point interconnection networks rather than buses, and more systems that have historically been bus-based switch to networks every year. This trend is due to non-uniform performance scaling. The demand for interconnection performance is increasing with processor performance (at a rate of 50% per year) and network bandwidth. Wires, on the other hand, aren't getting any faster. The speed of light and the attenuation of a 24-gauge copper wire do not improve with better semiconductor technology. As a result, buses have been unable to keep up with the bandwidth demand, and point-to-point interconnection networks, which both operate faster than buses and offer concurrency, are rapidly taking over.

Why are interconnection networks important? Because they are a limiting factor in the performance of many systems. The interconnection network between processor and memory largely determines the memory latency and memory bandwidth, two key performance factors, in a computer system.[1] The performance of the interconnection network (sometimes called the *fabric* in this context) in a communication switch largely determines the capacity (data rate and number of ports) of the switch. Because the demand for interconnection has grown more rapidly than the capability of the underlying wires, interconnection has become a critical bottleneck in most systems.

Interconnection networks are an attractive alternative to dedicated wiring because they allow scarce wiring resources to be shared by several low-duty-factor signals. In Figure 1.1, suppose each terminal needs to communicate one word with each other terminal once every 100 cycles. We could provide a dedicated word-wide channel between each pair of terminals, requiring a total of 30 unidirectional channels. However, each channel would be idle 99% of the time. If, instead, we connect the 6 terminals in a ring, only 6 channels are needed. (T1 connects to T2, T2 to T3, and so on, ending with a connection from T6 to T1.) With the ring network,

---

1. This is particularly true when one takes into account that most of the access time of a modern memory chip is communication delay.

the number of channels is reduced by a factor of five and the channel duty factor is increased from 1% to 12.5%.

1.2 Uses of Interconnection Networks

To understand the requirements placed on the design of interconnection networks, it is useful to examine how they are used in digital systems. In this section we examine three common uses of interconnection networks and see how these applications drive network requirements. Specifically, for each application, we will examine how the application determines the following network parameters:

1. The number of terminals
2. The *peak bandwidth* of each terminal
3. The *average bandwidth* of each terminal
4. The required *latency*
5. The *message size* or a distribution of message sizes
6. The *traffic pattern(s)* expected
7. The required *quality of service*
8. The required reliability and availability of the interconnection network

We have already seen that the number of terminals, or *ports*, in a network corresponds to the number of components that must be connected to the network. In addition to knowing the number of terminals, the designer also needs to know how the terminals will interact with the network.

Each terminal will require a certain amount of *bandwidth* from the network, usually expressed in bits per second (bit/s). Unless stated otherwise, we assume the terminal bandwidths are *symmetric* — that is, the input and output bandwidths of the terminal are equal. The *peak bandwidth* is the maximum data rate that a terminal will request from the network over a short period of time, whereas the *average bandwidth* is the average date rate that a terminal will require. As illustrated in the following section on the design of processor-memory interconnects, knowing both the peak and average bandwidths becomes important when trying to minimize the implementation cost of the interconnection network.

In addition to the rate at which messages must be accepted and delivered by the network, the time required to deliver an individual message, the message *latency*, is also specified for the network. While an ideal network supports both high bandwidth and low latency, there often exists a tradeoff between these two parameters. For example, a network that supports high bandwidth tends to keep the network resources busy, often causing *contention* for the resources. Contention occurs when two or more messages want to use the same shared resource in the network. All but one of the these messages will have to wait for that resource to become free, thus increasing the latency of the messages. If, instead, resource utilization was decreased by reducing the bandwidth demands, latency would be also lowered.

Message size, the length of a message in bits, is another important design consideration. If messages are small, overheads in the network can have a larger impact on performance than in the case where overheads can be amortized over the length of a larger message. In many systems, there are several possible message sizes.

How the messages from each terminal are distributed across all the possible destination terminals defines a network's *traffic pattern*. For example, each terminal might send messages to all other terminals with equal probability. This is the *random* traffic pattern. If, instead, terminals tend to send messages only to other nearby terminals, the underlying network can exploit this spatial *locality* to reduce cost. In other networks, however, it is important that the specifications hold for arbitrary traffic patterns.

Some networks will also require *quality of service* (QoS). Roughly speaking, QoS involves the *fair* allocation of resources under some service policy. For example, when multiple messages are contending for the same resource in the network, this contention can be resolved in many ways. Messages could be served in a first-come, first-served order based on how long they have been waiting for the resource in question. Another approach gives priority to the message that has been in the network the longest. The choice of between these and other allocation policies is based on the services required from the network.

Finally, the reliability and availability required from an interconnection network influence design decisions. *Reliability* is a measure of how often the network correctly performs the task of delivering messages. In most situations, messages need to be delivered 100% of time without loss. Realizing a 100% reliable network can be done by adding specialized hardware to detect and correct errors, a higher-level software protocol, or using a mix of these approaches. It may also be possible for a small fraction of messages to be dropped by the network as we will see in the following section on packet switching fabrics. The *availability* of a network is the fraction of time it is available and operating correctly. In an Internet router, an availability of 99.999% is typically specified — less than five minutes of total downtime per year. The challenge of providing this level availability of is that the components used to implement the network will often fail several times a minute. As a result, the network must be designed to detect and quickly recover from these failures while continuing to operate.

1.2.1 Processor-Memory Interconnect

Figure 1.2 illustrates two approaches of using an interconnection network to connect processors to memories. Figure 1.2(a) shows a *dance-hall* architecture[2] in which P processors are connected to M memory banks by an interconnection network. Most modern machines use the integrated-node configuration shown in Figure 1.2(b),

---

2.   This arrangement is called a dance-hall architecture because the arrangement of processors lined up on one side of the network and memory banks on the other resembles men and women lined up on either side of an old-time dance hall.

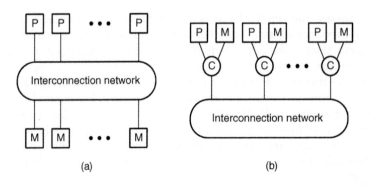

Figure 1.2 Use of an interconnection network to connect processor and memory. (a) *Dance-hall* architecture with separate processor (P) and memory (M) ports. (b) Integrated-node architecture with combined processor and memory ports and local access to one memory bank.

Table 1.1 Parameters of processor-memory interconnection networks.

Parameter	Value
Processor ports	1–2,048
Memory ports	0–4,096
Peak bandwidth	8 Gbytes/s
Average bandwidth	400 Mbytes/s
Message latency	100 ns
Message size	64 or 576 bits
Traffic patterns	arbitrary
Quality of service	none
Reliability	no message loss
Availability	0.999 to 0.99999

where processors and memories are combined in an integrated node. With this arrangement, each processor can access its local memory via a communication switch C without use of the network.

The requirements placed on the network by either configuration are listed in Table 1.1. The number of processor ports may be in the thousands, such as the 2,176 processor ports in a maximally configured Cray T3E, or as small as 1 for a single processor. Configurations with 64 to 128 processors are common today in high-end servers, and this number is increasing with time. For the combined node configuration, each of these processor ports is also a memory port. With a dance-hall configuration, on the other hand, the number of memory ports is typically much larger than the number of processor ports. For example, one high-end vector processor has 32 processor ports making requests of 4,096 memory banks. This large ratio maximizes memory bandwidth and reduces the probability of *bank conflicts* in which two processors simultaneously require access to the same memory bank.

A modern microprocessor executes about 10^9 instructions per second and each instruction can require two 64-bit words from memory (one for the instruction itself and one for data). If one of these references misses in the caches, a block of 8 words is usually fetched from memory. If we really needed to fetch 2 words from memory each cycle, this would demand a bandwidth of 16 Gbytes/s. Fortunately, only about one third of all instructions reference data in memory, and caches work well to reduce the number of references that must actually reference a memory bank. With typical cache-miss ratios, the average bandwidth is more than an order of magnitude lower — about 400 Mbytes/s.[3] However, to avoid increasing memory latency due to *serialization*, most processors still need to be able to fetch at a peak rate of one word per instruction from the memory system. If we overly restricted this peak bandwidth, a sudden burst of memory requests would quickly clog the processor's network port. The process of squeezing this high-bandwidth burst of requests through a lower bandwidth network port, analogous to a clogged sink slowly draining, is called serialization and increases message latency. To avoid serialization during bursts of requests, we need a peak bandwidth of 8 Gbytes/s.

Processor performance is very sensitive to memory latency, and hence to the latency of the interconnection network over which memory requests and replies are transported. In Table 1.1, we list a latency requirement of 100 ns because this is the basic latency of a typical memory system without the network. If our network adds an additional 100 ns of latency, we have doubled the effective memory latency.

When the *load* and *store* instructions miss in the processor's cache (and are not addressed to the local memory in the integrated-node configuration) they are converted into read-request and write-request packets and forwarded over the network to the appropriate memory bank. Each read-request packet contains the memory address to be read, and each write-request packet contains both the memory address and a word or cache line to be written. After the appropriate memory bank receives a request packet, it performs the requested operation and sends a corresponding read-reply or write-reply packet.[4]

Notice that we have begun to distinguish between *messages* and *packets* in our network. A message is the unit of transfer from the network's clients — in this case, processors and memories — to the network. At the interface to the network, a single message can create one or more *packets*. This distinction allows for simplification of the underlying network, as large messages can be broken into several smaller packets, or unequal length messages can be split into fixed length packets. Because

---

3.   However, this average demand is *very* sensitive to the application. Some applications have very poor locality, resulting in high cache-miss ratios and demands of 2 Gbytes/s or more bandwidth from memory.
4.   A machine that runs a cache-coherence protocol over the interconnection network requires several additional packet types. However, the basic constraints are the same.

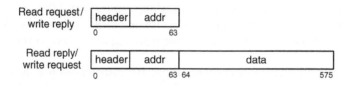

Figure 1.3 The two packet formats required for the processor-memory interconnect.

of the relatively small messages created in this processor-memory interconnect, we assume a one-to-one correspondence between messages and packets.

Read-request and write-reply packets do not contain any data, but do store an address. This address plus some header and packet type information used by the network fits comfortably within 64 bits. Read-reply and write-request packets contain the same 64 bits of header and address information plus the contents of a 512-bit cache line, resulting in 576-bit packets. These two packet formats are illustrated in Figure 1.3.

As is typical with processor-memory interconnect, we do not require any specific QoS. This is because the network is inherently *self-throttling*. That is, if the network becomes congested, memory requests will take longer to be fulfilled. Since the processors can have only a limited number of requests outstanding, they will begin idle, waiting for the replies. Because the processors are not creating new requests while they are idling, the congestion of the network is reduced. This stabilizing behavior is called self-throttling. Most QoS guarantees affect the network only when it is congested, but self-throttling tends to avoid congestion, thus making QoS less useful in processor-memory interconnects.

This application requires an inherently reliable network with no packet loss. Memory request and reply packets cannot be dropped. A dropped request packet will cause a memory operation to *hang* forever. At the least, this will cause a user program to crash due to a timeout. At the worst, it can bring down the whole system. Reliability can be layered on an unreliable network—for example, by having each network interface retain a copy of every packet transmitted until it is acknowledged and retransmitting when a packet is dropped. (See Chapter 21.) However, this approach often leads to unacceptable latency for a processor-memory interconnect. Depending on the application, a processor-memory interconnect needs availability ranging from three nines (99.9%) to five nines (99.999%).

1.2.2 I/O Interconnect

Interconnection networks are also used in computer systems to connect I/O devices, such as disk drives, displays, and network interfaces, to processors and/or memories. Figure 1.4 shows an example of a typical I/O network used to attach an array of disk drives (along the bottom of the figure) to a set of host adapters. The network operates in a manner identical to the processor-memory interconnect, but with different

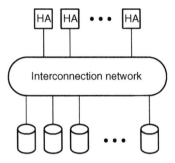

Figure 1.4 A typical I/O network connects a number of host adapters to a larger number of I/O devices — in this case, disk drives.

granularity and timing. These differences, particularly an increased latency tolerance, drive the network design in very different directions.

Disk operations are performed by transferring *sectors* of 4 Kbytes or more. Due to the rotational latency of the disk plus the time needed to reposition the head, the latency of a sector access may be many milliseconds. A disk read is performed by sending a control packet from a host adapter specifying the disk address (device and sector) to be read and the memory block that is the target of the read. When the disk receives the request, it schedules a head movement to read the requested sector. Once the disk reads the requested sector, it sends a response packet to the appropriate host adapter containing the sector and specifying the target memory block.

The parameters of a high-performance I/O interconnection network are listed in Table 1.2. This network connects up to 64 host adapters and for each host adapter there could be many physical devices, such as hard drives. In this example, there are up to 64 I/O devices per host adapter, for a total of 4,096 devices. More typical systems might connect a few host adapters to a hundred or so devices.

The disk ports have a high ratio of peak-to-average bandwidth. When a disk is transferring consecutive sectors, it can read data at rates of up to 200 Mbytes/s. This number determines the peak bandwidth shown in the table. More typically, the disk must perform a head movement between sectors taking an average of 5 ms (or more), resulting in an average data rate of one 4-Kbyte sector every 5 ms, or less than 1 Mbyte/s. Since the host ports each handle the aggregate traffic from 64 disk ports, they have a lower ratio of peak-to-average bandwidth.

This enormous difference between peak and average bandwidth at the device ports calls for a network topology with *concentration*. While it is certainly sufficient to design a network to support the peak bandwidth of all devices simultaneously, the resulting network will be very expensive. Alternatively, we could design the network to support only the average bandwidth, but as discussed in the processor-memory interconnect example, this introduces serialization latency. With the high ratio of peak-to-average bandwidth, this serialization latency would be quite large. A more efficient approach is to *concentrate* the requests of many devices onto an

Table 1.2 Parameters of I/O interconnection networks.

Parameter	Value
Device ports	1–4,096
Host ports	1–64
Peak bandwidth	200 Mbytes/s
Average bandwidth	1 Mbytes/s (devices)
	64 Mbytes/s (hosts)
Message latency	10 μs
Message size	32 bytes or 4 Kbytes
Traffic patterns	arbitrary
Reliability	no message loss[a]
Availability	0.999 to 0.99999

[a] A small amount of loss is acceptable, as the error recovery for a failed I/O operation is much more graceful than for a failed memory reference.

"aggregate" port. The average bandwidth of this aggregated port is proportional to the number of devices sharing it. However, because the individual devices infrequently request their peak bandwidth from the network, it is very unlikely that more than a couple of the many devices are demanding their peak bandwidth from the aggregated port. By concentrating, we have effectively reduced the ratio between the peak and average bandwidth demand, allowing a less expensive implementation without excessive serialization latency.

Like the processor-memory network, the message payload size is bimodal, but with a greater spread between the two modes. The network carries short (32-byte) messages to request read operations, acknowledge write operations, and perform disk control. Read replies and write request messages, on the other hand, require very long (8-Kbyte) messages.

Because the intrinsic latency of disk operations is large (milliseconds) and because the quanta of data transferred as a unit is large (4 Kbyte), the network is not very latency sensitive. Increasing latency to 10 μs would cause negligible degradation in performance. This relaxed latency specification makes it much simpler to build an efficient I/O network than to build an otherwise equivalent processor-memory network where latency is at a premium.

Inter-processor communication networks used for fast message passing in cluster-based parallel computers are actually quite similar to I/O networks in terms of their bandwidth and granularity and will not be discussed separately. These networks are often referred to as system-area networks (SANs) and their main difference from I/O networks is more sensitivity to message latency, generally requiring a network with latency less than a few microseconds.

In applications where disk storage is used to hold critical data for an enterprise, extremely high availability is required. If the storage network goes down, the business

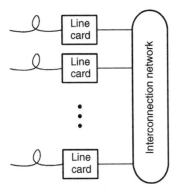

Figure 1.5 Some network routers use interconnection networks as a switching *fabric*, passing packets between line cards that transmit and receive packets over network channels.

goes down. It is not unusual for storage systems to have availability of 0.99999 (five nines) — no more than five minutes of downtime per year.

1.2.3 **Packet Switching Fabric**

Interconnection networks have been replacing buses and crossbars as the switching *fabric* for communication network switches and routers. In this application, an interconnection network is acting as an element of a router for a larger-scale network (local-area or wide-area). Figure 1.5 shows an example of this application. An array of *line cards* terminates the large-scale network channels (usually optical fibers with 2.5 Gbits/s or 10 Gbits/s of bandwidth).[5] The line cards process each packet or *cell* to determine its destination, verify that it is in compliance with its service agreement, rewrite certain fields of the packet, and update statistics counters. The line card then forwards each packet to the fabric. The fabric is then responsible for forwarding each packet from its source line card to its destination line card. At the destination side, the packet is queued and scheduled for transmission on the output network channel.

Table 1.3 shows the characteristics of a typical interconnection network used as a switching fabric. The biggest differences between the switch fabric requirements and the processor-memory and I/O network requirements are its high average bandwidth and the need for quality of service.

The large packet size of a switch fabric, along with its latency insensitivity, simplifies the network design because latency and message overhead do not have to be highly optimized. The exact packet sizes depend on the protocol used by the

5.   A typical high-end IP router today terminates 8 to 40 10 Gbits/s channels with at least one vendor scaling to 512 channels. These numbers are expected to increase as the aggregate bandwidth of routers doubles roughly every eighteen months.

Table 1.3 Parameters of a packet switching fabric.

Parameter	Value
Ports	4–512
Peak Bandwidth	10 Gbits/s
Average Bandwidth	7 Gbits/s
Message Latency	$10\,\mu s$
Packet Payload Size	40–64 Kbytes
Traffic Patterns	arbitrary
Reliability	$< 10^{-15}$ loss rate
Quality of Service	needed
Availability	0.999 to 0.99999

router. For Internet protocol (IP), packets range from 40 bytes to 64 Kbytes,[6] with most packets either 40, 100, or 1,500 bytes in length. Like our other two examples, packets are divided between short control messages and large data transfers.

A network switch fabric is *not* self-throttling like the processor-memory or I/O interconnect. Each line card continues to send a steady stream of packets regardless of the congestion in the fabric and, at the same time, the fabric must provide guaranteed bandwidth to certain classes of packets. To meet this service guarantee, the fabric must be *non-interfering*. That is, an excess in traffic destined for line-card a, perhaps due to a momentary overload, should not interfere with or "steal" bandwidth from traffic destined for a different line card b, even if messages destined to a and messages destined to b share resources throughout the fabric. This need for non-interference places unique demands on the underlying implementation of the network switch fabric.

An interesting aspect of a switch fabric that can potentially simplify its design is that in some applications it may be acceptable to drop a very small fraction of pack-ets — say, one in every 10^{15}. This would be allowed in cases where packet dropping is already being performed for other reasons ranging from bit-errors on the input fibers (which typically have an error rate in the 10^{-12} to 10^{-15} range) to overflows in the line card queues. In these cases, a higher-level protocol generally handles dropped packets, so it is acceptable for the router to handle very unlikely circumstances (such as an internal bit error) by dropping the packet in question, as long as the rate of these drops is well below the rate of packet drops due to other reasons. This is in contrast to a processor-memory interconnect, where a single lost packet can lock up the machine.

---

6.   The Ethernet protocol restricts maximum packet length to be less than or equal to 1,500 bytes.

1.3 **Network Basics**

To meet the performance specifications of a particular application, such as those described above, the network designer must work within technology constraints to implement the *topology*, *routing*, and *flow control* of the network. As we have said in the previous sections, a key to the efficiency of interconnection networks comes from the fact that communication resources are shared. Instead of creating a dedicated channel between each terminal pair, the interconnection network is implemented with a collection of shared router nodes connected by shared channels. The connection pattern of these nodes defines the network's *topology*. A message is then delivered between terminals by making several *hops* across the shared channels and nodes from its source terminal to its destination terminal. A good topology exploits the properties of the network's packaging technology, such as the number of pins on a chip's package or the number of cables that can be connected between separate cabinets, to maximize the bandwidth of the network.

Once a topology has been chosen, there can be many possible paths (sequences of nodes and channels) that a message could take through the network to reach its destination. *Routing* determines which of these possible paths a message actually takes. A good choice of paths minimizes their length, usually measured as the number of nodes or channels visited, while balancing the demand placed on the shared resources of the network. The length of a path obviously influences latency of a message through the network, and the demand or *load* on a resource is a measure of how often that resource is being utilized. If one resource becomes over-utilized while another sits idle, known as a load imbalance, the total bandwidth of messages being delivered by the network is reduced.

Flow control dictates which messages get access to particular network resources over time. This influence of flow control becomes more critical as the utilization of resource increases and good flow control forwards packets with minimum delay and avoids idling resources under high loads.

1.3.1 **Topology**

Interconnection networks are composed of a set of shared router nodes and channels, and the *topology* of the network refers to the arrangement of these nodes and channels. The topology of an interconnection network is analogous to a roadmap. The channels (like roads) carry packets (like cars) from one router node (intersection) to another. For example, the network shown in Figure 1.6 consists of 16 nodes, each of which is connected to 8 channels, 1 to each neighbor and 1 from each neighbor. This particular network has a *torus* topology. In the figure, the nodes are denoted by circles and each pair of channels, one in each direction, is denoted by a line joining two nodes. This topology is also a *direct network*, where a terminal is associated with each of the 16 nodes of the topology.

A good topology exploits the characteristics of the available packaging technology to meet the bandwidth and latency requirements of the application at minimum

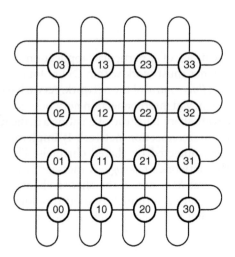

Figure 1.6 A network topology is the arrangements of nodes, denoted by circles numbered 00 to 33 and channels connecting the nodes. A pair of channels, one in each direction, is denoted by each line in the figure. In this 4 × 4, 2-D torus, or 4-ary 2-cube, topology, each node is connected to 8 channels: 1 channel to and 1 channel from each of its 4 neighbors.

cost. To maximize bandwidth, a topology should saturate the *bisection bandwidth*, the bandwidth across the midpoint of the system, provided by the underlying packaging technology.

For example, Figure 1.7 shows how the network from Figure 1.6 might be packaged. Groups of four nodes are placed on vertical printed circuit boards. Four of the circuit boards are then connected using a backplane circuit board, just as PCI cards might be plugged into the motherboard of a PC. For this system, the bisection bandwidth is the maximum bandwidth that can be transferred across this backplane. Assuming the backplane is wide enough to contain 256 signals, each operating at a data rate of 1 Gbit/s, the total bisection bandwidth is 256 Gbits/s.

Referring back to Figure 1.6, exactly 16 unidirectional channels cross the midpoint of our topology — remember that the lines in the figure represent two channels, one in each direction. To saturate the bisection of 256 signals, each channel crossing the bisection should be $256/16 = 16$ signals wide. However, we must also take into account the fact that each node will be packaged on a single IC chip. For this example, each chip has only enough pins to support 128 signals. Since our topology requires a total of 8 channels per node, each chip's pin constraint limits the channel width to $128/8 = 16$ signals. Fortunately, the channel width given by pin limitations exactly matches the number of signals required to saturate the bisection bandwidth.

In contrast, consider the 16-node ring network shown in Figure 1.8. There are 4 channels connected to each node, so pin constraints limit the channel width to $128/4 = 32$ signals. Four channels cross the bisection, so we would like to design these channels to be $256/4 = 64$ signals wide to saturate our bisection, but the

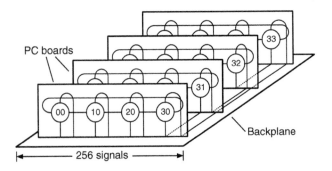

Figure 1.7 A packaging of a 16-node torus topology. Groups of 4 nodes are packaged on single printed circuit boards, four of which are connected to a single backplane board. The backplane channels for the third column are shown along the right edge of the backplane. The number of signals across the width of the backplane (256) defines the bisection bandwidth of this particular package.

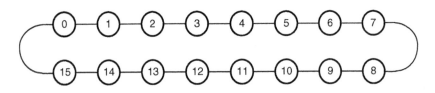

Figure 1.8 For the constraints of our example, a 16-node ring network has lower latency than the 16-node, 2-D torus of Figure 1.6. This latency is achieved at the expense of lower throughput.

pins limit the channel width to only half of this. Thus, with identical technology constraints, the ring topology provides only half the bandwidth of the torus topology. In terms of bandwidth, the torus is obviously a superior choice, providing the full 32 Gbits/s of bandwidth per node across the midpoint of the system.

However, high bandwidth is not the only measure of a topology's performance. Suppose we have a different application that requires only 16 Gbits/s of bandwidth under identical technology constraints, but also requires the minimum possible latency. Moreover, suppose this application uses rather long 4,096-bit packets. To achieve a low latency, the topology must balance the desire for a small average distance between nodes against a low *serialization latency*.

The distance between nodes, referred to as the *hop count*, is measured as the number of channels and nodes a message must traverse on average to reach its destination. Reducing this distance calls for increasing the node *degree* (the number of channels entering and leaving each node). However, because each node is subject to a fixed pin limitation, increasing the number of channels leads to narrower channel widths. Squeezing a large packet through a narrow channel induces serialization latency. To see how this tradeoff affects topology choice, we revisit our two 16-node topologies, but now we focus on message latency.

First, to quantify latency due to hop count, a traffic pattern needs to be assumed. For simplicity, we use *random traffic*, where each node sends to every other node with equal probability. The average hop count under random traffic is just the average distance between nodes. For our torus topology, the average distance is 2 and for the ring the average distance is 4. In a typical network, the latency per hop might be 20 ns, corresponding to a total hop latency of 40 ns for the torus and 80 ns for the ring.

However, the wide channels of the ring give it a much lower serialization latency. To send a 4,096-bit packet across a 32-signal channel requires 4,096/32 = 128 cycles of the channel. Our signaling rate of 1 GHz corresponds to a period of 1 ns, so the serialization latency of the ring is 128 ns. We have to pay this serialization time only once if our network is designed efficiently, which gives an average delay of 80 + 128 = 208 ns per packet through the ring. Similar calculations for the torus yield a serialization latency of 256 ns and a total delay of 296 ns. Even though the ring has a greater average hop count, the constraints of physical packaging give it a lower latency for these long packets.

As we have seen here, no one topology is optimal for all applications. Different topologies are appropriate for different constraints and requirements. Topology is discussed in more detail in Chapters 3 through 7.

1.3.2 Routing

The routing method employed by a network determines the path taken by a packet from a source terminal node to a destination terminal node. A route or path is an ordered set of channels $P = \{c_1, c_2, \ldots, c_k\}$, where the output node of channel c_i equals the input node of channel c_{i+1}, the source is the input to channel c_1, and the destination is the output of channel c_k. In some networks there is only a single route from each source to each destination, whereas in others, such as the torus network in Figure 1.6, there are many possible paths. When there are many paths, a good routing algorithm balances the load uniformly across channels regardless of the offered traffic pattern. Continuing our roadmap analogy, while the topology provides the roadmap, the roads and intersections, the routing method steers the car, making the decision on which way to turn at each intersection. Just as in routing cars on a road, it is important to distribute the traffic — to balance the load across different roads rather than having one road become congested while parallel roads are empty.

Figure 1.9 shows two different routes from node 01 to node 22 in the network of Figure 1.6. In Figure 1.9(a) the packet employs *dimension-order routing*, routing first in the x-dimension to reach node 21 and then in the y-dimension to reach destination node 22. This route is a *minimal* route in that it is one of the shortest paths from 01 to 22. (There are six.) Figure 1.9(b) shows an alternate route from 00 to 22. This route is non-minimal, taking 5 hops rather than the minimum of 3.

While dimension-order routing is simple and minimal, it can produce significant load imbalance for some traffic patterns. For example, consider adding another dimension-order route from node 11 to node 20 in Figure 1.9(a). This route also uses the channel from node 11 to node 21, doubling its *load*. A channel's load is the average amount of bandwidth that terminal nodes are trying to send across it. Normalizing the load to the maximum rate at which the terminals can inject data into the network, this channel has a load of 2. A better routing algorithm could reduce the normalized channel load to 1 in this case. Because dimension-order routing is placing twice the necessary load on this single channel, the resulting bandwidth of the network under this traffic pattern will be only half of its maximum. More generally, all routing algorithms that choose a single, fixed path between each source-destination pair, called *deterministic* routing algorithms, are especially subject to low bandwidth due to load imbalance. These and other issues for routing algorithm design are described in more detail in Chapters 8 to 10.

1.3.3 **Flow Control**

Flow control manages the allocation of resources to packets as they progress along their route. The key resources in most interconnection networks are the *channels* and the *buffers*. We have already seen the role of channels in transporting packets between nodes. Buffers are storage implemented within the nodes, such as registers or memories, and allow packets to be held temporarily at the nodes. Continuing our analogy: the topology determines the roadmap, the routing method steers the car, and the flow control controls the traffic lights, determining when a car can advance over the next stretch of road (channels) or when it must pull off into a parking lot (buffer) to allow other cars to pass.

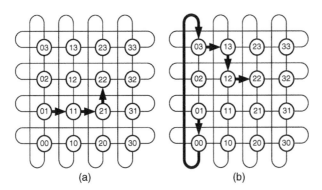

(a)   (b)

Figure 1.9 Two ways of routing from 01 to 22 in the 2-D torus of Figure 1.6. (a) Dimension-order routing moves the packet first in the *x* dimension, then in the *y* dimension. (b) A non-minimal route requires more than the minimum path length.

To realize the performance potential of the topology and routing method, the flow-control strategy must avoid resource conflicts that can hold a channel idle. For example, it should not block a packet that can use an idle channel because it is waiting on a buffer held by a packet that is blocked on a busy channel. This situation is analogous to blocking a car that wants to continue straight behind a car that is waiting for a break in traffic to make a left turn. The solution, in flow control as well as on the highway, is to add a (left turn) lane to *decouple* the resource dependencies, allowing the blocked packet or car to make progress without waiting.

A good flow control strategy is *fair* and avoids *deadlock*. An unfair flow control strategy can cause a packet to wait indefinitely, much like a car trying to make a left turn from a busy street without a light. Deadlock is a situation that occurs when a cycle of packets are waiting for one another to release resources, and hence are blocked indefinitely—a situation not unlike gridlock in our roadmap analogy.

We often describe a flow control method by using a time-space diagram such as the ones shown in Figure 1.10. The figure shows time-space diagrams for (a) store-and-forward flow control and (b) cut-through flow control. In both diagrams, time is shown on the horizontal axis and space is shown on the vertical axis. Time is expressed in cycles. Space is shown by listing the channels used to send the packet. Each packet is divided into five fixed-size *flits*. A flit, or flow control digit, is the smallest unit of information recognized by the flow control method. Picking a small, fixed size simplifies router design without incurring a large overhead for packets whose length are not a multiple of the flit size. Each flit of a single packet, denoted by labeled boxes, is being sent across four channels of a network. A box is shown in the diagram during the cycle that a particular flit is using the bandwidth of a channel. As seen in Figure 1.10, the choice of flow control techniques can significantly affect the latency of a packet through the network.

Flow control is described in more detail in Chapters 12 and 13. The problems of deadlock, livelock, tree saturation, and quality of service that arise in conjunction with flow control and routing are dealt with in Chapters 14 and 15.

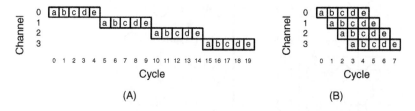

Figure 1.10 Time-space diagrams showing two flow control methods. The vertical axis shows space (channels) and the horizontal axis shows time (cycles). (a) With store-and-forward flow control, a packet, in this case containing 5 *flits*, is completely transmitted across one channel before transmission across the next channel is started. (b) With cut-through flow control, packet transmission over the channels is pipelined, with each flit being transmitted across the next channel as soon as it arrives.

1.3.4 **Router Architecture**

Figure 1.11 shows a simplified view of the internals of one of the 16 nodes in the network of Figure 1.6. A buffer is associated with each of the four input channels and these buffers hold arriving flits until they can be allocated the resources necessary for departure. Once a flit can be ensured buffer space in the next router along its path, the *downstream* router, it can begin to vie for access to the *crossbar switch*. The crossbar switch can be configured to connect any input buffer of the router to any output channel, but under the constraints that each input is connected to at most one output, and each output is connected to at most one input. The tasks of resolving all the potential requests to the crossbar and other shared resources of the router fall onto the *allocators*. To advance to the next router, a flit in one of the input buffers must be *allocated* space in a buffer on the next node of its route, bandwidth on the next channel of the route, and it must win the *allocation* to traverse the crossbar switch. Router architecture is covered in detail in Chapters 16 through 21.

1.3.5 **Performance of Interconnection Networks**

Performance of an interconnection network is described primarily by a latency vs. offered traffic curve like the one shown in Figure 1.12. The figure shows the average latency of a packet, the time from when the first bit of the packet arrives at the source terminal to when the last bit of the packet arrives at the destination terminal, as a function of *offered traffic*, the average amount of traffic (bits/s) generated by each

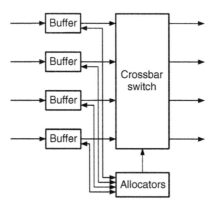

Figure 1.11   Simplified block diagram of a router. Flits arriving over the input channels are stored in buffers associated with each input. A set of allocators assigns buffers on the next node and channel bandwidth to pending flits. When a flit has been allocated the resources it needs, it is forwarded by the crossbar switch to an output channel.

source terminal of the network. To draw a particular latency vs. offered traffic curve, the traffic pattern (for example, random traffic) must also be specified.

Although latency vs. offered traffic curves give the most accurate view of the ultimate performance of an interconnection network, they do not have simple, closed-form expressions and are generally found by discrete-event simulation. To guide our understanding of the tradeoff in the early design stages of an interconnection network, we take an incremental approach to network performance that follows our exploration of topology, routing, and flow control.

Zero-load latency gives a lower bound on the average latency of a packet through the network. The zero-load assumption is that a packet never contends for network resources with other packets. Under this assumption, the average latency of a packet is its serialization latency plus its hop latency. For example, consider the torus network shown in Figure 1.6, with packets of length $L = 512$ bits, channels with a bandwidth of $b = 16$ Gbits/s, and random traffic. In this case, the serialization latency is $L/b = 32$ ns. The lowest possible hop latency for random traffic occurs with minimal routing and gives an average hop count of $H_{min} = 2$. For a router latency of $t_r = 10$ ns, the minimum hop latency is $H_{min}t_r = 20$ ns. This gives a lower bound of $32 + 20 = 52$ ns on the average latency of a packet through the network, based solely on the topology, packaging, and traffic pattern of the network.

Incorporating the average hop count H_{avg} of the actual routing algorithm used in the network gives a tighter bound on packet latency because $H_{avg} \geq H_{min}$. Finally, the flow control employed by the network can further reduce the performance over the bounds given by the topology and the routing. For example, if our network employs store and forward flow control (Figure 1.10[a]) the zero-load latency will be $H_{avg}t_r \times$

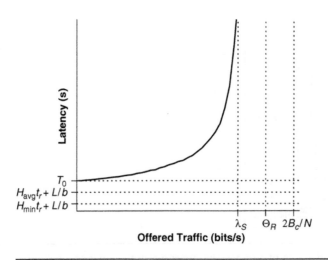

Figure 1.12   Latency vs. offered traffic curve for an interconnection network. At low offered traffic, latency approaches the zero-load latency T_0. Latency goes to infinity at the saturation throughput λ_s. Saturation throughput is bounded by a topology limit $2B_c/N$ and a routing limit Θ_R.

L/b rather than $H_{avg}t_r + L/b$. The actual zero-load latency, T_0, incorporates the constraints of topology along with the actual performance, routing, and flow control. These successively tighter bounds on latency are shown as horizontal asymptotes to the curve in Figure 1.12.

A similar approach gives a set of upper bounds on the *throughput* of the network. While each source offers a particular amount of traffic to the network, the through-put, or *accepted traffic*, is the rate that traffic (bits/s) is delivered to the destination terminals. For our example with random traffic, half of the traffic must cross the bisection of the network. The bisection consists of 16 channels with a total bandwidth of $B_c = 256$ Gbits/s. Hence, the traffic per node cannot exceed $2B_c/N$ or 32 Gbits/s. This bound assumes traffic is perfectly balanced across the bisection. Therefore, a particular routing algorithm R cannot exceed this bound and the throughput tak-ing into account the routing algorithm Θ_R may actually be lower if load imbalance occurs ($\Theta_R \leq 2B_c/N$). Finally, if our flow control results in idle channels due to re-source dependencies, the *saturation throughput* of the network λ_s can be significantly less than the bound of Θ_R. These three bounds on throughput are shown as vertical asymptotes in Figure 1.12.

Taking this incremental approach to performance — deriving successively tighter bounds due to topology, routing, and flow control — enables us to explore how each of the design decisions we consider affects performance without complicating our analysis with unnecessary details. For example, we can see how a topology choice affects latency independent of routing and flow control. In contrast, trying to deal with performance all at once makes it difficult to see the effects of any one design choice.

After developing our incremental model of performance, we will consider per-formance in its entirety in Chapters 23 through 25. In these chapters we discuss some subtle points of performance measurement, introduce analytic methods of estimating performance (based on queueing theory and probability theory), dis-cuss simulation methods for measuring performance, and give a number of example measurements.

1.4 **History**

Interconnection networks have a rich history that spans many decades. Networks developed along at least three parallel threads: telephone switching networks, inter-processor communication, and processor-memory interconnect.

Telephone switching networks have been around as long as the telephone. Early telephone networks were built from electro-mechanical crossbars or electro-mechanical step-by-step switches. As late as the 1980s, most local telephone switches were still built from electro-mechanical relays, although toll (long-distance) switches were completely electronic and digital by that time. Key developments in telephone switching include the non-blocking, multistage Clos network in 1953 [37] and the Beneš network in 1962 [17]. Many large telephone switches today are still built from Clos or Clos-like networks.

The first inter-processor interconnection networks were connections between the registers of neighboring processors connected in 2-D arrays. The 1962 Solomon machine [172] is an example of a processor array of this type. These early networks performed no routing. Thus, the processors had to explicitly relay communications to non-neighbors, making for poor performance and considerable programming complexity. By the mid-1980s, router chips, such as the torus routing chip [56], were developed to forward messages through intermediate nodes without processor intervention.

Inter-processor interconnection networks have gone through a series of topology *fads* over the years — largely motivated by packaging and other technology constraints. The early machines, like Solomon [172], Illiac [13], and MPP, were based on simple 2-D mesh or torus networks because of their physical regularity. Starting in the late 1970s, binary *n*-cube or *hypercube* networks became popular because of their low diameter. Many machines designed around the hypercube networks emerged, such as the Ametek S14, Cosmic Cube [163], the nCUBE computers [134, 140], and the Intel iPSC series [38, 155]. In the mid-1980s, it was shown that under realistic packaging constraints low-dimensional networks outperformed hypercubes [2, 46] and most machines returned to 2-D or 3-D mesh or torus networks. Consequently, most machines built over the last decade have returned to these networks, including the J-machine [138], Cray T3D [95] and T3E [162], Intel DELTA [117], and Alpha 21364 [131], to mention a few. Today, the high pin bandwidth of router chips relative to message length motivates the use of networks with much higher node degree, such as butterfly and Clos networks. We can expect a switch to such networks over the next decade.

Processor-memory interconnection networks emerged in the late 1960s when parallel processor systems incorporated alignment networks to allow any processor to access any memory bank without burdening the other processors [110]. The smallest machines employed crossbar switches for this purpose, whereas larger machines used networks with a butterfly (or equivalent) topology, in a dance-hall arrangement. Variations on this theme were used through the 1980s for many shared-memory parallel processors.

The three threads of interconnection network evolution recently merged. Since the early 1990s, there has been little difference in the design of processor-memory and inter-processor interconnection networks. In fact, the same router chips have been used for both. A variant of the Clos and Beneš networks of telephony has also emerged in multiprocessor networks in the form of the *fat tree* topology [113].

Our discussion of history has focused on topology because it is the most visible attribute of a network. Of course, routing and flow control methods evolved in parallel with topology. Early routing chips employed simple deterministic routing and either circuit-switching or store-and-forward packet switching. Later routers employed adaptive routing with sophisticated deadlock avoidance schemes and virtual-channel flow control.

1.5 **Organization of this Book**

We start in the next chapter with a complete description of a simple interconnection network from the topology down to the logic gates to give the reader the "big picture" view of interconnection networks before diving into details. The remaining chapters cover the details. They are organized into five main sections: topology, routing, flow control, router architecture, and performance. Each section is organized into chapters, with the first chapter of each section covering the basics and the later chapters covering more involved topics.

CHAPTER 2

A Simple Interconnection Network

In this chapter, we examine the architecture and design of a simple interconnection network to provide a global view. We will examine the simplest possible network: a butterfly network with dropping flow control. Although the resulting network is costly, it emphasizes many of the key aspects of interconnection network design. In later chapters, we will learn how to produce more efficient and practical networks.

2.1 Network Specifications and Constraints

Like all engineering design problems, network design starts with a set of specifications that describe what we wish to build and a set of constraints that limit the range of potential solutions. The specifications for the example network in this chapter are summarized in Table 2.1. These specifications include the size of the network (64 ports) and the bandwidth required per port. As shown in the table, the peak and average bandwidths are equal, implying that inputs inject messages continuously at a rate of 0.25 Gbyte/s. Random traffic, where each input sends to each output with equal probability, and message sizes from 4 to 64 bytes are expected. Also, the quality of service and reliability specifications allow for *dropped* packets. That is, not every packet needs to be successfully delivered to its destination. As we will see, the ability to drop packets will simplify our flow control implementation. Of course, an actual set of specifications would be quite a bit longer and more specific. For example, a QoS specification would indicate what fraction of packets could be dropped and under what circumstances. However, this set suffices to illustrate many points of our design.

The constraints on our example network design are illustrated in Table 2.2. These constraints specify the capacity and cost of each level of packaging. Our network is composed of chips that are assembled on circuit boards that are in turn connected

25

Table 2.1 Specifications for our example network. Only a portion of the possible design parameters is given to simplify the design.

Parameter	Value
Input ports	64
Output ports	64
Peak bandwidth	0.25 Gbyte/s
Average bandwidth	0.25 Gbyte/s
Message latency	100 ns
Message size	4–64 bytes
Traffic pattern	random
Quality of service	dropping acceptable
Reliability	dropping acceptable

Table 2.2 Constraints for our example network.

Parameter	Value
Port width	2 bits
Signaling rate	1 GHz
Signals per chip	150
Chip cost	$200
Chip pin bandwidth	1 Gbit/s
Signals per circuit board	750
Circuit board cost	$200
Signals per cable	80
Cable cost	$50
Cable length	4 m at 1 Gbit/s

via cables. The constraints specify the number of signals[1] that can be passed across a module interface at each level, and the cost of each module. For the cable, the constraints also specify the longest distance that can be traversed without reducing cable bandwidth.[2]

---

1. Note that *signals* does not necessarily imply pins. For example, it is quite common to use differential signaling that requires two pins per signal.
2. As bandwidth times distance squared, Bd^2, is a constant for a given type of cable, the bandwidth must be reduced by a factor of four, to 250 Mbits/s, to run the cable twice as far.

2.2 **Topology**

For simplicity our example network has a butterfly topology. From the point of view of a single input port, the butterfly looks like a tree. (See Figure 2.1.) Each level of the tree contains switching nodes, which, unlike the terminal nodes, do not send or receive packets, but only pass packets along. Also, each of the channels is unidirectional, as indicated by the arrows, flowing from the input to the output nodes (left to right in the figure). Choosing the butterfly, however, does not complete the job of topology design. We must also decide on the *speedup* of our network, determine the *radix* of our butterfly, and determine how the topology is mapped onto the packaging levels.

The *speedup* of a network is the ratio of the total input bandwidth of the network to the network's ideal *capacity*. The capacity is defined as the best possible throughput, assuming perfect routing and flow control, that could be achieved by the network under the given traffic pattern. Designing with a speedup of 1 means the demands of the inputs are exactly matched to the *ideal* ability of the network to deliver traffic. Providing more speedup increases the design's margin and allows for non-idealities in the implementation. In some sense, speedup is analogous to a civil engineer's notion of a structure's safety factor. A building with a safety factor of 4, for example, is designed to handle stresses 4 times greater than its specifications.

For the butterfly, sizing each of the network's channels to have the same bandwidth as a single input port gives a speedup of 1. To see this, consider the demand placed on any particular channel under random traffic — summing the fraction of

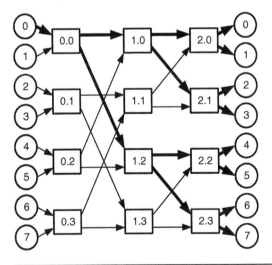

Figure 2.1 An 8-node butterfly network. Data flows from the input nodes on the left (circles) through three stages of switch nodes (rectangles) to the output nodes on the right (circles). The switch nodes are labeled with their stage and address. All channels are unidirectional, as indicated by the arrows.

traffic that each input sends over this channel always gives a demand equal to the input port bandwidth. In our simple network, this corresponds to designing our channels with a bandwidth of 0.25 Gbyte/s. However, based on the fact that we will be opting for simplicity rather than efficiency in our subsequent design choices, we choose a speedup of 8. While this speedup is quite large (the Brooklyn Bridge was designed with a safety factor of only 6!), we will quickly learn how to reduce our speedup requirements over the course of the book.

Our choice of speedup, along with our packaging constraints, determines the number of inputs and outputs of each switching node, referred to as the *radix* of the butterfly. For example, the butterfly in Figure 2.1 is designed with a radix of 2. Each of the switching nodes is implemented on a single chip, so the total number of channels (inputs and outputs) times the channel width must not exceed the limit of 150 signals per chip. To give a speedup of 8, we need a network channel bandwidth of $8 \times 0.25 = 2$ Gbytes/s, which takes 16 signals operating at 1 Gbit/s each. Allowing for 2 additional overhead signals, the channels are 18 signals wide and we can fit only $150/18 \approx 8$ channels on a chip. We therefore choose a radix-4 butterfly, which has 4 input channels and 4 output channels per switching node, for a total of 8 channels.

To connect each input port to all 64 output ports, our butterfly requires $\log_4 64 = 3$ levels or *stages* of switching nodes. Thus, our network will be a radix-4, 3-stage butterfly, or a 4-ary 3-fly for short. The full topology of this network is illustrated in Figure 2.2. While this diagram may seem daunting at first, it is an extension of the smaller butterfly we introduced in Figure 2.1. The channels that connect input 1 to each of the 64 outputs again form a tree (shown in bold), but now the degree of the tree is 4 — the radix of our network.

The last step in designing our topology is packaging it. We have already made one packaging decision by placing one switching node per chip. By choosing the radix of the network to meet the per chip constraints, we know that the switching nodes are packaged within our design constraints. These chips must be mounted on a circuit board, and, in order to minimize cost, we would like to mount as many switching chips on a circuit board as possible. We are constrained, however, not to exceed the maximum pinout of 750 signals entering or leaving a circuit board.[3] This constraint is driven by the maximum number of signals that can be routed through a connector on one edge of the board — the connector density (signals/m) times the length of the connector (m).

A valid partitioning of switch nodes between circuit boards is shown in Figure 2.2. The boundary of each of the 8 boards is denoted by a dashed box. As shown, the network is packaged by placing the first stage of switches on 4 circuit boards with 4 chips per board. The next 2 stages are packaged on 4 boards, each containing 8 chips connected as a 16-port butterfly network. We verify that each circuit board's pinout constraint is met by observing that 32 channels each containing 18 signals enter and leave each board. This gives us a total pinout of $32 \times 18 = 576$

---

3.   In a real system there would also be a constraint on the signal density on the circuit board.

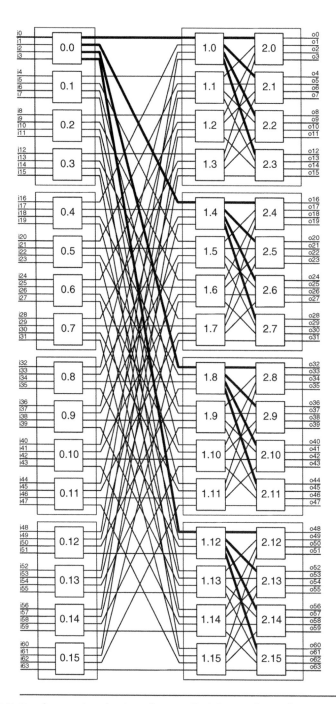

Figure 2.2 Topology and packaging of our radix-4 3-stage butterfly network. Channels are unidirectional and data flows from left (inputs) to right (outputs).

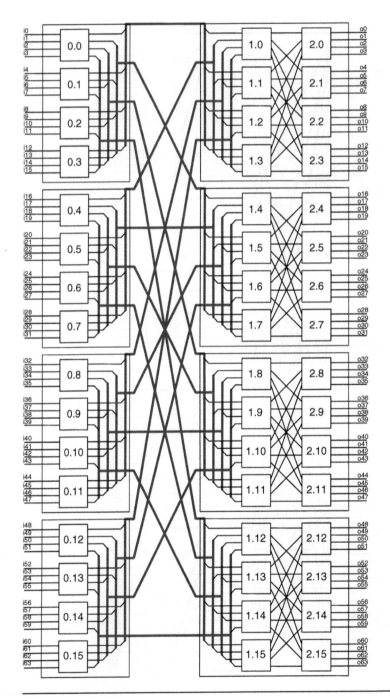

Figure 2.3 Cabling of the board-to-board connections four our radix-4 3-stage butterfly network.

signals, which comfortably fits within our constraint of 750 signals. The astute reader will notice that we could put 5 router chips on the first stage boards (40 channels or 720 signals total), but this would not be efficient because we would still need 4 first-stage boards. Also, we cannot put 10 router chips on the second-stage boards because the 46 channels (828 signals) required would exceed the pinout of our board.

Finally, the connections between boards are carried on cables, as illustrated in Figure 2.3. Each thick gray line in the figure corresponds to a single cable carrying four 18-bit channels from one circuit board to another. The 8 circuit boards are cabled together with 16 of these cables: one from each board in the first stage to each board in the second and third stages. With the 8 circuit boards arranged in a single chassis, these cables are all well within the maximum length.

Taking a step back, we see how the switching nodes in our topology can connect any input to any output. The first stage of switches selects between the 4 circuit boards holding the remaining stages. The second stage of switches selects 1 of 4 chips making up the third stage on the selected circuit board. Finally, the last stage of switches selects the desired output port. We exploit this *divide-and-conquer* structure when routing packets through the network.

2.3 Routing

Our simple butterfly network employs *destination-tag routing*[4] in which the bits of the destination address are used to select the output port at each stage of the network. In our 64-node network, the destination address is 6 bits. Each stage of switches uses 2 address bits to select 1 of the 4 switch outputs, directing the packet to the proper quarter of the remaining nodes. For example, consider routing a packet from input 12 to output node $35 = 100011_2$. The most significant dibit of the destination (10) selects the third output port of switch 0.3, taking the packet to switch 1.11. Then, the middle dibit (00) selects the first output port of switch 1.11. From switch 2.8, the least-significant dibit (11) selects the last output port, delivering the packet to output port 35.

Notice that this sequence of switch output selections was completely independent of the input port of the packet. For example, routing from node 51 to output 35 follows the same sequence of selections: the third switch port in the first stage, the first port in the second stage, and the last port in the third stage. So, we can implement the routing algorithm by storing just the destination address with each packet.

For uniformity, all of our switch nodes operate on the most significant dibit of the destination address field. Then, before the packet leaves the node, the address field is shifted two bits to the left, discarding the bits that have just been used and exposing the next dibit in the most significant position. After the first stage of routing

---

4. See Section 8.4.1 for a more complete description of destination-tag routing.

to node 35, for example, the original address 100011_2 is shifted to 001100_2. This convention allows us to use the same switching node at each position in the network without the need for special configuration. It also facilitates expanding the network to larger numbers of nodes, limited only by the size of the address field.

2.4 Flow Control

The channels of our network transport 16-bit-wide physical digits, or *phits*, of data per cycle. However, we have specified that the network must deliver entire packets of data that contain from 32 to 512 bits of data. Thus, we use a simple protocol, illustrated in Figure 2.4, to assemble phits into packets. As shown in the figure, each packet consists of a *header* phit followed by zero or more *payload* phits. The header phit signifies the beginning of a new packet and also contains the destination address used by our routing algorithm. Payload phits hold the actual data of the packet, split into 16-bit chunks. The phits of a given packet must be contiguous, without interruption. However, any number of *null* phits may be transported on the channel between packets. To distinguish header words from payload words and to denote the end of a packet, we append a 2-bit type field to each channel. This field describes each 16-bit word as a header (H), payload (P), or null (N) word. Packets may be of any length but always consist of a single H word followed by zero or more P words, followed in turn by zero or more N words. Using regular expression notation, the zero or more packets flowing on a link may be described as $(HP^*N^*)^*$.

Now that we have assembled our phits into packets, we can get to the main business of flow control: allocating resources to packets. For simplicity, our butterfly network uses dropping flow control. If the output port needed by a packet is in use when the packet arrives at a switch, the packet is dropped (discarded). The flow control assumes that some higher-level, end-to-end error control protocol will eventually

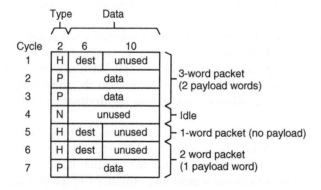

Figure 2.4 Packet format for our simple network. Time, in cycles, is shown in the vertical direction, while the 18 signals of a channel are shown in the horizontal direction. The leftmost signals contain the phit type, while the 16 remaining signals contain either a destination address or data, or are unused in the case of a null phit.

Table 2.3 Phit type encoding for our example network.

Type	Code
H	11
P	10
N	00

resend the dropped packet. Dropping packets is among the worst of flow-control methods because it has a high rate of packet loss and it wastes channel bandwidth on packets that are ultimately dropped. As we shall see in Chapter 12, there are much better flow-control mechanisms. Dropping flow-control is ideal for our present purposes, however, because it is extremely simple both conceptually and in implementation.

2.5 **Router Design**

Each of the switching nodes in our butterfly network is a *router*, capable of receiving packets on its inputs, determining their destination based on the routing algorithm, and then forwarding packets to the appropriate output. The design decisions we have made up to this point result in a very simple router. The block diagram of a single router is shown in Figure 2.5. The datapath of the router consists of four 18-bit input registers, four 18-bit 4:1 multiplexers, four shifters (for shifting the route field of header phits), and four 18-bit output registers. The datapath consists of 144 bits of register and about 650 gates (2-input NAND-equivalent).

Phits arrive each clock cycle in the input register and are routed to all four multiplexers. At each multiplexer, the associated *allocator* examines the type of each phit and the next hop field of each head phit, and sets the switch accordingly. Phits from the selected input are next routed to a shifter. Under control of the allocator, the shifter shifts all head phits left by two bits to discard the current route field and expose the next route field. Payload phits are passed unchanged.

The control of the router resides entirely in the four allocators associated with each output that control the multiplexers and shifters. Each allocator, as the name suggests, allocates an output port to one of the four input ports. A schematic diagram of one of the allocators is shown in Figure 2.6. The allocator consists of four nearly identical bit slices, each divided into three sections: decode, arbitrate, and hold. In the decode section, the upper four bits of each input phit is decoded. Each decoder generates two signals. Signal request$_i$ is true if the phit on input i is a head phit and the upper two bits of the route field match the output port number. This signal indicates that the input phit *requests* use of the output port to route the packet starting with this head phit. The decoder also generates signal payload$_i$ which is true if the phit on input i is a payload phit. The hold logic of the allocator uses this signal to hold a channel for the duration of a packet (as long as payload phits are on the selected input port).

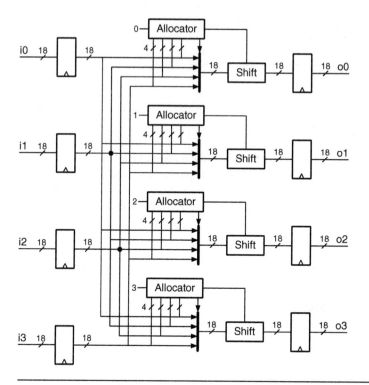

Figure 2.5 Block diagram of router for simple network.

The second stage of the allocator is a four-input fixed-priority[5] arbiter. The arbiter accepts four `request` signals and generates four `grant` signals. If the output port is available (as indicated by signal `avail`), the arbiter grants the port to the first (uppermost) input port making a request. Asserting a `grant` signal causes the corresponding `select` signal to be asserted, selecting the corresponding input of the multiplexer to be passed to the output. If any of the grant signals is asserted, it indicates that a header is being passed through the multiplexer and the `shift` signal is asserted to cause the shifter to shift the routing field.

The final stage of the allocator holds the assignment of the output port to an input port for the duration of a packet. Signal $last_i$ indicates that input port i was selected on the last cycle. If the port carries a payload phit this cycle, then that payload is part of the same packet and the channel is held by asserting signal $hold_i$. This signal causes $select_i$ to be asserted and `avail` to be cleared, preventing the arbiter from assigning the port to a new header.

---

5. In practice, one would never use a fixed-priority arbiter in an application like this, as it results in an extremely unfair router that can lead to livelock or starvation problems when used in a network. In Chapters 18 and 19 we will examine better methods of arbitration and allocation.

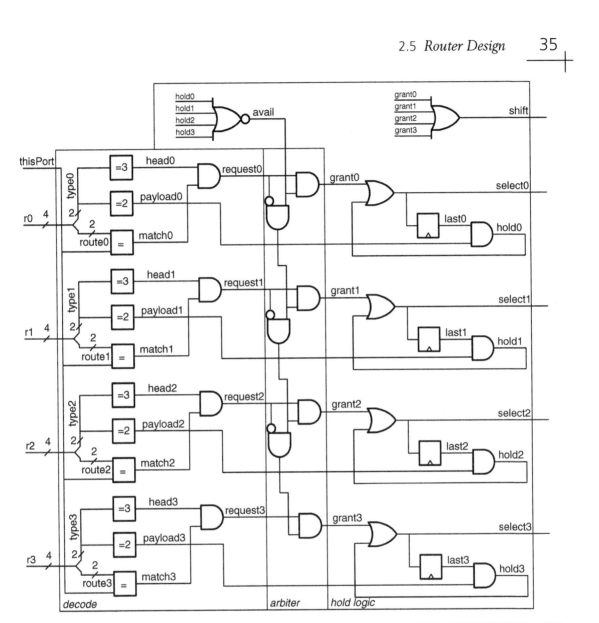

Figure 2.6 Allocator for the router of Figure 2.5.

In this book, we will often describe hardware by using Verilog register-transfer language (RTL) models. A Verilog model is a textual description of a module that describes its inputs, outputs, and internal functions. A Verilog description of the allocator of Figure 2.6 is given in Figure 2.7. The module declaration begins, after a comment, with the module declaration that gives the module name, alloc, and declares its eight inputs and outputs. The next five lines declare these inputs and outputs in terms of dimension and width. Next, internal wires and registers are declared. The real logic starts with the ten assign statements. These describe the combinational logic of the allocator and correspond exactly to the schematic of

```
// allocator: assigns output port to input port based on type
//            of input phit and current field of routing header
//            once assigned, holds a port for the duration of a packet
//            (as long as payload phits are on input).
//            uses fixed priority arbitration (r0 is highest).
module alloc(clk, thisPort, r0, r1, r2, r3, select, shift) ;
  input clk ;                    // chip clock
  input [1:0] thisPort ;         // identifies this output port
  input [3:0] r0,r1,r2,r3 ;      // top four bits of each input phit
  output [3:0] select ;          // radial select to multiplexer
  output shift ;                 // directs shifter to discard upper two bits
  wire [3:0] grant, select, head, payload, match, request, hold ;
  wire [2:0] pass ;
  reg [3:0] last ;
  wire avail ;

  assign head =     {r3[3:2]==3,r2[3:2]==3,r1[3:2]==3,r0[3:2]==3} ;
  assign payload =  {r3[3:2]==2,r2[3:2]==2,r1[3:2]==2,r0[3:2]==2} ;
  assign match =    {r3[1:0]==thisPort,r2[1:0]==thisPort,
                     r1[1:0]==thisPort,r0[1:0]==thisPort} ;
  assign request = head&match ;
  assign pass =     {pass[1:0],avail}&~request[2:0] ;
  assign grant =    request&{pass,avail} ;
  assign hold =     last&payload ;
  assign select =   grant|hold ;
  assign avail =    ~(|hold) ;
  assign shift =    |grant ;

  always @(posedge clk) last = select ;
endmodule
```

Figure 2.7 Verilog code for the allocator.

Figure 2.6. Finally, the `always` statement defines the flip-flops that hold the state `last[3:0]`.

In addition to being a convenient textual way to describe a particular piece of hardware, Verilog also serves as a simulation input language and as a synthesis input language. Thus, after describing our hardware in this manner, we can simulate it to verify proper operation, and then synthesize a gate-level design for implementation on an ASIC or FPGA. For your reference, a Verilog description of the entire router is given in Figure 2.8. The descriptions of the multiplexer and shifter modules are omitted for brevity.

2.6 **Performance Analysis**

We judge an interconnection network by three measures: cost, latency, and through-put. Both latency and throughput are performance metrics: latency is the time it takes a packet to traverse the network and throughput is the number of bits per second the network can transport from input to output. For our example network

```
// simple four-input four output router with dropping flow control
module simple_router(clk,i0,i1,i2,i3,o0,o1,o2,o3) ;
   input clk ;              // chip clock
   input [17:0]   i0,i1,i2,i3 ;   // input phits
   output [17:0] o0,o1,o2,o3 ;    // output phits

   reg [17:0] r0,r1,r2,r3 ;  // outputs of input registers
   reg [17:0] o0,o1,o2,o3 ;  // output registers
   wire [17:0] s0,s1,s2,s3 ; // output of shifters
   wire [17:0] m0,m1,m2,m3 ; // output of multiplexers
   wire [3:0] sel0, sel1, sel2, sel3 ; // multiplexer control
   wire shift0, shift1, shift2, shift3 ; // shifter control

   // the four allocators
   alloc a0(clk, 2'b00, r0[17:14], r1[17:14], r2[17:14], r3[17:14], sel0, shift0) ;
   alloc a1(clk, 2'b01, r0[17:14], r1[17:14], r2[17:14], r3[17:14], sel1, shift1) ;
   alloc a2(clk, 2'b10, r0[17:14], r1[17:14], r2[17:14], r3[17:14], sel2, shift2) ;
   alloc a3(clk, 2'b11, r0[17:14], r1[17:14], r2[17:14], r3[17:14], sel3, shift3) ;

   // multiplexers
   mux4_18 mx0(sel0, r0, r1, r2, r3, m0) ;
   mux4_18 mx1(sel1, r0, r1, r2, r3, m1) ;
   mux4_18 mx2(sel2, r0, r1, r2, r3, m2) ;
   mux4_18 mx3(sel3, r0, r1, r2, r3, m3) ;

   // shifters
   shiftp sh0(shift0, m0, s0) ;
   shiftp sh1(shift1, m1, s1) ;
   shiftp sh2(shift2, m2, s2) ;
   shiftp sh3(shift3, m3, s3) ;

   // flip flops
   always @(posedge clk)
     begin
       r0=i0 ; r1=i1 ; r2=i2 ; r3=i3 ;
       o0=s0 ; o1=s1 ; o2=s2 ; o3=s3 ;
     end
endmodule
```

Figure 2.8 Verilog code for the router.

with dropping flow control, these performance metrics are heavily influenced by the probability that a packet will be dropped.

Our analysis begins with a simple model of the network for the case where dropped packets are resent by the network. (See Figure 2.9.) First, because of symmetry between the inputs and outputs of the network and the random traffic pattern, it is sufficient to consider the packets from a single input of the network. As shown in Figure 2.9, packets are injected into the network at a rate of λ. Instead of expressing λ in bits per second, it is normalized to the channel bandwidth of 2 Gbytes/s, so that $\lambda = 1$ corresponds to injecting packets at the maximum rate allowed by the channel. Before packets enter the network, they are merged with packets that are being resent through the network. The sum of these two rates p_0 is

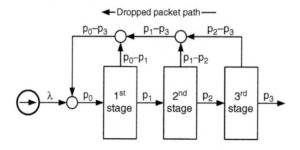

Figure 2.9 A simple analytical model of a 3-stage butterfly network with dropping flow control and reinjection of dropped packets.

the total rate of packets injected into the first stage of the network. In the first stage, some collisions of packets may occur and a smaller fraction of packets p_1 will pass through without being dropped. The difference in rates $p_0 - p_1$ represents the packets that have been dropped. If dropped packets are to be resent, these packets flow back to the input and are reinjected. Similarly, the output rates of the second stage p_2 and third stage p_3 will continue to decrease due to more collisions and dropped packets.

These output rates can be calculated iteratively by starting at the input of the network and working toward the output. A single stage of our butterfly network is built from 4×4 crossbar switches and, by symmetry, the rate of incoming packets at each input of the switch is equal. So, for stage $i + 1$ of the network, the input rate at each port of the crossbar is p_i. Because the rates have been normalized, they can also be interpreted as the probability of a packet arriving at an input during any particular cycle. Then, the probability that a packet leaves a particular output p_{i+1} is one minus the probability that no packet wants that output. Since the traffic pattern is random, each input will want an output with probability $p_i/4$, and therefore the probability that no input wants a particular output is just

$$\left(1 - \frac{p_i}{4}\right)^4 . \tag{2.1}$$

Therefore, the output rate p_{i+1} at stage $i + 1$ is

$$p_{i+1} = 1 - \left(1 - \frac{p_i}{4}\right)^4 . \tag{2.2}$$

Applying Equation 2.2 $n = 3$ times, once for each stage of the network, and momentarily ignoring resent packets ($p_0 = \lambda$), we calculate that with an input duty factor of $\lambda = 0.125$ (corresponding to a speedup of 8), the duty factors at the outputs of the three switch stages are 0.119, 0.114, and 0.109, respectively. That is, with an *offered traffic* of 0.125 of link capacity at the input of the network, the *accepted traffic* or *throughput* of the network is only 0.109. The remaining 0.016 (12.6% of the packets) was dropped due to collisions in the network. Once dropped packets are reinjected into the network, an interesting dynamic occurs. Resending dropped packets increases the effective input rate to the network p_0. This in turn increases the amount of dropped packets, and so on. If this feedback loop stabilizes such that the network can support the resulting input rate ($p_0 \leq 1$), then the amount of traffic injected into the network will equal the amount ejected ($p_3 = \lambda$).

Figure 2.10 plots the relationship between offered traffic and throughput for our example network. Both axes are normalized to the ideal capacity of the network. We see that at very low loads almost all of the traffic gets through the network, and throughput is equal to offered traffic. As the offered traffic is increased, however, dropping quickly becomes a major factor and if packets are not resent, the throughput of the network drops well below the offered traffic. Eventually throughput *saturates*, reaching an asymptote at 43.2%. No matter how much traffic is offered to the network, we cannot achieve a throughput greater than 43.2% of the channel capacity whether packets are resent or not. Note that we could operate this network with a speedup as low as 2.5, effectively limiting the maximum injection rate to 0.4. However, the original choice of a speedup of 8 will have benefits in terms of latency, as we will see. Also, the fact our network is achieving less than half its capacity is the main reason dropping flow control is almost never used in practice. In Chapter 12, we shall see how to construct flow control mechanisms that allow us to operate networks above 90% of channel capacity without saturating.

Throughout this discussion of throughput, we have expressed both offered traffic and throughput as a fraction of channel capacity. We follow this convention through-

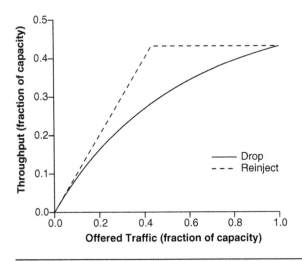

Figure 2.10 Throughput as a function of offered traffic (injection rate) for our simple network. Both the throughput for when packets are dropped and not reinjected (drop) and when they are reinjected (reinject) are shown.

out the book because it lends considerably more insight than expressing these figures in terms of bits per second. For example, we get far more insight into the relative performance of the flow control method by stating that the network saturates at 24% capacity than stating that it saturates at 480 Mbytes/s. In the discussion below, we normalize latency in a similar manner.

The latency of a packet is determined by the number of times it must be retransmitted before it successfully traverses the network. With no other packets in the network, the header of a packet traverses the network in 6 clock cycles. The packet is clocked through two flip-flop registers in each of the three stages. For clarity, we will refer to this 6-cycle delay as a relative latency of 1.0 and express network latency at higher loads relative to this number. In this section we are concerned only with *header latency*. The overall latency of the packet also includes a *serialization* term equal to the length of the packet divided by the bandwidth of the channel, L/b, that reflects the time required for the tail of the packet to catch up with the head.

As the load is increased, a fraction of packets p_D is dropped and must be retransmitted.

$$P_D = \frac{p_0 - p_3}{p_0}$$

The same fraction of these retransmitted packets is dropped again and must be retransmitted a second time, and so on. We calculate the average latency of a packet by summing the latencies for each of these cases weighted by the probability of the case. Assuming, unrealistically, that the source discovers immediately that a packet is dropped and retransmits it after 6 cycles, then the latency of a packet that is dropped i times is $i + 1$. The probability that a packet is dropped exactly i times is $P_D^i(1 - P_D)$. Summing the weighted latencies, we calculate the average latency as

$$T = \sum_{i=0}^{\infty} (i + 1) P_D^i (1 - P_D) = \frac{1}{1 - P_D} = \frac{p_0}{p_3}. \tag{2.3}$$

This latency is plotted as a function of offered traffic in Figure 2.11. With no load on the network, latency starts at unity (6 clock cycles). As the throughput is increased, some packets are dropped and the average latency increases, doubling at a throughput of about 0.39. Finally, the network reaches saturation at a throughput of 0.43. At this point, no greater throughput can be realized at any latency.

For modest loads, Equation 2.3 gives a reasonable model for latency, but as the network approaches saturation the assumption that a packet can be immediately resent is brought into question. This is because there is an increasing chance that a resent packet and a newly injected packet will simultaneously be introduced into the network. In a real implementation, one packet would have to wait for the other to go ahead, thus incurring an additional *queueing latency*. Figure 2.11 also shows another latency curve that incorporates a model for queueing time. The shape of this curve, with latency growing to infinity as throughput approaches saturation, is more typical in an interconnection network. Both curves are compared to a simulation in Exercise 2.9.

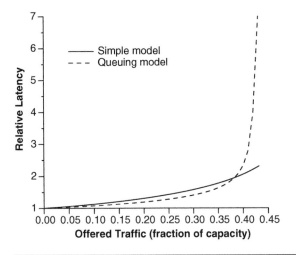

Figure 2.11 Relative latency as a function of offered traffic (injection rate) for our simple network. The solid curve shows the simple model presented in the text, while the dashed curve incorporates an additional queueing delay.

It is important to keep in mind that Equation 2.3 and Figure 2.11 give the *average* latency of a packet. For many applications, we are interested not only in the average latency, but also in the probability distribution of latencies. In particular, we may be concerned about the worst case latency or the variation in latency (sometimes called *jitter*). For example, in a video playback system, the size of the buffer required to hold packets before playback is determined by the jitter, not the average latency. For our example network, the probability that a packet is received with a relative latency of i is given by

$$P(T = i) = P_D^{(i-1)} (1 - P_D) .$$

This exponential distribution of latency results in an infinite maximum latency and, hence, an infinite jitter. More realistically, we might express the jitter in terms of the bound of delays achieved by a given fraction (for example, 99%) of the packets.

The performance measures discussed above are all for *uniform random traffic*. For a butterfly network, this is the best case. As we shall see, for certain traffic patterns, like bit-reversal,[6] the performance of the network is far worse than described here. The sensitivity of the butterfly network to bad traffic patterns is largely due to the

---

6. In a bit-reversal traffic pattern, the node with binary address $\{b_{n-1}, b_{n-2}, \ldots, b_0\}$ sends a packet to the node with address $\{b_0, b_1, \ldots, b_{n-1}\}$.

fact that there is just a single path from each input of the network to each output. We shall see that networks with *path diversity* fare far better under difficult loads.

2.7 **Exercises**

2.1 *Cost of the simple network.* Compute the cost of our simple network using the data in Tables 2.1 and 2.2.

2.2 *Incorporating a power constraint.* Limit the number of chips per board to six to ensure enough power can be delivered to these chips, and their heat can be properly dissipated. Suggest a packaging that meets this new constraint in addition to the original set of constraints. What is the cost of this packaging?

2.3 *Fair allocators.* Modify the Verilog code for the allocator given in Figure 2.7 to implement a more fair type of arbitration. Verify your new allocator via simulation and describe your design.

2.4 *Increasing the degree of the router.* If the simple router is extended to be a 5×5 rather than a 4×4 switch, it can be used to implement a 2-D mesh or torus network. Describe how the router and packet format can be extended to add an additional input and output port.

2.5 *Reducing multiple drops of the same packet.* Change the Verilog code to give priority to retransmitted packets. That is, if two head phits are requesting the same output at the same switch in the butterfly, the allocator should always give the resource to a retransmitted packet first. Add a priority field to the phit header to include this information and assume it is appropriately set when a packet is reinjected.

2.6 *Multiple drops and average latency.* Does introducing a scheme to reduce multiple drops of a packet, such as the one in Exercise 2.5, reduce the *average* packet latency? Explain why or why not.

2.7 *Effects of a larger butterfly on dropping.* If we add more stages to our example butterfly network (and increase the number of nodes accordingly), will the fraction of packets dropped increase? How will the fraction dropped change if the degree of the switches is increased instead? Considering only the dropping probability, is it more efficient to expand the degree of the switches or the number of stages when adding nodes?

2.8 *Realistic drop delays.* In real networks there is usually a significant delay before a dropped packet is retransmitted. First, time must be allowed for an acknowledgment to reach the source, then a timeout is allowed to account for delay in the acknowledgment. Modify Equation 2.3 to reflect this delay.

2.9 **Simulation.** Write a simple computer program to simulate our example network. Using this program, experimentally determine the network's latency as a function of offered traffic. How do your results compare to the analytical results from Equation 2.3 and Figure 2.11? Also compare against the relative delay given by the queueing model

$$T = \frac{T_0}{2} + \frac{p_0}{p_3}\left(\frac{T_0}{2} + \frac{p_0}{2(1 - p_0)}\right)$$

where T_0 is the zero-load latency — the latency of a packet that is never dropped by the network.

2.10 Simulation. Add a timeout mechanism to the simulator from Exercise 2.9 and compare the results from the model developed in Exercise 2.8. Comment on any major differences.

CHAPTER 3

Topology Basics

Network topology refers to the static arrangement of channels and nodes in an interconnection network — the roads over which packets travel. Selecting the network topology is the first step in designing a network because the routing strategy and flow-control method depend heavily on the topology. A roadmap is needed before a route can be selected and the traversal of that route scheduled. As illustrated by the example in Chapter 2, the topology specifies not just the type of network (for example, butterfly), but also the details, such as the radix of the switch, the number of stages, and the width and bit-rate of each channel.

Selecting a good topology is largely a job of fitting the requirements of the network to the available packaging technology. On one hand, the design is driven by the number of ports and the bandwidth and duty factor per port, and on the other hand, by the pins available per chip and board, by wire density, by the available signaling rate, and by the length requirements of cables.

We choose a topology based on its cost and performance. The cost is determined by the number and complexity of the chips required to realize the network, and the density and length of the interconnections, on boards or over cables, between these chips. Performance has two components: bandwidth and latency. Both of these measures are determined by factors other than topology — for example, flow control, routing strategy, and traffic pattern. To evaluate just the topology, we develop measures, such as bisection bandwidth, channel load, and path delay, that reflect the impact of the topology on performance.

A common pitfall for network designers is to try to match the topology of the network to the data communication of the problem at hand. On the surface, this seems like a good idea. After all, if a machine performs a divide-and-conquer algorithm with a tree-structured communication pattern, shouldn't a tree network be the optimum to handle this pattern? The answer is usually no. For a variety of

45

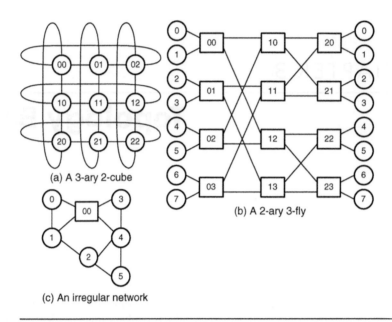

(a) A 3-ary 2-cube

(b) A 2-ary 3-fly

(c) An irregular network

Figure 3.1 Example network topologies: (a) a 3-ary 2-cube, (b) a 2-ary 3-fly, and (c) an irregular network.

reasons, a special purpose network is usually a bad idea. Due to dynamic load imbalance in the problem, or a mismatch between problem size and machine size, the load on such networks is usually poorly balanced. If data and threads are relocated to balance load, the match between the problem and the network is lost. A problem-specific network often does not map well to available packaging technology, requiring long wires or a high node degree. Finally, such networks are inflexible. If the algorithm changes to use a different communication pattern, the network cannot be easily changed. It is almost always better to use a good general purpose network than to design a network with a topology matched to the problem.

Figure 3.1 shows three example topologies. A 2-D torus with three nodes per dimension, a 3-ary 2-cube, is shown in Figure 3.1(a). Each node in the cube network is both a terminal and a switching node. Figure 3.1(b) shows a three-stage radix-two butterfly, a 2-ary 3-fly. The butterfly network makes a clear distinction between terminal nodes and switch nodes with terminal-only nodes at either end and switch-only nodes (rectangles) in the middle. An irregular network is shown in Figure 3.1(c).

3.1 **Nomenclature**

3.1.1 **Channels and Nodes**

The topology of an interconnection network is specified by a set of nodes N^* connected by a set of channels C. Messages originate and terminate in a set of terminal nodes N where $N \subseteq N^*$. In a network where all nodes are terminals, we simply refer to the set of nodes as N. Each channel, $c = (x, y) \in C$, connects a source node, x, to a destination node, y, where $x, y \in N^*$. We denote the source node of a channel c as s_c and the destination as d_c. Note that each edge in Figure 3.1 denotes a pair of channels, one in each direction. This definition of a topology is equivalent to a directed graph and, not surprisingly, much of the terminology used to describe a topology borrows heavily from graph theory. For notational convenience, we will often refer to the number of nodes in a network as simply N^* instead of $|N^*|$ and likewise for the number of channels.

A channel, $c = (x, y)$, is characterized by its width, w_c or w_{xy}, the number of parallel signals it contains; its frequency, f_c or f_{xy}, the rate at which bits are transported on each signal; and its latency, t_c or t_{xy}, the time required for a bit to travel from x to y. For most channels, the latency is directly related to the physical length of the channel, $l_c = vt_c$, by a propagation velocity v. The bandwidth of channel c is $b_c = w_c f_c$. In the common case where the bandwidths of all the channels are the same, we drop the subscript and refer to the network channel bandwidth as b.

Each switch node, x, has a channel set $C_x = C_{Ix} \cup C_{Ox}$. Where $C_{Ix} = \{c \in C | d_c = x\}$ is the input channel set, and $C_{Ox} = \{c \in C | s_c = x\}$ is the output channel set. The degree of x is $\delta_x = |C_x|$ which is the sum of the in degree, $\delta_{Ix} = |C_{Ix}|$, and the out degree, $\delta_{Ox} = |C_{Ox}|$. Where the degree of all $x \in N^*$ is the same, we drop the subscript and denote degree by δ.

3.1.2 Direct and Indirect Networks

A network node may be a terminal node that acts as a source and sink for packets, a switch node that forwards packets from input ports to output ports, or both. In a direct network, such as the torus of Figure 3.1(a), every node in the network is both a terminal and a switch. In an indirect network, such as the butterfly of Figure 3.1(b), on the other hand, a node is either a terminal (round nodes) or a switch (rectangular nodes). It cannot serve both functions. In a direct network, packets are forwarded directly between terminal nodes, while in an indirect network they are forwarded indirectly by means of dedicated switch nodes. Some networks, like the random network of Figure 3.1(c), are neither direct nor indirect. Every direct network can be redrawn as an indirect network by splitting each node into separate terminal and switch nodes, as illustrated in Figure 3.2. With such networks, the distinction between direct and indirect networks is largely academic.

A potential advantage of a direct network is that the resources of a terminal (which usually include a computer) are available to each switch. In some early networks, the switching function was implemented in software running on the terminal CPU, and buffering was performed using the terminal computer's memory [163, 192]. Software switching is, however, both very slow and demanding of the terminal's resources. Thus, it is rarely used today.

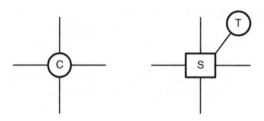

Figure 3.2 A combined node consists of a terminal node and a switch node.

3.1.3 **Cuts and Bisections**

A cut of a network, $C(N_1, N_2)$, is a set of channels that partitions the set of all nodes N^* into two disjoint sets, N_1 and N_2. Each element of $C(N_1, N_2)$ is a channel with a source in N_1 and destination in N_2, or vice versa. The number of channels in the cut is $|C(N_1, N_2)|$ and the total bandwidth of the cut is

$$B(N_1, N_2) = \sum_{c \in C(N_1, N_2)} b_c.$$

A bisection of a network is a cut that partitions the entire network nearly in half, such that $|N_2| \le |N_1| \le |N_2| + 1$, and also partitions the terminal nodes nearly in half, such that $|N_2 \cap N| \le |N_1 \cap N| \le |N_2 \cap N| + 1$. The channel bisection of a network, B_C, is the minimum channel count over all bisections of the network.

$$B_C = \min_{\text{bisections}} |C(N_1, N_2)|$$

The bisection bandwidth of a network, B_B, is the minimum bandwidth over all bisections of the network.

$$B_B = \min_{\text{bisections}} B(N_1, N_2)$$

For networks with uniform channel bandwidth b, $B_B = b B_C$. In the cost model developed later in this chapter (Section 3.4), we use the bisection bandwidth of a network as an estimate of the amount of global wiring required to implement it.

3.1.4 **Paths**

A path in a network is an ordered set of channels $P = \{c_1, c_2, \ldots, c_n\}$, where $d_{c_i} = s_{c_{i+1}}$ for $i = 1 \ldots (n - 1)$. Paths are also referred to as *routes*. The source of a path is, $s_P = s_{c_1}$. Similarly, the destination of a path is $d_P = d_{c_n}$. The length or *hop count* of a path is $|P|$. If, for a particular network and its routing function, at least one path exists between all source-destination pairs, it is said to be *connected*.

A *minimal path* from node x to node y is a path with the smallest hop count connecting these two nodes. The set of all minimal paths from node x to node y is denoted R_{xy}. $H(x, y)$, is the hop count of a minimal path between x and y. The *diameter* of a network H_{max} is the largest, minimal hop count over all pairs of terminal nodes in the network.

$$H_{max} = \max_{x, y \in N} H(x, y)$$

For a fully-connected network with N terminals built from switches with out degree δ_O, H_{max} is bounded by

$$H_{max} \geq \log_{\delta_O} N. \tag{3.1}$$

or for symmetric switches where $\delta_I = \delta_O = \delta/2$,

$$H_{max} \geq \log_{\delta/2} N.$$

Each terminal can reach at most δ_O other terminals after one hop, at most δ_O^2 after two hops, and at most δ_O^H after H hops. If we set $\delta_O^H = N$ and solve for H, we get Equation 3.1, which provides a lower bound on network diameter. Networks for which this bound is tight, such as butterfly networks, have no path diversity. All of the decisions are *used up* selecting the destination node and no decisions are left to select between alternate paths.

The average minimum hop count of a network H_{min} is defined as the average hop count over all sources and destinations.

$$H_{min} = \frac{1}{N^2} \sum_{x, y \in N} H(x, y)$$

While H_{min} represents the smallest possible average hop count, a specific implementation may choose to incorporate some non-minimal paths. In this case, the actual average hop count H_{avg} is defined over the paths used by the network, not just minimal paths, and $H_{avg} \geq H_{min}$.

The physical distance of a path is

$$D(P) = \sum_{c \in P} l_c$$

and the delay of a path is $t(P) = D(P)/v$. Distance and delay between node pairs and average and maximum distances and delays for a network are defined in the same manner as hop counts.

3.1.5 **Symmetry**

The symmetry of a topology plays an important role in load-balance and routing as we will discuss in later sections. A network is *vertex-symmetric* if there exists an automorphism that maps any node a into another node b. Informally, in a vertex-symmetric network, the topology looks the same from the point-of-view of all the nodes. This can simplify routing, because all nodes share the same roadmap of the network and therefore can use the same directions to route to the same relative position.

In an *edge-symmetric* network, there exists an automorphism that maps any channel a into another channel b. Edge symmetry can improve load balance across the channels of the network since there is no reason to favor one channel over another.

3.2 Traffic Patterns

Before introducing the performance metrics for topologies, it is useful to consider the spatial distribution of messages in interconnection networks. We represent these message distributions with a traffic matrix Λ, where each matrix element $\lambda_{s,d}$ gives the fraction of traffic sent from node s destined to node d. Table 3.1 lists some common static traffic patterns used to evaluate interconnection networks. Historically, several of these patterns are based on communication patterns that arise in particular applications. For example, matrix transpose or corner-turn operations induce the

Table 3.1 Network traffic patterns. Random traffic is described by a traffic matrix, Λ, with all entries $\lambda_{sd} = 1/N$. Permutation traffic, in which all traffic from each source is directed to one destination, can be more compactly represented by a permutation function π that maps source to destination. Bit permutations, like transpose and shuffle, are those in which each bit d_i of the b-bit destination address is a function of one bit of the source address, s_j where j is a function of i. In digit permutations, like tornado and neighbor, each (radix-k) digit of the destination address d_x is a function of a digit s_y of the source address. In the two digit permutations shown here, $x = y$. However, that is not always the case.

Name	Pattern
Random	$\lambda_{sd} = 1/N$
Permutation	$d = \pi(s)$
Bit permutation	$d_i = s_{f(i)} \oplus g(i)$
Bit complement	$d_i = \neg s_i$
Bit reverse	$d_i = s_{b-i-1}$
Bit rotation	$d_i = s_{i+1 \bmod b}$
Shuffle	$d_i = s_{i-1 \bmod b}$
Transpose	$d_i = s_{i+b/2 \bmod b}$
Digit permutations	$d_x = f(s_{g(x)})$
Tornado	$d_x = s_x + (\lceil k/2 \rceil - 1) \bmod k$
Neighbor	$d_x = s_x + 1 \bmod k$

transpose pattern, whereas fast Fourier transform (FFT) or sorting applications might cause the shuffle permutation [175], and fluid dynamics simulations often exhibit neighbor patterns. Temporal aspects of traffic also have important effects on network performance and are discussed in Section 24.2.

Random traffic, in which each source is equally likely to send to each destination is the most commonly used traffic pattern in network evaluation. Random traffic is very benign because, by making the traffic uniformly distributed, it balances load even for topologies and routing algorithms that normally have very poor load balance. Some very bad topologies and routing algorithms look very good when evaluated only with random traffic.

To stress a topology or routing algorithm, we typically use permutation traffic in which each source s sends all of its traffic to a single destination, $d = \pi(s)$. The traffic matrix Λ for a permutation is a permutation matrix Π where each row and each column contains a single entry with all other entries zero. Because they concentrate load on individual source-destination pairs, permutations stress the load balance of a topology and routing algorithm.

Bit permutations are a subset of permutations in which the destination address is computed by permuting and selectively complementing the bits of the source address. For example, if the four-bit source address is $\{s_3, s_2, s_1, s_0\}$, the destination for a bit-reversed traffic pattern is $\{s_0, s_1, s_2, s_3\}$, for a bit-complement traffic pattern the destination is $\{\neg s_3, \neg s_2, \neg s_1, \neg s_0\}$, and for a shuffle, the destination is $\{s_2, s_1, s_0, s_3\}$.

Digit permutations are a similar subset of permutations in which the digits of the destination address are calculated from the digits of the source address. Such permutations apply only to networks in which the terminal addresses can be expressed as n-digit, radix-k numbers, such as k-ary n-cube (torus) networks (Chapter 5) and k-ary n-fly (butterfly) networks (Chapter 4). The tornado pattern is designed as an adversary for torus topologies, whereas neighbor traffic measures a topology's ability to exploit locality.

3.3 **Performance**

We select a topology for a network based on its cost and performance. In this section, we address the three key metrics of performance: throughput, latency, and path diversity. These measures are revisited in Section 3.4, where they are tied to the implementation cost of a network.

3.3.1 **Throughput and Maximum Channel Load**

The *throughput* of a network is the data rate in bits per second that the network accepts per input port. Throughput is a property of the entire network and depends on routing and flow control (as we have seen in Chapter 2) as much as on the topology. However, we can determine the *ideal* throughput of a topology by measuring the throughput that it could carry with perfect flow control and routing. This is the throughput that would result if the routing perfectly balanced the load over alternative paths in the network and if the flow control left no idle cycles on the bottleneck channels. For simplicity, we present only the throughput and load equations for networks where all the channel bandwidths are b in this section. This limitation is removed as part of Exercise 3.5.

Maximum throughput occurs when some channel in the network becomes saturated. If no channels are saturated, the network can carry more traffic and is thus not operating at maximum throughput. Thus, to compute throughput, we must consider channel load. We define the load on a channel c, γ_c, as the ratio of the bandwidth demanded from channel c to the bandwidth of the input ports. Equivalently, this ratio is the amount of traffic that must cross channel c if each input injects one unit of traffic according to the given traffic pattern. Because it is a ratio, channel load is a dimensionless quantity. Unless otherwise specified, we consider channel loads under uniform traffic.

Under a particular traffic pattern, the channel that carries the largest fraction of the traffic determines the *maximum channel load* γ_{max} of the topology, $\gamma_{max} = \max_{c \in C} \gamma_c$. When the offered traffic reaches the throughput of the network, the load on this bottleneck channel will be equal to the channel bandwidth b. Any additional traffic with the specified pattern would overload this channel. Thus, we define the ideal throughput of a topology Θ_{ideal} as the input bandwidth that saturates the bottleneck channel

$$\Theta_{ideal} = \frac{b}{\gamma_{max}} \qquad (3.2)$$

where b is in bits per second and γ_{max} is unitless.

Computing γ_{max} for the general case of an arbitrary topology and an arbitrary traffic pattern requires solving a multicommodity flow problem as described below. For uniform traffic, however, we can compute some upper and lower bounds on γ_{max} with much less effort.

The load on the bisection channels of a network gives a lower bound on γ_{max} that in turn gives an upper bound on throughput. For uniform traffic, we know on average that half of the traffic, $N/2$ packets, must cross the B_C bisection channels. The best throughput occurs when these packets are distributed evenly across the bisection channels. Thus, the load on each bisection channel γ_B is at least

$$\gamma_{max} \geq \gamma_B = \frac{N}{2B_C}. \qquad (3.3)$$

Combining Equations 3.2 and 3.3 gives us an upper bound on ideal throughput

$$\Theta_{ideal} \leq \frac{2bB_C}{N} = \frac{2B_B}{N}. \qquad (3.4)$$

For example, consider a k node ring under uniform traffic. $B_C = 4$ channels cross the network bisection, two in each direction. (Figure 3.3 shows an 8-node ring.) Thus, from Equation 3.4 we know that $\Theta_{ideal} \leq 8b/k$. For the ring, it turns out that this bound is exact.

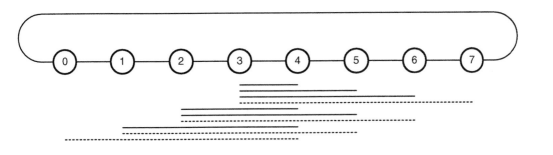

Figure 3.3 An 8-node ring.

Another useful lower bound on channel load can be computed in a similar manner. The product $H_{\min}N$ gives the channel demand — the number of channel traversals required to deliver one round of packets for a given traffic pattern. If we assume the best case in which all the channels are loaded equally, dividing this demand by the number of channels bounds the load on every channel in the network:

$$\gamma_{c,LB} = \gamma_{\max,LB} = \frac{H_{\min}N}{C}. \tag{3.5}$$

These lower bounds can be complemented with a simple upper bound on maximum channel load by considering a routing function that balances load across all minimal paths equally. That is, if there are $|R_{xy}|$ paths, $1/|R_{xy}|$ is credited to each channel of each path. The maximum load $\gamma_{\max,UB}$ is the largest $\gamma_{c,UB}$ over all channels. Mathematically, we define these loads as

$$\gamma_{c,UB} = \frac{1}{N}\sum_{x\in N}\sum_{y\in N}\sum_{P\in R_{xy}} \begin{cases} 1/|R_{xy}| & \text{if } c \in P \\ 0 & \text{otherwise} \end{cases}$$

$$\gamma_{\max,UB} = \max_{c\in C}\gamma_{c,UB}. \tag{3.6}$$

For any topology, $\gamma_{\max,LB} \le \gamma_{\max} \le \gamma_{\max,UB}$ and in the case of an edge-symmetric topology (e.g., tori), both bounds exactly equal $\gamma_{\max}$.

To see how channel load can be used to estimate ideal throughput, consider the case of an eight-node ring network (an 8-ary 1-cube) as shown in Figure 3.3. This topology is edge-symmetric; therefore, our simple bounds are equal to the maximum channel load and we also know that the load on all channels is the same, so it suffices to compute the load on a single channel. We apply the upper bound approach to channel (3,4), the right-going channel from node 3 to node 4. The summation of Equation 3.6 is illustrated by the lines below the network in the figure. Each line denotes a path that uses the channel. The dotted lines represent paths that count as half. For these paths of length four, there is a second minimum path that traverses the network in the clockwise direction. Performing the summation, we see that there are six solid lines (six node pairs that always use (3,4)) and four dotted lines (four node pairs that use (3,4) half the time) for a total summation of eight. Dividing this sum by $N = 8$, we see that the channel load is $\gamma_{\max} = 1$.

We can also verify that the lower bound gives the same maximum channel load. In this case, the average packet travels $H_{\min} = 2$ hops. Each node also contributes two channels, one to the left and one to the right; therefore, the channel load from this approach is also $\gamma_{\max} = H_{\min} N/C = 2 \cdot 8/16 = 1$.

To compute $\gamma_{\max}$ in the general case, we must find an optimal distribution of the packets across the network that minimizes channel load. This task can be formulated as a convex optimization problem. While determining the solutions to these problems is beyond the scope of this book, we will present the problem formulation.

For each destination d in the network, we denote the average distribution of packets destined for d over the channels of the network with a vector x_d of length $|C|$. A valid distribution is maintained by adding flow balance equations at each node: the sum of the incoming distributions minus the sum over the outgoing channels must equal the average number of packets that the node is sourcing (positive values) or sinking (negative values). These balance equations are maintained for each of the distributions x_d for all $d \in N^*$. For a distribution x_d under uniform traffic, all terminal nodes (including the destination d) source $1/N$ units of traffic and the destination d sinks 1 unit. This is represented by using an $|N^*|$ element balance vector f_d:

$$f_{d,i} = \begin{cases} 1/N - 1 & \text{if } d = i \text{ and } d \in N, \\ 1/N & \text{if } d \neq i \text{ and } d \in N, \\ 0 & \text{otherwise} \end{cases} \qquad (3.7)$$

where $f_{d,i}$ is i^{th} element of f_d. We then express the topology by using an $N^* \times C$ node-arc incidence matrix A, where

$$A_{n,c} = \begin{cases} +1 & \text{if } s_c = n, \\ -1 & \text{if } d_c = n, \\ 0 & \text{otherwise.} \end{cases} \qquad (3.8)$$

Then, the objective is to minimize the maximum load on any one channel, and the overall optimization problem is written as

$$\text{minimize} \quad \max_{c \in C} \sum_{d \in N} x_{d,c}$$

$$\text{subject to} \quad A x_d = f_d \quad \text{and} \qquad (3.9)$$

$$x_d \geq 0 \quad \text{for all } d \in N^*.$$

The solution to this convex optimization problem gives the optimal maximum channel load γ_{max}.[1]

Equations 3.6 and 3.9 estimate the throughput of the network under uniform traffic: every node is equally likely to send to every node. Arbitrary traffic patterns can be incorporated into Equation 3.9 by adjusting the balance vectors f_d. Similarly, Equation 3.6 can be adapted by weighting the final summation by λ_{xy}, the probability that x sends to y

$$\gamma_c(\Lambda) = \sum_{x \in N} \sum_{y \in N} \lambda_{xy} \sum_{P \in R_{xy}} \begin{cases} 1/|R_{xy}| & \text{if } c \in P \\ 0 & \text{otherwise.} \end{cases}$$

Finally, Equation 3.5 is modified by substituting the average hop count for an arbitrary traffic pattern:

$$H_{min}(\Lambda) = \sum_{x \in N} \sum_{y \in N} \lambda_{xy} H(x, y).$$

We often refer to the ideal throughput of a network on uniform traffic $\Theta(U)$ as the *capacity* of the network. It is often useful to express throughput of the network on an arbitrary non-uniform traffic pattern Λ as a *fraction of capacity*: $\Theta(\Lambda)/\Theta(U)$. If the channels of a network all have equal bandwidth, it is also equivalent to express the fraction of capacity as $\gamma_{max}(U)/\gamma_{max}(\Lambda)$.

3.3.2 **Latency**

The *latency* of a network is the time required for a packet to traverse the network, from the time the head of the packet arrives at the input port to the time the tail of the packet departs the output port. We separate latency, T, into two components

$$T = T_h + \frac{L}{b}.$$

The *head latency*, T_h, is the time required for the head of the message to traverse the network, and the *serialization latency*,

$$T_s = L/b \qquad (3.10)$$

is the time required for the tail to catch up — that is, the time for a packet of length L to cross a channel with bandwidth b.

Like throughput, latency depends not only on topology but also on routing, flow control, and the design of the router. As above, however, we focus here on the contribution of topology to latency and will consider the other effects later.

---

1. This type of network optimization problem is commonly called a multicommodity flow problem in optimization literature. See [6] for example.

In the absence of contention, head latency is the sum of two factors determined by the topology: router delay, T_r, and time of flight, T_w. Router delay is the time spent in the routers, whereas time of flight is the time spent on the wires. The average router delay is $T_r = H_{\min}t_r$ for a network with an average hop count of $H_{\min}$ and a delay of t_r through a single router. The average time of flight is $T_w = D_{\min}/v$ for a network with an average distance of $D_{\min}$ and a propagation velocity of v.

Combining these components gives the following expression for average latency in the absence of contention

$$T_0 = H_{\min}t_r + \frac{D_{\min}}{v} + \frac{L}{b}. \tag{3.11}$$

The three terms correspond to the three components of total latency: switch delay, time of flight, and serialization latency. We refer to this latency as T_0 as it represents the latency at zero load where no contention occurs. As we increase the load, a fourth term, T_c, is added to the equation, which reflects the time spent waiting for resources (i.e., contention).

The topology of a network and its mapping onto the physical packaging largely determine the three critical parameters in this equation. The average hop count, $H_{\min}$ is entirely a property of the topology. The average distance, $D_{\min}$, is affected by packaging as well as topology. Finally, the bandwidth b is set by the node degree (a property of the topology) and the packaging constraints.

Figure 3.4 shows a Gantt chart of a packet propagating along a two-hop route from node x to node z, via intermediate node y, in a manner that illustrates the three terms of Equation 3.11. The first row of the chart shows each phit of the packet arriving at node x. One hop routing delay, t_r, is incurred before the first phit of this packet leaves node x. Then, a link latency, t_{xy}, occurs before the phit arrives at node y. A second routing delay and link latency are incurred before the first phit arrives at node z. The complete packet is received a serialization delay after the first phit of the message arrives at z. The bar at the bottom of the figure summarizes the sources

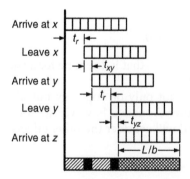

Figure 3.4 Gantt chart showing latency of a packet traversing two channels in the absence of contention.

of latency. The single hatched areas correspond to routing delay, the light gray areas to link latency, and the double-hatched areas to serialization latency.

Consider a 64-node network with $H_{avg} = 4$ hops and 16-bit wide channels. Each channel, c, operates at $f_c = 1$ GHz and takes $t_c = 5$ ns to traverse. If the delay of a single router is $t_r = 8$ ns (4 2 ns clocks), then the total routing delay is $8 \cdot 4 = 32$ ns (16 clocks). For our example, the 4 wire delays are each 5 ns adding $T_w = 20$ ns (10 clocks) to the total. If the packet length is $L = 64$ bytes and the channels all have a uniform 2 Gbytes/s bandwidth, the serialization delay is $64/2 = 32$ ns (16 clocks). Thus, for this network, $T_0 = 32 + 20 + 32 = 84$ ns.

3.3.3 Path Diversity

A network with multiple minimal paths between most pairs of nodes, $|R_{xy}| > 1$ for most $x, y \in N$, is more robust than a network with only a single route from node to node, $|R_{xy}| = 1$. This property, which we call *path diversity*, adds to the robustness of our network by balancing load across channels and allowing the network to tolerate faulty channels and nodes.

So far, we have been primarily concerned with the throughput of a network under random traffic, where each node is equally likely to send a message to any other node. For many networks, this random traffic is a best case load because by uniformly distributing traffic between node pairs, it also uniformly balances the load across network channels. A more challenging case for many networks is arbitrary permutation traffic, where each node, x, sends all of its traffic to exactly one other node, $\pi(x)$, for some permutation, π. Without path diversity, some permutations focus a considerable fraction of traffic on a single bottleneck channel, resulting in substantial degradation of throughput.

As an example of the importance of path diversity in balancing load, consider sending bit-rotation traffic over both a 2-ary 4-fly with unit bandwidth channels and a 4-ary 2-cube with half-unit bandwidth channels. Both of these networks have $\gamma_{max} = 1$ for random traffic. For bit-rotation (BR) traffic, the node with address $\{b_3, b_2, b_1, b_0\}$ sends packets only to the node with address $\{b_2, b_1, b_0, b_3\}$. Expressed differently, this traffic corresponds to the shuffle permutation

$$\{0, 2, 4, 6, 8, 10, 12, 14, 1, 3, 5, 7, 9, 11, 13, 15\}.$$

That is, node 0 sends to itself, node 1 sends to node 2, and so on. Figure 3.5 illustrates the concentration of traffic that occurs when this permutation is applied to a 2-ary 4-fly. As illustrated by the bold lines in the figure, all of the packets from nodes 0, 1, 8, and 9 must traverse channel $(10, 20)$. Similarly, all traffic from nodes 2, 3, 10, and 11 must traverse channel $(11, 23)$; 4, 5, 12, 13 concentrate on $(16, 24)$; and 6, 7, 14, 15 concentrate on $(17, 27)$. Thus, for this traffic pattern, $\gamma_{max,BR} = 4$, and the throughput is therefore $\gamma_{max}/\gamma_{max,BR} = 25\%$ of capacity.

Figure 3.6 shows how this permutation traffic maps onto a 4-ary 2-cube. Two routes traverse no channels, four traverse one channel, four traverse two channels (each with two possible minimal routes, $H(5, 10) = 2$ and $|R_{5,10}| = 2$ for

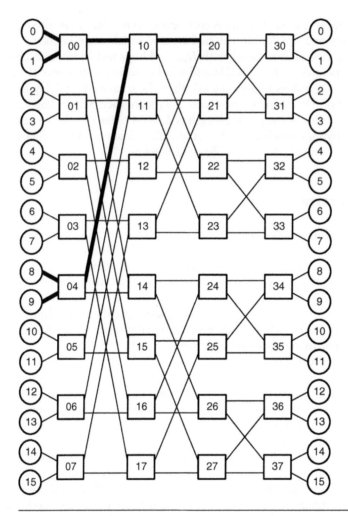

Figure 3.5 Routing a shuffle permutation on a 2-ary 4-fly results in all of the traffic concentrating on one quarter of the channels at the center of the network, degrading performance by a factor of 4.

example), four traverse three channels (with three alternate routes), and four traverse four channels (with 24 alternate routes). With minimal routing, the one-hop channels become the bottleneck. There are no alternative minimal routes and thus $\gamma_{max, BR} = 1$. For this topology, uniform traffic gives $\gamma_{max} = 0.5$ and therefore the throughput under the bit-reversal pattern is $\gamma_{max}/\gamma_{max,BR} = 50\%$ of capacity. However, if we allow the four one-hop routes to route half of their traffic non-minimally and in the opposite direction, the traffic is spread uniformly. For example, half of the traffic from node 1 to 2 takes the route $\{(1, 2)\}$, while the remaining half takes $\{(1, 0), (0, 3), (3, 2)\}$. The resulting throughput is increased to 89% of capacity, and the 11% degradation is because of the extra hops taken by the non-minimal routes.

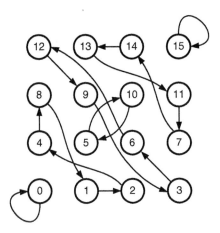

Figure 3.6 Routing a shuffle permutation on a 4-ary 2-cube. With minimal routing, throughput is degraded by a factor of 2, since all of the one-hop traffic must traverse a single channel with bandwidth of 0.5. With non-minimal routing, the traffic is spread uniformly and throughput is 0.89 of capacity.

This example shows how non-minimal routing is often required to balance the load across the channels. Even though more total work is done with non-minimal routing (more channels are traversed), overall performance improves because of better load balancing. However, non-minimal routing can increase implementation complexity. For example, networks that use non-minimal routing require special attention to avoid deadlock, as we shall see in Chapter 14.

Another important advantage of increased path diversity is a network's ability to handle faults, such as a failed node or link. For example, for the butterfly network in Figure 3.5, if the link from switch 07 to 17 fails, there is no possible path from source 14 to destination 15. Since the mean time to failure of an interconnection network decreases with the number of components in the system, it is critical for large interconnection networks to tolerate one or more faulty nodes or links.

One measure of a network's ability to handle faults is the number of *edge-disjoint* or *node-disjoint* paths allowed by the routing function between each source-destination pair. A set of paths is edge-disjoint if the paths do not share any common links. So, if one link in a network fails, it is guaranteed to affect at most one path in an edge-disjoint set of paths. In general, if the minimum number of edge-disjoint paths between all nodes for a given routing function is j, the network is guaranteed to be connected as long as there are fewer than j link failures.

Node-disjoint paths are a set of paths that share no common nodes, excluding the source and destination. Analogous to edge-disjoint paths, if one node fails, it can only affect one path in a set of node-disjoint paths, unless, of course, the failed node is the source or destination of the paths. A node-disjoint set of paths is also edge-disjoint, so a network and routing function that has at least j node-disjoint paths between each source-destination pair can tolerate up to j total link plus node failures.

An unlucky group of faults could all affect the neighbors of a particular node in a network. For this unlucky node, if all its neighboring nodes, all its incoming links, all its outgoing links, or any equivalent combination fails, there is no way to route either to or from this node. So, it is possible that a network will no longer be connected after $\min_x [\min\{|C_{Ix}|, |C_{Ox}|\}]$ failures.

3.4 Packaging Cost

When constructing a network, the nodes of a topology are mapped to packaging modules, chips, boards, and chassis, in a physical system. The properties of the topology and the packaging technology, along with the placement of nodes, determine the constraints on the channels' bandwidth. Without these constraints, one cannot evaluate a topology or perform a fair comparison between topologies. In this section, we develop a simple model of packaging cost based on a typical two-level packaging hierarchy where the channel width w is constrained by both the number of pins per node and the total amount of global wiring. We also discuss how frequency f, the other factor in a channel's bandwidth, is affected by packaging choices.

At the first level of hierarchy in the packaging model, individual routers are connected by local wiring. Compared to global wiring, local wiring is inexpensive and abundant, so a topology must be arranged to take advantage of spatial locality — neighboring nodes should be placed physically close to one another. This arrangement allows local wiring to be used in place of global wiring. For example, consider a system in which 16 nodes can be fit on a single printed circuit (PC) board. By packaging a 4×4 array of nodes on a PC board as shown in Figure 3.7, three quarters of all pins remain local to the PC board and only 32 channels need to cross the module boundary.

Once an efficient local arrangement of nodes is found, the main constraint on channel width becomes the available number of pins on each node. For a node with a maximum pin count of W_n per node, the channel width w is constrained by

$$w \leq \frac{W_n}{\delta}. \tag{3.12}$$

The second level of the hierarchy connects local groups of nodes via global wiring. A typical realization of this global wiring is a backplane connecting several individual boards of nodes, as shown in Figure 3.8. Any global signal travels from one board across an electrical connector, onto the backplane, and across another connector to another board. At this level, the number of available global wires, W_s, limits the width of the individual channels. For example, a network constructed on a backplane has a wire bisection limited by the wire density of the backplane. Allowing space for through-hole vias, a typical PC board can support a wire density of 1 wire/mm, or 0.5 signals/mm for differential signals, on each signal layer. A moderate-cost PC board may have a total of 8 wiring layers, 4 in the x direction and 4 in the y direction, for a total wire density of 2 signals/mm in each direction.

To estimate the number of global channels required for a particular topology, we use the minimum channel bisection B_C of the topology. The minimum bisection is

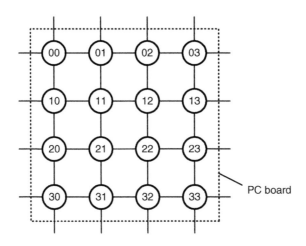

Figure 3.7 A 4 × 4 array of nodes is packaged on a PC board so that $\frac{3}{4}$ of all node pins are connected to other pins on the board using local wiring.

a cut of the network that partitions the network almost in half while cutting as few wires as possible. Therefore, the two sets of nodes created by a bisection represent a good partitioning of nodes into local groups for packaging. While this model is limited in that many networks may have to be partitioned into more than two local groups to meet the constraints of the packaging technology, it is generally a good estimate in these cases.

Using the minimum bisection, the available global wiring constrains the channel width to

$$w \leq \frac{W_s}{B_C}. \tag{3.13}$$

While our discussion has focused on a two-level packaging hierarchy, Equation 3.13 can also be applied to additional levels of packaging necessary in larger systems. (See Exercise 5.3.)

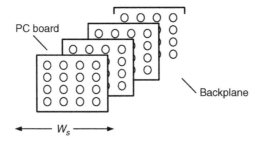

Figure 3.8 The connection of several boards of nodes with a backplane.

By combining Equations 3.12 and 3.13, we get an overall constraint on channel width

$$w \leq \min \left(\frac{W_n}{\delta}, \frac{W_s}{B_C} \right).$$ (3.14)

The first term of Equation 3.14 tends to dominate networks with low degree (e.g., rings). These networks are node-pin limited. Networks with high degree (e.g., binary n-cubes), on the other hand, tend to be bisection limited with the second term of Equation 3.14 dominating.

We can also express our packaging constraints in terms of bandwidth rather than channel width. The maximum bandwidth of a node is $B_n = f W_n$ and the maximum bandwidth across the bisection of the system is $B_s = f W_s$. Using bandwidth, we can rewrite Equation 3.14 to give the maximum bandwidth per channel as

$$b \leq \min \left(\frac{B_n}{\delta}, \frac{B_s}{B_C} \right).$$

In addition to the width of wiring available at different levels of the packaging hierarchy, another important consideration is the length of these wires. The length of the network channels must be kept short because, above a critical length, the frequency of a signal falls off quadratically with wire length:[2]

$$f = \min \left(f_0, f_0 \left(\frac{l_w}{l_c} \right)^{-2} \right).$$

The critical length of a wire, l_c, is a function of the nominal signaling rate[3] of the wire, f_0, the physical properties of the wire, and the amount of frequency-dependent attenuation that can be tolerated by the system.

Table 3.2 shows the critical length for several common types of wires at a signaling rate of 2 GHz, assuming no more than 1 dB of attenuation can be tolerated.[4] Density-cost constraints lead most networks to be constructed from stripguides in PC boards, fine wire cables, or both. These interconnects, which correspond to the first two rows of the table, can be run only a short distance, less than a meter, before the critical wire length is reached and data rate begins to drop off quadratically with length.

By inserting repeaters into the lines, one can build long channels and operate them at high bit rates. However, inserting a repeater is about the same cost as inserting

---

2.   This bandwidth limitation is caused by the skin-effect resistance of the wire. See Section 3.3.4 of [55] for a detailed treatment.
3.   Note that a signaling rate of f_0 corresponds to a maximum frequency of $f_0/2$.
4.   Much higher amounts of frequency-dependent attenuation can be tolerated, and hence much longer wires can be driven, by equalizing the signal [54].

Table 3.2 Critical length of common wires at 2 GHz (without equalization).

Wire Type	l_c
5 mil stripguide	0.10 m
30 AWG pair	0.56 m
24 AWG pair	1.11 m
RG59U coax	10.00 m

a switch, so there is little point in this exercise. One would be better off using a topology that keeps the channels under the critical length, inserting switches rather than repeaters into long routes. The relationship between wire speed and wire length makes it impractical to build high-speed electrical networks using topologies that require long channels.

One can also build long channels that operate at high bit rates by using optical signaling. While optical fibers also attenuate and disperse signals, limiting distance, they do so at a much lower rate than electrical transmission lines. Single-mode optical fibers can transmit signals tens to hundreds of kilometers before they need to be regenerated. The downside of optical channels is cost. An optical channel in 2003 costs over ten times as much as an electrical channel of the same bandwidth.

A comparison between two different 6-node topologies is shown in Figure 3.9. The first topology is a simple 6-ring, with $\delta = 4$ and $B_C = 4$. The second topology falls under a large class of graphs known as Cayley graphs and has $\delta = 6$ and $B_C = 10$. Our packaging technology allows $W_n = 140$ pins per node and a backplane that is $W_s = 200$ signals wide. So, for a fair comparison between these topologies, the channel width w is chosen so both topologies meet the packaging constraints. First, for the ring,

$$w \leq \min\left(\frac{W_n}{\delta}, \frac{W_s}{B_C}\right) = \min\left(\frac{140}{4}, \frac{200}{4}\right) = 35. \tag{3.15}$$

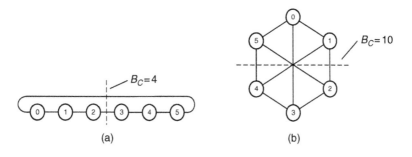

(a)   (b)

Figure 3.9 The bisection and degree for (a) a 6-ring and (b) a Cayley graph. Each edge represents two unidirectional channels going in opposite directions.

And for the Cayley graph,

$$w \le \min\left(\frac{W_n}{\delta}, \frac{W_s}{B_C}\right) = \min\left(\frac{140}{6}, \frac{200}{10}\right) = 20. \tag{3.16}$$

Using these widths and a signaling frequency of $f = 1\,\text{GHz}$, the channel bandwidth of the ring is 35 Gbits/s and 20 Gbits/s for the Cayley graph. If the message length is $L = 1024\text{bits}$, the router delay $t_r = 20\,\text{ns}$, and given the maximum channel load and average hop count, the ideal throughput and zero-load latency of both networks can be compared. (See Table 3.3.) As shown, the Cayley graph gives better ideal throughput, while the ring has a lower zero-load latency for the given packaging constraints. Examining Equations 3.15 and 3.16 shows that the ring's channel width is limited by pin bandwidth, whereas the Cayley graph is able to take advantage of the full bisection width. This results in better bisection utilization and hence better ideal throughput for the Cayley graph. However, the higher degree of the Cayley graph limits the size of an individual channel, resulting in higher serialization delay and a better zero-load latency for the ring. One might conclude that this latency result is counterintuitive by examining the topologies alone, because the Cayley graph has a smaller average hop count. However, by considering the limits on the channel width imposed by a particular package, we see that the actual latency of the Cayley graph is in fact higher than that of the ring— in this case, the reduction in hop count is overwhelmed by the increase in serialization latency.

3.5 Case Study: The SGI Origin 2000

To give you a concrete example of a topology, we will take a look at the interconnection network of the SGI Origin 2000 [108]. The Origin 2000 system, shown in Figure 3.10, was first announced in 1997. The Origin 2000 supports up to 512 nodes with 2 MIPS R10000 [197] processors on each node. Because it is a shared-memory multiprocessor, the requirements on the network are both low latency and high throughput.

Table 3.3 Example performance for the packaged ring and Cayley graph networks.

	Ring	Cayley
b	35 Gbits/s	20 Gbits/s
H_{avg}	3/2	7/6
γ_{max}	3/4	7/18
Θ_{ideal}	≈ 46.7 Gbits/s	≈ 51.4 Gbits/s
T_h	30 ns	≈ 23.3 ns
T_s	≈ 29.3 ns	≈ 51.2 ns
T_0	≈ 69.3 ns	≈ 74.5 ns

Figure 3.10   SGI Origin Servers. A single cabinet (16-processor) and a deskside (8-processor) configuration are shown.

By examining the network of this machine, we will see a number of issues involved in mapping an interconnection network to a packaging hierarchy. We will also explore issues in building an interconnection network from a fixed set of components that supports scalable machine sizes from a few nodes to hundreds of nodes.

The Origin 2000 network is based on the SGI SPIDER routing chip [69]. The SPIDER chip provides 6 bidirectional network channels. Each channel is 20 bits wide and operates at 400 MHz for a channel bandwidth of 6.4 Gbits/s. Thus, each node of the network has six 6.4 Gbits/s links for a total node bandwidth of 38.4 Gbits/s. All these channels may be driven across a backplane and three of the channels have a physical interface capable of driving up to five meters of cable.

Figure 3.11 illustrates how the Origin 2000 network topology changes as the number of nodes is increased. In all configurations, the network attaches 2 processing nodes (4 processors) to each terminal router.[5] These terminal connections use 2 of the 6 channels on the router, leaving 4 channels to connect to other routers. Systems with up to 16 routers (32 nodes, 64 processors) are configured as binary n-cubes, with each router connecting to neighboring routers in up to 4 dimensions, as illustrated in Figure 3.11(a). If all 4 dimensions are not used (for example, in an 8-router system), then the unused channels may be connected across the machine to reduce network diameter.

---

5.   This is an example of concentration, which we will discuss in Section 7.1.1.

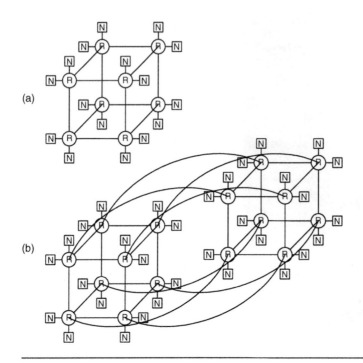

Figure 3.11   For up to 16 routers (R) (32 nodes [N]) the Origin 2000 has a binary n-cube topology, $n \leq 4$. (a) An 8-router (16-node) machine has a binary 3-cube topology. (b) A 16-router (32-node) machine has a binary 4-cube topology.

Machines with more than 16 routers are implemented with hierarchical networks, as shown in Figure 3.12. These larger configurations are composed of 8-router (16-node, 32-processor) local sub-networks that are configured as binary 3-cubes with one channel on each node left open. Eight router-only global subnetworks are then used to connect the 8-router subnetworks together. For a machine with 2^n routers, each global network is a binary cube with $m = n - 3$ dimensions. For example, a maximal 256-router (512-node, 1024-processor) configuration uses 8 32-node binary 5-cubes for global subnetworks. The open channel of router i in local subnetwork j is connected to router j of global subnetwork i. For the special case of a 32-node machine ($n = 5$), each 4-port global subnetwork is realized with a single router chip. This structure is, in effect, a Clos network (see section 6.3) with the individual switches constructed from binary n-cubes.

The Origin 2000 is packaged in a hierarchy of boards, modules, and racks, as shown in Figure 3.13. Each node (2 processors) is packaged on a single 16-inch × 11-inch circuit board. Each router chip is also packaged on a separate circuit board. Four node boards (8 processors) and 2 router boards are packaged in a chassis and connected by a midplane. The remaining room in each chassis is used for the I/O subsystem. Two chassis (four routers) are placed in each cabinet. A system may have up to 64 cabinets (256 routers). In large systems, additional router-only cabinets are used to realize the global subnetworks.

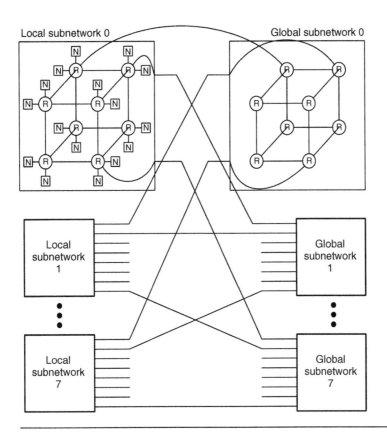

Figure 3.12   Origin 2000 machines with more than 32 nodes use a hierarchical topology where 8-router (16-node) binary 3-cube subnetworks are connected by 1 link per node to 8 central binary *n*-cube, $n \le 5$, networks. The figure shows a 128-node machine in which there are 8 local subnetworks connected to 8 binary 3-cube global subnetworks.

Of the 6 channels on each router board, 2 connect to 2 of the 4 node boards on the midplane, 1 connects to the other router on the midplane, and the remaining 3 are brought out to back-panel connectors for connection to other chassis. One of these back-panel connections on each router is used to connect to the other chassis in the same cabinet, a second connects to the second cabinet in a local subnet, and the third connects either to a global sub-network or to a second pair of cabinets in a 16-router, 4-cabinet system.

Table 3.4 shows how this hierarchical topology meets the requirements of a shared-memory multiprocessor given the constraint of constructing the network entirely from 6-port routers. As shown in Equation 3.11, zero-load latency grows with average hop count and distance, both of which grow with diameter, and serialization latency, Equation 3.10, which is fixed by the 20-bit width of the router channel. To keep latency low, the Origin 2000 uses a topology in which diameter, and hence hop count, increases logarithmically with the size of the machine. Going to a hierarchical

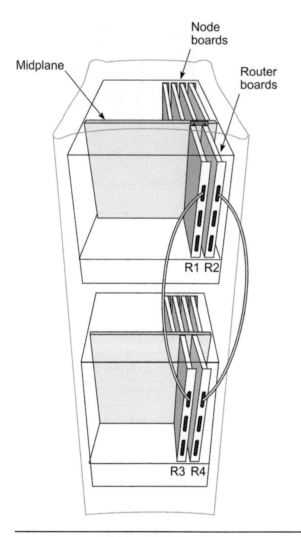

Figure 3.13   Packaging of a 4-router (16-processor) Origin 2000 system. Reprinted by permission from Laudon/Lenoski [ISCA '97] © [1997] ACM, Inc.

network allows the designers to continue the logarithmic scaling at machine sizes larger than can be realized with a degree-4 (2 ports are used by the nodes) binary *n*-cube. We leave the exact calculation of latency as Exercise 3.8.

The Origin 2000 topology meets the bandwidth requirements of a shared-memory multiprocessor by providing a flat bisection bandwidth per node as the machine scales. For each machine configuration, a bisection of the machine cuts N channels where N is the number of routers ($N/2$ in each direction). For the small machines, a binary n cube network has 2^n routers and 2^n channels across the bisection, 2^{n-1} in each direction. For the larger machines, each node has one channel to a global sub-network and each global sub-network has a bisection bandwidth equal to its input bandwidth.

Table 3.4 Configuration and performance of Origin 2000 network as a function of the number of nodes.

Size (nodes)	Topology	Chassis	Diameter	B_C
4	binary 1-cube	1	3	2
8	binary 2-cube	2	4	4
16	binary 3-cube	4	5	8
32	binary 4-cube	8	6	16
64	4 3-cubes × 8 4 × 4 switches	16	7	32
128	8 3-cubes × 8 3-cubes	32	9	64
256	16 3-cubes × 8 4-cubes	64	10	128
512	32 3-cubes × 8 5-cubes	128	11	256

3.6 Bibliographic Notes

Although we focus on the two most common interconnection networks in the next chapters of this book, there are several other notable topologies. Cube-connected cycles [153] maintain the low hop count (diameter) of hypercube networks, but with constant-degree nodes. Fat trees [113] have been shown to be a universal network topology, in that they can emulate the behavior of any other topology in poly-logarithmic time. A 4-ary fat tree was used in the Connection Machine CM-5's network [114]. (See Section 10.6.) Cayley graphs [7], a family of topologies that subsume the cube-connected cycle, offer simple routing and lower degree than hypercube networks of equivalent size.

3.7 Exercises

3.1 *Tornado traffic in the ring.* Consider an 8-node ring network in which each node sends traffic to the node 3 hops around the ring. That is, node i sends traffic to $i + 3$ (mod 8). Each channel has a bandwidth of 1 Gbit/s and each input offers traffic of 512 Mbits/s.

What is the channel load, ideal throughput, and speedup if minimum routing is used on this network? Recalculate these numbers for the case where non-minimal routing is allowed and the probability of taking the non-minimal route is weighted by its distance so that a packet takes the three-hop route with probability 5/8 and the five-hop route with probability 3/8.

3.2 *A worst-case channel load bound.* Derive a lower bound for the maximum channel load under *worst-case* traffic, assuming that the bisection channels are the most heavily loaded channels and all channels have an equal bandwidth b.

3.3 *Limitations of channel load bounds.* Find a topology where the upper bound on maximum channel load given by Equation 3.6 is not tight. Does the topology you have found require non-minimal routing to optimally balance traffic? Explain your answers.

3.4 *Tightness of channel load bounds for symmetric topologies.* Prove that the maximum channel load bounds in Equations 3.5 and 3.6 equal the optimal load for any edge-symmetric topology.

3.5 *Throughput with asymmetric channel bandwidths.* Derive an expression for the ideal throughput of a network when the channel bandwidths are not equal. If necessary, change the definitions of γ_c and γ_{max} to be consistent with this new expression for throughput.

3.6 *Impact of serialization latency on topology choice.* A system designer needs to build a network to connect 64 processor nodes with the smallest possible packet latency. To minimize cost, each router is placed on the same chip as its corresponding processor (a direct network) and each processor chip has 128 pins dedicated to the network interface. Each pin's bandwidth is 2 Gbits/s and the average packet length is $L = 512$ bits. The hop latency of the routers is $t_r = 15$ ns. Ignore wire latency ($T_w = 0$).

(a) The designer first considers a fully connected topology. That is, each node has a dedicated channel to every other node. What is the average router latency $T_{r_{min}}$ and serialization latency T_s of this network? What is the average, zero-load message latency T_0?

(b) Recompute the latencies for a ring topology. H_{min} for this ring is 16. (See Section 5.2.2.)

3.7 *Latency under non-random traffic.* For the torus and ring networks described in Section 1.3.1, random traffic was assumed when computing latency. This led to the conclusion that the ring had superior latency performance. Is there *any* traffic pattern where the torus would have lower latency? If not, explain why not. For simplicity, assume node (i, j) of the torus is mapped to node $(4i + j)$ of the ring. (For example, torus 00 maps to ring 0, torus 13 maps to ring 7, and so on.)

3.8 *Latency in the Origin 2000.* Each of the racks in the Origin 2000 (Section 3.5) is 19 inches wide and 72 inches high. Assume that all cables within a rack are 48 inches long, all cables between racks must be routed under a raised floor, and that the propagation velocity of the cables is 2×10^8 m/s. Also, assume that all messages are 512 bits long (16 32-bit words). Compute the average zero-load message latencies, including wire delay T_w, for an Origin 2000 configured with 16 nodes and one configured with 512 nodes for uniform traffic.

3.9 *Diameter improvements from partially random wiring.* Random topologies, where nodes are randomly connected by channels, are known to have several good graph-theoretic properties such as low diameter. This comes at the cost of regularity of packaging and routing, which makes most purely random networks impractical. However, a hybrid approach can still realize some of the benefits of randomization [191] without sacrificing ease of packaging. Consider the mesh network in Figure 3.14(a), where the left and right sides of the network are packaged in separate cabinets.

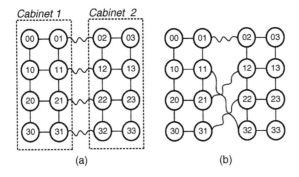

Figure 3.14 The packaging of a 16-node network across 2 cabinets (a) using coaxial cables to create a mesh topology, and (b) with a random permutation of the cables between cabinets.

Coaxial cables (wavy lines in the graph) connect corresponding nodes between the two cabinets.

(a) What is the diameter of this mesh network?

(b) How does the diameter of this network change if the cable connections are randomly permuted? (Figure 3.14[b] shows one such permutation.) What are the minimum and maximum diameters of the network over all permutations? Give a permutation that realizes the minimum diameter.

3.10 *Performance of a fat tree network.* Figure 3.15 shows a 16-node, radix-2 fat tree topology.

(a) Assuming all channel bandwidths equal the injection and ejection rates of the terminal nodes, what is the capacity of this network?

(b) Consider a randomized approach for routing on this network: for each packet, route first from the source "up" to the top of the tree (these are the 8 small, center nodes in the figure), along the way randomly choosing between the two possible "up" channels, and then finish by routing along the "down" channels to the destination. What is the maximum channel load you can reach by using this routing algorithm for *any* traffic pattern?

3.11 *Performance and packaging of a cube-connected cycles topology.* The topology in Figure 3.16 is called the 3rd-order cube-connected cycles.

(a) What is the channel bisection B_C of this topology?

(b) If minimal routing is used, what is the maximum hop count H_{max}? What is the average hop count H_{min}?

(c) Now we want to package this topology under a constraint of $W_n = 128$ signals per node and $W_s = 180$ across the backplane. Assume a packet size of $L = 200$ bits and a signaling frequency of 800 MHz, and also ignore wire length. What is maximum channel width w under these constraints? Is this network

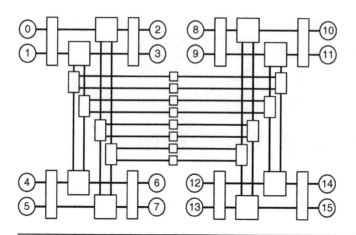

Figure 3.15   A 16-node, radix-2 fat tree. All rectangular nodes are switch nodes.

pin or bisection bandwidth limited? What does the router latency t_r need to be to ensure a zero-load latency of 75 ns?

3.12 *Physical limits of performance.* Using simple ideas, it is possible to compute realizable bounds on the diameter of a network and the physical distance between nodes once it has been packaged. First, if radix k switches (out degree k) are used, what is the smallest possible diameter of an N node network? Now assume that each node has a volume V. What packaging shape in three-dimensions gives the smallest maximum distance between nodes? What is this distance, assuming a large number of nodes?

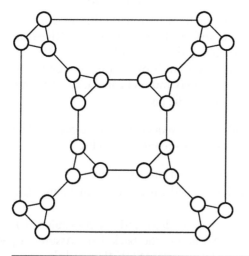

Figure 3.16 3rd-order cube-connected cycles.

3.13 *Capacity of arbitrary networks.* Write a program to determine the capacity of an arbitrary network topology on uniform traffic using the optimization problem from Equation 3.9. The program's input is simply the node-arc incidence matrix corresponding to the topology, and the output is the capacity of the network. For simplicity, assume all channels have equal bandwidth and that the ejection and injection rates of all nodes are also equal. Finally, it may be simpler to work with this alternative formulation of the optimization problem:

$$\text{minimize} \quad t$$

$$\text{subject to} \quad Ax_d = f_d \quad \text{and}$$

$$\sum_{d \in N} x_{d,c} \leq t \quad \text{for all } c \in C$$

$$x_d \geq 0 \quad \text{for all } d \in N^*$$

By introducing an extra variable t, the optimization is expressed as a linear program, which can be solved by specialized tools or by the linear programming routines included in software packages such as MATLAB.

CHAPTER 4

Butterfly Networks

While numerous topologies have been proposed over the years, almost all networks that have actually been constructed use topologies derived from two main families: butterflies (k-ary n-flies) or tori (k-ary n-cubes). In this chapter, we will define the family of butterfly networks and explore its properties. Torus networks are examined in Chapter 5.

A butterfly network is the quintessential indirect network. The butterfly topology has the minimum diameter for an N node network with switches of degree $\delta = 2k$, $H = \log_k N + 1$. Although this optimal diameter is an attractive feature, butterfly networks have two main drawbacks. First, the basic butterfly network has no path diversity: there is exactly one route from each source node to each destination node. As we shall see, this problem can be addressed by adding extra stages to the butterfly. These extra stages improve the path diversity of the network, while keeping the diameter of the network within a factor of two of optimal.

Second, the butterfly cannot be realized without long wires that must traverse at least half the diameter of the machine. Because the speed of a wire decreases quadratically with distance over the critical length, these long wires make butterflies less attractive for moderate-sized and larger interconnection networks. However, the logarithmic diameter and simple routing of the butterfly network has made it and its variants some of the most popular of interconnection networks for many applications.

4.1 The Structure of Butterfly Networks

We have already seen many examples of k-ary n-flies. The simple network shown in Figure 2.2 is a 4-ary 3-fly (three stages of radix-four switches), Figure 3.1(b) shows a

2-ary 3-fly, and Figure 3.5 shows a 2-ary 4-fly. Many other "flies" are possible; k need not be a power of 2, for example.

A k-ary n-fly network consists of k^n source terminal nodes, n stages of k^{n-1} $k \times k$ crossbar switch nodes, and finally k^n destination terminal nodes. We adopt the convention of drawing the source nodes of the butterfly at the left and the destinations at the right. All channels in a butterfly are *unidirectional* and flow from left to right, unless otherwise stated. In most realizations of the butterfly, the source and destination terminal nodes are physically colocated, although they are often drawn as logically separate. We will count each of these source and destination node pairs as a single terminal node so that a k-ary n-fly has a total of $N = k^n$ terminals.

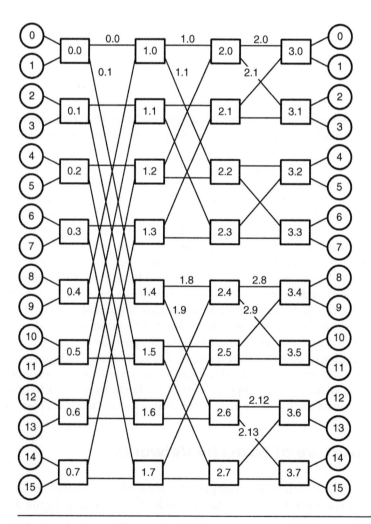

Figure 4.1 Labeling of a 2-ary 4-fly. A node is labeled with its stage number concatenated with its address. A subset of the channels is also labeled with its stage and address.

We label each terminal node and the outgoing channels from each switch node with an n-digit radix-k number, $\{d_{n-1}, d_{n-2}, \ldots, d_0\}$. The first $n-1$ digits, $\{d_{n-1}, d_{n-2}, \ldots, d_1\}$ identify the switch, and the last digit, d_0, identifies the terminal on the switch. The wiring between the stages permutes the terminal address. Between stages $i-1$ and i (numbering starts with 0), the wiring exchanges digits d_{n-i} and d_0. Both the node and a partial channel labeling of a 2-ary 4-fly are shown in Figure 4.1. To distinguish nodes and channels from different stages, the stage number is appended to their label separated by a period. So, for example, node $\{1, 0, 1\} = 5$ in stage 1 is labeled 1.5 in the figure.

Routing on the k-ary n-fly is understood easily in terms of these permutations. Switch stage i sets the current low-order digit of the terminal address, d_0, to an arbitrary value. The wiring stage between switch stage $i-1$ and stage i then places this value into position d_{n-i}. Thus, the first stage sets d_{n-1}, the second stage d_{n-2}, and so on, with the final stage setting d_0, which is already in the right position, so no further wiring is needed.

4.2 Isomorphic Butterflies

Over the years, many multistage networks have been proposed: the shuffle-exchange, the butterfly, the data manipulator, and the flip, to name a few. It turns out, however, that they are all the same network [195] with a few of the switches renumbered. That is, they are isomorphic.

A network, K, is defined by its node and channel sets: $K = (N^*, C)$. Two networks, $K_1 = (N_1, C_1)$ and $K_2 = (N_2, C_2)$ are *isomorphic* if there exists a permutation π of the vertices such that an edge $\{u, v\} \in C_1$ iff $\{\pi(u), \pi(v)\} \in C_2$. As an example of this type of isomorphism, Figure 4.2 shows a 2-ary 3-fly drawn two ways: as a shuffle-exchange network and as a butterfly. The shuffle-exchange network has identical wiring between each stage that performs a shuffle permutation.[1] This permutation connects the output terminal of stage i with address $\{d_2, d_1, d_0\}$ to the input terminal of stage $i+1$ with address $\{d_1, d_0, d_2\}$.

The two networks shown in Figure 4.2 are isomorphic with a very simple mapping: the position of switch nodes 11 and 12 are simply swapped. To make the mapping clear, these switch nodes are numbered with a butterfly numbering in both networks. The simplest way to see the isomorphism is to consider routing from input terminal $\{a_2, a_1, a_0\}$ to output terminal $\{b_2, b_1, b_0\}$. The sequence of switch ports visited in the two networks is illustrated in Table 4.1. The sequences are identical except that the address of the stage 1 switch is reversed: it is $\{b_2, a_1\}$ in the butterfly and $\{a_1, b_2\}$ in the shuffle exchange. Thus, routing in the two networks visits the identical switches modulo this simple relabeling of the first stage switches.

---

1. This is the permutation performed on the cards of a deck when shuffling. Card 0 remains in position 0, card 1 goes to position 2, card 2 goes to position 4, and so on.

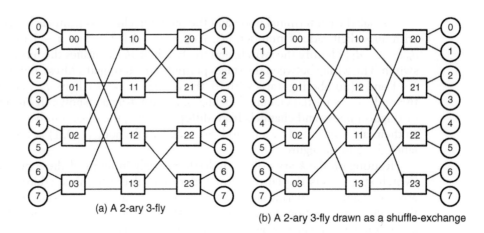

(a) A 2-ary 3-fly

(b) A 2-ary 3-fly drawn as a shuffle-exchange

Figure 4.2 A 2-ary 3-fly drawn two ways (a) as a conventional butterfly, and (b) as a shuffle exchange network. The only difference is the position of switch nodes 11 and 12.

Table 4.1 Routing from $\{a_2, a_1, a_0\}$ to $\{b_2, b_1, b_0\}$ in 2-ary 3-flies wired as a butterfly and a shuffle exchange. The only difference is in the address of the stage 1 switch: $\{b_2, a_1\}$ in the butterfly and $\{a_1, b_2\}$ in the shuffle exchange.

Stage	Butterfly	Shuffle Exchange
Stage 0 input	$\{a_2, a_1, a_0\}$	$\{a_2, a_1, a_0\}$
Stage 0 output	$\{a_2, a_1, b_2\}$	$\{a_2, a_1, b_2\}$
Stage 1 input	$\{b_2, a_1, a_2\}$	$\{a_1, b_2, a_2\}$
Stage 1 output	$\{b_2, a_1, b_1\}$	$\{a_1, b_2, b_1\}$
Stage 2 input	$\{b_2, b_1, a_1\}$	$\{b_2, b_1, a_1\}$
Stage 2 output	$\{b_2, b_1, b_0\}$	$\{b_2, b_1, b_0\}$

4.3 Performance and Packaging Cost

As we saw in Chapter 3, a topology is characterized by its throughput, latency, and path diversity. We discuss throughput and latency of butterflies in this section, while addressing path diversity in Section 4.4. All performance comparisons use the two-level packaging model introduced in Section 3.4.

A k-ary n-fly network has k^n input terminals, k^n output terminals, and nk^{n-1} switch nodes. The degree of each switch node is

$$\delta_{\text{fly}} = 2k. \tag{4.1}$$

The channel bisection is

$$B_{C,\text{fly}} = \frac{N}{2} \tag{4.2}$$

for N even. The case for N odd is addressed in Exercise 4.7.

Although butterflies are not edge-symmetric networks, the channels between each stage are symmetric and this is sufficient for the bounds on maximum channel load developed in Section 3.3.1 to be tight (that is, all channels are loaded equally under uniform traffic). Then a formulation for channel load only requires the average hop count.

The hop count for any packet sent in a butterfly network is the same regardless of source and destination and is simply the number of stages plus one,[2] or

$$H_{\text{min,fly}} = n + 1.$$

Using this hop count and Equation 3.5, the channel load is

$$\gamma_{U,\text{fly}} = \frac{N H_{\text{min,fly}}}{C} = \frac{k^n (n + 1)}{k^n (n + 1)} = 1.$$

As we have just shown for uniform traffic, half of the traffic crosses the network bisection, and each bisection channel is loaded with $\gamma_{\text{fly,uniform}} = 1$. For a *reverse* traffic pattern, where node i sends a packet to node $N - i - 1$, or any other pattern where each node in one half of the network sends to a node in the other half, all traffic will cross a bisection channel and load increases to $\gamma_{\text{fly,rev}} = 2$. This difference between average and worst-case bisection traffic is not unique to butterfly networks. However, as we shall see in Section 4.4, the lack of path diversity in butterflies can increase γ to as high as $\sqrt{N}$ by concentrating a large fraction of the traffic on a single channel.

The channel width of the butterfly network under the two-level packaging hierarchy is calculated by substituting Equations 4.1 and 4.2 into Equation 3.14 giving

$$w_{\text{fly}} \leq \min \left(\frac{W_n}{\delta_{\text{fly}}}, \frac{W_s}{B_{C,\text{fly}}} \right) = \min \left(\frac{W_n}{2k}, \frac{2W_s}{N} \right).$$

Then with uniform loading, and thus $\gamma = 1$, the ideal throughput is

$$\Theta_{\text{ideal,fly}} = \frac{f w_{\text{fly}}}{\gamma} = \min \left(\frac{B_n}{2k}, \frac{2B_s}{N} \right) \tag{4.3}$$

where B_n and B_s are the node and bisection bandwidths, respectively.

For most butterfly networks the goal is first to get maximum throughput, and second to minimize message latency. To achieve this goal, one chooses the largest k for which the network is bisection bandwidth limited. This occurs when

$$k = \left\lfloor \frac{N B_n}{4 B_s} \right\rfloor. \tag{4.4}$$

---

2. This assumes that the source and destination are terminal nodes distinct from the switching nodes. If the source and destination nodes are switching nodes, then the hop count is $n - 1$.

This value of k gives the smallest diameter, which minimizes the H portion of latency while maximizing channel bandwidth. This also maximizes ideal throughput and minimizes serialization latency. Any k less than this value does not improve throughput and only increases latency due to additional hop count[3].

For a k-ary n-fly, all three components of latency (T_s, T_r, and T_w) are influenced by the choice of k and n. Serialization latency for a butterfly is $T_s = L/b$ and is determined by the bandwidth given in Equation 4.3. T_r is simply $t_r H_{\min} = t_r(n+1)$. The exact value of wire latency T_w depends on how the k-ary n-fly is packaged. All k-ary n-flies of any size require many long channels. Half of the channels in the first stage cross to the other half of the machine, half the channels in the second stage cross to a different quadrant, and so on. In many butterflies these channels are longer than the critical wire length and repeaters must be inserted in the wires if they are to operate at full speed.

To illustrate the butterfly performance measures described in this section, consider an $N = 2^{12}$ node k-ary n-fly that is to route $L = 512$-bit packets and is to be packaged in a technology with a wire bisection of $W_s = 2^{14}$, a node pinout of $W_n = 2^8$, and a channel frequency of $f = 1$ GHz. We also assume router latency is $t_r = 10$ ns and ignore wire delay. From Equation 4.4 we calculate $k = \frac{2^{12} \times 2^8}{4 \times 2^{14}} = 16$. Thus, for best performance, this will be a 16-ary 3-fly. With $\delta = 2k = 32$, each channel is at most $w = \frac{W_n}{\delta} = 8$ bits wide. This gives a network with a hop count of $H = 4$, an ideal throughput of $\Theta_{\text{ideal}} = b/\gamma = 8$ Gbits/s, and an average latency of $T = 104$ ns. The performance of this butterfly over all choices of k and n is summarized in Table 4.2.

Table 4.2 Throughput and latency of a 4,096-node butterfly network. w_s and w_n are the channel widths due to bisection and node pinout constraints, respectively. Values of w limited by node pinout are shown in italics, and the optimal value of k is bolded.

n	k	w_s	w_n	w	Θ_{ideal} (Gbits/s)	T_h(ns)	T_s(ns)	T (ns)
1	4,096	8	*0.03125*	*0.03125*	0.03125	20	16,384	16,404
2	64	8	2	*2*	2	30	256	286
3	**16**	8	8	8	8	40	64	104
4	8	8	16	8	8	50	64	114
6	4	8	32	8	8	70	64	134
12	2	8	64	8	8	130	64	194

---

3.   In Chapter 7 we will introduce the option of bit slicing or channel slicing the butterfly network, which allows the diameter to be reduced while keeping the network bisection bandwidth limited. Channel slicing, in particular, allows serialization latency to be traded off against diameter.

4.4 **Path Diversity and Extra Stages**

There is no path diversity in a k-ary n-fly network: $|R_{xy}| = 1 \; \forall x, y \in N$. This can lead to significant degradation in throughput due to load imbalance across the channels when the traffic pattern is non-uniform.

This load imbalance can be mitigated by adding extra stages to the network [166]. To see how adding stages to the network reduces the problem, we will first see how traffic becomes concentrated on the bottleneck channels.

Let $k = 2$, n be even, and $m = n/2$. Using the address format described in Section 4.1, suppose packets are sent from node $\{x_m, \ldots, x_2, a_m, \ldots, a_1, x_1\}$ to node $\{b_m, \ldots, b_1, y_m, \ldots, y_0\}$ for all $x \in \mathcal{Z}(2^m)$, where $y = \pi(x)$ is any permutation of the elements of $\mathcal{Z}(2^m)$.[4] For brevity of notation, we will write an m-digit address using a single character, so $x = \{x_m, \ldots, x_1\}$, for example.

In the first stage of routing, a packet from node x moves through switch $\{x_m, \ldots, x_2, a\}$. During the next stage of routing, the same packet moves through switch $\{b_m, x_{m-1}, \ldots, x_2, a\}$. This continues for the mth stage, where the packet is at switch $\{b_m, \ldots b_2, a\}$, and $(m + 1)^{th}$ stage, where the packet is at switch $\{b, a_{m-1}, \ldots, a_1\}$. Because the mth stage switch is not a function of x, all $2^m = \sqrt{N}$ packets in our pattern travel through this switch. The same can be said about the $(m + 1)^{th}$ stage switch, and since there is only one channel between these two switches, the load on this channel is $\sqrt{N}$. Finding the address of this overloaded channel is left as Exercise 4.4.

This traffic concentration is easy to visualize if we draw our butterfly network in two dimensions, as illustrated in Figure 4.3. The figure shows that a k-ary n-fly can be thought of as 2^m y-axis switching planes, each with 2^m switches per stage followed by an equal number of z-axis switching planes with the same number of switches

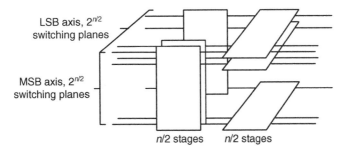

Figure 4.3 A 2-D view of a butterfly network. The first m stages resolve the most significant m switch address bits, routing the packet to the correct position along the y axis in this figure. The next m stages resolve the remaining m switch address bits, routing the packet to the correct position along the z axis.

---

4. $\mathcal{Z}(i)$ is shorthand for the set of integers $\{0, \ldots, i - 1\}$.

per stage. This is because the first m stages adjust only the most significant m switch address bits and, hence, move the packet only in the y dimension (vertically). Then the next m stages adjust only the least significant address bits, moving the packet in the z dimension (into the page).[5]

To see how a butterfly can become congested, consider what happens when all of the input ports connected to the first y-axis network attempt to send packets to all of the output ports connected to the first z-axis network. All 2^m of these packets must route through the single channel connecting these two networks, which corresponds exactly to the permutation traffic we described above.

Suppose now that we add m stages to the front of the network, and that these stages are wired identically to the last m stages in the network — that is, they adjust the m least significant bits of the switch address. These extra stages give a path diversity of 2^m. That is $|R_{xy}| = 2^m \ \forall x, y \in N$. To see this, consider routing a packet from $\{s, t\}$ to $\{u, v\}$. The new m stages can route from $\{s, t\}$ to $\{s, i\}$ for an arbitrary $i \in \mathcal{Z}(2^m)$. With an appropriate routing strategy,[6] i will be chosen uniformly and the traffic from $\{s, t\}$ to $\{u, v\}$ will be uniformly distributed over the 2^m intermediate channels $\{s, i\}$ from stage $m - 1$. Now, the next $\frac{n}{2}$ stages (the first half of the original network) route the packet from $\{s, i\}$ to $\{u, i\}$. Finally, the last $\frac{n}{2}$ stages route from $\{u, i\}$ to $\{u, v\}$. Note that our pathological traffic pattern, from $\{x_m, \ldots, x_2, a, x_1\}$ to $\{b, y\}$, does not concentrate on a single channel in this network, since the additional stages distribute this traffic uniformly over the channels $\{x_m, \ldots, x_2, a_m, i\}$ and then $\{b, i\}$.

Figure 4.4 shows how adding an extra stage to a 2-ary 3-fly gives the network a path diversity of 2. In the unmodified network, bit-reversal traffic — that is, from $0 \to 0, 4 \to 1$, and so on — will concentrate all of the traffic on half the links out of the second stage. Adding the additional input stage eliminates this problem, giving a uniform channel load of $\gamma = 1$ on bit-reversal traffic.

The optimistic reader may assume at this point that the problem is solved: that by adding m stages to a butterfly network, we can perfectly load-balance the network. Unfortunately, this is not the case. Two problems remain. First, because the channel bisection of the network, $B_{C, \text{fly}}$ is less than N, the network cannot handle traffic in which all nodes transmit across the bisection (for example, where node i sends to $N - i - 1$). Second, for larger networks, $n > 3$, the problem of traffic concentration still exists, but with a concentration factor of $2^{n/4}$ rather than $2^{n/2}$.

To see the problem with concentration, we will express our switch addresses in terms of $r = n/4$ bit strings (assume $n = 2m = 4r$ for some integers m and r). Consider a route from $\{x_r, \ldots, x_2, a, x_1, s, t\}$ to $\{b, y, u, v\}$, where $y = \pi(x)$ is a permutation of the elements of $\mathcal{Z}(2^r)$. After our extra input stages, we are at channel $\{x_r, \ldots, x_2, a, x_1, i, j\}$. Our middle stages then take us to $\{b, y, i, j\}$. Since the two

---

This 2-D arrangement of a butterfly has been used to efficiently package small butterfly networks. With this approach, the first half of the stages is packaged on a set of vertical circuit cards, while the remaining stages are packaged on horizontal circuit cards. The cards are connected together as shown in Figure 4.3 by a midplane

6. The output port for a packet traveling through the redundant stages may either be chosen randomly or by selecting any idle output port. This latter strategy is often called *deflection routing*.

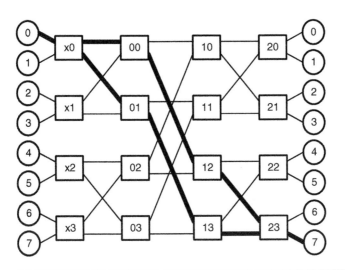

Figure 4.4 A 2-ary 3-fly with one extra stage provides a path diversity of 2 on routes between all pairs of nodes. The two paths between nodes 0 and 7 are shown in bold.

least significant digits of the address are fixed during the middle stages, we have recreated the same concentration as in the original adversary. However, in this case the total load is $2^r = 2^{n/4}$. To see this intuitively, note that the middle stages of the network are equivalent to 2^m separate networks (indexed by the low address bits), each of which has 2^m ports. We have just moved the problem from the whole network into these multiple middle networks.

Figure 4.5 gives a 2-D view of a k-ary n-fly with m extra stages. The extra stages form a set of 2^m m-stage z-axis networks, each with 2^{m-1} switches per stage. These networks, properly routed, uniformly distribute the traffic across the z-dimension, preventing congestion at the crossing from the y-axis networks to the z-axis networks.

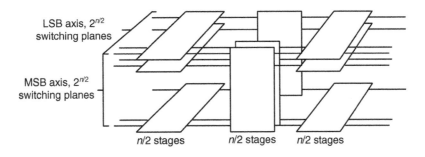

Figure 4.5 A 2-D view of a butterfly network with m extra stages. Properly routed, the extra stages act to balance traffic in the z direction. Load imbalance may still occur, however, routing in the y direction through the center networks.

However, congestion can still occur entirely within the *y*-axis. We can, for example, divide each *y*-axis network into two *m*/2 stage networks, and concentrate traffic at the junction of these two networks.

The load balance problem for butterflies can be solved by duplicating all *n* stages of the network. The resulting 2*n*-stage network, equivalent to two back-to-back *n*-stage butterflies, is called a Beneš (pronounced Ben-ish) network. It has a channel bisection of *N* and thus can handle patterns like reversal traffic. It also perfectly load-balances the traffic by distributing traffic between each pair of nodes over *N* alternate paths. In fact, the Beneš network is *non-blocking*, as we shall discuss in Section 6.4.

4.5 Case Study: The BBN Butterfly

Bolt, Beranek, and Neumann (BBN) Advanced Computer Systems built a series of shared-memory parallel processors by using interconnection networks with *k*-ary *n*-fly topologies. The first of these machines, offered in 1981, was the BBN Butterfly. The Butterfly connected up to 256 processing nodes, each composed of an 8 MHz Motorola 68000 microprocessor and 512 Kbytes of memory. In 1987, the Butterfly GP-1000's processors were upgraded to faster Motorola 68020s and memory was increased to 4 Mbytes per node. In 1989, the TC-2000 was upgraded again with Motorola 88100 RISC processors with 88200 cache and memory management chips and up to 16 Mbytes of memory per node. Each processor of the Butterfly was able to read and write the memory of any other node via an interconnection network. This was not a cache-coherent machine. The original Butterfly and the GP-1000 had no caches, and the TC-2000 allowed only caching of local data. The Monarch (Section 23.4) was designed as a follow-on to the TC-2000, but was never completed.

Remote memory accesses were made over a 4-ary *n*-fly network with one extra stage, as shown in Figure 4.6[7] for a 64-node machine. Each channel in the figure is 4 bits wide and operated at 8 MHz (also the clock rate of the 68000 processor) for a channel bandwidth of $b = 32$ Mbits/s. The network contained 4 stages of 4×4 switches, rather than the $\log_4(64) = 3$ stages that would be required without the extra stage. The extra stage provided path diversity and improved throughput on adversarial traffic patterns, as discussed in Section 4.4. With radix-4 switches, the extra stage provided four edge disjoint paths between any source and destination. When a packet was transmitted, the network interface randomly choose one of the four paths through the switch.[8] A 4-bit mask enabled the system to disable any combination of the 4 paths if one or more became faulty.

The 4×4 by 4-bit wide switches were realized on a single 12-inch-high PC board using discrete TTL logic, and the channel interfaces used emitter-coupled logic

---

7. The TC-2000 used a slightly different network with an 8-ary *n*-fly topology. Here we will confine our discussion to the radix-4 network used on the original Butterfly and the GP-1000.

8. This is an example of oblivious routing, which we will discuss in Chapter 9.

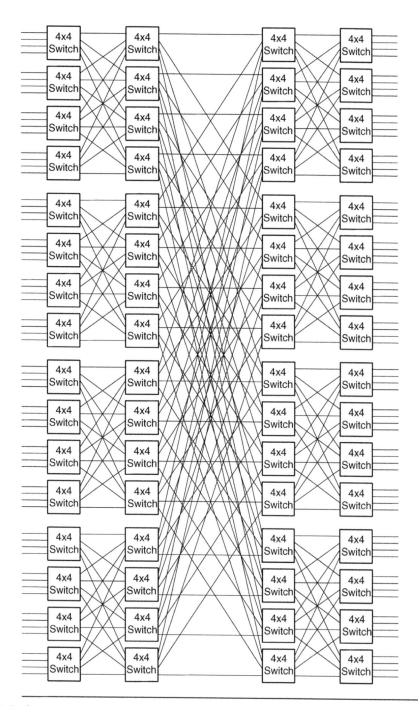

Figure 4.6 The BBN Butterfly (1981) used a 4-ary *n*-fly network with one extra stage of switches to connect processors to memories. Each channel had a bandwidth of 32 Mbits/s and the switches dropped messages to resolve contention. The figure shows a 4-ary 3+1-fly that connects 64 nodes with 4 stages of 4 × 4 switches.

(ECL) line drivers and receivers. The switch cards and processor/memory cards were inserted into 19-inch racks in an arbitrary pattern. All network wiring was realized with ribbon cable connections on the front side of the cards.

The latency through the switch was three 125 ns clocks, one of which was used to consume the routing nybble that specified the switch output port, so the 4-stage network of Figure 4.6 had an end-to-end zero-load latency of twelve 125 ns clocks or $1.5\mu s$. The throughput on uniform traffic (neglecting routing overhead) was 32 Mbits/s per node. Adversarial traffic patterns, however, could result in much lower throughput. (See Exercise 4.6.)

The network employed dropping flow control similar to that used in our simple network of Chapter 2. When multiple packets request the same output port, one packet is granted the port and the remaining packets are dropped. Dropped packets are retransmitted after a short timeout. Each time a packet is retransmitted, it again randomly selects one of the four paths to the destination. This makes multiple collisions between the same set of packets unlikely.

4.6 Bibliographic Notes

Butterfly topologies have appeared in many forms over the years. Early examples include Stone's shuffle-exchange [175], Feng's data manipulator [65], Lawrie's omega [110], Batcher's flip [15], and Pease's indirect binary cube topology [144]. Wu and Feng subsequently showed that all of these networks are isomorphic to the butterflies we studied in this chapter [195]. A further comparison of butterflies and butterfly-like topologies is presented by Kruskal and Snir [106]. In addition, several commercial and academic systems have been built around butterflies, such as the NYU Ultracomputer [75], the BBN Butterfly studied in this chapter [85], and more recently, NEC's Cenju-4 computer [133].

4.7 Exercises

4.1 *Isomorphism between butterflies and shuffle-exchange networks.* A radix-2 shuffle-exchange network consists of $n = \log_2 N$ switch stages where each stage is connected in a *perfect shuffle* pattern. A perfect shuffle connects the output terminal of stage i with address $\{a_{n-1}, a_{n-2}, \ldots, a_0\}$ to the input terminal of stage $i+1$ with address $\{a_{n-2}, \ldots, a_0, a_{n-1}\}$. Show that 2-ary n-flies are isomorphic to radix-2 shuffle exchange networks with the same number of nodes. (An example of this is shown in Figure 4.2.)

4.2 *Throughput under bit-reversal.* What fraction of capacity does a 4-ary 2-fly achieve when routing the bit-reversal traffic pattern?

4.3 *Packaging a butterfly topology.* You need to connect 2^{10} nodes in a packaging technology with $W_n = 128$ and $W_s = 1024$. Choose a butterfly topology that first maximizes the throughput of the network and then minimizes latency. What is this throughput and the corresponding latency? Assume $L = 512$ bits, $f = 1$ GHz, and $t_r = 10$ ns. Also, ignore wire latency.

4.4 *Overloaded channel under adversarial traffic.* In Section 4.4 a permutation traffic pattern is found to load a particular channel to $\sqrt{N}$. By using the channel address notation for butterflies described in Section 4.1, determine the address of this highly overloaded channel.

4.5 *Identical extra stages.* Consider the network of Figure 4.4. Suppose you add a second redundant stage wired identically to the first stage of the network. Does this increase the path diversity? Is the maximum channel load reduced?

4.6 *Worst-case traffic on the BBN Butterfly.* Describe a traffic pattern that results in worst-case throughput — that is, one that maximizes γ_{max} — on the network of the BBN Butterfly shown in Figure 4.6. What is the worst-case throughput on a butterfly with an odd number of stages in general?

4.7 *Packaging odd radix butterflies.* What is the minimum number of bisection channels B_C for a 3-ary 2-fly network? Remember that the sizes of the two sets of nodes separated by the bisection will differ by one because $N = 9$ in this case. Does this bisection represent a good packaging of the network? If not, suggest an alternative cut or set of cuts that partitions the network into approximately equal size sections and does correspond to an efficient packaging.

4.8 *Mixed radix butterflies.* Sketch the topology of a 12-node butterfly, first using 2 switch stages and then using 3 switch stages. Each stage should contain switches that have the same radix, but the radix may differ from stage to stage. Also, do not leave any switch ports unused. (For example, do not implement a 16-node butterfly with 4 unconnected terminals.) How are these two topologies related to factorizations of 12?

4.9 *Wounded butterfly networks.* Consider a 2-ary n-fly network with one faulty switch node in one of its stages. The error recovery mechanism of the network removes this faulty switch from the network and messages are no longer allowed to route through it. If there are no extra stages, this fault will obviously create a disconnected network — some source-destination pairs will not have a path between them. For a butterfly with x extra stages, where $0 \le x \le n$, what is the probability that the network will remain connected after a single switch failure? For $x = 1$, what is the probability the network is still connected after two switch failures?

4.10 *Wire length in the layout of a butterfly.* Calculate the average wire length traversed by a message in a 2-ary butterfly network laid out in a plane. Assume that all nodes are aligned on a 2-D grid and the inter-node spacing is 10 cm. So, for example, the first stage of the butterfly is in the first column of the grid and requires $10 \times 2^{n-1}$ cm of vertical space. The next stage is laid out in the next grid column with 10 cm of horizontal separation from the first stage, and so on. For simplicity, measure distances from the center of each switch node and ignore the wiring to and from the terminal nodes. Compare the wire length for the layout in the plane versus the layout used in Figure 4.3, assuming n is even. How much shorter are the wires for a large network?

CHAPTER 5

Torus Networks

Torus and mesh networks, k-ary n-cubes, pack $N = k^n$ nodes in a regular n-dimensional grid with k nodes in each dimension and channels between nearest neighbors. They span a range of networks from rings ($n = 1$) to binary n-cubes ($k = 2$), also know as *hypercubes*.

These networks are attractive for several reasons. This regular physical arrangement is well matched to packaging constraints. At low dimensions, tori have uniformly short wires allowing high-speed operation without repeaters. Logically minimal paths in tori are almost always physically minimal as well. This physical conservation al lows torus and mesh networks to exploit physical locality between communicating nodes. For local communication patterns, such as each node sending to its neighbor in the first dimension, latency is much lower and throughput is much higher than for random traffic. Butterfly networks, on the other hand, are unable to exploit such locality.

Tori have good path diversity and can have good load balance even on permutation traffic. Also, since all channels in a torus or mesh network are bidirectional, they can exploit bidirectional signaling, making more efficient use of pins and wires.

One disadvantage of torus networks is that they have a larger hop count than logarithmic networks. This gives them a slightly higher latency than the minimum bound and increases the pin cost of the network. Note, however, that some increase in hop count is required for path diversity.

A designer can determine the properties of a torus network by choosing the dimension, n, of the network. As we shall see in Section 5.2.1, the throughput of the network increases monotonically with dimension until the point where the network becomes bisection limited. Beyond that point, there is no further increase in throughput. Network latency is high at either extreme. With low dimension, latency is dominated by the high hop count, H, whereas with high dimension, serialization latency, T_s, dominates. Minimum latency is usually achieved at a relatively low dimension, typically between 2 and 4. To minimize both latency and wire length, we typically choose n to be the smallest dimension that makes the network bisection limited.

5.1 **The Structure of Torus Networks**

An n-dimensional, radix-k torus, or k-ary n-cube, consists of $N = k^n$ nodes arranged in an n-dimensional cube with k nodes along each dimension. Being a direct network, each of these N nodes serves simultaneously as an input terminal, output terminal, and a switching node of the network. Each node is assigned an n-digit radix-k address $\{a_{n-1}, \ldots, a_0\}$ and is connected by a pair of channels (one in each direction) to all nodes with addresses that differ by $\pm 1 (\bmod k)$ in exactly one address digit. This requires 2 channels in each dimension per node or $2nN$ channels total. Tori are regular (all nodes have the same degree) and are also edge-symmetric, which helps to improve load balance across the channels.

We have already seen many examples of k-ary n-cubes. Figure 3.1(a) shows a 3-ary 2-cube (a 2-D torus with three nodes per dimension), and an 8-ary 1-cube (an 8-node ring) is shown in Figure 3.3.

In general, an arbitrary k-ary n-cube can be constructed by adding dimensions iteratively, as illustrated in Figure 5.1. A k-ary 1-cube (Figure 5.1[a]) is simply a k-node ring. Connecting k of these 1-cubes in a cycle adds a second dimension, forming a k-ary 2-cube (Figure 5.1[b]). The process continues one-dimension at a time, combining k k-ary $(n-1)$-cubes to form a k-ary n-cube (Figure 5.1[c]).

A mesh network is a torus network with the connection from address a_{k-1} to address a_0 omitted in each direction. For example, Figure 5.2 compares a 4-ary 2-cube (torus) with a 4-ary 2-mesh. A mesh network has the same node degree, but half the number of bisection channels as a torus with the same radix and dimension. Although the mesh has a very natural 2-D layout that keeps channel lengths short, it gives up the edge symmetry of the torus. This can cause load imbalance for many traffic patterns, as the demand for the central channels can be significantly higher than for the edge channels. As we shall see in Section 5.3, the maximum channel length of the torus can be reduced to twice that of the mesh by using a folded layout.

A torus may be unidirectional, with channels in only one direction (from a_i to a_{i+1}) in each dimension, or bidirectional with channels in both directions between connected nodes. Meshes can also be made either unidirectional or bidirectional; however, unidirectional meshes must alternate channel direction between rows of the network to keep the network fully connected. Unless they employ simultaneous bidirectional signaling, bidirectional networks require twice the pin count and twice the wire bisection as unidirectional networks. Even with this increase in cost, bidirectional networks are generally preferred, as they have a lower hop count H and greater path diversity. We consider a torus or mesh to be bidirectional unless otherwise specified.

Each dimension of a torus network may have a different radix. For example, Figure 5.3 illustrates a 2,3,4-ary 3-mesh that has a radix of 2 in the y dimension, a

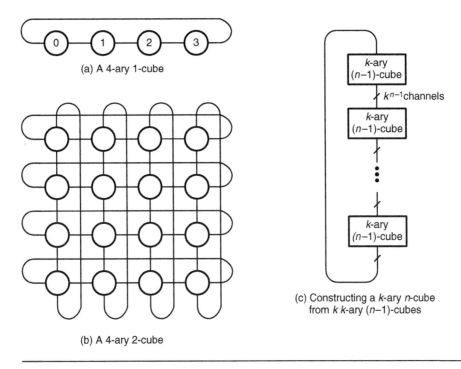

(a) A 4-ary 1-cube

(b) A 4-ary 2-cube

(c) Constructing a *k*-ary *n*-cube
from *k* *k*-ary (*n*–1)-cubes

Figure 5.1 Three *k*-ary *n*-cube networks: (a) a 4-ary 1-cube; (b) a 4-ary 2-cube that is constructed from 4 1-cubes by connecting like elements in a ring; and (c) a *k*-ary *n*-cube that is constructed by connecting like nodes of *k* *k*-ary (*n* − 1)-cubes in a cycle.

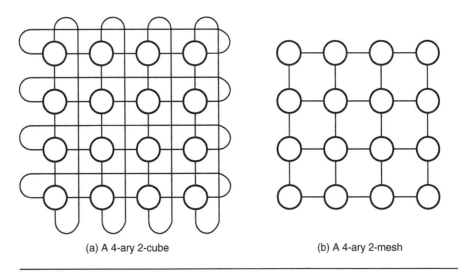

(a) A 4-ary 2-cube

(b) A 4-ary 2-mesh

Figure 5.2 Torus and mesh networks: (a) a torus network (4-ary 2-cube) includes the connection from node 3 to node 0 in both dimensions, but (b) a mesh network (4-ary 2-mesh) omits this connection.

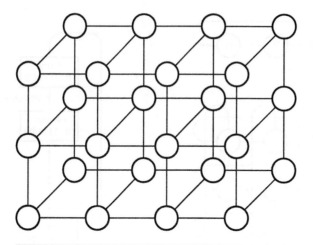

Figure 5.3 A mixed-radix 2,3,4-ary 3-mesh has a different radix in each of its three dimensions.

radix of 3 in the z dimension, and a radix of 4 in the x dimension. Mixed-radix tori and meshes are often built for practical reasons of packaging and modularity. However, a mixed-radix tori is no longer edge-symmetric and a mixed-radix mesh has further asymmetries compared to a single radix mesh. These asymmetries introduce load imbalance, and for many traffic patterns, including uniform traffic, the channel load on the longer dimensions is larger than the load on the shorter dimensions. With uniform traffic — for example, γ_x — the load on the x dimension in Figure 5.3 will be twice γ_z.

5.2 **Performance**

The performance of a torus network, as with any network, is characterized by its throughput, latency, and path diversity.

5.2.1 **Throughput**

In the two-level packaging model, throughput is limited by either pin bandwidth or bisection bandwidth. We first consider the bisection limit and calculate the channel bisection of the networks as

$$B_{C,T} = 4k^{n-1} = \frac{4N}{k} \tag{5.1}$$

$$B_{C,M} = 2k^{n-1} = \frac{2N}{k} \tag{5.2}$$

for k even.

The minimum bisection that yields Equation 5.1 can be visualized by using Figure 5.1(c). When k is even, there is an even number of k-ary $(n-1)$-cubes in the loop of the outermost dimension. The minimum bisection divides this loop at its center point, cutting 2 sets of k^{n-1} bidirectional channels or $4k^{n-1}$ channels total. The minimum bisection is similar for the mesh, except the wraparound channel does not exist, thus halving the number of channels cut.

Since a torus is both node- and edge-symmetric, the channel load under uniform traffic can be determined from the bisection channel load by substituting Equation 5.1 into Equation 3.3.

$$\gamma_{T,U} = \frac{N}{2B_C} = \frac{N}{2} \times \frac{k}{4N} = \frac{k}{8} \tag{5.3}$$

Due to the edge symmetry of the torus, we can find the same result by substituting the hop count for a torus from Equation 5.10 into Equation 3.5 and get

$$\gamma_{T,U} = \frac{N H_{min}}{C} = \frac{N(nk/4)}{2nN} = \frac{k}{8} \tag{5.4}$$

for k even.

Because the mesh is asymmetric, it cannot achieve the lower bound due to hop count used above for the torus. Rather, the channel load for the mesh under uniform traffic is

$$\gamma_{M,U} = \frac{k}{4} \tag{5.5}$$

for k even. Initutively, this result comes from the fact that the mesh can saturate its bisection channels under uniform traffic. Half of the traffic, $N/2$ packets per cycle, will cross the bisection on average, and balancing load across the $B_{C,mesh} = 2N/k$ channels gives an average load of $(N/2)/(2N/k) = k/4$. For the non-bisection channels, the average load drops off further from the center of the array.[1]

For worst-case traffic, all traffic crosses the bisection and the channel loadings are doubled.[2]

$$\gamma_{T,W} = \frac{N}{B_C} = \frac{k}{4}$$

$$\gamma_{M,W} = \frac{N}{2B_C} = \frac{k}{2}$$

---

1. Equations 5.3 through 5.5 are for bidirectional networks. The channel load for unidirectional tori is studied in Exercise 5.4.
2. In a bidirectional mesh or torus, non-minimal routing is required to prevent load imbalance between the two directions from increasing the peak channel load (in one direction) to almost $4\gamma_{uniform}$.

To consider the pin bandwidth limit, we first compute the degree of each node. Each bidirectional cube node has 4 connections in each of its n dimensions.[3,4]

$$\delta_T = \delta_M = 4n \tag{5.6}$$

Substituting Equations 5.1 and 5.6 into Equation 3.14 gives the maximum channel width of a torus.

$$w_T \leq \min \left(\frac{W_n}{4n}, \frac{kW_s}{4N} \right) \tag{5.7}$$

Similarly, for a mesh we get

$$w_M \leq \min \left(\frac{W_n}{4n}, \frac{kW_s}{2N} \right).$$

From channel width and channel load, we then compute throughput with uniform loading as

$$\Theta_{\text{ideal,T}} = \frac{f w_T}{\gamma} = \frac{8}{k} \min \left(\frac{B_n}{4n}, \frac{kB_s}{4N} \right),$$

$$\Theta_{\text{ideal,M}} = \frac{f w_M}{\gamma} = \frac{4}{k} \min \left(\frac{B_n}{4n}, \frac{kB_s}{2N} \right).$$

In general, the best throughput for a torus occurs when the dimension is high enough to keep the network bisection limited and small enough to keep all wires below the critical wire length. Ignoring wire length for the moment, maximum throughput is achieved when

$$nk \leq \frac{N B_n}{B_s}. \tag{5.8}$$

To keep wires short and to minimize serialization latency (see below) we typically choose the smallest n for which Equation 5.8 holds.

Suppose, for example, that you must build an $N = 2^{12}$ node torus where $W_s = 2^{14}$ and $W_n = 2^8$ and $f = 1\,\text{GHz}$. Table 5.1 compares the alternative tori networks of this size. For each row of the table, the dimension n and radix k determine the remaining properties. The table shows the product, nk, which must be kept smaller than $\frac{NW_n}{W_s} = 64$ to meet Equation 5.8. This value is shown in bold at the point where it falls below this limit. Three channel widths are shown: w_n, the width limit due to node pinout; w_s, the width limit due to bisection; and w, the actual channel width (the smaller of w_n and w_s). The channel width is shown in italics for cases that are node pin limited. Finally, the table shows the throughput Θ in Gbit/s.

---

3.   Strictly speaking, some nodes in the mesh have degree less than $4n$, but in practice it is not worth designing special parts for these edge cases. Rather, the degree of all nodes is constant and channels are simply left unconnected at the edge of the mesh.

4.   If simultaneous bidirectional signaling is used, pins can be shared between channels in opposite directions and the effective node degree is halved. (See Exercise 5.11.)

Table 5.1 Throughput of 4,096-node torus networks.

n	k	nk	W_s	W_n	W	Θ_{ideal} (Gbits/s)
1	4,096	4,096	4,096	64	64	0.125
2	64	128	64	32	32	4
3	16	48	16	21	16	8
4	8	32	8	16	8	8
6	4	24	4	10	4	8
12	2	24	2	5	2	8

Table 5.1 shows that throughput increases monotonically with dimension until the network becomes bisection limited at $n = 3$. Beyond that point, the throughput is flat because the design uses all the available bisection wiring. Most likely, we would choose a 16-ary 3-cube topology for this network because it gives us the maximum throughput with a minimum dimension. We will see how the low dimension benefits us in latency and wire length in the next section.

5.2.2 Latency

The latency of a torus network depends strongly on dimension. At the low extreme of dimension, latency is dominated by the high hop count; at the high extreme of dimension, serialization latency dominates due to the narrow channel width, w. The optimal latency typically occurs at a low, intermediate dimension.

The serialization latency of a torus network is given by substituting Equation 5.7 into Equation 3.10.

$$T_{s,\mathrm{T}} = \frac{L}{b} = \frac{L}{f \min \left(\frac{W_n}{4n}, \frac{kW_s}{4N} \right)} = \frac{1}{f} \max \left(\frac{4nL}{W_n}, \frac{4NL}{kW_s} \right).$$

Similarly, for a mesh,

$$T_{s,\mathrm{M}} = \frac{1}{f} \max \left(\frac{4nL}{W_n}, \frac{2NL}{kW_s} \right). \tag{5.9}$$

With large n, and hence small k, both the pin and bisection terms of Equation 5.9 become large. However, the bisection term usually dominates. With small n, and hence large k, T_s is small. However, in this regime hop count dominates latency.

The average minimum hop count in a torus network is determined by averaging the shortest distance over all pairs of nodes, giving

$$H_{\mathrm{min,T}} = \begin{cases} \frac{nk}{4} & k \text{ even} \\ n \left(\frac{k}{4} - \frac{1}{4k} \right) & k \text{ odd} \end{cases}. \tag{5.10}$$

Table 5.2 Latency of 4,096-node torus networks.

n	k	w	Θ_{ideal}	T_h (ns)	T_s(ns)	T(ns)
1	4,096	64	0.125	10,240	8	10,248
2	64	32	4	320	16	336
3	16	16	8	120	32	152
4	8	8	8	80	64	144
6	4	4	8	60	128	188
12	2	2	8	60	256	316

With uniform traffic and even radix, a packet travels, on average, one quarter of the way, $\frac{k}{4}$ hops, around each of n dimensions. Hence, the hop count is $\frac{nk}{4}$. For odd k, the average distance includes a small additional factor.

Similarly, for a mesh the hop count is

$$H_{\min,M} = \begin{cases} \frac{nk}{3} & k \text{ even} \\ n\left(\frac{k}{3} - \frac{1}{3k}\right) & k \text{ odd} \end{cases}. \tag{5.11}$$

To see the combined effect of T_s and H on latency, consider our example $N = 2^{12}$ node torus and let $L = 512$ bits, $t_r = 8$ ns, and $t_c = 2$ ns, a total of 10 ns per hop. Table 5.2 and Figure 5.4 compare the header latency T_h, serialization latency, T_s, and overall latency, T, as dimension is varied. The table shows that the minimum latency, 144 ns, occurs at $n = 4$. However, the latency of 152 ns when $n = 3$ is close enough that it would probably be chosen to gain the packaging and wire length advantages of a lower dimension.

5.2.3 Path Diversity

Many distinct paths exist between every pair of nodes in a torus network. By selectively dividing traffic over these paths, load can be balanced across the network channels, even for very irregular traffic patterns. This path diversity also enables the network to be quickly reconfigured around faulty channels, by routing traffic along alternative paths.

How many paths exist between a source node, a, and a destination node, b? To answer this question, we will first examine the simple case in which we consider only minimal paths and in which all paths take the same direction in each dimension (one-way routes). The number of possible paths increases rapidly with dimension. For a one-dimensional network, there is only a single path. For a 2-D network where a and b are separated by Δ_x hops in the x dimension and Δ_y hops in the y dimension, the number of paths is given by

$$|R_{ab}| = \begin{pmatrix} \Delta_x + \Delta_y \\ \Delta_x \end{pmatrix}.$$

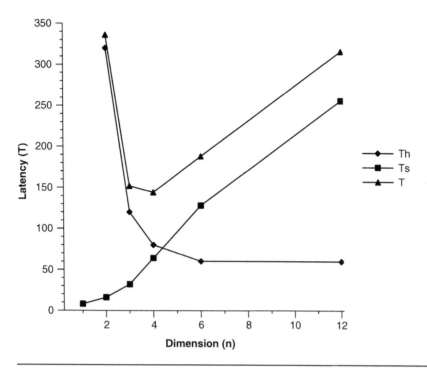

Figure 5.4 Latency vs. dimension for 4,096-node torus networks. The graph plots header latency, T_h, serialization latency, T_s, and overall latency, T.

There are this many ways of choosing where to take the Δ_x hops out of the $\Delta_x + \Delta_y$ total hops. Similarly, in three dimensions, with Δ_z hops in the z dimension, the formula is

$$|R_{ab}| = \binom{\Delta_x + \Delta_y + \Delta_z}{\Delta_x} \binom{\Delta_y + \Delta_z}{\Delta_y} = \frac{(\Delta_x + \Delta_y + \Delta_z)!}{\Delta_x!\Delta_y!\Delta_z!}.$$

As above, the first term gives the number of ways to pick where to take the Δ_x x hops out of the total. The second term gives the number of ways to choose where to take the Δ_y y hops out of the remaining y and z hops. These two terms are multiplied together to give the total number of unique paths.

In general, if there are n dimensions numbered 0 to $n-1$ and there are Δ_i hops from a to b in the ith dimension, then the total number of minimal one-way routes from a to b is given by

$$|R_{ab}| = \prod_{i=0}^{n-1} \binom{\sum_{j=i}^{n-1} \Delta_j}{\Delta_i} = \frac{\left(\sum_{i=0}^{n-1} \Delta_i\right)!}{\prod_{i=0}^{n-1} \Delta_i!}.$$

Each term of the product gives the number of ways of choosing where to take the hops in dimension i out of all of the remaining hops. As expected, the number of minimal paths grows rapidly with dimension and distance. For example, in a 3-D network with $\Delta_x = 3$, $\Delta_y = 3$, and $\Delta_z = 3$, the number of paths is 1,680. Some routing algorithms will also incorporate non-minimal paths and, in these cases, the path diversity can be nearly unbounded.

5.3 Building Mesh and Torus Networks

To construct a network, and to determine properties such as channel length, l, the abstract nodes of a network must be mapped to real positions in physical space. Depending on the packaging technology, the mapping may be to one-dimensional space (e.g., boards along a linear backplane), 2-D space (e.g., chips on a PC board), or 3-D space (e.g., modules placed in arbitrary locations in a volume). While the world we live in has three dimensions, some packaging systems constrain networks to live in one or two dimensions.

Torus and mesh networks are particularly easy to map to physical space with uniformly short wires. The simplest case is when the network is a mesh with the same number of dimensions as the physical dimensions of the packaging technology. In this case, the n-digit address of a node $\{a_1, \ldots, a_n\}$ is also its position in n-dimensional Cartesian space. Mathematically, the node's position in dimension i equals its address in dimension i: $p_i = a_i$. All channels connect physically adjacent nodes and hence are all of unit length[5].

If a torus is packaged in the same manner, with $p_i = a_i$, the long end-around channels, from node $k - 1$ to and from node 0 in each dimension, will have length k. Such long channels may result in excessive latency or require slower signaling rates. This problem can be avoided by *folding* a torus network as shown in Figures 5.5 and 5.6. In a folded torus, the mapping to physical space is

$$p_i = \begin{cases} 2a_i & \text{if } a_i < \frac{k}{2} \\ 2k - 2a_i - 1 & \text{if } a_i \geq \frac{k}{2} \end{cases}.$$

Folding the torus eliminates the long end-around channel, but at the expense of doubling the length of the other channels. As shown in Figure 5.5, a one-dimensional torus or ring network is folded by interleaving the first half of the nodes, $0, \ldots, \frac{n}{2} - 1$ in ascending order with the second half of the nodes, $n - 1, \ldots, \frac{n}{2}$ in descending order. In general, a folded k-ary n-cube is constructed by combining k folded k-ary $n - 1$-cubes in folded order, as shown in Figure 5.6.

---

5. Here we assume that each node has unit diameter in physical space and thus channels connecting physically adjacent nodes in each dimension are of unit length.

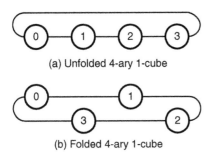

(a) Unfolded 4-ary 1-cube

(b) Folded 4-ary 1-cube

Figure 5.5 Folding a ring: (a) a 4-ary 1-cube unfolded, and (b) a 4-ary 1-cube folded to shorten the long connection from 3 to 0.

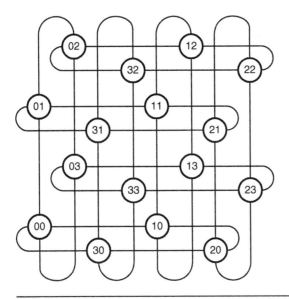

Figure 5.6 A folded 4-ary 2-cube.

When the number of logical dimensions exceeds the number of physical dimensions, several logical dimensions must be mapped into each physical dimension. If the number of physical dimensions is q, a straightforward mapping is to fold $\frac{n}{q}$ of the logical dimensions into each physical dimension.

$$p_i = \sum_{j=0}^{\frac{n}{q}-1} k^j a_{i+jq}.$$

This mapping is illustrated in Figure 5.7 for the case of a 3-ary 2-mesh. Here, the three rows of the mesh are laid out in a line. The dimension 0 (logical x) channels

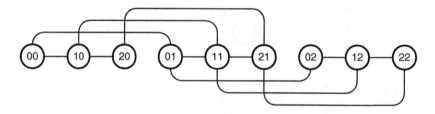

Figure 5.7 A 3-ary 2-mesh mapped to a single physical dimension. Logical x channels are unit length. Logical y channels have length 3.

have unit length while the dimension 1 (logical y) channels have length $l = k = 3$. In general, the channel length for dimension i of a mesh mapped into q physical dimensions in this manner is

$$l_i = k^{\lfloor \frac{i}{q} \rfloor}.$$

This projection of logical dimensions to physical dimensions can also be performed for a folded torus. In this case, the lengths are doubled.

$$l_i = 2k^{\lfloor \frac{i}{q} \rfloor}$$

5.4 **Express Cubes**

Because torus networks have short physical channels, the channel latency, t_c, is often dominated by the routing latency, t_r. This results in a larger header latency, T_h, than could be achieved by a network with a smaller diameter. The diameter can be decreased by increasing dimension. However, as illustrated in Figure 5.4, this narrows the channel width, resulting in an increased serialization latency, T_s.

An express cube network is a k-ary n-cube augmented with a number of long (express) channels [45]. By routing packets that must traverse a long distance in a dimension over the express channels, the header latency can be reduced to nearly the channel latency limit. Because the number of express channels can be controlled to match the bisection width of the network, this reduction in header latency can be achieved without increasing serialization latency.

A k-ary n-cube network is transformed into an express cube by inserting interchanges every i nodes in each dimension and connecting the interchanges along a dimension with long express channels, as illustrated for one dimension in Figure 5.8. If a message had a hop count of H_j in dimension j of the original network, and assuming that H_j is divisible by i, the hop count in dimension j of the express

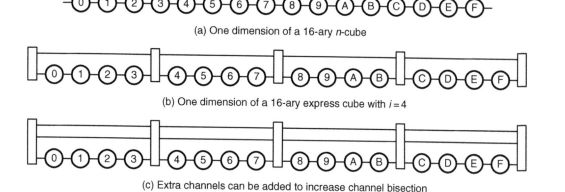

(a) One dimension of a 16-ary *n*-cube

(b) One dimension of a 16-ary express cube with $i = 4$

(c) Extra channels can be added to increase channel bisection

Figure 5.8 A flat express cube is created by inserting interchanges connected by long channels: (a) one dimension of a 16-ary *n*-cube, (b) this network is transformed into an express cube by inserting an interchange every *i* nodes and connecting these interchanges by long, *express*, channels, and (c) channel bisection can be controlled by varying the number of express channels between interchanges.

cube becomes[6]

$$H_{je} = \left(\frac{H_j}{i} + i \right).$$

For very long dimensions, the first term of this equation dominates. In this case, channel delay can be balanced against routing delay by choosing $i = \frac{t_r}{t_c}$. For networks with shorter dimensions, the two terms of the equation can be balanced against one another by choosing $i = \sqrt{H_{min}}$ (where H_{min} is the average hop count without the express channels).

A hierarchical express cube gives a network with the locality properties of a torus and logarithmic diameter approaching the limit of Equation 3.1. As shown in Figure 5.9, a hierarchical express cube is constructed recursively. First a level-0 interchange is inserted every *i* nodes. Then every i^{th} level-0 interchange is made a level-1 interchange, and so on for *l* levels. Each interchange at level *j* is spaced i^j nodes from other interchanges at level *j* and connects to adjacent interchanges at levels 0 through *j*.

Assuming that H_j is divisible by i^l, if a packet had a hop count of H_j in dimension *j* in a conventional cube, on a hierarchical express cube this hop count

6.   If H_j is not evenly divisible by *i*, the equation becomes more complicated, but the result is the same. See [45] for details.

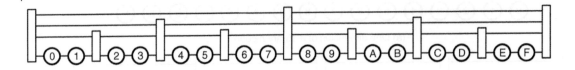

Figure 5.9 A hierarchical express cube with $i = 2$ and $l = 3$.

becomes[7]

$$H_{jh} = \left(\frac{H_j}{i^l} + (i - 1)l + 1 \right).$$

Choosing $i = 2$ minimizes delay due to local nodes. Choosing $l = \log_i \frac{t_r}{t_c}$ balances router and channel delay for long distances.

Overall, a hierarchical express cube has a latency that is within a small constant of the sum of two fundamental limits: (1) the minimum channel delay required to traverse the minimum distance between two nodes, and (2) the logarithmic delay required to make the routing decisions to select the destination node. When the number of express channels is varied, the throughput of an express cube can be made to saturate both pin and bisection bandwidth constraints. Despite these advantages, express cubes have seen very little use in practice. This is largely because most practical networks do not have a radix that is high enough to justify the complexity of adding interchanges. Also, as technology advances, both t_r and the critical wire length, l_c, get smaller, reducing the motivation for long channels.

5.5   Case Study: The MIT J-Machine

The MIT J-Machine [51, 136], shown in Figure 5.10, illustrates how torus and mesh networks map naturally into three dimensions. The J-Machine, constructed at MIT in collaboration with Intel in 1991, was a message-passing multicomputer. Each processing node contained a message-driven processor chip [53] and three external DRAM chips. At the time, the machine was notable for the performance of its network, which had the highest bisection bandwidth and lowest latency of any parallel computer of comparable size. The machine had a novel network interface that allowed short messages to be sent directly out of the processor registers in a single instruction.[8]

---

7.   This also assumes that H_j is large enough to use all l levels. Again, see [45] for the more general case in which H_j is not evenly divisible by i^l.

8.   We defer our discussion of processor-network interfaces to Section 20.4, where we discuss the network interface of the M-Machine, a descendant of the J-Machine.

Figure 5.10 A pair of J-Machines. The near machine has its skins removed. The upper portion of the machine houses up to 1,024 processing nodes connected in an $8 \times 8 \times 16$ mesh. This machine is half populated with eight 8×8 64-processor boards. Below the processor array are I/O cards for graphics, disk control, and LAN interface. The bottom of the machine houses an 80-disk disk array.

The J-Machine used the network both for inter-node message communication and for I/O. An array of I/O nodes were arranged along one edge of the machine, as shown in Figure 5.10. Any node in the machine could communicate with any I/O device by sending it a message. I/O devices for graphics (distributed frame buffer), disks (SCSI host adapter), and network (LAN interface) were developed for the J-Machine.

The J-Machine processing nodes were connected in a 3-D mesh network [138]. The natural mapping of the mesh to physical space enabled a very dense machine construction.[9] Each board contained 64 nodes connected in an 8×8 mesh. Each processing node measured 2×3 inches. Adjacent boards were connected vertically using *elastomeric* connectors distributed across the face of the board. Up to 16 boards could be connected in a single chassis (Figure 5.10) to form a 1,024-node $8 \times 8 \times 16$ mesh. Ribbon cable connectors were provided along the edges of the boards to allow multiple chassis to be connected together to form larger machines. Router addressing limited the maximum machine size to a 64 K-node $32 \times 32 \times 64$ mesh; however, no machine larger than 1,024 nodes was ever constructed.

Each network channel consisted of 15 lines (9 data signals[10] and 6 control signals) and operated at 32 MHz, twice the rate of the 16 MHz processor. This gave a channel payload bandwidth of 288 Mbits/s. A single physical channel was used for communication in both directions between a pair of adjacent nodes, with the nodes arbitrating for use of the channel on a flit-by-flit basis. Depending on the location in the machine, a network channel could be implemented on a PC board (internal connections between the 64 nodes on a card), over an elastomeric connector (vertical connections between two adjacent 64-node cards), or over a ribbon cable (horizontally between adjacent cards).

The J-Machine network employed dimension-order routing (Section 8.4.2) and wormhole flow control (Section 13.2). The router was integrated on the message-driven processor (MDP) chip (Figure 5.11). The dimension-order routing enabled the router to be cleanly partitioned into X, Y, and Z sections. The latency through the router, including channel traversal, was two 32-MHz router cycles (one 16 MHz processor cycle) per hop.

The zero-load latency of a 4-word message on an $8 \times 8 \times 16$ J-Machine network on uniform traffic can be calculated by extending Equation 5.11 to handle this mixed-radix case. The serialization latency is 16 31.25 ns network cycles, and the average hop count is 32/3 at two network cycles per hop for a total latency of 37 31.25 ns network cycles (18.5 processor cycles), or 1.16 μs.

The throughput of this network on uniform traffic can be calculated using the bisection bound. Applying Equation 5.2 in the long $k = 16$ dimension, we calculate that $B_C = 128$ channels cross the network bisection. However, because the physical channel is shared between the two directions, the actual channel bisection is half this number, $B_C = 64$.[11] With uniform traffic, half of the 1,024 nodes send packets across this bisection for a channel load of $\gamma = 512/64 = 8$. We then calculate the throughput as $\Theta = b/\gamma = 288/8 = 36$ Mbits/s per node.

---

9.   In Exercise 5.9, we will explore the difficulties of mapping other topologies to such dense packaging.
10. The machine had a 36-bit word length (32 data + 4 tag bits) and sent a word over the network in 4 network cycles.
11. We explore the advantages of this bidirectional channel in Exercise 5.10.

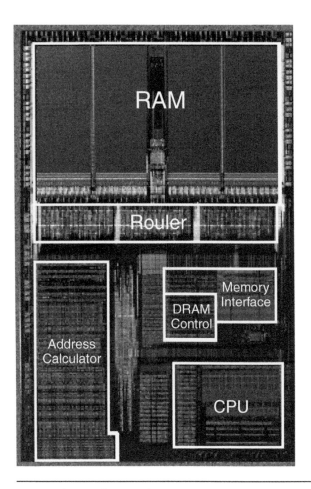

Figure 5.11   An MDP chip, implemented in 1.2 μm CMOS, includes a 32-bit RISC processor, a 3-D dimension-order router, a network interface, a DRAM interface, and 18 Kbytes of on-chip memory. The router, just below the RAM, is divided into X, Y, and Z sections.

For comparison with our calculated performance numbers, Figure 5.12 (from [136]) shows measured latency as a function of offered traffic for a 512-node J-Machine operating at a 12.5 MHz processor clock. The measured zero-load latency of 23 processor cycles is 4.5 cycles greater than our calculated value (of 18.5 cycles) due to a fixed 4-cycle latency through the network input and output interfaces and a half cycle of roundup. The saturation bisection traffic of 6 Gbits/s is only 42% of the calculated throughput of 14.4 Gbits/s (64 channels × 9 bits × 25 MHz). This is due to two principal factors. First, the four-word message includes a two-word header giving it a 50% header overhead. Second, the wormhole flow control employed in the machine is not capable of reaching 100% channel utilization.

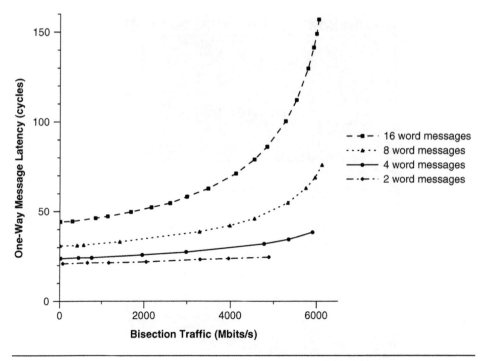

Figure 5.12   Measured latency (in processor cycles) vs. offered traffic (in bisection traffic) for a 512-node J-Machine.

5.6 **Bibliographic Notes**

Meshes and tori have long been popular network topologies and were employed on some of the earliest parallel computers, including the Solomon [172] and Illiac-IV [13]. Binary *n*-cubes or hypercube networks were popularized by Sullivan and Bashkow [179]. A hypercube was employed in the Caltech Cosmic Cube [163] and many commercial parallel computers were modeled after the Cosmic Cube, including the Intel iPSC computers [155, 38] and NCUBE systems [140, 134]. In 1985, Dally introduced the concept of comparing network topologies based on physical constraints (wire bisection and pinout) and showed the advantages of lower-dimensional tori under realistic constraints [46]. Agarwal later improved on this analysis [2]. After this point, low-dimensional torus and mesh networks became the standard for parallel computers [56, 164, 138, 95, 162]. More recently, torus networks have been applied to Internet router design [49]. Dally developed express cubes to address the large asymptotic routing latency of conventional torus networks [45]. However, practical networks rarely get large enough for asymptotics to matter. Many variants of torus networks have been published that involve twisting an end-around connection or varying the ordering of connections within a dimension. Some of these variants give a small improvement in diameter. See for example

[119, 165, 33]. You will have the opportunity to experiment with such topologies in Exercise 5.6.

5.7 **Exercises**

5.1 *Comparing butterfly and torus topologies.* Compare 1,024-node butterfly and torus networks where the node bandwidth is 300 Gbits/s and the bisection bandwidth is 4 Tbits/s. For both networks, choose the smallest values of k and n so that the bisection bandwidth is saturated (for simplicity, consider only combinations of k and n where $k^n = 1,024$). What is the serialization latency of these networks for a packet length of $L = 1,280$? What is the average hop count H_{min}? Ignoring wire latency, with a per-hop latency of $t_r = 12$ ns, what is the zero-load latency of both networks?

5.2 *Tradeoffs in a 4,096-node torus.* Examine the tradeoff between k and n for a 4,096-node torus. For each combination of k and n where $k^n = 4,096$, what is the ideal throughput and average zero-load message latency? Assume each node has 120 signal pins, the bisection width of the system is 1,500 signals, the signalling frequency is $f = 2.5$ GHz, the packet length is $L = 512$ bits, and the router hop delay is 20 ns. Ignore wire latency ($T_w = 0$).

5.3 *A three-level packaging hierarchy.* A 256-node torus needs to be packaged in a three-level hierarchy under the following constraints: each node has 384 signal pins, 1,200 signals may go off a board, and 6,000 signals can cross the midsection of the backplane. Determine k and n that maximize the network's bandwidth. If several values of k and n achieve this goal, choose the one with the minimum zero-load latency. Explain how you would package this network by using nodes, boards, and a backplane while minimizing the number of boards used.

5.4 *Channel load in unidirectional tori.* Calculate the average channel load in unidirectional tori under uniform traffic. Approximately how many times greater is this load than in bidirectional tori? Explain the sources of this increase.

5.5 *Number of slightly non-minimal routes.* As a function of the minimal hop counts in the x and y dimensions, Δ_x and Δ_y respectively, how many routes are there in a 2-D torus if routes that are at most 1 hop longer than minimal are allowed? How many routes are there that are at most 2 hops longer than minimal? Assume k is odd.

5.6 *Doubly twisted tori.* The topology shown in Figure 5.13 is a doubly-twisted torus [165]. Assuming minimal routing, compare the average hop count H_{min} of this topology with that of a standard 4-ary 2-cube. Why does twisting change the hop count?

5.7 *Wire length in the layout of a torus.* Calculate the average wire length traversed by a message in a k-ary 6-mesh network laid out in a plane. Assume that all nodes are aligned on a 2-D grid and the inter-node spacing is 10 cm. For simplicity, measure distances between node centers. How does the average distance change if the network is packaged in three dimensions and the inter-node spacing in third dimension is also 10 cm? Assume k is even in both cases.

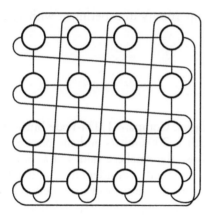

Figure 5.13   A doubly twisted torus topology.

5.8   *Cube connected cycles topologies.* While high dimensional tori offer low hop counts, their high degree requirements result in narrow channels that in turn increase serialization latency. The cube connected cycles (CCC) topologies address this issue in hypercubes (2-ary n-cubes). An nth-order CCC is constructed by taking a 2-ary n-cube and replacing each node with a cycle of n nodes (Figure 5.14). The resulting topology has a fixed node degree of 3 independent of the network size. Figure 3.16 shows a 3rd-order CCC. How many nodes are in an nth-order CCC and what is its average minimum hop count $H_{\min}$? (Computing the minimum hop count between nodes is difficult — approximate if necessary.) Consider the design of a 64-node network with a node limit of 120 pins, each operating at 2.5 Gbits/s. Compare the zero-load latency of this network using a CCC topology to an optimally sized k-ary n-cube. Make sure the channels are wide enough to saturate the pin bandwidth of each node. Use a message size of $L = 1{,}024$ bits, a per-hop latency of $t_r = 20$ ns, and ignore any wire latency.

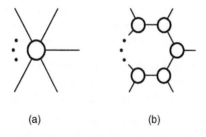

(a)   (b)

Figure 5.14   The node transformation used from (a) a 2-ary n-cube to (b) an nth-order CCC. As shown, each single node in the 2-ary n-cube is replaced with a cycle of n nodes in the CCC. Each node in the CCC has a fixed degree of 6: 2 bidirectional connections within the cycle and a bidirectional channel to another cycle.

5.9 *Packaging the J-Machine.* Each J-Machine (Section 5.5) 64-node circuit board was an 8-layer board with 2 power planes, 2 surface layers, 2 *x* routing layers, and 2 *y* routing layers. Each routing layer provided a signal density of 20 signals per linear inch. There were 16 elastomeric connectors between the boards that each carried 60 signals. What were the throughput and latency (on uniform traffic with 128-bit messages) of the highest performance 1,024-port butterfly network that you could realize in the same volume as the J-Machine using the same packaging technology?

5.10 *Benefits of bidirectional signaling in the J-Machine.* The J-Machine used bidirectional signaling, sharing a single physical channel between logical channels traveling in opposite directions between two nodes. To see why this was done, compare the throughput and latency of a 1,024-node J-Machine network with a network in which each 9-bit wide bidirectional channel is replaced with two 4.5-bit wide unidirectional channels. (Do not worry about the fractional bit width.)

5.11 *Benefits of simultaneous bidirectional signaling.* Compare the throughput and latency of a 1,024-node J-Machine with 9-bit wide sequentially bidirectional channels (one direction at a time) with the same network using *simultaneous* bidirectional channels (that can send information in both directions at once).

5.12 *The J-Machine network as a torus.* Using the same packaging density assumptions from Exercise 5.9, what is the throughput and latency (on uniform traffic with 128-bit long messages) of the highest-performance 1,024-node torus network that you can realize in the same volume using the same packaging?

CHAPTER 6

Non-Blocking Networks

Until this point, we have focused on packet-switched networks, but for this chapter, we shift our attention to circuit-switched networks. In circuit-switched networks, connections between a particular source and destination are first set up, then held as data flows continuously through the connection, and then finally torn down.[1] Historically, non-blocking, circuit-switched networks were first associated with the telephone system. A call placed from one person to another represented a single connection request. For the call to go through, an unused path through the network had to be found and then allocated for that call. If the telephone system was *non-blocking*, then the call would always succeed as long as the recipient's phone was not in use.

More precisely, a network is said to be *non-blocking* if it can handle all circuit requests that are a permutation of the inputs and outputs. That is, a dedicated path can be formed from each input to its selected output without any conflicts (shared channels). Conversely, a network is *blocking* if it cannot handle all such circuit requests without conflicts.[2]

In this chapter, we examine two types of non-blocking networks. First, a network is *strictly non-blocking* if any permutation can be set up incrementally, one circuit at a time, without the need to reroute (or rearrange) any of the circuits that are already set up. If any unused input can be connected to any unused output without altering the path taken by any other traffic, then the network is strictly non-blocking.

In contrast, a network is *rearrangeably non-blocking* (or simply *rearrangeable*) if it can route circuits for arbitrary permutations, but incremental construction of

---

1. Chapter 12 includes a detailed description of both packet and circuit switching.
2. The U.S. phone system is obviously *blocking* to any person who has ever received the "all circuits are busy" recording.

a permutation may require rearranging some of the early circuits to permit later circuits to be set up. A rearrangeble network can connect any unconnected input to any unconnected output, but it may need to reroute some unrelated traffic to make this connection.

All of the above definitions apply to *unicast traffic*, where each input is connected to at most one output. We also consider *multicast traffic*, where a connection from a single input may fan out to several outputs. A network is strictly non-blocking for multicast traffic if any unconnected output can be connected to any input (connected or unconnected) without affecting other traffic. Also, a network is rearrangeable for multicast traffic if it can connect any unconnected output to any input but may have to reroute other traffic to make the connection.

Crossbar switches and Clos networks with an expansion of 2:1 are examples of strictly non-blocking networks for unicast traffic. Crossbar switches and Clos networks with larger expansion are also strictly non-blocking for multicast traffic. Beneš networks and Clos networks without expansion are examples of rearrangeable networks for unicast traffic.

6.1 Non-Blocking vs. Non-Interfering Networks

For packet switching applications, the notion of a non-blocking network is largely irrelevant, or at least overkill. Because packet switches do not tie down a circuit for the duration of a connection or session, they can share channels between connections and packet flows without interference as long as two constraints having to do with bandwidth and resource allocation are met. First, there must be adequate channel bandwidth to support all of the traffic sharing the channel. This bandwidth constraint is met if the channel load γ_{max} is less than the channel bandwidth. Second, the allocation of resources (buffers and channel bandwidth) must be done in a manner such that no single flow denies service to another flow for more than a short, predetermined amount of time. This resource allocation constraint can be realized by a suitable flow control mechanism as described in Chapter 12.

We will call a packet-switched network that meets these criteria *non-interfering*. Such a network is able to handle arbitrary packet traffic with a guaranteed bound on packet delay. The traffic neither exceeds the bandwidth capacity of any network channel, nor does it result in coupled resource allocation between flows.

For almost all applications today, when people say they want a non-blocking network, what they really require is a non-interfering network, which can usually be realized with considerably less expense. For the sake of history, however, and for those cases in which true non-blocking is needed to support circuit switching, we give a brief survey of non-blocking networks in the remainder of this chapter.

6.2 Crossbar Networks

A $n \times m$ crossbar or crosspoint switch directly connects n inputs to m outputs with no intermediate stages. In effect, such a switch consists of m n:1 multiplexers, one for each output. Many crossbar networks are *square* in that $m = n$. Others are *rectangular* with $m > n$ or $m < n$.

Figure 6.1 shows a conceptual model of a 4×5 crossbar. There are 4 input lines and 5 output lines. At each point where an input line crosses an output line — that is, at each crosspoint — a switch optionally connects the input line to the output line. For correct operation, each output must be connected to at most one input. Inputs, however, may be connected to more than one output.[3] In Figure 6.1, for example, input 1 drives outputs 0 and 3.

At one point in time, much of the telephone network was implemented using crossbar switches composed of mechanical relays that did in fact have the structure of Figure 6.1. With such switch-based crossbars, there is no real distinction between inputs and outputs. The connections made by the switches are bidirectional.

Today, however, most crossbars are implemented using digital logic and have the structure shown in Figure 6.2. Each of the n input lines connect to one input of m n:1 multiplexers. The outputs of the multiplexers drive the m output ports. The multiplexers may be implemented with tri-state gates or wired-OR gates driving an output line, mimicking the structure in Figure 6.1, or with a tree of logic gates to realize a more conventional multiplexer.

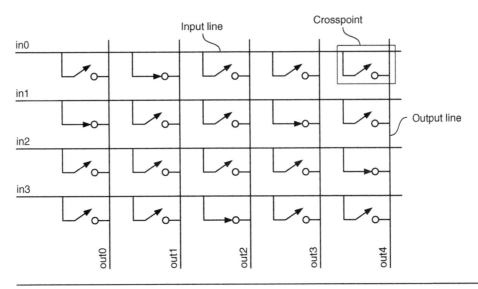

Figure 6.1 A 4×5 crossbar switch consists of 4 input lines, 5 output lines, and 20 crosspoints. Each output may be connected to at most one input, while each input, may be connected to any number of outputs. This switch has inputs 1,0,3,1,2 connected to outputs 0,1,2,3,4, respectively.

3. This is true of most, but not all, crossbar implementations. Some have limited fanout.

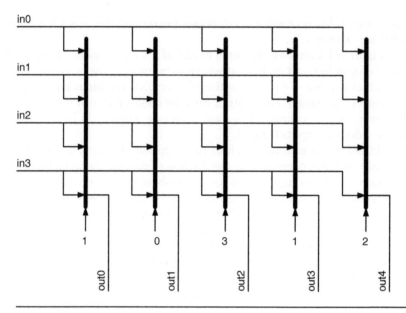

Figure 6.2 A 4×5 crossbar switch as implemented with 5 4:1 multiplexers. Each multiplexer selects the input to be connected to the corresponding output. The multiplexer settings here realize the same connection as in Figure 6.1: {1, 0, 3, 1, 2} → {0, 1, 2, 3, 4}.

However it is realized, when we depict crossbar switches in a system, to avoid drawing the entire schematic each time, we employ the symbol for a crossbar shown in Figure 6.3. Where it is clear from context that the box is a crossbar switch, we will omit the "X" and depict the crossbar as a simple box with inputs and outputs.

A crossbar switch is obviously strictly non-blocking for both unicast and multicast traffic. Any unconnected output can be connected to any input by simply closing the switch connecting the input and the output, or by setting that output's multiplexer appropriately. Many other networks work very hard to achieve this property that comes to the crossbar very easily.

If crossbars are trivially non-blocking, then why bother with any other non-blocking network? The reasons are cost and scalability. Although economical in small configurations, the cost of a square $n \times n$ crossbar increases as n^2. As n grows, the

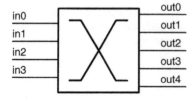

Figure 6.3 Symbol for a 4×5 crossbar switch.

cost of a crossbar quickly becomes prohibitive compared to networks whose cost increases as $n \log n$. Also, as we shall see in Chapter 19, the problem of scheduling a crossbar in a packet switch becomes increasingly difficult as n becomes larger.

The quadratic cost of the crossbar is evident both in the structure of the switch itself and in the construction used to combine several smaller crossbars into one larger crossbar. From Figures 6.1 and 6.2, it is clear that n^2 area is needed to lay out the grid of input and output lines, n^2 area is needed to contain the n^2 crosspoints, and n^2 area is needed to hold n multiplexers, each of which has area proportional to n.[4]

As shown in Figure 6.4, the cost of building a large crossbar from small crossbars is also quadratic. The figure shows how a $2n \times 2n$ crossbar can be constructed using a 2×2 array of $n \times n$ crossbars. Because high-speed signals require point-to-point distribution, 1:2 distributors are needed on the inputs to fan out the input signals to the crossbars in each row. The complementary operation is performed in each column where 2:1 multiplexers select the signal to be forwarded to each output. In general, a $jn \times jn$ crossbar can be constructed from a $j \times j$ array of j^2 $n \times n$ crossbars with n 1:j distributors to fan out the inputs and n j:1 multiplexers to select the outputs.

As with most things in interconnection networks, the cost of a real crossbar depends strongly on how it is packaged. A crossbar packaged entirely within a chip

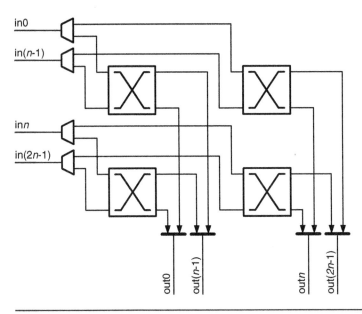

Figure 6.4 A $2n \times 2n$ crossbar can be constructed from 4 $n \times n$ crossbars, as shown. Each sub-crossbar handles one quarter of the crosspoint array. Distributors are needed on the inputs to maintain point-to-point connections and multiplexers are needed on the outputs to select between the upper and lower input sets. In general, a $jn \times jn$ crossbar can be constructed from j^2 $n \times n$ crossbars along with jn 1:j distributors and jn j:1 multiplexers.

---

4. $m \times n$ area for a rectangular crossbar.

that contains no other system components tends to be pin limited. That is, its size is limited by W_n, the number of pins on the chip, rather than the area required to implement the crossbar on the chip. In this case, the effective cost of the crossbar is linear up to the largest crossbar that can be packaged on a single chip. If, however, a crossbar fits on a chip but also connects together other components on that same chip, then its quadratic area becomes important, since it might no longer be pin limited. Once a crossbar must be implemented with multiple chips, as illustrated in Figure 6.4, then its cost increases quadratically with size.

6.3 Clos Networks

6.3.1 Structure and Properties of Clos Networks

A Clos network is a three-stage[5] network in which each stage is composed of a number of crossbar switches. A symmetric Clos is characterized by a triple, (m, n, r) where m is the number of middle-stage switches, n is the number of input (output) ports on each input (output) switch, and r is the number of input and output switches.[6] In a Clos network, each middle stage switch has one input link from every input switch and one output link to every output switch. Thus, the r input switches are $n \times m$ crossbars to connect n input ports to m middle switches, the m middle switches are $r \times r$ crossbars to connect r input switches to r output switches, and the r output switches are $m \times n$ crossbars to connect m middle switches to n output ports. For example, a (3,3,4) Clos network is shown in Figure 6.5 and a (5,3,4) Clos network is shown in Figure 6.6. In referring to the input and output ports of Clos networks, we denote port p of switch s as $s.p$.

It is often valuable to visualize the all-to-all connection between stages of the Clos network in three dimensions, as shown in Figure 6.7. The input and output switches can be thought of as moving the traffic horizontally to and from the vertical middle switches. The middle switches move the traffic vertically from a horizontal input switch to a horizontal output switch. This crossed arrangement is also a useful way to package small Clos networks, as it keeps all of the connections between the stages short.

The properties of an (m, n, r) Clos network with $N = rn$ terminals follow from the topology. All three-stage Clos networks have $H = 4$. The network can be bisected either horizontally or vertically, as drawn in Figure 6.7, giving a bisection of $B_C = mr$ for the horizontal cut through the middle switches or $B_C = 2nr = 2N$ for the vertical cut through the input and output switches. In practice, most networks are packaged by co-locating the input and output switches and cutting all of the inter-switch

---

5. Clos networks with any odd number of stages can be derived recursively from the three-stage Clos by replacing the switches of the middle stage with three-stage Clos networks.
6. In some situations, asymmetric Clos networks are used, in which r and n differ between the input and output stages. Asymmetric Clos networks are described by a 5-tuple (m, n_i, r_i, n_o, r_o).

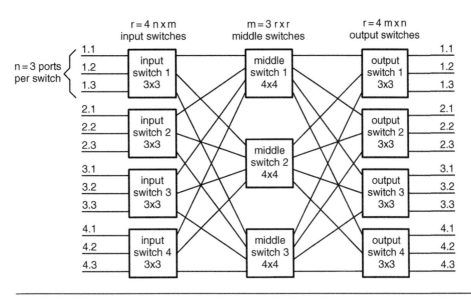

Figure 6.5 An $(m = 3, n = 3, r = 4)$ symmetric Clos network has $r = 4$ $n \times m$ input switches, $m = 3$ $r \times r$ middle-stage switches, and $r = 4$ $m \times n$ output switches. All switches are crossbars.

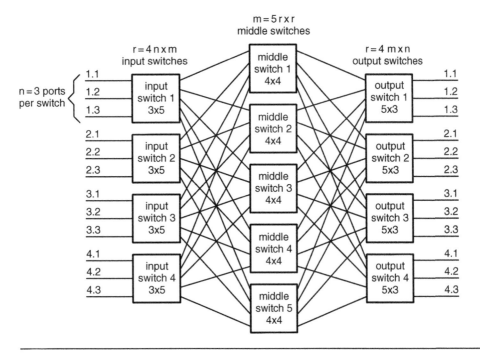

Figure 6.6 A $(5,3,4)$ Clos network. This network is strictly non-blocking for unicast traffic.

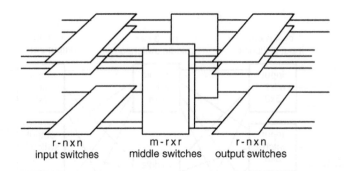

r-nxn   m-rxr   r-nxn
input switches   middle switches   output switches

Figure 6.7 The all-to-all connection between the stages of a Clos network is better understood by drawing the network in three dimensions. The input switches move traffic *horizontally* to select a middle switch, the middle switches move traffic *vertically* to select an output switch, and the output switches move traffic *horizontally* to select an output port.

channels giving $B_C = 2N$. The degree of the input and output switches is $\delta_{io} = n + m$ and of the middle switches is $\delta_m = 2r$.

The most interesting property of the Clos, and the one from which its non-blocking properties derive, is its path diversity. For a Clos network with m middle switches, there are $|R_{ab}| = m$ routes from any input a to any output b, one through each middle stage switch.

In routing a circuit in a Clos network, the only free decision is at the input switch, where any of the m middle switches can be chosen as long as the link to that middle switch is available. The middle switches must choose the single link to the output switch (and the route is not possible if this link is busy). Similarly, the output switch must choose the selected output port. Thus, the problem of routing in a Clos network is reduced to the problem of assigning each circuit (or packet) to a middle switch.

6.3.2 Unicast Routing on Strictly Non-Blocking Clos Networks

THEOREM 6.1

A Clos network is strictly non-blocking for unicast traffic iff $m \geq 2n - 1$.

Proof Consider an arrangement of $N - 1$ calls where all inputs are connected except for input $a.i$ and all outputs are connected except for output $b.j$. The previous calls in the network have been set up so that the middle switches being used by the calls on switch a are disjoint from the middle switches being used by the calls on switch b. Without loss of generality, we can assume that the calls on a are using the first $n - 1$ middle switches and the calls on b are using the last $n - 1$ middle switches. To connect $a.i$ to $b.j$ we must find a middle switch that is not currently being used by either a or b. Since $n - 1$ middle switches are being used by a and different $n - 1$ middle switches are being used by b, there must be at least $2n - 1$ middle switches for there to be a switch available to route this call.[7]   □

Thus, the (5,3,4) Clos network of Figure 6.6 is strictly non-blocking for unicast traffic, since $m = 2n - 1 = 5$. The (3,3,4) Clos network of Figure 6.5 is not strictly non-blocking because $m = 3 < 2n - 1 = 5$. We will see in Section 6.3.3 that the (3,3,4) network is rearrangeable.

The algorithm for routing unicast calls on a strictly non-blocking Clos network is shown in Figure 6.8. To route a unicast call from $a.i$ to $b.j$, the middle stage switch is found by intersecting the list of switches that are not used by a with the list of switches that are not used by b. By the counting argument above, there will always be at least one switch in this intersection.

Consider routing a circuit from input 1.1 (1) to output 3.3 (9) as shown in Figure 6.9. In this case, input switch 1 has already blocked the paths to middle switches 1 and 2 with routes from 1.2 (2) and 1.3 (3) to 4.1 (10) and 2.3 (6), respectively, as shown with dotted lines. The paths from middle switches 4 and 5 to output switch 3 are also blocked with routes to 3.1 (7) and 3.2 (8) from 3.1 (7) and 2.1 (4), respectively. Because the input switch and output switch can each block at most $n - 1 = 2$ middle switches, there can be at most $2n - 2 = 4$ middle switches blocked. With $m = 2n - 1 = 5$ middle switches, there is always guaranteed to be one middle switch available to handle the new call. In this case, middle switch 3 is available to route the new call as shown with the bold lines in the figure.

For a more complete routing example, consider the permutation {5, 7, 11, 6, 12, 1, 8, 10, 3, 2, 9, 4}. That is, input 1 (1.1) routes to output 5 (2.2), input 2 (1.2) to output 7 (3.1), and so on. We also simplify the permutation to {(2, 3, 4), (2, 4, 1), (3, 4, 1), (1, 3, 2)} by considering only switches, not ports. We need to consider only switches when finding the middle stage assignments, and once these assignments are found, scheduling the first and last stages is trivial because the crossbar elements are strictly non-blocking. Input switch 1 has connections to output switches 2, 3, and 4; input switch 2 to output switches 2, 4, and 1; and so on. Suppose the calls are applied in the order {9,6,7,8,3,12,10,5,1,11,2,4}. The call from input 9

```
For each call (a,b)
    freeab = free(a) ∧ free(b)          middle switches available on both a and b
    middle = select(freeab)                                          pick one
    assign((a,b),middle)                                            route (a,b)
```

Figure 6.8 Algorithm for routing unicast traffic on a strictly non-blocking Clos network.

---

7. This proof technique is commonly referred to as the pigeonhole principal and is useful in a wide range of scheduling problems.

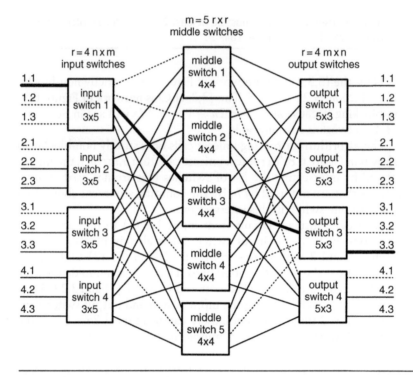

Figure 6.9 Routing from input 1.1 to output 3.3 on a (5,3,4) Clos network. Input switch 1 already has blocked paths (shown as dotted lines) to middle switches 1 and 2, and output switch 3 has blocked paths to middle switches 4 and 5. The only available path (shown as a bold line) is via middle switch 3.

(3.3) to output 3 (1.3) is applied first, then the call from 6 (2.2) to 1 (4.3), and so on.[8]

The process of routing this call set on the (5,3,4) Clos network is shown in Table 6.1. Each row of the table corresponds to one step in the routing process. The first three columns show the input switch from which the call is coming (In), the output switch to which the call is going (Out), and the middle switch allocated to the call (Middle). The remaining eight columns give bit vectors showing which middle switches are free from each input and output switch. For example, the first row of the table shows setting up the call from 9 (3.3) to 3 (1.3). The input switch is 3, the output switch is 1, and, since there are no paths blocked for this first call, it is assigned middle switch 1. The free vectors then show that after this call is set up, the path from input switch 3 to middle switch 1 is busy (Input Free 3 = 01111) and the path to output switch 1 from middle switch 1 is busy (Output Free 1 = 01111).

---

8.   This permutation and call ordering was chosen to be particularly nasty for the rearrangeable (3,3,4) network. Most call sequences are much easier to schedule.

Table 6.1 Example of routing a call set on a strictly non-blocking Clos network. The set of calls $\{(2, 3, 4), (2, 4, 1), (3, 4, 1), (1, 3, 2)\}$ is routed in the order $\{9,6,7,8,3,12,10,5,1,11,2,4\}$. Each row of the table shows the setup of one call. In each row, the columns indicate the input, output, and middle switches used by the call (In, Out, and Middle), and the free vector for each input and output switch.

In	Out	Middle	Input Free				Output Free			
			1	2	3	4	1	2	3	4
3	1	1	11111	11111	01111	11111	01111	11111	11111	11111
2	1	2	11111	10111	01111	11111	00111	11111	11111	11111
3	3	2	11111	10111	00111	11111	00111	11111	10111	11111
3	4	3	11111	10111	00011	11111	00111	11111	10111	11011
1	4	1	01111	10111	00011	11111	00111	11111	10111	01011
4	2	1	01111	10111	00011	01111	00111	01111	10111	01011
4	1	3	01111	10111	00011	01011	00011	01111	10111	01011
2	4	4	01111	10101	00011	01011	00011	01111	10111	01001
1	2	2	00111	10101	00011	01011	00011	00111	10111	01001
4	3	4	00111	10101	00011	01001	00011	00111	10101	01001
1	3	3	00011	10101	00011	01001	00011	00111	10001	01001
2	2	3	00011	10001	00011	01001	00011	00011	10001	01001

Using this bit-vector representation for the free sets, we select the middle switch for a new call from input switch a to output switch b, by AND-ing the input free vector a, which indicates the switches available from a, with output free vector b, which indicates the switches available from b. The resulting vector has a 1 for every middle switch that is available to handle the call. Because the two free vectors can have at most two 0s each, there will be at most four 0s, and hence at least a single 1, in the result of the AND. One of the 1s in this vector is selected to determine the middle switch to carry the new call.

For example, consider the last call set up in Table 6.1, from input 4 (2.1) to output 4 (2.1). Before this call, the free vectors are shown in the second to last row of the table. The free vector for input switch $a = 2$ is 10101 (middle switches 2 and 4 blocked), and the free vector for output switch $b = 2$ is 00111 (middle switches 1 and 2 blocked). We AND together these two free vectors to give a combined free vector of 00101, which indicates that switches 3 and 5 are available to handle the new circuit. Middle switch 3 is selected.

The procedure for routing a strictly non-blocking Clos network is very simple. However, this simplicity comes at a cost. A strictly non-blocking Clos network is nearly twice as expensive, requiring $\frac{2n-1}{n}$ times as many middle switches as a corresponding rearrangeable network. Hence, for most applications a rearrangeable network is preferred.

6.3.3 **Unicast Routing on Rearrangeable Clos Networks**

THEOREM
6.2

A Clos network with $m \geq n$ is rearrangeable.

To simplify our reasoning about routing Clos networks, we represent a set of circuits to be routed as a bipartite graph. The input switches are one set of vertices, the output switches the other set of vertices, and each call is represented by an edge in the graph. The routing problem becomes an edge coloring problem in this graph. Each middle switch corresponds to a color, and the problem is to assign each edge a color (each call to a middle switch) so that no vertex is incident to two edges of the same color (no input or output uses a middle switch more than once).

Consider, for example, the routing problem from Section 6.3.2. Recall that, in terms of switches, the set of circuits to be routed is {(2, 3, 4), (2, 4, 1), (3, 4, 1), (1, 3, 2)}. That is, input switch 1 has circuits to output switches 2, 3, and 4; input switch 2 connects to 2, 4, and 1; and so on. The bipartite graph representing this set of calls is shown in Figure 6.10.

An algorithm for routing a set of calls on a rearrangeable non-blocking network is shown in Figure 6.11. To route a call from input switch a to output switch b, the algorithm starts by looking for a middle switch that is free on both a and b. If such a free middle switch is found, it is assigned and the call is completed. Up to this point, the procedure is the same as the procedure for a strictly non-blocking network shown in Figure 6.8. If a common free middle switch cannot be found, a middle switch, **mida**, that is free on a is assigned to the call, and the call (c,b) that uses this switch on b is moved to a middle switch, **midb**, that is free on b. If **midb** is in use on c, then the algorithm loops, renaming (c,d) to (a,b) to reroute (c,d) on **mida**, rearranging further calls as needed.

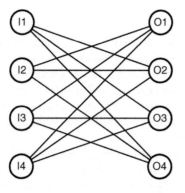

Figure 6.10   The set of circuits to be routed on a Clos network can be represented as a bipartite graph.

```
For each call (a,b)
    freeab = free(a) ∧ free(b)                    middle switches available on both a and b
    If freeab ≠ 0 then                            if there is one free on both a and b
        middle = select(freeab)                                            pick one
        assign((a,b),middle)                                             route (a,b)
    else
        mida = select(free(a))                              middle switch for (a,b)
        midb = select(free(b))                    middle switch for rearranged call
        do                                                          rearrange calls
            (c,b) = call(b,mida)                     rearrange call using mida on b
            if(c,b) then unassign((c,b),mida)                    disconnect (c,b)
            assign((a,b),mida)                                         setup (a,b)
            if(c,b) then                        if mida on b was used - rearrange (c,b)
                (c,d) = call(c,midb)                     get call using midb on c
                if(c,d) then unassign((c,d))                       disconnect call
                assign((c,b),midb)                             move (c,b) to midb
                if(c,d) then                               if midb on c was in use
                    (a,b) = (c,d)                         iterate to move this call
        while (a,b) ∧ (c,b)                         while calls remain to be moved
```

Figure 6.11   *Looping* algorithm for routing unicast traffic on a rearrangeable Clos network.

The application of the looping algorithm to the call set and ordering of Section 6.3.2 is shown in Table 6.2. The algorithm is able to set up calls without rearrangement until it attempts to set up the call (2,4) from input 2 to output 4 (eighth row of the table). The situation at this point is illustrated in Figure 6.12(a). Input switch 2 is using middle switch 2 to route to output 1, and output switch 4 is using middle switches 1 and 3 to route to input switches 1 and 3, respectively. Thus, there are no middle switches available to handle a call from switch 2 to switch 4. The looping algorithm routes ($a=2$, $b=4$) over switch mida = 1. To make this switch available, it rearranges ($c=1$, $b=4$) to switch midb = 2. Because input switch 1 has no call using middle switch 2, no further rearrangement is needed. The result is shown in Figure 6.12(b).

A larger rearrangement that demonstrates the *looping* action of the algorithm occurs when the program attempts to set up the call from $a=4$ to $b=3$ (row 11 of the table). The situation before attempting this call is shown in Figure 6.13(a). As there is no free middle switch common between I4 and O3, the call is set up on middle switch mida = 2, causing a conflict with the call (I3,O3) as shown in Figure 6.13(b). Call (I3,O3) is the first call of a chain of conflicts, (I3,O3), (I3,O1), (I2,O1), (I2,O4), and (I1,O4), illustrated in Figure 6.13(c) that must be *flipped* between mida=2 and midb=1. Flipping (I3,O3) to midb=1 conflicts with (I3,O1). This conflict is resolved, on a second iteration of the algorithm, by flipping (I3,O1) to mida=2, which in turn causes a conflict with (I2,O1), and so on. Flipping all of the calls in the chain resolves all conflicts, as shown in Figure 6.13(d).

Table 6.2 A routing problem.

In	Out	Middle		Input Free				Output Free			
		New	Old	1	2	3	4	1	2	3	4
3	1	1		111	111	011	111	011	111	111	111
2	1	2		111	101	011	111	001	111	111	111
3	3	2		111	101	001	111	001	111	101	111
3	4	3		111	101	000	111	001	111	101	110
1	4	1		011	101	000	111	001	111	101	010
4	2	1		011	101	000	011	001	011	101	010
4	1	3		011	101	000	010	000	011	101	010
2	4	1		011	001	000	010	000	011	101	010
1	4	2	1	101	001	000	010	000	011	101	000
1	2	3		100	001	000	010	000	010	101	000
4	3	2		100	001	000	000	000	010	101	000
3	3	1	2	100	001	010	000	000	010	001	000
3	1	2	1	100	001	000	000	100	010	001	000
2	1	1	2	100	011	000	000	000	010	001	000
2	4	2	1	100	001	000	000	000	010	001	100
1	4	1	2	010	001	000	000	000	010	001	000
1	3	2		000	001	000	000	000	010	001	000
4	3	3	2	000	001	000	010	000	010	000	000
4	1	2	3	000	001	000	000	001	010	000	000
3	1	3	2	000	001	010	000	000	010	000	000
3	4	2	3	000	001	000	000	000	010	000	001
2	4	3	2	000	010	000	000	000	010	000	000
2	2	2		000	000	000	000	000	000	000	000

THEOREM 6.3

Setting up a single call using the looping algorithm requires rearranging at most $2r - 2$ other calls.

Proof This is true because the looping algorithm, in setting up a single call, never revisits a vertex of the bipartite graph. Thus, it completes its iterations after visiting at most $2r$ vertices. Calls are rearranged starting with the visit to the third vertex. Thus, at most $2r - 2$ calls are rearranged.

To see that the path traveled by the algorithm is acyclic, consider unfolding the conflict chain of Figure 6.13(c) as shown in Figure 6.14. In Figure 6.14, a_i and b_i denote input switch a and output switch b on the ith iteration of the loop. On the first iteration of the loop, the algorithm sets up the original call (a_1, b_1), and then moves the call (a_2, b_1) (if it exists) from **mida** to **midb**. On each subsequent iteration, i, of the loop, the algorithm moves call (a_i, b_i) from **midb** to **mida** and (if it exists) call (a_{i+1}, b_i) from **mida** to **midb**.

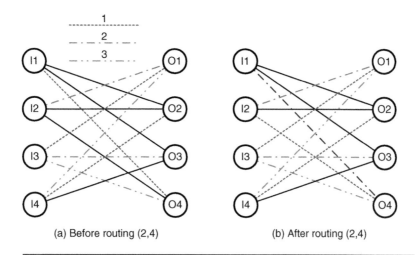

(a) Before routing (2,4)   (b) After routing (2,4)

Figure 6.12 Stages of routing the call set of Figure 6.10. Assignment to middle stages is denoted by the *dotting* of the edges. (a) Before routing call (I2,O4), there are no middle switches free on both I2 and O4. (b) To route this call, (I1,O4) is moved to middle switch 2, allowing (I2,O4) to use middle switch 1.

We show that no vertex can be revisited by induction. We know vertices a_1 and b_1 are not the same, since a is an input switch and b is an output switch. Suppose vertices a_i and b_i are unique for $i < j$. Then when we visit node a_j, it must also be unique because it is found by following the **mida** link from b_{j-1} and the **mida** link on all $a_i, i < j$ is accounted for. On a_1, **mida** is originally unconnected. For $a_i, 1 < i < j$, **mida** originally connected to b_{i-1}, which is distinct from b_{j-1}. Similarly, b_j is guaranteed to be unique because it is identified by following the **midb** link from a_j and the **midb** link on all $b_i, i < j$ is accounted for. Thus, the chain is guaranteed to be acyclic since no vertex will be revisited. Because there are a total of $2r$ vertices, the algorithm is guaranteed to terminate after visiting at most $2r$ vertices and rearranging at most $2r - 2$ calls, starting with the call (a_2, b_1). ☐

Proof   *(Theorem 6.2)* When the algorithm terminates after scheduling $N = rn$ calls, it has produced a schedule for arbitrary unicast traffic. Thus, we have shown that a Clos network with $m \geq n$ is rearrangeable. ☐

When rearranging an existing call, (c, d), to set up a new call, (a, b), it is desirable to switch call (c, d) to its new route without affecting the traffic being carried. Such *hitless* switching can be realized by synchronizing the switching of the input, middle, and output switches carrying call (c, d) so they all pass unit i of the call along the old path and then switch to pass unit $i + 1$ of the call along the new path. The unit here could be a bit, byte, or frame. What is important is not the granularity of switching, but rather that it be synchronized.

Finally, the previous examples have considered only bipartite graphs with at most one edge between each pair of nodes. However, in general, we need to allow

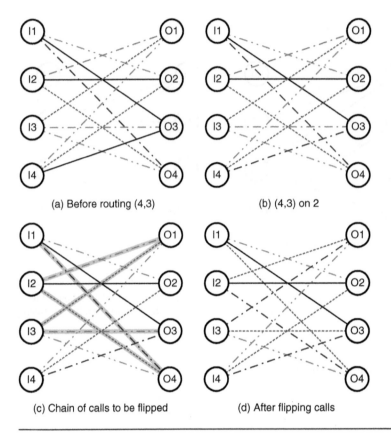

Figure 6.13   Stages of routing the call set of Figure 6.10. Assignment to middle stages is denoted by the *dotting* of the edges. (a) Before routing call (I4,O3), there are no middle switches free on both I4 and O3. (b) The call is routed on middle switch **mida=2**, creating a conflict with call (I3,O3). (c) Moving (I3,O3) to switch **midb=1** starts a chain of conflicting calls (I3,O3),(I3,O1),(I2,O1),(I2,O4), and (I1,O4) that must be *flipped* between switches 1 and 2 to resolve the conflict. (c) After flipping these calls, no further conflicts exist.

for bipartite multigraphs. A multigraph is a graph that can contain parallel edges. This would arise, for example, if a circuit from input 1.1 to output 3.3 along with another circuit from 1.2 to 3.1 were required. Then, in terms of switches, we would have two circuits from switch I1 to switch O3, corresponding to two parallel edges from I1 to O3 in the bipartite multigraph. Fortunately, this complication does not affect the results of this section, and the looping algorithm works equally well on multigraphs.

6.3.4 **Routing Clos Networks Using Matrix Decomposition**

A rearrangeable Clos network can also be routed using matrix decomposition. The set of calls to be routed can be represented as a matrix R where each entry x_{ij} indicates

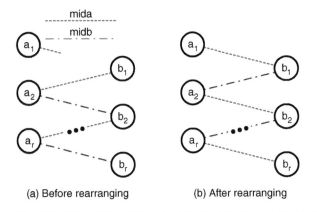

(a) Before rearranging   (b) After rearranging

Figure 6.14 Sequence of visiting vertices during the looping algorithm.

the number of calls from input swtch i to output switch j. For example, the matrix representing our routing example is

$$R = \begin{bmatrix} 0 & 1 & 1 & 1 \\ 1 & 1 & 0 & 1 \\ 1 & 0 & 1 & 1 \\ 1 & 1 & 1 & 0 \end{bmatrix}.$$

This matrix can be decomposed into the sum of a set of m positive matrices where each row and column sum to at most one. Each matrix in this set corresponds to the setting of a middle stage switch. For example, the solution shown in Table 6.2 corresponds to the following matrix decomposition:

$$\begin{bmatrix} 0 & 1 & 1 & 1 \\ 1 & 1 & 0 & 1 \\ 1 & 0 & 1 & 1 \\ 1 & 1 & 1 & 0 \end{bmatrix} = \begin{bmatrix} 0 & 0 & 0 & 1 \\ 1 & 0 & 0 & 0 \\ 0 & 0 & 1 & 0 \\ 0 & 1 & 0 & 0 \end{bmatrix} + \begin{bmatrix} 0 & 0 & 1 & 0 \\ 0 & 1 & 0 & 0 \\ 0 & 0 & 0 & 1 \\ 1 & 0 & 0 & 0 \end{bmatrix} + \begin{bmatrix} 0 & 1 & 0 & 0 \\ 0 & 0 & 0 & 1 \\ 1 & 0 & 0 & 0 \\ 0 & 0 & 1 & 0 \end{bmatrix}$$

$$\tag{6.1}$$

In this case the three terms on the right side of Equation 6.1 are permutation matrices that correspond to the settings of the three middle-stage switches in Figure 6.5. As mentioned in the previous section, it is possible to have multiple circuit requests between an input-output switch pair. This corresponds to elements greater than one in the original request matrix.

Matrix decomposition can be used to route unicast or multicast traffic and can be applied to both rearrangeable and strictly non-blocking Clos networks. The derivation of decomposition algorithms is left as an exercise.

6.3.5 **Multicast Routing on Clos Networks**

In multicast call set $C = \{c_1, \ldots, c_n\}$, each multicast call $c_i = (a_i, \{b_{i1}, \ldots, b_{if}\})$ specifies the connection between an input port, a_i, and f output ports, $b_{i1}, \ldots, b_{if}$. The *fanout*, f, of the call is the number of output ports connected. The call set C is constrained so that each input port and output port can be included in at most one call. Equivalently, if we express the calls in terms of switches rather than ports, each input switch and output switch can be included in at most n calls.

Routing multicast traffic on a Clos network is a harder problem than routing unicast traffic for three reasons. First, more middle switches are required. To route a multicast call set with a maximum fan out of f it is sufficient[9] to have $m(f)$ middle switches, as given in Equation 6.5. Second, the looping algorithm cannot be applied to multicast traffic unless all of the fan out is done in the input switches. Any fan out in the middle switches will result in cycles in our conflict graphs and cause the algorithm to fail. Finally, with multicast we have an additional degree of freedom. We can choose where to fan out a call — in the input switch, in a middle switch, or partially in both places — in addition to choosing particular middle switches. As we shall see, properly exploiting this additional degree of freedom is the key to efficient multicast routing in Clos networks.

To illustrate the multicast problem, consider the problem of routing the fanout-of-two call set $C = \{(1, \{1, 2\}), (1, \{3, 4\}), (1, \{1, 3\}), (2, \{1, 4\}), (2, \{2, 4\}), (2, \{2, 3\})\}$ on a Clos network with $n = 3$ and $r = 4$. As before, we show only the switch numbers corresponding to the circuits, not the specific ports.

Table 6.3 gives a vector representation of this call set. The table has a column for each input switch and a column for each output switch. Row i of the table represents call $c_i = (a_i, \{b_{i1}, b_{i2}\})$ by placing a 1 in the positions corresponding to a_i, b_{i1}, and

Table 6.3 Conflict vector representation for call set $C =$
$\{(1, \{1, 2\}), (1, \{3, 4\}), (1, \{1, 3\}), (2, \{1, 4\}), (2, \{2, 4\}), (2, \{2, 3\})\}$.

Call	Input				Output			
	1	2	3	4	1	2	3	4
c_1	1				1	1		
c_2	1						1	1
c_3	1				1	1		
c_4		1			1			1
c_5		1				1		1
c_6		1				1	1	

---

9. Although we do not have a bound on the necessary number of middle switches, there are several examples in which n is not enough.

b_{i2}. For example, the first row of the table has 1s in the column for input switch 1 and 1s in the columns for output switches 1 and 2, and hence represents $(1, \{1, 2\})$.

Two calls, c_i and c_j, can share the same middle switch only if their vectors have no 1s in the same column. That is, c_i and c_j can be mapped to the same middle stage iff $c_i \wedge c_j = 0$. An examination of Table 6.3 shows that every pair of rows has a non-zero intersection. Hence, six different middle-stage switches are needed to route these six calls unless they are split by performing some of the fanout in the input switches.

Fanout performed in the input switch allows the output set to be split across multiple middle stages. In effect, k-way input fanout splits a single call $c_i = (a, B)$ into k calls $c_{i1} = (a, B_1), \ldots, c_{ik} = (a, B_k)$, where $B_1, \ldots, B_k$ are disjoint and $\bigcup_{j=1}^{k} B_j = B$.

For example, the call set of Table 6.3 is shown in Table 6.4 with input fanout applied to calls c_3 and c_6. Splitting calls in this manner eliminates many of the pairwise conflicts and allows the call set to be routed on four middle-stage switches, as shown in Table 6.5 and Figure 6.15.

Table 6.4 Call set of Table 6.3 with input fanout to split calls c_3 and c_6. The resulting set of calls can be routed on four middle stage switches rather than six.

Call	Input				Output			
	1	2	3	4	1	2	3	4
c_1	1				1	1		
c_2	1						1	1
c_{3a}	1				1			
c_{3b}	1						1	
c_4		1			1			1
c_5		1				1		1
c_{6a}		1				1		
c_{6b}		1					1	

Table 6.5 Call set of Table 6.4 packed into 4 middle switches.

Middle Switch	Calls	Input				Output			
		1	2	3	4	1	2	3	4
1	c_1, c_{6b}	1	1			1	1	1	
2	c_2, c_{6a}	1	1				1	1	1
3	c_{3a}, c_5	1	1			1	1		1
4	c_{3b}, c_4	1	1			1		1	1

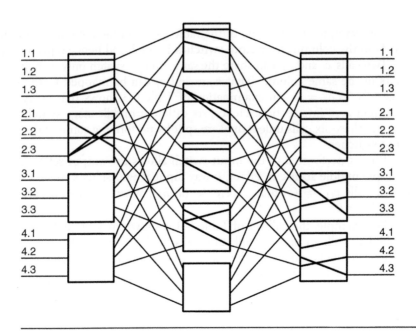

Figure 6.15   Routing specified by Table 6.5. Fanout for calls c_3 and c_6 is performed in input switches 1 and 2, respectively. All other fanout is performed in the middle switches. All six dual-cast calls are routed on 4 middle stage switches.

If we have m middle switches, then our *speedup* is $S = \frac{m}{n}$. To route a general multicast call set, we can always fan out by S in the input stage. This leaves a fanout of at most $g = \left\lceil \frac{f}{S} \right\rceil$ to be performed in the middle stages. Each of the split calls to be assigned to middle stage switches has at most $g + 1$ 1s in its vector, one for the input switch, and g for the g output switches. This call can conflict with at most $(g + 1)(n - 1)$ other calls, $n - 1$ calls in each of the $g + 1$ columns where it has a 1. Thus, we are guaranteed to be able to route an arbitrary fanout of f call set if

$$m \geq (g + 1)(n - 1) + 1. \tag{6.2}$$

If f is evenly divisible by S, we can rewrite this as

$$S = \frac{f}{g} \geq \frac{(g + 1)(n - 1) + 1}{n}. \tag{6.3}$$

After splitting calls for fanout in the first-stage switches, a given input column may have as many as Sn 1s. However, we are still guaranteed that a given call can conflict with at most n calls in this column. This is because two sub-calls that are derived from the same main call do not conflict.

Table 6.6 Fanout as a function of the number of middle stages.

n	S	m	f
2	2	4	4.0
2	3	6	12.0
2	4	8	24.0
2	5	10	40.0
3	2	6	3.0
3	3	9	9.0
3	4	1	18.0
3	5	1	30.0
4	2	8	2.7
4	3	1	8.0
4	4	1	16.0
4	5	2	26.7
48	2	96	2.0
48	3	144	6.1
48	4	192	12.3
48	5	24	20.4

Thus, from Equation 6.3, given a Clos network with m middle switches, we are guaranteed to be able to route calls with a fanout of

$$f \le \frac{m(m-n)}{n(n-1)}. \tag{6.4}$$

Table 6.6 gives values for the maximum fanout that can be handled for several values of n and S. Note that because of the $n-1$ term in Equation 6.4, the fanout that can be handled for a given speedup drops considerably as n is increased. For example, with a speedup of $S = 2$, the fanout drops from 4 when $n = 2$ to just over 2 when $n = 48$.

If we solve Equation 6.4 for m, we can derive the following expression for the number of middle switches required for a fanout of f:

$$m \ge \frac{n + \sqrt{n^2 + 4n(n-1)f}}{2} = \frac{n + \sqrt{(4f+1)n^2 - 4fn}}{2}. \tag{6.5}$$

Our argument about the number of conflicting calls, Equations 6.2 through 6.5, works only if g is an integer. In cases where $\frac{f}{S}$ is not an integer, we must round up to $g = \left\lceil \frac{f}{S} \right\rceil$.

$C_s = \phi$ *initialize split call set empty*
for each call $c_i \in C$ *first split calls for first stage fanout*
    split c_i into $c_{i1}, \ldots, c_{iS}$ each with fanout $\leq g = \left\lceil \frac{f}{S} \right\rceil$
    add $c_{i1}, \ldots, c_{iS}$ to C_s
for each call $c_i \in C_s$
    for each middle switch $m_j \in M$ *find first middle stage that doesn't conflict*
       if c_i doesn't conflict with calls already assigned to m_j
          assign c_i to m_j
          break   *skip to next call c_i*

Figure 6.16   Greedy algorithm to route multicast calls on a Clos network.

The bounds calculated above can be achieved by the simple greedy algorithm shown in Figure 6.16. The algorithm starts by splitting the calls in call set C, creating a split call set C_s with a maximum fanout of g. The calls in C_s are then assigned to middle switches in a greedy manner. For each call c_i, the algorithm scans through the middle switches. For each middle switch m_j, c_i is checked to see if it conflicts with calls already assigned to m_j. If there is no conflict, c_i is assigned to m_j and the algorithm moves on to the next call.

To determine if c_i conflicts with a call assigned to m_j, we use an $m \times 2r$ assignment matrix A. Each row of A corresponds to a middle switch and each column of A corresponds to an input or output switch. Each element of A is initialized to ϕ. When we assign $c_k = (a_k, \{b_{k1}, \ldots, b_{kg}\})$ to m_j, we set $A_{m_j, a_k} = A_{m_j, b_{k1}} = \cdots = A_{m_j, b_{kg}} = p(c_k)$ where $p(c_k) \in C$ is the parent of c_k, the call from which this call was split. To check a call, $c_i = (a_i, \{b_{i1}, \ldots, b_{ig}\})$ for a conflict with m_j, we check that each relevant entry of A, $A_{m_j, a_i}, A_{m_j, b_{i1}}, \ldots, A_{m_j, b_{ig}}$ is either ϕ or equal to $p(c_j)$.

THEOREM 6.4   *The greedy algorithm (Figure 6.16) meets the bound of Equation 6.2 on the number of middle stages required to route multicast traffic with a fanout of f.*

Proof   We prove that the algorithm meets the bound of Equation 6.2 by contradiction. Assume we had a call in C_s that cannot be assigned to any of the $m \geq (g+1)(n-1)+1$ middle switches. Then this call must conflict with at least one call assigned to each of these switches, or m calls altogether. However, this is not possible, since the call can conflict with at most $(g+1)(n-1)$ calls.   □

The ability of the algorithm to handle an arbitrary fanout of f callset without moving any calls leads directly to the following theorem.

THEOREM 6.5   *An (m, n, r) Clos network is strictly non-blocking for multicast traffic with fanout f if Equation 6.4 holds.*

6.3.6 **Clos Networks with More Than Three Stages**

Using $n \times n$ crossbar switches as a building block, we can build a rearrangeably non-blocking (n, n, n) three-stage Clos network with n^2 ports. If we need more than n^2 ports, we can use this $n^2 \times n^2$ port Clos network for each of the middle switches of a Clos network, using the $n \times n$ crossbars for input and output switches. This gives us a five-stage Clos network with n^3 ports. In general, for any integer i we can construct a $2i + 1$ stage rearrangeable Clos network with n^{i+1} ports by using n^i $n \times n$ crossbars for the first stage, n $2i - 1$ stage Clos networks for the middle stage, and n^i $n \times n$ crossbars for the last stage. A similar construction can be used to create a $2i + 1$-stage strictly non-blocking Clos network, with n^i $n \times (2n - 1)$ input switches, and $2n - 1$ n^i-port $2i - 1$-stage Clos networks for middle switches.

For example, a five-stage $(2,2,4)$ Clos network composed of 2×2 crossbar switches is shown in Figure 6.17. The two 4×4 middle switches in this network are each themselves $(2,2,2)$ Clos networks.

Traffic is routed on a Clos network with more than three stages by working from the outside in using the algorithm of Figure 6.11 or Figure 6.16. We start by assigning each call to one of n^i middle switches. This divides one $(2i + 1)$-stage scheduling problem into n $(2i - 1)$-stage scheduling problems. We repeat this process until we get to the point of scheduling n^i $n \times n$ crossbars, after i steps. At the jth step, we schedule n^{j-1} $(2i - 2j + 3)$-stage networks.

Consider scheduling unicast calls on a five-stage Clos network ($i = 2$) constructed from $n \times n$ crossbars. The first step is to assign each call to one of the n middle switches. This is done using the algorithm of Figure 6.11 to schedule to an (n, n, n^2) Clos. This gives the switch settings for the two outside stages and poses

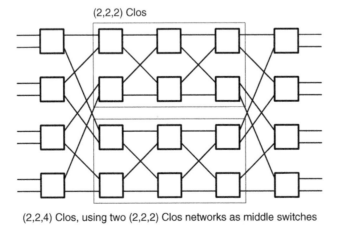

(2,2,2) Clos

(2,2,4) Clos, using two (2,2,2) Clos networks as middle switches

Figure 6.17 A (2,2,4) rearrangeable Clos network is constructed using two (2,2,2) Clos networks as 4×4 middle switches. A Clos network such as this composed of 2×2 switches is also referred to as a Beneš network.

routing problems for each of the n three-stage Clos networks used as middle switches. We then move inward and apply the same procedure to each of the Clos networks used as a middle switch. For each of the n middle switches, we route the calls assigned to this (n, n, n) Clos middle switch using the algorithm of Figure 6.11. This assigns each call to one of the n crossbars in this Clos (one of n^2 middle-stage crossbars overall).

The five-stage Clos network of Figure 6.17 is rearrangeably non-blocking for unicast traffic. The outer (2,2,4) network is rearrangeable and thus can assign each call to one of the two middle stage switches. Also, each of the (2,2,2) middle stage networks are rearrangeable, and thus can route all of the calls assigned to them.

For a five-stage Clos network to be strictly non-blocking for unicast traffic, expansion is required in each input stage. For example, a strictly non-blocking network comparable to the network of Figure 6.17 would require a (3,2,4) outer network with four 2×3 switches in the input stage, four 3×2 switches in the output stage, and three 4×4 middle stage subnetworks. To be strictly non-blocking, each of the four middle stage subnetworks must be realized as (3,2,2) Clos networks with two 2×3 input switches, two 3×2 output switches, and three 2×2 middle switches. The five-stage strictly non-blocking network has nine 2×2 switches down its midpoint, as opposed to four 2×2 switches for the five-stage rearrangeable network. This is because expansion is needed in both the first and second stages. If we use a factor of two expansion in each input stage,[10] a $2n + 1$-stage strictly non-blocking Clos network requires a total expansion of 2^n.

To route multicast traffic in a Clos network that has more than three stages, one must decide where to perform the fanout. Fanout can be performed at any stage of the network. A complete discussion of this problem is beyond the scope of this book. However, when routing a fanout of f multicast call on a $2n + 1$-stage Clos network, a good heuristic is to perform a fanout of $f^{\frac{1}{n+1}}$ on each of the first $n + 1$ stages.

6.4 Beneš Networks

A Clos network constructed from 2×2 switches — for example, the network of Figure 6.17 — is also called a Beneš network. These networks are notable because they require the minimum number of crosspoints to connect $N = 2^i$ ports in a rearrangeably non-blocking manner. As discussed above, to connect $N = 2^i$ ports with 2×2 switches requires $2i - 1$ stages of 2^{i-1} 2×2 switches. This gives a total of $(2i - 1)2^{i-1}$ switches. With 4 crosspoints per 2×2 switch, the total number of crosspoints is $(2i - 1)2^{i+1}$.

The astute reader will have noticed that a $2i - 1$ stage Beneš network is equivalent to two 2-ary i-fly networks back-to-back with the abutting stages fused. Very often, Beneš networks and other multistage Clos networks are implemented by *folding* the

---

10.   Strictly speaking, the expansion required in each stage is $\frac{2n-1}{n}$, which is slightly less than 2.

network along this middle stage of switches. In a $2i - 1$-stage folded Clos network, stages j and $2i - j$ are co-located and share common packaging. This arrangement takes advantage of the symmetry of the network to reduce wiring complexity.

6.5 **Sorting Networks**

An N-input *sorting network* accepts a set of N records tagged with unique sorting keys on its N input terminals and outputs these records with the keys in sorted order on its N output terminals. A sorting network can be used as a non-blocking network

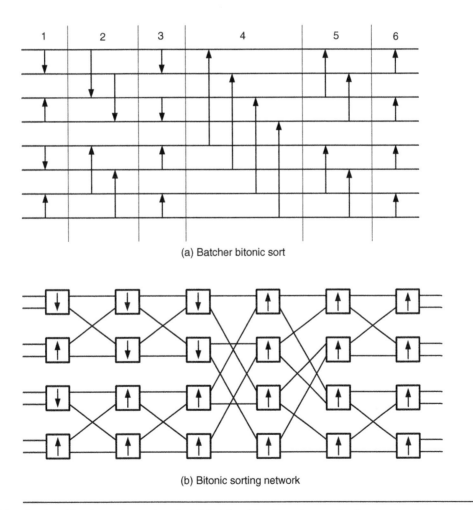

(a) Batcher bitonic sort

(b) Bitonic sorting network

Figure 6.18 Batcher bitonic sorting network. (a) Batcher bitonic sort operates by treating two sorted sequences of length 2^i as a bitonic sequence of length 2^{i+1} and then merging this into a sorted sequence of length 2^{i+1}. (b) A sorting network that realizes this sort is constructed from 2×2 switches that route based on the relative values of the sorting keys.

by using the output address for each record as its sorting key.[11] Because the sorting network is able to sort any permutation of the sorting keys into sorted order, it is able to route any permutation.

Figure 6.18 shows a Batcher bitonic sorting network and illustrates the bitonic sorting algorithm on which this network is based. The sorting algorithm takes advantage of the properties of a bitonic sequence, a sequence that, when viewed cyclically, has at most one ascending subsequence and one descending subsequence (two monotonic subsequences form a bitonic sequence). As shown in Figure 6.18(a), the bitonic sorting algorithm starts with 2^n (in this case 8) unsorted search keys on the left side and generates a sorted sequence on the right side. In this figure, each arrow denotes an operation that compares the keys of the two records on the left side and swaps them if necessary so that the record with the highest key is at the head of the arrow on the right side. The first stage sorts even and odd keys in alternating directions so that the input to stage 2 consists of two bitonic sequences, one from inputs 0–3, and the other from inputs 4–7. Stages 2 and 3 then merge each of these 4-long bitonic sequences into 4-long monotonic sequences. The result is an 8-long bitonic sequence. Stages 4, 5, and 6 then merge this bitonic sequence into a monotonic sequence, and the sort is complete. For example, Table 6.7 illustrates the sort on a sequence of eight numbers.

In general, an $N = 2^n$ input bitonic sort takes $S(N) = \frac{(\log_2 N)(1+\log_2 N)}{2}$ stages consisting of $S(N/2)$ stages to sort the input into a bitonic sequence consisting of two 2^{n-1}-long monotonic sequences, and $\log_2 N$ stages to merge this bitonic sequence into an N-long monotonic sequence.

The sorting network, shown in Figure 6.18(b) follows directly from the algorithm. Here each arrow in Figure 6.18(a) is replaced by a 2×2 switch that routes by comparing keys.

Table 6.7 Example of a bitonic merge sort.

Row	0	1	2	3	4	5	6	7
Input	4	6	2	7	3	1	5	0
Stage 1	4	6	7	2	1	3	5	0
Stage 2	4	2	7	6	5	3	1	0
Stage 3	2	4	6	7	5	3	1	0
Stage 4	2	3	1	0	5	4	6	7
Stage 5	1	0	2	3	5	4	6	7
Stage 6	0	1	2	3	4	5	6	7

---

11.   Note that dummy records must be sent to idle outputs to make this work.

Some asynchronous transfer mode (ATM) switches have used a Batcher sorting network in combination with a butterfly (this is usually called a Batcher-banyan network, using the banyan alias for butterfly). The Batcher network sorts the input cells by destination address. At the output of the Batcher, a *trap* stage detects cells destined for the same output and removes all but one from consideration. The remaining cells, in sorted order and all destined for different outputs, are then shuffled and input into the butterfly. Because of their ordering, the butterfly is guaranteed to be able to route them. (See Exercise 6.4.)

6.6 Case Study: The Velio VC2002 (Zeus) Grooming Switch

The Velio VC2002 is a single-chip time-domain-multiplexing (TDM) circuit switch (Figure 6.19). A VC2002 accepts 72 synchronous optical network (SONET) *STS-48* 2.488 Gbits/s serial input streams and produces 72 *STS-48* output streams. Each of these streams contains 48 51.83 Gbits/s *STS-1* streams multiplexed on a byte-by-byte basis. That is, each 3.2-ns byte time or *time slot* one byte from a different *STS-1* is carried on the line—first a byte from channel 0, then channel 1, and so on up to channel 47. After 48 time slots, the pattern repeats. A TDM switch, also called

Figure 6.19   The Velio VC2002 is a single-chip 72×72 STS-48 grooming switch. The chip is packaged in a 37.5mm×37.5mm 1296-ball ball-grid-array package.

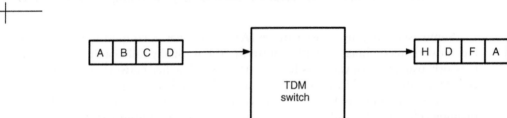

Figure 6.20   A TDM switch accepts and generates time-domain multiplexed streams and switches the contents of input time slots into the contents of output time slots. For example, channel *C* on 0.2 (input 0, time slot 2) is switched to 1.3.

a *cross connect* or a *grooming switch*, switches in time as well as space to map input STS-1 time slots onto output STS-1 time slots.[12]

Figure 6.20 shows an example of TDM switching for the simplified case of two inputs and outputs with four time slots each. This example includes both unicast connections—for example, the *B* on 0.1 (input 0, time slot 1) is switched to 1.0 (output 1, time slot 0)—and multicast connections—the *A* on 0.0 is switched to both 0.3 and 1.1. The VC2002 performs a similar function, but on 72 inputs with 48 time slots each.

TDM grooming switches form the bulk of most metropolitan and long-haul voice and data communication networks.[13] Most of these networks are implemented as interconnected sets of SONET rings (because of the restoration properties of rings). Grooming switches act as cross-connects that link multiple rings together and as add-drop multiplexers that multiplex slower feeds onto and off of the ring.

The configuration of a TDM switch typically changes slowly except during *protection events*. New STS-1 circuits are provisioned at a rate of at most a few per minute. However, when a fault occurs in the network (for example, a backhoe cuts a fiber bundle), a large fraction of the 3,456 connections in a VC2002 may need to be switched within a few milliseconds to avoid interruption of service.

A TDM switch can take advantage of the fact that the input and output streams are multiplexed by implementing the first and third stages of a Clos network in the time domain, as shown in Figure 6.21. The figure shows how the 2-input by 4-time slot example of Figure 6.20 can be realized without the complexity of an 8 × 8 crossbar switch. Instead, a three-stage Clos network in which m=4, n=4 and

---

12.   A single 51.83-Gbits/s STS-1 may contain 28 1.5-Mbits/s T1 or DS1 channels, each of which in turn carries 24 64-Kbits/s DS0 channels, each of which carries one phone call.

13.   Although people think of the Internet as being implemented with packet routers, these routers exist only at the edges and are connected by TDM circuits between these endpoints.

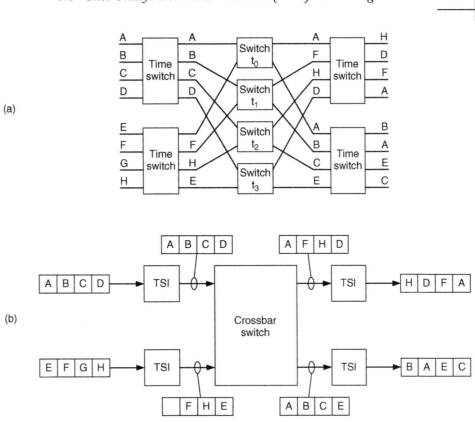

Figure 6.21 A time-space-time (TST) TDM switch implementation for the example of Figure 6.20. (a) Conceptually, the inputs are switched by two 4 × 4 switches in the time domain, then, in each of four time slots t_0 to t_3, the outputs of the two input switches are switched by a 2 × 2 space switch. Finally, a pair of output time switches reorders the outputs. (b) The input and output time switches are implemented by TSIs. The four 2 × 2 switches in (a) are implemented by a single 2 × 2 switch operated with a different configuration on each time slot.

$r = 2$ is realized. The $r = 2$ 4 × 4 ($n \times m$) input and output switches are implemented as *time-slot interchangers* (TSIs) and the $m = 4$ 2 × 2 ($r \times r$) middle switches are implemented as a single time-multiplexed 2 × 2 switch. A TSI is a $2n$ entry memory. While half of the entries (n entries) are being filled by the input stream in order, an output stream is created by reading the other half of the entries in an arbitrary order (as required to realize the input switch configuration). Each time n entries are filled by the input stream, the process repeats, with the output stream reading out these entries, while the input stream fills the other half — which were just read out. Because the outputs of the TSIs in Figure 6.21(b) are time multiplexed, the four 2 × 2 switches of Figure 6.21(a) can be realized by a single switch. In each of the four time slots, the switch acts as the middle stage for that time slot.

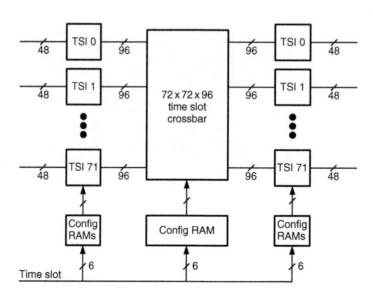

Figure 6.22 Block diagram of the VC2002. The switch consists of 72 48 × 96 input TSIs, a 72 × 72 space switch that is reconfigured for each of the 96 logical time slots, and 72 96 × 48 output TSIs. The configuration of the crossbar switch and the time slot output from each interchanger are controlled by configuration RAMs addressed by the current time slot.

Figure 6.22 shows a block diagram of the VC2002.[14] Like the example of Figure 6.21, the VC2002 uses a three-stage TST Clos network. However, in this case, the Clos has $(m, n, r) = (96, 48, 72)$. There are 72 input channels with 48 time slots each. To make the switch strictly non-blocking for unicast traffic and rearrangeably non-blocking for dual-cast traffic (fanout = 2), the switch has a speedup of two. There are 96 logical middle stages for 48 inputs to each input switch.[15] The switch is controlled by a set of configuration memories addressed by the current timestep. For each timestep, the configuration memories specify which location (time slot) to read from each input TSI, which input port should be connected to each output port of the switch, and which location to read from each output TSI.

The VC2002 is scheduled by assigning a middle stage switch (time slot) to each STS-1 *call* that must traverse the switch. Because the VC2002 is strictly non-blocking for unicast traffic, the middle-stage switch assignments can be performed using the intersection method of Figure 6.8 or by the looping algorithm of Figure 6.11. Dual-cast traffic may be scheduled using the greedy algorithm of Figure 6.16. Alternatively, dual-cast traffic may be scheduled using the looping algorithm by splitting each

---

14. The VC2002 also includes a SONET framer on each input and each output. The input framers align the input streams and monitor and terminate SONET transport overhead. The output framers generate transport overhead.

15. To keep clock rate manageable, the 96 time slot 72 × 72 switch is implemented as two 48 time-slot switches, each running at the STS-48 byte rate (311 MHz).

dual-cast call into two unicast calls and splitting the middle stage switches into two equal subsets. One unicast call is scheduled on each subset of middle switches. Since each subset has $m/2 = n$ switches, the subset is rearrangeble for these unicast calls.

In practice, the VC2002 can handle multicast calls of arbitrary fanout with very low probability of blocking by splitting the calls as described in Section 6.3.5. We leave the calculation of blocking probability for high fanout calls as Exercise 6.5.

A single VC2002 provides 180 Gbits/s of STS-1 grooming bandwidth, enough to switch 3,456 STS-1s. If a larger grooming switch is needed, multiple VC2002s can be connected in a Clos network as illustrated in Figure 6.23. The figure shows how 120 VC2002s can be connected in a folded Clos configuration to switch up to 4.3 Tbits/s (1,728 STS-48s or 82,944 STS-1s). Each of the 72 chips in the first rank acts as both a 24 × 48× STS-48 first-stage input switch and as a 48 × 24× STS-48 third-stage output switch. Each of the 48 chips in the second rank form 72 × 72× STS-48 middle stage switch.

The system of Figure 6.23 is logically a five-stage Clos network. Since each chip is internally a three-stage switch itself, it may appear that we have created a nine-stage

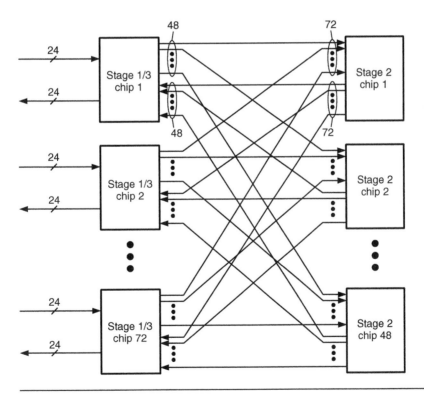

Figure 6.23 A folded Clos network using 120 VC2002s provides 4.3 Tbits/s of STS-1 grooming bandwidth. A first rank contains 72 chips that each split their inputs and outputs to provide both a 24 × 48× STS-48 input switch and a 48 × 24× STS-48 output switch. A second rank of 48 chips forms 48 72 × 72× STS-48 middle switches.

network. However, we can ignore all but the outermost TSIs and view the network as a TSSST network with a 48 × 48 TSI followed by a 24 × 48 space switch, then a 72 × 72 space switch, a 48 × 24 space switch, and a 48 × 48 TSI. The other four TSIs along each path are left in the *straight-through* configuration. Exercise 6.6 examines the possibilities of treating the network as a seven-stage TSTSTST.

To schedule our logical five-stage network, we start by considering the middle three space-switch stages as 48 3,456 × 3,456 space switches, one for each time slot. We assign each call to one of the 48 time slots by scheduling this $(m, n, r) = (48, 24, 3456)$ Clos. This solution then leaves 48 subproblems, one per time slot, of assigning each call in that time slot to one of the 48 physical middle-stage switches. For each subproblem, we schedule all calls in that middle-stage time slot on the $(m, n, r) = (48, 24, 72)$ Clos formed by the three space switches for a single time slot.

6.7 Bibliographic Notes

The seminal paper on non-blocking networks was published by Clos in 1953 [37]. This paper introduced Clos networks and described the requirements for them to be strictly non-blocking. Beneš, Slepian, and Duguid discovered that much smaller Clos networks were rearrangeably non-blocking [17, 171, 64]. The looping algorithm presented to find these arrangements can be traced back to König, who proved a bipartite graph of maximum degree Δ can be edge-colored using Δ colors [103]. Several more efficient algorithms for edge-coloring are due to Cole [39, 40] and matrix decomposition is addressed by Waksman [190]. Beneš's classic 1965 book [18] derives a bound for the number of crosspoints required to realize a rearrangeable non-blocking network and introduces the Beneš network, a special form of a Clos network, to realize this bound. Batcher's 1968 paper introduced the bitonic sort and the Batcher sorting network [14]. A detailed treatment of sorting networks can be found in Knuth [102]. Multicast in Clos networks was first considered by Masson and Jordan [120] and has since been studied by many researchers. More recent results include the near-optimal multicast scheduling algorithm of Yang and Masson [196].

6.8 Exercises

6.1 *A 27-port Clos.* Sketch a rearrangeable Clos network using 3 × 3 crossbar switches that has exactly 27 ports.

6.2 *Maximum size Clos networks.* Using $n × n$ crossbar switches as a building block, how large a Clos can you build with k stages (k odd)? Assume that your Clos need only be rearrangeable for unicast traffic.

6.3 *Performance of the looping algorithm.* Implement the looping algorithm described in Figure 6.11. Run the algorithm for a (5,5,5) Clos network: start with an empty network (no connections) and add one randomly chosen connection at a time until all input and output ports are occupied. We know that the maximum number of rearrangements necessary for each connection is $2r - 2 = 8$, but what is the average number of rearrangements required during your experiment? Now modify the original algorithm and try to further reduce the average number of rearrangements. Explain your changes and report the average number of rearrangements required using the original and modified algorithms.

6.4 *Batcher-banyan networks.* Batcher-banyan networks are non-blocking networks built from a Batcher stage, in which calls are are sorted by their output port, followed by a banyan (butterfly) stage. Between the stages, a shuffle of the ordered calls is performed — the call leaving port $\{a_{n-1}, a_{n-2}, \ldots, a_0\}$ of the Batcher network enters port $\{a_{n-2}, \ldots, a_0, a_{n-1}\}$ of the banyan network. Prove that 2-ary n-flies (banyan stages) can always route a shuffle of a set of ordered addresses without blocking.

6.5 *Probabilistically non-blocking multicast.* Estimate the probability of blocking when routing a full set of randomly distributed calls with a fanout of four on a VC2002 (Section 6.6) when using the greedy algorithm.

6.6 *A seven-stage time-space switch.* In the 120-chip network of Figure 6.23, we initially considered the network as a five-stage TSSST switch and ignored the internal speedup of the VC2002. Now consider the network as a seven-stage TSTSTST switch in which both TSIs on the first and third stages are used. How does this affect what traffic patterns the network is able to route? How does this affect how you schedule the network?

6.7 *Proof of rearrangeability.* Use Hall's Theorem (below) to prove that a Clos network with $m \geq n$ is rearrangeably non-blocking.

THEOREM   **6.6**

(**Hall's Theorem**) *Let A be a set and $A_1, A_2, \ldots, A_r$ be any r subsets of A. There exists a set of distinct representatives $a_1, \ldots, a_r$ of $A_1, \ldots, A_r$ such that $a_i \in A_i$ and $a_i \neq a_j \; \forall i \neq j$, iff $\forall k, 1 \leq k \leq r$, the union of any k of the sets $A_1, \ldots, A_r$ has at least k elements.*

6.8 *Graph coloring with Euler partitions.* The idea of an *Euler partition* [40] can be used to improve the performance of the looping algorithm to find a bipartite edge-coloring. Euler partitioning begins by finding an *Euler circuit* through the bipartite graph — an Euler circuit is a path through the graph that starts and ends at the same node and visits each edge exactly once. Then, the path is followed and alternate edges are partitioned between two new bipartite graphs with the same number of nodes as the original. For example, for a path that visits nodes $\{L_1, R_6, L_3, R_4, \ldots\}$, where R and L denote the left and right sides of the bipartite graph, the edge (L_1, R_6) is placed in the first new graph, (R_6, L_3) in the second, (L_3, R_4) in the first, and so on. Notice that the degree of the two new graphs is exactly half of the original. These new graphs are then partitioned again, and this continues until we are left with graphs of degree one, each of which is assigned a unique color.

(a) For a bipartite graph in which the degree of all nodes is 2^i, for some integer i, prove that an Euler circuit always exists. Describe a simple algorithm to find such a circuit.

(b) Find the asymptotic running time for an edge-coloring algorithm based on Euler partitions for a graph with V nodes, each with degree 2^i, for some integer i. Compare this with the running time of the looping algorithm.

CHAPTER 7

Slicing and Dicing

In this closing chapter of the section on topology, we briefly explore some pragmatics of packaging a topology. We start by looking at concentrators and distributors. Concentrators combine the traffic of several terminal nodes onto a single network channel. They can be used when the traffic from any one terminal is too small to make full use of a network channel. They are also effective in combining traffic from many bursty terminals. When the ratio of peak to average traffic is large, using concentrators results in lower serialization latency and a more cost-effective network.

Distributors are the opposite of concentrators. They take traffic from a single node and distribute it across several network channels on a packet-by-packet basis. Distributors are used when the traffic from a node is too large to be handled by a single network channel. They increase serialization latency and reduce load balance, but in some cases are still useful to connect nodes that otherwise could not be served.

There are three ways to *slice* a network node across multiple chips or modules: bit slicing, dimension slicing, and channel slicing. With bit slicing, a w-bit-wide node is divided across k $\frac{w}{k}$-bit-wide slices, each slice packaged in a separate module. Each slice contains a $\frac{w}{k}$-bit-wide portion of the router datapath. Control information must be distributed to all slices so they act in unison. This requires both additional pins to distribute the control information and latency to allow time for this distribution.

With dimension slicing, the network node is sliced so that entire channels, such as those associated with a dimension, are contained on each slice. With this approach, additional data channels must be added between the slices to carry traffic that enters on one slice but must leave on another.

Finally, with channel slicing, a w-bit-wide node is divided into k independent nodes, each of which has $\frac{w}{k}$-bit-wide channels. Unlike the bit-sliced approach, however, there are no connections between these sub-nodes. They form completely separate networks. Often, a distributor is used at each terminal to distribute traffic over these network slices.

7.1   Concentrators and Distributors

7.1.1   Concentrators

In some applications it is preferable to combine the load from several — say, M — network terminals into a single channel. The M terminals then appear to the network as a single large terminal. The device that performs this combining is called a *concentrator*. As illustrated in Figure 7.1 for the case where $M = 4$, a concentrator accepts M bidirectional channels on the terminal side and combines these into a single bidirectional channel on the network side. In addition to reducing the number of ports, the concentrator may also reduce the total bandwidth. The *concentration factor* of a concentrator is the ratio of bandwidth on the terminal side to bandwidth on the network side and is given by $k_C = \frac{Mb_T}{b_N}$ where b_T is the terminal channel bandwidth and b_N is the bandwidth of the channel on the network side of the concentrator.

Figure 7.2 shows how a concentrator is applied to a network. An 8-node ring could directly serve 8 terminals, as shown in Figure 7.2(a). Placing 2:1 concentrators between the terminals and the network, as shown in Figure 7.2(b), allows the same 8 terminals to be handled by a 4-node ring.

Concentrators are often used to combine multiple terminals that have bursty traffic characteristics. Sharing the bandwidth on the network side of the concentrator smooths the load from bursty sources and makes more efficient use of channel bandwidth.

Consider, for example, an $N=512$-node multicomputer in which each node presents an average 100 Mbits/s load to the network. However, when a node takes a cache miss, it presents an instantaneous load of 1 Gbit/s to the network for a 128-ns

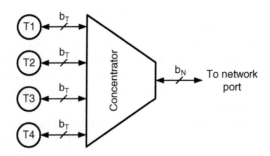

Figure 7.1 A 4:1 concentrator combines traffic to and from 4 terminal nodes, $T_1, \ldots, T_4$, each with bandwidth b_T, onto a single channel to the network with bandwidth b_N.

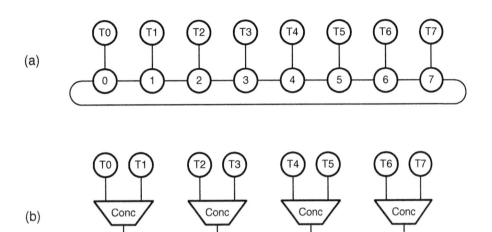

Figure 7.2 Application of a 2:1 concentrator: (a) An 8-ary 1-cube (ring) network connects 8 terminals. (b) The same 8 terminals are connected by a 4-ary 1-cube (ring) network preceded by four 2:1 concentrator.

period ($L = 128$ bits). To prevent serialization latency from increasing the memory access time of the network, we must provide a 1-Gbit/s channel from each processor into the network. Without concentrators, we would build an 8-ary 3-cube with 1-Gbit/s channels to handle this load. The total pin bandwidth of the routers in this network, a good estimate of cost, is 3 Tbits/s. If we size the network instead to handle the average bandwidth with worst-case traffic, we would use 200 Mbits/s channels and incur a 5× increase in serialization latency, from 128 ns to 640 ns.

A more efficient approach to this 512-node network is to combine groups of 8 nodes with 8:1 concentrators that feed a 64-node 4-ary 3-cube network with 1-Gbit/s channels. A 2-Gbits/s channel connects each concentrator to one node of this network to prevent this link from becoming a bottleneck.[1] The concentrated node now has an average bandwidth of 800 Mbits/s. Although the peak load of the concentrated node is 8 Gbits/s, this peak is rarely attained.[2] In only a tiny fraction of the time are there more than two nodes transmitting, and the additional delay during this small fraction of time is more than offset by the reduced diameter of the network and the reduced serialization latency. If we size our network with 1-Gbit/s channels, the total pin bandwidth is 384 Gbits/s (a factor of 8 less than the

---

1. If pinout permits, we could combine the concentrator with the node router and eliminate this channel entirely.
2. If we assume that the access patterns of the 8 nodes are independently distributed Poisson processes, the probability of all 8 nodes transmitting simultaneously is 10^{-8}.

unconcentrated network). If, on the other hand, we size the network to handle average bandwidth, we would use 800-Mbits/s channels and have a serialization latency of 160 ns (a factor of 4 fewer than the unconcentrated network).

Concentrators are also used to facilitate packaging of a network. For example, if 4 terminals are packaged together on a module (for example, a chip), it is often convenient from a packaging perspective to combine the traffic from the 4 terminals together using a concentrator and treat the module as a single network terminal.

7.1.2 Distributors

A distributor is the opposite of a concentrator. As illustrated in Figure 7.3, a *distributor* takes one high bandwidth channel and distributes its packets over several lower bandwidth channels. Although at first glance it may appear that a distributor is just a concentrator inserted backward, the functionality is different. Each packet traversing a concentrator in the reverse direction must be delivered to a particular terminal, the one to which it is addressed. A distributor, on the other hand, can arbitrarily distribute packets to network channels. The distribution may be random, round-robin, or load balancing. In some cases, to maintain ordering, packets of the same class[3] are always distributed to the same network channel.

Distributors are used in several applications. They may be used to interface a higher bandwidth module (such as a faster processor or higher-rate line card) with an existing network that has a lower bandwidth by using multiple network ports. In other applications, distributors are used to provide fault tolerance. Distributing the load across two half-speed channels allows the network to gracefully degrade to half performance if one channel fails.

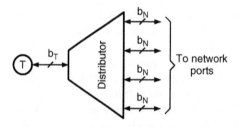

Figure 7.3 A 1:4 distributor accepts packets from a single terminal node with bandwidth b_T and distributes the packets over 4 network channels, each with bandwidth b_N. The distributor may deal out the packets evenly over the 4 network channels or may load balance them based on packet length or queue length.

---

3.   For example, packets from the same flow in an IP router or packets that address the same cache block in a multiprocessor.

Distributors are also used when the bandwidth of a channel is too high to easily handle in a router. For example, if the clock rate is 500 MHz (2 ns period), packets are 8 bytes, and the router can handle at most 1 packet every 4 cycles, then the maximum bandwidth that can be handled by this router is 8 bytes every 8 ns, or 1 Gbyte/s. If we wish to attach a 4-Gbyte/s port to this network, we need to use a 1:4 distributor to break up this traffic into lower bandwidth channels that are within the capabilities of the router. After we break up the high-bandwidth traffic, we can insert the lower bandwidth channels into multiple ports of one network or into parallel networks. (This is called channel slicing — see Section 7.2.3.)

Adding distributors to a network adversely affects performance in two ways. First, distribution increases serialization latency. The serialization latency of a packet is $\frac{L}{b}$ at the bottleneck link, where b is the lowest. Thus, distributing a packet from a terminal link with bandwidth b_T to a network channel with bandwidth b_N increases the serialization latency by $\frac{b_T}{b_N}$. Queueing delays, which are related to serialization latency, are also increased proportionally. Second, distribution reduces load balance. The balance on the output channels of the distributor is never perfect. Also, if the distributor is used to feed parallel networks, at any given point in time, a link of one network may be overloaded, while the corresponding link on the parallel network may be idle. In general, from a performance perspective, it is always better to share resources. Thus, distribution hurts performance, but may aid implementability or fault tolerance.

7.2 Slicing and Dicing

Occasionally, a network node does not fit entirely on a single module (chip or board). This is most often due to pin limitations, but area can also be a consideration (especially for memory). In such cases, we need to divide the router across several chips, a process that is called *slicing*. There are three approaches to slicing the router: bit slicing, dimension slicing, and channel slicing.

7.2.1 Bit Slicing

Bit slicing, illustrated in Figure 7.4, is the most straightforward method of dividing a network node across multiple chips. If each channel is w bits wide and we wish to divide the node across m chips, we simply put $\frac{w}{m}$ bits of each channel on each chip.[4] Figure 7.4, for example, shows an 8-bit-wide, 2-D node being sliced into two 4-bit-wide modules. Each of the 8 channels is divided into a pair of 4-bit channels. Several control lines (`ctl`) are used to pass information between the two router bit slices.

---

4. Of course, this assumes that w is evenly divisible by m.

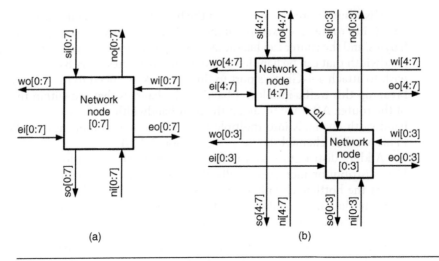

Figure 7.4 Bit slicing: (a) A network node for a 2-D torus network with $w = 8$-bit-wide channels. Each direction (n-north, s-south, w-west, e-east) has both an 8-bit input channel (for example, `ni[0:7]` is the north input) and an 8-bit output channel (for example, `wo[0:7]` is the west output). (b) The same node sliced into 2 packages each with $w = 4$-bit-wide channels

The difficulty with bit slicing comes with control distribution and fault recovery. Half of the bits of a flit arrive on network node `[4:7]`, and the other half arrive on network node `[0:3]`; yet the entire flit must be switched as a unit. To switch the flit as a unit, both slices must have identical and complete control information. The first problem encountered is distribution of header information. One approach is to distribute all of the relevant bits of the header, which specify the route, destination, virtual channel, and so on, to both slices. With this approach, all relevant bits of the header must cross the control channel between the two chips. The cost of this distribution is both pin overhead and latency. The pin overhead is exactly the ratio of header bits to total flit bits. For small flit sizes, this overhead can be quite large, 25% or more. The latency overhead is also significant. Transmitting the header from one chip to another can easily add two clocks to the packet latency in a typical router.

The pin bandwidth overhead can be reduced somewhat by performing all of the control on one of the bit slices and distributing the control decisions, and not the header information, to the other slices. This approach, however, does not reduce the latency penalty. The chip crossing still costs several clocks.

Another issue with bit slicing is error detection and recovery. Many routers perform error detection by computing a function, such as a cyclic redundancy check (CRC) across all bits of the flit. In a bit-sliced router, such a check must be partitioned across the slices and intermediate results exchanged between slices.

Problems also occur if the control state that is replicated across the bit slices diverges — for example, due to a transient fault. A robust sliced router must constantly

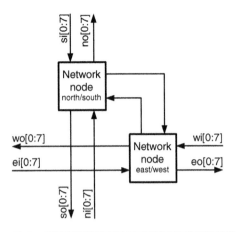

Figure 7.5 Dimension slicing: a 2-D node is partitioned into 2 one-dimensional nodes, one carrying north-south traffic and one carrying east-west traffic. An inter-slice channel handles traffic that crosses between these directions.

check to make sure this state remains synchronized and must be able to recover in the event that the state diverges.

Despite the complexity of control distribution and error handling, bit slicing works well in routers that have large flits over which this overhead can be amortized. Bit slicing is particularly attractive with flow control methods that allow the control to be pipelined ahead of the data, such as flit-reservation flow control (Section 13.4).

7.2.2 Dimension Slicing

The control and error handling complexity of bit slicing can be avoided if a router is sliced in a manner that keeps an entire flit together in a single slice. Dimension slicing achieves this goal by dividing the router across its ports[5] while keeping each port intact on a single slice. In effect, each network node of degree d is partitioned into a subnetwork of m nodes of degree $\frac{d}{m} + p$, where the p additional ports per node are required for inter-slice communication.

For example, Figure 7.5 shows the 2-D $w = 8$-bit-wide router of Figure 7.4(a) sliced into 2 one-dimensional routers: one that handles north-south traffic and one that handles east-west traffic. If further partitioning is required, this router could

---

5. It would be more accurate to call this method *port slicing* since it partitions across ports. The first applications of this method were applied to cube and mesh networks across dimensions — hence the name dimension slicing.

be partitioned to run each of the four directions on a separate one-dimensional, unidirectional router.

The channel or channels between the 2 one-dimensional routers must be sized with enough bandwidth to handle all traffic that is switching from north-south to east-west or vice versa. The amount of traffic needing to switch directions depends greatly on the routing algorithm (Chapter 8). Routing algorithms, like dimension-ordered routing, that favor straight-line travel over direction switching require much less inter-router bandwidth than algorithms that permit arbitrary switching of directions.

7.2.3 Channel Slicing

Both of the previous approaches to node partitioning require communication between the partitions. Bit slicing requires control information to be exchanged and dimension slicing requires a datapath between the partitions. Channel slicing eliminates all communication between the partitions by splitting the entire network into two completely separate networks. The only communication between the parallel networks is at the terminals where distributors are used.

Channel slicing is illustrated in Figure 7.6. The $w = 8$-bit-wide network of Figure 7.4(a) is replaced by two completely separate 4-bit-wide networks. There are no connections between the two 4-bit-wide subnetworks except at the terminal links where distributors are used to divide traffic across the two networks for transmission and recombine the traffic at the destination.

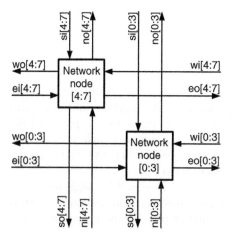

Figure 7.6 Channel slicing: a $w = 8$-bit-wide 2-D network node is partitioned into two completely separate $w = 4$-bit-wide nodes. There are no connections between the two 4-bit-wide nodes except on the terminal links (not shown).

7.3 **Slicing Multistage Networks**

Slicing of multistage networks can be used to trade off serialization latency T_s (which depends on channel width) against head latency T_h (which depends on diameter), allowing us to optimize the latency of the network by balancing these two components. The technique is similar to that discussed in Section 5.2.2 for minimizing the latency of a torus by selecting the dimension of the network to balance T_s and T_h. Channel slicing a multistage network also reduces total cost by reducing the number of routing components and the total number of pins.

An $N = k^n$ node k-ary n-fly with w-bit-wide channels can be channel sliced into x xk-ary n'-fly networks with w/x-bit-wide channels. For example, Figure 7.7 shows that cost and diameter can be reduced by replacing a binary 2-fly network that has 2-bit wide channels with two 4-ary 1-fly networks that have serial (1-bit wide) channels. The two networks have equal throughput and require switches with equal pin count (eight signals). However, the channel-sliced 4-ary fly has a smaller diameter and requires half the number of switch components as the binary fly.

Channel slicing a multistage k-ary n-fly network reduces the diameter of the network from $n + 1$ to $n' + 1$ where

$$n' = \log_{xk} N = \log_{xk} k^n = \frac{n}{1 + \log_k x}.$$

As a result of this slicing, T_s increases by a factor of x and T_h decreases by a factor of $\frac{1}{1+\log_k x}$. Using this expression for n' and lumping our wire delay into t_r gives

$$T_s = \frac{xL}{b},$$

$$T_h = t_r \left(\frac{n}{1 + \log_k x} \right).$$

For example, consider an $N = 4{,}096$ node binary 12-fly with $w = 32$ bit-wide channels, b=1 Gbit/s, t_r=20 ns, and L=256-bit-long messages. Table 7.1 shows how diameter n', channel width w, and the two components of latency T_s and T_h vary as the slicing factor x is increased from 1 (no slicing, 32-bit-wide channels) to 32 (network sliced into serial channels) in powers of two. The latency data is plotted in Figure 7.8. With no slicing, latency T of the binary cube is dominated by T_h — 240 ns to propagate through $n' = 12$ stages at 20 ns each. When we slice the network into two networks of 16-bit channels ($x = 2$), we can now realize radix-4 switches with the same pin count and hence can use two $n' = 6$-stage networks. This halves T_h to 120 ns while doubling T_s to 16 ns — a good trade. Slicing into 4 networks with 8-bit channels ($x = 4$) gives the minimum latency of $T = 112$ ns with T_h reduced further to 80 ns ($n' = 4$ stages). Beyond this point, the incremental increase in T_s, which doubles each time we double x, is larger than the incremental decrease in T_h, which decreases as the inverse log of x, and overall latency increases. The extreme point of $x = 32$ parallel bit-serial $n' = 2$ stage networks with radix $k = 64$ switches

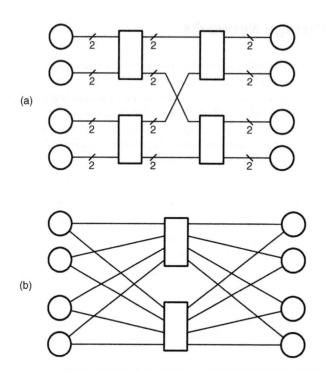

Figure 7.7 Channel slicing a multistage network allows switches of a higher radix to be realized with the same pin count. This reduces network diameter and cost at the expense of higher serialization latency. (a) A binary 2-fly with $w=2$ channels has a diameter of 3 and uses switches that have 8 signals. (b) Two parallel 4-ary 1-flies with serial ($w=1$) channels have the same total bandwidth and the same switch signal count, but a diameter of 2 and half the number of switches as the binary 2-fly.

Table 7.1 Channel slicing to minimize latency in an $N = 4{,}096$ node butterfly network. With no slicing, a binary 12-fly with 32-bit-wide channels is limited by header latency. At the other extreme, slicing into 32 parallel 64-ary 2-flies with serial channels gives a network that is dominated by serialization latency. Minimum latency is achieved when the two components of latency are comparable, with 4 parallel 8-ary 4-flies with 8-bit-wide channels.

x	k	n'	w	T_s	T_h	T
1	2	12	32	8 ns	240 ns	248 ns
2	4	6	16	16 ns	120 ns	136 ns
4	8	4	8	32 ns	80 ns	112 ns
8	16	3	4	64 ns	60 ns	124 ns
16	32	3	2	128 ns	60 ns	188 ns
32	64	2	1	256 ns	40 ns	296 ns

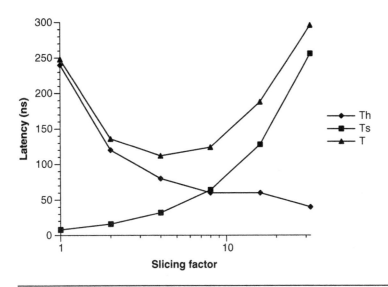

Figure 7.8 Plot of total latency T and its two components: serialization latency T_s and head latency T_h as a function of slicing factor x.

gives the largest latency of all configurations, dominated by $T_s = 256$. Although this configuration has the highest latency, it has the lowest pin cost, since it has the lowest diameter n'.

We have described the use of channel slicing in the context of butterfly networks, but it can also be applied to other multistage networks, including Clos and Batcher networks. We explore the slicing of Clos networks in Exercise 7.5. Also, while we have described multistage network slicing using channel slicing, we can also slice these networks using bit slicing. We explore this alternative in Exercise 7.6.

7.4 **Case Study: Bit Slicing in the Tiny Tera**

The Tiny Tera is a fast packet switch originally designed at Stanford University and later commercialized by Abrizio [125].[6] As indicated by its name, it provides 1 Tbit/s of aggregate bandwidth.[7] The high-level architecture of the Tiny Tera switch is illustrated in Figure 7.9. As shown, the Tiny Tera is organized around a 32 × 32, 8-bit-wide crossbar. Port cards provide both input and output buffering between the crossbar and the physical interface to the outside network. Each port card communicates with a central scheduler unit responsible for computing configurations for the crossbar.

---

6.   Our discussion of the details of the Tiny Tera is based on the academic design from Stanford.
7.   The bandwidth of a packet switch is often advertised as the sum of the input and output bandwidth (aggregate bandwidth). It would be more accurate to say that this switch has 500 Gbits/s of throughput.

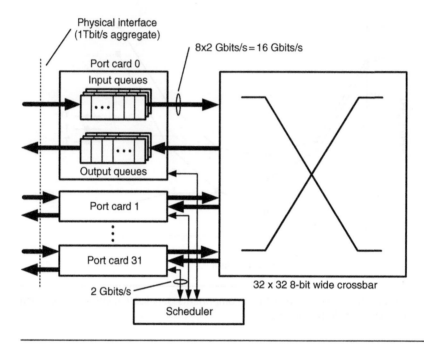

Figure 7.9 The Tiny Tera architecture, which uses 32 port cards, each with 32 Gbits/s of bandwidth to feed a 32 × 32 8-bit-wide crossbar, achieves 1 Tbit/s of aggregate bandwidth. A centralized scheduler communicates with the port cards to compute crossbar schedules, which are relayed back to the port cards.

The configurations are relayed back to the port cards and appended to the packets as they are passed to the crossbar.

The challenges of building the crossbar become apparent by considering the number of pins it requires. First, the signals between the port cards and the crossbar are differential and operate at 2 Gbits/s. Each port of the switch sends and receives 16 Gbits/s of data, so this requires a total of 16 signals, or 32 pins per port. Since there are 32 ports to the crossbar, it requires a total of 1,024 high-speed pins. While packages with this number of pins are available, we have not yet accounted for the power and ground pins or the amount of power that each high-speed link dissipates. So, based on its demanding requirements, the designers chose to bit slice the Tiny Tera crossbar.

Figure 7.10 shows the bit slicing approach selected by the switch designers. The crossbar is sliced into eight 32 × 32 1-bit-wide crossbars. Slicing reduces the pin count to a more manageable 128 high-speed pins per chip and power dissipation is also spread across these 8 chips. Although not shown here, the slicing is carried through the port cards themselves where the input and output queues are bit sliced and spread across several SRAMs.

Not only does the sliced design alleviate the many packaging issues in the Tiny Tera, but it also improves flexibility. For example, additional crossbar slices can be

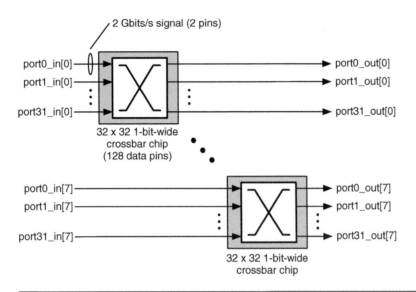

Figure 7.10 The bit sliced implementation of the Tiny Tera crossbar. The 8-bit-wide interface is cut into 8 slices, requiring only 128 pins per crossbar chip.

added to provide a fractional speedup to the switch or to improve reliability. If a single chip crossbar had been employed, any increase in speedup or redundancy would have required another entire crossbar.

7.5 **Bibliographic Notes**

Slicing, concentration, and distribution have a long history. Concentration has long been used in the telephone network to account for the fact that most phones are on the hook most of the time. The J-Machine router [138] (Section 5.5) was dimension sliced with all three slices on a single chip. The Cray T3D [95] used a similar organization with each slice on a separate ECL gate array. A bit sliced crossbar was used for the network of the MARS accelerator [3] as well as for the Tiny Tera (Section 7.4) [125]. A bit sliced 3-D torus with six 4-bit slices was used in the Avici TSR [49].

7.6 **Exercises**

7.1 *Concentration in a torus.* You need to connect 4,096 nodes that have a peak bandwidth of 10 Gbits/s and an average bandwidth of 500 Mbits/s. Limiting yourself to torus networks with the bisection bandwidth fixed at 2.56 Tbits/s, compare alternative networks in terms of concentrators, dimension, and radix. Assume each concentrator and router node is packaged in a separate chip and each chip can have a total pin bandwidth of up to 100 Gbits/s. Which topology offers the lowest pin bandwidth? Which offers the lowest pin bandwidth without incurring additional serialization latency?

7.2 *Distributing traffic from a line card.* Using distributors, suggest how you might connect 64 40-Gbits/s line cards using a torus network composed of channels that do not exceed 10 Gbits/s. The underlying network should support worst-case traffic in which all nodes send across the bisection. Does this arrangement change the bisection bandwidth compared to a network in which the channels could operate at 40 Gbits/s?

7.3 *Slicing a butterfly.* Consider the partitioning of a radix-4 butterfly node with $w = 2$-bit-wide channels into two modules using bit slicing, dimension (port) slicing, and channel slicing. Flits are 64-bits, 16-bits of which are header information. Sketch each partitioning, labeling all channel widths and the number of signals required per chip, and qualitatively compare the latency in each case.

7.4 *Sub-signal slicing.* It it possible to choose a channel slicing factor such that the resulting channels are less than one signal wide. To implement these channels, several could be multiplexed onto a single physical signal. For example, if the channels are sliced to one-half a signal, two channels would share one physical signal. Can this level of slicing ever reduce zero-load latency? Explain why or why not.

7.5 *Channel slicing a Clos network.* Find the channel slicing factor x that gives the minimum latency for an $N = 256$ rearrangeable $(m = n)$ Clos network built from switches with 32 signals. Each node requires 8 Gbits/s of bandwidth, $L = 128$ bits, $t_r = 20$ ns, and $f = 1$ Gbit/s.

7.6 *Bit slicing a 4,096 node butterfly.* Consider a network that has the parameters of the network of the first row of Table 7.1. Find the *bit* slicing of this network that gives minimum zero-load latency. Assume that a 32-bit header is repeated in each bit slice. (No control signals are required between slices.) Also, the full header must be received before any of the router latency t_r is experienced.

7.7 *Header distribution in the Tiny Tera.* In the bit-sliced Tiny Tera switch, crossbar configurations are computed by the centralized scheduler and then redistributed to the input ports. This header information is then duplicated and attached to each outgoing packet slice. However, for a port A, the packet leaving A does not contain its destination port, but rather the address of the port writing to A (that is, configurations are not described by where to write a packet, but rather from where to receive one). Why does this result in a more efficient encoding? Hint: The Tiny Tera supports multicast traffic.

CHAPTER 8

Routing Basics

Routing involves selecting a path from a source node to a destination node in a particular topology. Once we have a topology, or road map for our network, routing is the next logical step: picking a route on the map that gets us to our destination. Whereas a topology determines the ideal performance of a network, routing is one of the two key factors that determine how much of this potential is realized. We discuss the other key factor, flow control, in Chapters 12 and 13.

The routing algorithm used for a network is critical for several reasons. A good routing algorithm balances load across the network channels even in the presence of non-uniform traffic patterns such as permutation traffic. The more balanced the channel load, the closer the throughput of the network is to ideal. Surprisingly, many routers that have been built and are in use today do a poor job of balancing load. Rather, the traffic between each pair of nodes follows a single, predetermined path. As you would expect, non-uniform traffic patterns can induce large load misbalances with this type of routing algorithm, giving suboptimal throughput. However, these routing choices can be at least partially explained because most of these routers have been designed to optimize a second important aspect of any routing algorithm: short path lengths.

A well-designed routing algorithm also keeps path lengths as short as possible, reducing the number of hops and the overall latency of a message. What might not be immediately obvious is that, often, routing minimally (always choosing a shortest path) is at odds with balancing load and maximizing throughput. In fact, for *oblivious routing algorithms*, to improve load balance over *all* traffic patterns, we are forced to increase the average path length of *all* messages. The converse is also true. This tradeoff exists for oblivious algorithms because they do not factor the current traffic pattern into the routing algorithm. We explore these algorithms in more detail in Chapter 9.

On the other hand, a clever designer might suggest an approach to offer the "best of both worlds." Instead of picking an algorithm independent of the traffic pattern, as in oblivious algorithms, why not *adapt* to the current traffic conditions? Then we could send traffic minimally for an "easy" traffic pattern such as uniform traffic, but then could resort to non-minimal routing for "hard" non-uniform traffic patterns. This simple idea forms the basis of *adaptive routing algorithms*, which we explore in Chapter 10. The potential advantage of these algorithms is realizing both load balance and locality (short path lengths). However, we will see that practical design issues make achieving this goal challenging.

Another important aspect of a routing algorithm is its ability to work in the presence of faults in the network. If a particular algorithm is hardwired into the routers and a link or node fails, the entire system fails. However, if an algorithm can be reprogrammed or adapt to the failure, the system can continue to operate with only a slight loss in performance. Obviously, this is critical for systems with high-reliability demands. Finally, routing interacts with the flow control of the network and careful design of both is often required to avoid deadlocks and/or livelocks (Chapter 14).

Our discussion of routing begins below with a short example and a discussion of routing taxonomy and an introduction to deterministic routing algorithms. We continue in Chapter 9 with a discussion of deterministic and oblivious routing, and then adaptive routing in Chapter 10. We conclude with a discussion of routing mechanics in Chapter 11.

8.1　**A Routing Example**

Consider the problem of routing on the 8-node ring network shown in Figure 8.1. If we rule out *backtracking*, or revisiting a node in the network, the routing decision here is binary. For each packet being sent from s to d, we can either send the packet clockwise or counterclockwise around the ring starting at s and ending at d. Even with this simple topology and only a binary decision to make, there are many possible routing algorithms. Here are a few:

Greedy: Always send the packet in the shortest direction around the ring. For example, always route from 0 to 3 in the clockwise direction and from 0 to 5 in the counterclockwise direction. If the distance is the same in both directions, pick a direction randomly.

Uniform random: Randomly pick a direction for each packet, with equal probability of picking either direction.

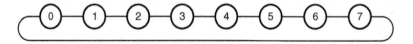

Figure 8.1　An 8-node ring network.

Weighted random: Randomly pick a direction for each packet, but weight the short direction with probability $1 - \Delta/8$ and the long direction with $\Delta/8$, where Δ is the (minimum) distance between the source and destination.

Adaptive: Send the packet in the direction for which the local channel has the lowest *load*. We may approximate load by either measuring the length of the queue serving this channel or recording how many packets it has transmitted over the last T slots. Note that this decision is applied once at the source because we have disallowed backtracking.

Which algorithm gives the best worst-case throughput? The vast majority of people pick the greedy algorithm.[1] Perhaps this says something about human nature. However, it turns out that the greedy algorithm does not give the best worst-case throughput on this topology.

To see how greedy routing can get us into trouble, consider a tornado traffic pattern in which each node i sends a packet to $i + 3$ mod 8, as shown in Figure 8.2. The performance of the 4 routing algorithms described above on tornado traffic on an 8-node ring is summarized in Table 8.1. With the greedy routing algorithm, all

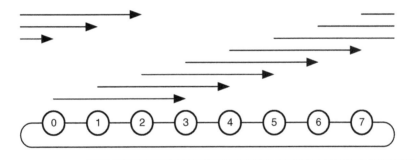

Figure 8.2 Tornado traffic on an 8-node ring network. With greedy routing, all traffic moves in a clockwise direction around the ring, leaving the counterclockwise channels idle.

Table 8.1 The throughput of several example routing algorithms (as a fraction of capacity) for an 8-node ring on the tornado traffic pattern.

Algorithm	Throughput on Tornado
Greedy	0.33
Random	0.40
Weighted random	0.53
Adaptive	0.53

---

1. This problem of routing on a ring was given as a Ph.D. qualifying exam question in 2002, and over 90% of the examinees initially picked the greedy algorithm.

of the traffic routes in the clockwise direction around the ring, leaving all of the counterclockwise channels idle and loading the clockwise channels with 3 units of traffic — that is, $\gamma = 3$ —, which gives every terminal a throughput of $\Theta = b/3$. With random routing, the counterclockwise links become the bottleneck with a load of $\gamma = 5/2$, since half of the traffic traverses 5 links in the counterclockwise direction. This gives a throughput of $2b/5$. Weighting the random decision sends 5/8 of the traffic over 3 links and 3/8 of the traffic over 5 links for a load of $\gamma = 15/8$ in both directions giving a throughput of $8b/15$. Adaptive routing, with some assumptions on how the adaptivity is implemented, will match this perfect load balance in the steady state, giving the same throughput as weighted random routing.

This example has shown how the choice of routing function can significantly affect load balance. However, worst-case throughput is only one of several possible metrics a designer may wish to optimize. And as one would expect, different metrics can lead to different conclusions about which of these four algorithms would be the most appropriate. We explore some of these in Exercise 8.1.

8.2 Taxonomy of Routing Algorithms

We classify routing algorithms in terms of how they select between the set of possible paths R_{xy} from source node x to destination node y.

Deterministic routing algorithms always choose the same path between x and y, even if there are multiple possible paths ($|R_{xy}| > 1$). These algorithms ignore path diversity of the underlying topology and hence do a very poor job of balancing load. Despite this, they are quite common in practice because they are easy to implement and easy to make deadlock-free.

Oblivious algorithms, which include deterministic algorithms as a subset, choose a route without considering any information about the network's present state. For example, a random algorithm that uniformly distributes traffic across all of the paths in R_{xy} is an oblivious algorithm.

Adaptive algorithms, adapt to the state of the network, using this state information in making routing decisions. This information may include the status of a node or link (up or down), the length of queues for network resources, and historical channel load information.

The tornado example from the previous section includes examples of all three of these types of routing. The greedy algorithm on the ring is an example of deterministic routing. All packets between s and d travel in the same direction around the ring. The uniform and weighted random routing schemes are examples of oblivious routing. They choose between directions around the ring without taking into account the state of the network. Finally, the adaptive algorithm makes its decision based on channel load of the initial hop.

In these definitions, we described each type of routing algorithm over the set of routes in R_{xy} — the minimal, or shortest path, routes from source to destination.

Therefore these algorithms are referred to as *minimal*. As we have already seen, it's often important to include non-minimal routes and in this case, routing functions choose paths from the set of all minimal and non-minimal routes R'_{xy}. These algorithms are refered to as *non-minimal*. Again, from our simple example on the ring, the greedy algorithm is minimal, while randomized and adaptive algorithms are non-minimal.

8.3 **The Routing Relation**

It is useful to represent the routing algorithm as a routing relation R and a selection function ρ. R returns a set of paths (or channels for an incremental routing algorithm), and ρ chooses between these paths (or channels) to select the route to be taken. With this division of the algorithm, issues relating to channel dependencies and deadlock deal with the relation R while issues relating to adaptivity deal with the selection function ρ. We address deadlock in detail as part of Chapter 14.

Depending on whether our algorithm is incremental, and whether it is node-based or channel-based, we define R in three different ways:

$$R : N \times N \mapsto \mathcal{P}(P) \tag{8.1}$$

$$R : N \times N \mapsto \mathcal{P}(C) \tag{8.2}$$

$$R : C \times N \mapsto \mathcal{P}(C) \tag{8.3}$$

where $\mathcal{P}(X)$ denotes the power set, or the set of all subsets, of the set X. This notation allows us to reflect the fact that a routing relation may return multiple paths or channels, one of which is chosen by the selection function.

When the output of the routing relation is an entire path, as in Relation 8.1 — the first of our three routing relations, the routing algorithm is referred to as *all-at-once*. This name reflects exactly how the routing algorithm is used. When a packet is injected into the network at the source node x destined to node y, the routing relation is evaluated: $U = R(x, y)$. Since U may be a set of routes, one is selected and assigned to the packet. Of course, U does not have to include all possible routes R'_{xy} or even all minimal routes R_{xy}, and in the case of a deterministic routing algorithm, it returns only one ($|U| = 1$). Once the route is chosen, it is stored along with the packet. As we will see in Chapter 11, all-at-once routing minimizes the time spent evaluating the routing relation for each packet, but this advantage comes with the overhead of carrying the routes inside the packets.

An alternate approach is *incremental* routing, where the relation returns a set of possible channels. Instead of returning an entire path at once, the routing relation is evaluated once per hop of the packet. The output of the relation is used to select the next channel the packet follows. In Relation 8.2, the second form of the routing relation, for example, the inputs to the relation are the packet's current node w and its destination y. Evaluating the relation gives us a set of channels $D = R(w, y)$, where each element of D is an outgoing channel from w or $D \subseteq C_{Ow}$. The selection function is then used to choose the next channel used by the packet from D. This incremental process is repeated until the packet reaches its final destination.

Relation 8.3, the third relation, is also incremental and is used in a similar way. The only difference is that the inputs to the function are the previous channel used by the packet and its destination.

Compared to all-at-once routing, there is no overhead associated with carrying the route along with a packet, but the routing relation may have to be evaluated many times, which potentially increases the latency of a packet. Another important point is that incremental algorithms cannot implement every routing strategy possible with all-at-once routing. This is because we are using little or no history from a packet to compute its next hop. For example, with an all-at-once algorithm, we could design a routing algorithm for a 2-D mesh where packets are only routed vertically or horizontal through a particular node. (No packets turn from a horizontal to vertical dimension at this node.) However, this would be impossible with the second routing relation (Relation 8.2) because there is no way to distinguish a packet that has arrived from a vertical channel from one that arrived from a horizontal channel. Of course, the third relation could alleviate this problem, but it still does not cover many all-at-once algorithms. (See Exercise 8.2.)

The third form of the routing relation (Relation 8.3) is also incremental, but bases the routing decision on a packet's current channel c rather than its current node, w. In this case, the routing relation takes the current channel c and the destination node y and returns a set of channels $D = R(c, y)$. Basing this decision on the channel c over which a packet arrived at node w rather than on w itself provides just enough history to decouple dependencies between channels, which is important for avoiding deadlock (Chapter 14).

Whichever form of R we use, unless the routing is deterministic, it returns a set of possible paths or channels. The selection function ρ is used to choose the element of this set that will be used. If ρ uses no information about the network state in making this choice, the routing will be oblivious. If, on the other hand, ρ bases the choice on output channel availability, the routing will be adaptive.

8.4 Deterministic Routing

The simplest routing algorithms are deterministic — they send every packet from source x to destination y over exactly the same route. The routing relation for a deterministic routing algorithm is a function — $R : N \times N \mapsto P$, for example. As we saw in Section 8.1, this lack of path diversity can create large load imbalances in the network. In fact, there is a traffic pattern that causes large load imbalance for *every* deterministic routing algorithm. So, for a designer interested in the worst case, these algorithms would not be a first choice. However, deterministic algorithms still have their merits.

Many early networks adopted deterministic routing because it was so simple and inexpensive to implement. What may be surprising is that deterministic routing continues to appear in networks today. This is especially true for irregular topologies, where designing good randomized or adaptive algorithms is more difficult. For almost[2] any topology, it only makes sense to choose a minimal deterministic routing function. So, at least the path lengths will be short. For some topologies, simple deterministic approaches actually load balance as well as any other *minimal* routing algorithm, including adaptive (Exercise 9.2). Finally, for networks in which the ordering of messages between particular sort-destination pairs is important, deterministic routing is often a simple way to provide this ordering. This is important, for example, for certain cache coherence protocols.

In this section, we describe two of the most popular deterministic routing algorithms: destination-tag routing on the butterfly and dimension-order routing for tori and meshes.

8.4.1 **Destination-Tag Routing in Butterfly Networks**

In a k-ary n-fly network (see Section 4.1), the destination address interpreted as an n-digit radix-k number is used directly to route a packet. Each digit of the address is used in turn to select the output port at each step of the route, as if the address itself was the routing header determined from a source-routing table. This is the routing that we employed in the simple router of Chapter 2.

Figure 8.3 shows two examples of destination-tag routing. A route from source 3 to destination 5 is shown as a thick line in the binary 3-fly of Figure 8.3(a). Working from left to right, each stage of the network uses one bit of the binary destination

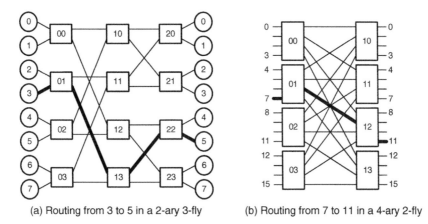

(a) Routing from 3 to 5 in a 2-ary 3-fly   (b) Routing from 7 to 11 in a 4-ary 2-fly

Figure 8.3 Two examples of destination-tag routing: (a) Routing from source 3 to destination 5 in a 2-ary 3-fly. The destination address in binary, $5 = 101_2 =$ down, up, down, selects the route. (b) Routing from 7 to 11 in a 4-ary 2-fly. The destination address interpreted as quaternary digits, $11 = 1011_2 = 23_4$, selects the route.

----

2.   One exception is the strange group of topologies in which minimal routing is not optimal under uniform traffic, as in Exercise 3.3.

address 101 to select an output. The most significant 1 selects the lower output at the first stage of switching, the 0 selects the upper output in the second stage, and the least significant 1 selects the lower output in the final stage.

If we look back over our description of how we routed from 3 to 5, we never actually used the source node's address. In fact, starting from any source and using the same 101 pattern of switch ports routes to destination 5 regardless of the source node. It is also not difficult to convince yourself that the same fact holds for all of the possible destinations. Therefore, destination-tag routing in k-ary n-fly networks depends on the destination address only, not the starting position.

Figure 8.3(b) shows an example route in a higher-radix butterfly. The thick line in this figure shows a route from node 7 to node 11 in a quaternary (radix-4) 2-fly network. As with the binary network, working from left to right the digits of the destination address determine the output port at each stage of the network. With the quaternary network, however, the destination address is interpreted as a quaternary number $11 = 1011_2 = 23_4$. The output ports of each router are numbered from the top starting at zero. The destination address 23_4 selects port 2 (third from the top) of the first router and port 3 (bottom) of the second router. As in the previous example, this selection of ports selects destination 11 regardless of the starting point.

8.4.2 Dimension-Order Routing in Cube Networks

Dimension-order or e-cube routing is the analog of destination-tag routing for direct k-ary n-cube networks (tori and meshes). Like destination-tag routing, the digits of the destination address, interpreted as a radix-k number, are used one at a time to direct the routing. Rather than selecting an output port at a given stage, however, each digit is used to select a node in a given dimension. Unlike butterfly networks, cube networks may require several hops to resolve each address digit before moving on to the next digit.

As an example of dimension-order routing, consider a packet traveling from node $s = 03$ to node $d = 22$ in the 6-ary 2-cube shown in Figure 8.4. Because each dimension of a torus can be traversed in either the clockwise or counterclockwise direction, the first step in e-cube routing is to compute the shortest or *preferred direction* in each dimension. To find the preferred directions, we first compute a relative address Δ_i for each digit i of our source and destination addresses:

$$m_i = d_i - s_i \bmod k$$

$$\Delta_i = m_i - \begin{cases} 0 & \text{if } m_i \leq k/2, \\ k & \text{otherwise} \end{cases}.$$

This can then be used to compute our preferred directions:

$$D_{T,i} = \begin{cases} 0 & \text{if } |\Delta_i| = k/2 \\ \text{sign}(\Delta_i) & \text{otherwise} \end{cases} \tag{8.4}$$

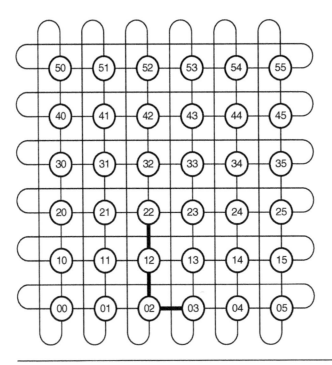

Figure 8.4 An example of dimension-order routing in a 6-ary 2-cube. A packet is routed from node $s = 03$ to node $d = 22$ by first routing in the x dimension and then in the y dimension.

where the T indicates the function is for tori. Before discussing the zero case for preferred direction, we return to our example.

Following the above formulas, the relative address is

$$\begin{aligned} m &= (2, 2) - (0, 3) \bmod 6 = (2, 5) \\ \Delta &= (2, 5) - (0, 6) = (2, -1). \end{aligned}$$

Thus, our preferred directions are

$$D = (+1, -1).$$

Once the preferred direction vector is computed, the packet is routed one dimension at a time. Within each dimension, the packet travels in the preferred direction until it reaches the same coordinate as the destination in that dimension. In the example in Figure 8.4, the packet starts at node $s = 03$ and moves in the negative direction (decreasing addresses) in the x dimension. After one hop, at node 02, it has reached the proper coordinate in the x dimension and hence starts routing in the positive direction in the y dimension. The packet takes two more hops to reach destination node 22.

Now consider the same routing problem with the destination moved slightly to $d = 32$. Following the same procedure, we find that $D = (0, -1)$. Routing in the

x dimension remains the same, but for the y dimension the preferred direction is $D_y = 0$. How do we route the packet in this case? By moving the destination node to 32, routing in either the positive or negative direction in the y dimension requires three hops. So, to balance load, it is important that traffic be evenly distributed in the two directions. A simple approach for doing this is to abandon a deterministic algorithm and randomly split the traffic equally between the positive and negative y directions.[3] It can also be easily verified by intuition or from Equation 8.4 that a preferred direction of zero can occur only when k is even.

We have focused on the torus up to this point, but dimension-order routing works similarly in the mesh. The lack of wraparound channels simplifies the choice of the preferred directions, which in this case are also the only *valid* directions:

$$D_{M,i} = \begin{cases} +1 & \text{if } d_i > s_i \\ -1 & \text{otherwise.} \end{cases}$$

Despite its generally poor load balancing properties, dimension-order routing has been widely used in mesh and torus networks for two reasons. First, it is very simple to implement. In particular, it allows the router to be *dimension-sliced* or partitioned across dimensions. Second, it simplifies the problem of deadlock avoidance by preventing any cycles of channel dependency between dimensions. However, deadlock can still occur within a dimension. (See Chapter 14.)

8.5 Case Study: Dimension-Order Routing in the Cray T3D

Figure 8.5 shows a Cray T3D [95, 161], which connects up to 2,048 DEC Alpha processing elements in a 3-D torus. The T3D is a shared-memory multiprocessor. Each processing element includes a local memory but can access the local memory of all other processing elements by forwarding load and store operations over the torus network. Each pair of processing elements shares a single router via a network interface.

The T3D network uses dimension-order routing and is implemented using a dimension-sliced (Section 7.2.2) router, as shown in Figure 8.6. The router is realized on three identical ECL gate arrays that route in the x, y, and z dimensions, respectively. The overall design closely follows the organization of the J-Machine router (Section 5.5). This partitioning is possible because of the dimension-order routing. We consider a different partitioning in Exercise 8.6.

When a packet arrives from the network interface, the x router examines the packet to determine if it must route in the $+x$ dimension, the $-x$ dimension, or, if it is already at the destination x coordinate, proceed to the y router. Suppose the

---

3.   In Exercise 8.9, we explore the cost of not balancing this load and deterministic ways to achieve this balance.

Figure 8.5 A Cray T3D connects up to 2,048 DEC Alpha processors in a 3-D torus with shared memory.

packet is forwarded in the $+x$ direction (along the xpOut channel). At each subsequent x router, the router checks the packet to see if it is at the proper x coordinate. The packet is forwarded to the y router when it reaches the proper coordinate. Otherwise, it continues to move in the $+x$ direction.

Each T3D router channel has a bandwidth of 300 Mbytes/s and is carried over a *wire mat* between modules. The channels each have 16 data bits and 8 control bits and operate at 150 MHz, the same frequency as the original Alpha 21064 processors. The wire mat is a harness of wires that are manually connected to the board edge connectors to implement the torus topology. It is called a *mat* because it resembles an irregularly woven fabric. Each data and control signal is carried as a differential ECL signal over a twisted pair of wires in the wire mat.

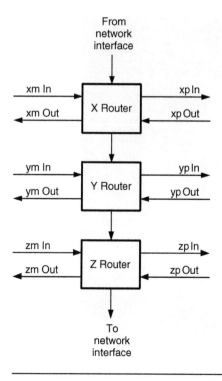

Figure 8.6 A T3D router is partitioned onto three identical ECL gate array chips, one for each of the x, y, and z dimensions.

The Cray T3D includes a set of I/O nodes that connect to the network in only the x and z dimensions. This makes the torus network a bit irregular. This seems plausible since messages always start in x and end in z, but what happens when the node sending the message differs from the address of the I/O node only in y? To make the dimension-order routing still work, these nodes are given two addresses. We examine this issue in Exercise 8.8.

8.6 **Bibliographic Notes**

The problem of routing tornado traffic and the weighted random solution is described by Singh et al. [168]. The different forms of the routing relation and their importance in analyzing deadlock have been described by Dally [57] and Duato [60, 61, 62]. Destination-tag routing in butterfly networks was first described by Lawrie [110] and e-cube routing in torus networks is due to Sullivan and Bashkow [179]. For degree δ networks, Borodin and Hopcroft [28] and Kaklamanis et al. [91] show that some traffic pattern can induce a channel load of at least $\Omega(\sqrt{N}/\delta)$ for any deterministic routing algorithm.

8.7 **Exercises**

8.1 *Tradeoffs between routing algorithms.* Reconsider the routing algorithms and network from Section 8.1. Which algorithm would you choose to optimize for the following:

(a) Minimum message latency.
(b) Best throughput under uniform traffic.
(c) Highest average throughput over many permutation traffic patterns.

Limit your choice to one algorithm for each of the different criteria and defend your choices.

8.2 *Limitations of incremental routing.* Describe a routing algorithm that could be specified using the the path-based relation from Relation 8.1, but could not be expressed with either of the incremental forms from Relations 8.2 and 8.3.

8.3 *Header bits for incremental and all-at-once routing.* Destination tag routing can be implemented as either an incremental or all-at-once algorithm. Compute the number of bits that need to be stored along with the packet to implement each approach. Does one approach require fewer bits? Does this relationship hold for minimal routing in a general topology? How is this related to the path diversity of the topology?

8.4 *Backtracking in the ring.* Suppose we allow backtracking in the routing example of Section 8.1. Is it possible to develop an algorithm that gives better worst-case throughput than the weighted random algorithm? If so, give such an algorithm; otherwise, explain why no such algorithm exists.

8.5 *Routing in a butterfly with extra stages.* Describe a deterministic extension for destination-tag routing to handle a k-ary n-fly with one or more extra stages. Suggest a simple way to introduce randomization into this algorithm to improve load balance.

8.6 *Direction-order routing in the Cray T3D.* Suppose you rearrange the labels on the Cray T3D router's channels in Figure 8.6 so that the first router handles $+x$ and $+y$, the second router handles $+z$ and $-x$, and the third router handles $-y$ and $-z$. Describe a routing algorithm that can work with this partitioning. Remember that once a packet reaches each of the three routers, it can never return to the previous routers.

8.7 *Advantages of direction-order routing.* Consider the routing algorithm you derived for Exercise 8.6. What advantages does this algorithm have over dimension-order routing?

8.8 *Routing to and from I/O nodes in the T3D.* I/O nodes are added to a T3D network only in the x and z dimensions by adding an additional x and/or z coordinate to a 3-cube. For example, suppose you have a 64-node 4-ary 3-cube with nodes addressed from (0,0,0) to (3,3,3). An I/O node might be added with address (4,0,0) or (0,0,4). Explain how you can assign each I/O node a pair of addresses so that it is always possible to route from any node in the interior of the machine to the I/O node and from the I/O node to any interior node using dimension-order routing.

8.9 *Balancing "halfway around" traffic in tori.* For dimension-order routing, we discussed the load balance issues that arise when a node is exactly halfway around a ring of a torus. If we always chose the positive direction in this halfway case instead of load-balancing, how would this affect the throughput of uniform traffic on a k-ary n-cube with k even? What's the worst-case throughput in this case? Express your results in terms of fraction of capacity. Suggest a way to improve load balance for the halfway case while maintaining a *deterministic* algorithm. Recalculate the uniform and worst-case throughputs.

8.10 *Minimal routing in CCCs.* Design a near minimal routing algorithm for a general CCC topology described in Exercise 5.8. Opt for simplicity rather than finding exact minimal routes in all cases, but be sure that no path generated by the algorithm is greater than the diameter $H_{\max} = 2n + \lfloor n/2 \rfloor - 2$, as shown in [128]. Comment on the load balance of your routing algorithm under uniform traffic.

CHAPTER 9

Oblivious Routing

Oblivious routing, in which we route packets without regard for the state of the network, is simple to implement and simple to analyze. While adding information about network state can potentially improve routing performance, it also adds considerable complexity and if not done carefully can lead to performance degradation.

The main tradeoff with oblivious routing is between locality and load balance. By sending each packet first to a random node and from there directly to its destination, Valiant's randomized routing algorithm (Section 9.1) exactly balances the load of any traffic pattern. However, this load balance comes at the expense of destroying any locality in the traffic pattern — even nearest neighbor traffic gives no better performance than worst-case traffic. Minimal oblivious routing (Section 9.2), on the other hand, preserves locality and generally improves the average case throughput of a network over all traffic patterns. However, on torus networks, any minimal algorithm gives at most half the worst-case throughput of Valiant's algorithm. We also introduce load-balanced oblivious algorithms, which provide a middle point between the minimal algorithm's and Valiant's approaches.

It is straightforward to analyze oblivious routing algorithms because they give linear channel load functions. This linearity makes it easy to compute the channel loads γ from the traffic pattern Λ and, hence, to compute the ideal throughput of the routing algorithm on any traffic pattern. We shall also see in Section 9.4 that this linearity makes it relatively easy to compute the worst-case traffic pattern for an oblivious routing algorithm.

9.1   **Valiant's Randomized Routing Algorithm**

Load can be balanced for any traffic pattern on almost any topology[1] using Valiant's algorithm, in which a packet sent from s to d is first sent from s to a randomly chosen intermediate terminal node x and then from x to d.[2] An arbitrary routing algorithm can be used for each of the two phases, but in general a routing algorithm that balances load under uniform traffic works best. So, for tori and mesh networks, dimension-order routing is an appropriate choice, and for butterflies, destination-tag routing works well. Then, regardless of the original traffic pattern, each phase of Valiant's algorithm appears to be uniform random traffic. Thus, Valiant's algorithm reduces the load of any traffic pattern to twice the load of random traffic or half the capacity of a network.

9.1.1   **Valiant's Algorithm on Torus Topologies**

Valiant's algorithm gives good worst-case performance on k-ary n-cube networks at the expense of locality. Each of the two random phases sends each packet an average distance of $k/4$ in each of n dimensions for a total hop count over both phases of $nk/2$. Thus, each link, on average, has a load of $\gamma = k/4$ and throughput of $4b/k$.

This throughput is almost optimal, as demonstrated by the tornado traffic pattern. Under tornado traffic, each packet must travel $H = n(k/2-1)$ hops (for k even). The channel load for this pattern under *any* routing algorithm must be at least

$$\gamma \geq \frac{H_{\min} N}{C} = \frac{n(k/2-1)N}{2nN} = \frac{k}{4} - \frac{1}{2}.$$

As k increases, this ideal channel load approaches the channel load induced by Valiant's algorithm and therefore Valiant's algorithm is asymptotically optimal.

In contrast, any minimal algorithm will give a channel load of at least $\gamma = k/2-1$ on the clockwise channels and 0 on the counterclockwise channels under tornado traffic. Because of this poor load balance, the throughput is at most $\frac{b}{(k/2-1)} \approx 2b/k$. Thus, no minimal routing algorithm can achieve a worst-case throughput of more than about half of Valiant's algorithm.

This good performance on worst-case traffic patterns, like tornado, comes at the expense of locality, which is destroyed by randomization, and overhead, which is added by routing in two passes. Nearest neighbor traffic, for example, which normally requires one hop per packet, is reduced to two passes of random traffic, increasing hop count from $H = 1$ to $nk/2$ and decreasing throughput. Applying Valiant's algorithm

---

1.   Valiant's algorithm can be applied to any *connected* topology — a topology with at least one path between each terminal pair. For example, butterflies are connected if the output terminals are connected to the corresponding input terminals.
2.   This is actually a simplification of Valiant's original work, which required specialization per topology and allowed a more rigorous analysis than presented here.

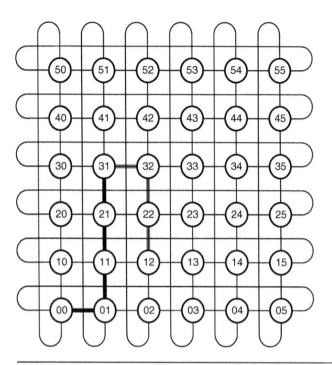

Figure 9.1 An example of randomized routing (Valiant's algorithm) on a 6-ary 2-cube. A packet is routed from $s = 00$ to $d = 12$ in two phases. In the first phase, the packet is routed to randomly-selected intermediate node $x = 31$, as shown with the dark bold lines. The second phase delivers the packet from $x = 31$ to $d = 12$, as shown with the gray bold lines. Both phases use dimension-order (e-cube) routing.

to random traffic halves throughput and doubles latency by replacing one phase of random traffic with two phases.

Figure 9.1 shows an example of using Valiant's algorithm to deliver a packet from node $s = 00$ to node $d = 12$ in a 2-D radix-6 torus (6-ary 2-cube). The packet is routed via randomly-selected intermediate node $x = 31$. During the first phase, the packet uses dimension-order routing to travel from $s = 00$ to $x = 31$, taking 4 hops as shown by the dark bold lines. Then, during the second phase, the packet routes from $x = 31$ to $d = 12$ taking an additional 3 hops. Randomized routing takes 7 hops to reach the destination, which could have been reached in 3 hops by a minimal routing algorithm.

9.1.2 Valiant's Algorithm on Indirect Networks

Applying Valiant's algorithm to a k-ary n-fly eliminates the bottlenecks caused by certain traffic patterns, as described in Section 4.4. In fact, the two-pass routing of Valiant's algorithm is equivalent to logically duplicating the butterfly network — resulting in a Beneš network in which the first n stages share the hardware with the last n stages. In this case Valiant's algorithm is a method for routing the resulting Beneš or Clos (if $k \neq 2$) network by randomly selecting the middle stage switch to be used by each packet. This random routing of Clos networks results in well-balanced *average* load, but can result in instantaneous imbalance due to variation in the number of packets picking each intermediate node. (See Exercise 9.5.) The flow control mechanism must have deep enough buffers to average out these transient variations in load. Alternatively, adaptive routing can avoid this statistical variation in load.

If oblivious routing is to be used on an indirect network, it is advantageous to *fold* the network so the first and second passes of Valiant's algorithm passes over the nodes of the butterfly network in opposite directions. This folded arrangement eliminates the need to connect corresponding source and destination nodes together to forward packets between passes. More importantly, it enables efficient minimal oblivious routing by allowing a packet to terminate the first phase of routing as soon as it reaches a switch from which it can reach the ultimate destination node.

9.2 Minimal Oblivious Routing

Minimal oblivious routing attempts to achieve the load balance of randomized routing without giving up the locality by restricting routes to be minimal (shortest path). While a non-minimal oblivious routing algorithm may choose any path in R'_{xy} to route a packet from x to y, minimal oblivious routing restricts its choice to paths in R_{xy}. For hierarchical topologies, minimal oblivious routing works extremely well — it gives good load balance while preserving locality. We explore an example of this for a fat-tree network in Section 9.2.1. In other networks, however, non-minimal routing is required to balance load, as we have seen with tornado traffic in tori.

9.2.1 Minimal Oblivious Routing on a Folded Clos (Fat Tree)

Figure 9.2 shows an example of minimal oblivious routing on a 16-node folded Beneš network with concentration (Section 6.4). This type of network is often called a fat tree. The figure shows 16 terminal nodes along the left edge. Each terminal connects to a 2:1 concentrator (Section 7.1.1) labeled with an address *template* that matches its two terminal nodes. Each concentrator then connects to a terminal of an 8-port radix-2 folded Clos network (Beneš network). Each node of the Beneš network is labeled with the address template that matches all nodes reachable from the left terminals of that node. Template bits set to X indicate don't care and match with both 0 and 1. For example, the node 00XX matches 0000 through 0011, nodes 0 through 3.

To route a packet from node s to node d, the packet is routed to a randomly selected, nearest common ancestor x of s and d, and then from x to d. For example,

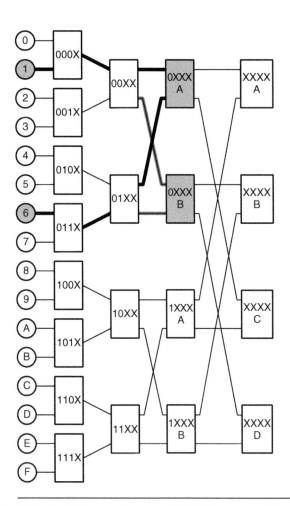

Figure 9.2 Minimal oblivious routing on a folded Clos network (fat tree). A packet from source node $s = 1$ to destination node $d = 6$ is routed first to a randomly-selected, nearest common ancestor of s and d, switch 0XXX-A or B, and then from this common ancestor switch to $d = 6$. Because the network is folded, all channels are bidirectional.

to route from $s = 1$ to $d = 6$, switch nodes 0XXX-A and 0XXX-B are the two nearest common ancestors of s and d. The packet may be routed either over the dark bold path via 0XXX-A or over the gray bold path via 0XXX-B. In either case, the route is a minimal route — taking six hops from $s = 1$ to $d = 6$.

This route can also be constructed incrementally. Initially, the packet proceeds to the right, making random routing decisions between the upper right and lower right ports of each switch node until it reaches a switch node with an address template that matches d. Revisiting the example, the nodes labeled 0XXX are the first switch nodes that match $d = 6 = 0110$. Once a match is made, the packet reverses direction and routes to the left using partial destination-tag routing to select the unique path to d. The destination-tag routing uses only those bits of the destination address corresponding to Xs in the switch template, since the other address bits are already resolved at the common ancestor. In this case, the second phase of the route uses the low three bits of d, 110 giving a route of down, down, up from 0XXX-A or B to terminal $d = 6$.

By randomly choosing a common ancestor node, and hence routing randomly during the first phase of the route, traffic from a given source/destination pair is exactly balanced over the common ancestors and over the paths to these ancestors. There is nothing to be gained by routing outside of this minimal subnetwork, bounded on the right by the common ancestor nodes. Every route from s to d must start with a path from s to a common ancestor x_1 and must end with a path from a common ancestor x_2 (possibly the same as x_1) to d. Any routing outside of this subnetwork simply consumes bandwidth on other switches and channels without improving the load balance within the subnetwork.

The Thinking Machines CM-5 (Section 10.6) uses an algorithm nearly identical to the one described here, except that the first phase of routing is adaptive rather than oblivious.

9.2.2   Minimal Oblivious Routing on a Torus

A minimal version of Valiant's algorithm can be implemented on k-ary n-cube topologies by restricting the intermediate node, x, to lie in the minimal *quadrant* between s and d. The *minimal quadrant* is the smallest n-dimensional subnetwork that contains s and d as corner nodes.

Figure 9.3 gives an example of minimal oblivious routing from $s = 00$ to $d = 21$ (the same s and d as in Figure 9.1 on a 6-ary 2-cube). The first step is to compute the relative address $\Delta = (2, 1)$ as in Section 8.4.2. The magnitude of Δ gives the size of the minimal quadrant (in this example, 2 hops in the x dimension and 1 in the y dimension). The preferred direction vector gives the position of the minimal quadrant with respect to node s. In this case, $D = (+1, +1)$, so the minimal quadrant is above and to the right of $s = 00$.

Once the minimal quadrant has been identified, an intermediate node x is selected from within the quadrant. The packet is then routed from s to x and then from x to d using e-cube routing. In this case, there are six possibilities for x and each is illustrated by the shaded nodes in Figure 9.3(b). The portion of each route before the intermediate node is illustrated with bold solid lines and the portion after the intermediate node is illustrated with gray solid lines. Note that the source and destination themselves may be selected as intermediate nodes.

While there are six possible intermediate nodes, there are only three possible routes in this case — corresponding to the three points at which the hop in the y direction can be taken. The load is not distributed evenly across these routes. The route that takes the y hop from 02 appears four times while the route that takes the y hop from 00 appears only once. This imbalance can be partially corrected by randomizing the order in which dimensions are traversed. The dashed gray lines in Figure 9.3(b) show the paths taken when the routing is performed first in the y

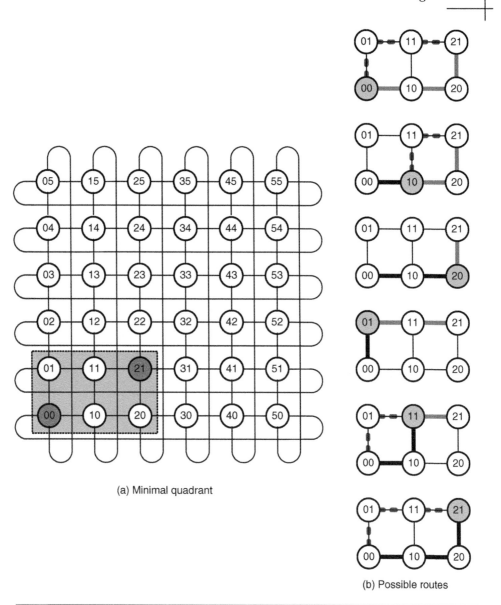

(a) Minimal quadrant

(b) Possible routes

Figure 9.3 Minimal oblivious routing on a 6-ary 2-cube from $s = 00$ to $d = 21$. (a) The route is restricted to remain within the minimum quadrant — shown shaded. (b) With x-first routing there are six possible routes, corresponding to the six possible intermediate nodes. Bold solid lines show the route to the intermediate node and gray solid lines show the route from the intermediate node to the destination. Dashed gray lines show the route taken if y-first routing is used.

direction for each of the two phases when it differs from x-first routing. Randomly selecting between x-first and y-first routing balances the load on the two end y channels of the minimal quadrant (each are used in 5 of 12 routes), but the middle channel is used only in 2 of the 12 routes.

Minimal oblivious routing on a torus does a great job of preserving locality. Nearest neighbor traffic remains local with this routing algorithm, and random traffic is not burdened with a halving of throughput. Unfortunately this locality is achieved at the expense of worst-case performance. As explained above, patterns such as tornado traffic will result in severe load imbalance.

9.3 Load-Balanced Oblivious Routing

A compromise can be struck between completely randomized routing (Valiant's algorithm) and minimal oblivious routing on a torus by randomly choosing the quadrant to route in. By weighting the choice of quadrants by distance (as in Section 8.1) we can exactly balance load for tornado traffic. In each dimension, i, we select the short direction $D_i' = D_i$ with probability $\frac{k-\Delta_i}{k}$ and the long direction $D_i' = -D_i$ with probability $\frac{\Delta_i}{k}$. Once we have chosen direction vector D', we route in the quadrant it selects as in minimal oblivious routing. We randomly select an intermediate node in the quadrant, route to that intermediate node, and then route from the intermediate node to the destination. For each of the routing phases, the dimension order of the routing is randomly selected.[3] Routing during both phases always proceeds in the direction specified by D'.

This load-balanced oblivious routing method is a compromise between locality and worst-case throughput. On local traffic, it outperforms Valiant's algorithm, because it routes in the short direction more frequently, but it does not perform as well as minimal routing because it routes some packets through the non-minimal quadrants. Although load-balanced oblivious routing performs well for tornado traffic, its worst-case throughput is much lower than Valiant's algorithm.

9.4 Analysis of Oblivious Routing

Because an oblivious routing algorithm chooses a path independent of network state, the load $\gamma_{c(sd)}$ induced on channel c by one unit of load being sent from s to d (that is, $\lambda_{sd} = 1$) is independent of the load being sent between any other node pairs. This property allows us to solve for the load on a channel for a given algorithm and traffic pattern by summing the contributions of each element of the traffic matrix:

$$\gamma_c = \sum_{i,j} \lambda_{ij} \gamma_{c(ij)}. \tag{9.1}$$

---

3.   It suffices to randomly choose one of n rotations of the dimensions rather than to choose over all of the $n!$ permutations of dimensions.

For example, in Figure 9.3, when routing from $s_1 = 00$ to $d_1 = 21$ using x-first routing in both phases, channel $(10, 20)$ is used in two of the six possible routes, so we have $\gamma_{(10,20)(00,21)} = 1/3$. When routing from $s_2 = 10$ to $d_2 = 31$ channel $(10, 20)$ is used by four of the six possible routes so $\gamma_{(10,20)(10,31)} = 2/3$.

Now, consider a traffic matrix with only two non-zero entries $\lambda_{(00,21)} = \lambda_{(10,31)} = 1$. We can calculate $\gamma_{(01,02)}$ by summing the components due to the two non-zero elements of the traffic matrix:

$$\gamma_{(10,20)} = \gamma_{(10,20)(00,21)} + \gamma_{(10,20)(10,31)} = 1/3 + 2/3 = 1.$$

With oblivious routing, channel loads are *linear*. We compute the total load on a channel due to a traffic pattern as the superposition of the loads due to the individual elements of the traffic pattern. This *linearity* of channel load enables us to solve for the worst-case traffic pattern by finding the traffic matrix that gives the highest maximum channel load.

To compute the throughput achieved by a routing algorithm on a given traffic pattern, we normalize the traffic matrix so the rows and columns sum to unity[4] and then compute a scale factor Θ so that a traffic matrix of $\Lambda' = \Theta\Lambda$ exactly saturates the critical channel. Our original traffic matrix Λ induces a maximal channel load $\gamma_{\max}(\Lambda)$, which may be larger than the channel bandwidth b. Thus, the throughput and our scale factor is given by $\Theta = \frac{b}{\gamma_{\max}(\Lambda)}$. By linearity, this scale factor gives a maximum channel load equal to the channel capacity: $\gamma_{\max}(\Lambda') = \Theta\gamma_{\max}(\Lambda) = b$.

To simplify our search for the worst-case traffic pattern, we first observe that all worst-case traffic patterns are permutations. To see this, note that all normalized traffic matrices can be expressed as the superposition of permutations:

$$\Lambda = \sum_i w_i \Pi_i, \quad \text{s.t.} \sum_i w_i = 1. \tag{9.2}$$

One of the permutations in this sum, say $\Pi_{\max}$, generates the largest maximum channel load compared to the other permutations in the sum: $\forall i, \ \gamma_{\max}(\Pi_{\max}) \geq \gamma_{\max}(\Pi_i)$. By linearity, we then know

$$\gamma_{\max}(\Lambda) = \sum_i w_i \gamma_{\max}(\Pi_i) \leq \gamma_{\max}(\Pi_{\max}). \tag{9.3}$$

Hence, a permutation can always be a worst-case traffic pattern.

We can find the permutation that gives the highest channel load for a given channel, c, by constructing a bipartite graph, as shown in Figure 9.4. On the left side of the graph there is a vertex for every source node and on the right side of the graph is a vertex for every destination node. There is an edge from every source

---

4. With a normalized traffic matrix, each source sends one unit of traffic and each destination receives one unit of traffic. Such a matrix with unit row and column sums is often called a doubly stochastic matrix.

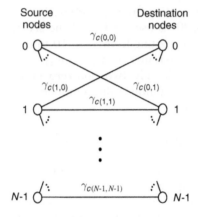

Figure 9.4 To find the permutation that gives the maximum load on a channel c, a bipartite graph is constructed with a left vertex for every source node and a right vertex for every destination node. The edge from each source node s to each destination node d is labeled with $\gamma_{c(sd)}$. A matching of this graph describes a permutation Π and the weight of the matching is the load on c due to the permutation $\gamma_c(\Pi)$. Thus, the maximum-weight matching of the graph gives the permutation that gives the maximum load on c.

node s to every destination node d labeled with $\gamma_{c(sd)}$, the load induced on c by unit traffic from s to d. Any permutation Π corresponds to a matching of this graph where there is an edge from s to d iff $\pi_{sd} = 1$. By linearity, the total load on c for this permutation, $\gamma_c(\Pi)$, is the weight of the matching, the sum of the edge weights over the matching. Thus, by computing a maximum-weight matching on this graph, we can find the permutation that generates the maximum load on c. We then repeat this procedure for each channel in the network to find the permutation that generates the maximum load over all of the channels.

Calculating the worst-case traffic pattern for a routing algorithm as described above often gives worst-case throughput that is substantially lower than the worst of the standard traffic patterns described in Section 3.2. Table 9.1 shows the throughput

Table 9.1 Throughput (as a fraction of capacity) of four routing algorithms on an 8-ary 2-cube on six traffic patterns. The worst-case pattern is different for each routing algorithm.

Load	e-cube	Valiant	Minimal	Load-Balanced
Nearest neighbor	4.00	0.50	4.00	2.33
Uniform	1.00	0.50	1.00	0.76
Bit complement	0.50	0.50	0.40	0.42
Transpose	0.25	0.50	0.54	0.57
Tornado	0.33	0.50	0.33	0.53
Worst-case	0.25	0.50	0.21	0.31

of four routing algorithms: e-cube, Valiant, minimal oblivious (Section 9.2.2), and load-balanced oblivious (Section 9.3) on five standard traffic patterns and on the worst-case traffic pattern for that particular algorithm. The worst-case traffic patterns are different for each algorithm and reveal substantially lower throughputs than the standard patterns for both the minimal oblivious and load-balanced routing algorithms.

Testing routing algorithms by simulating them on several *suspect* traffic patterns, as was the practice before the method described above was developed, can give very misleading results for worst-case throughput. Unfortunately, because channel load functions for adaptive routing algorithms are not linear, this method cannot be applied to adaptive routing. At present, simulation on specific traffic patterns is the only method we have for evaluating adaptive routing algorithms.

9.5 Case Study: Oblivious Routing in the Avici Terabit Switch Router (TSR)

The Avici Terabit Switch Router (TSR) (Figure 9.5) is a scalable Internet router that uses a 3-D torus interconnection network as a switch fabric to connect input *line cards* to output line cards [34, 49, 50]. Each cabinet contains 40 5-Gbits/s line cards connected as a $2 \times 4 \times 5$ folded torus network. Up to 8 cabinets can be combined to give machines with a total size of up to 320 nodes ($8 \times 8 \times 5$) for an aggregate bandwidth of 1.6 Tbits/s. Each line card provides 5 Gbits/s (full duplex — 5 Gbits/s in and 5 Gbits/s out) of interface bandwidth typically to an OC-48 or OC-192 (2 cards) packet over SONET (POS) links.

A line card with a single OC-48 POS interface is shown in Figure 9.6. The 6 chips (with heatsinks) along the bottom of the card comprise a bit-sliced torus router. The router is 24 bits wide, 4-bits on each of the 6 chips. Each network link is 28 bits wide (24 data plus 3 control and 1 clock) and operates at 400 MHz for a data rate of 9.6 Gbits/s (1.2 Gbytes/s).[5] Each router connects to 12 such links, one input and one output in each of 6 directions.

The Avici TSR is a notable router in many respects. In Section 15.7 we shall examine how it uses a virtual network for each output to provide service guarantees — in particular to guarantee that traffic destined for one output does not interfere with traffic destined for a different output. In this section, we will restrict our attention to the oblivious routing algorithm employed by the TSR to balance load on the network links.

The TSR network must provide non-stop routing even on irregular networks that occur in partially populated routers or when one or more line cards fail or are removed. To facilitate routing in such irregular, partial torus topologies, the TSR employs source routing (Section 11.1.1). The sending node or source determines the exact route in the form of a string of routing directions (such as $+x, -y, +x, +z,$

---

5.   Later TSR systems have higher bandwidth line cards and higher bandwidth fabric links.

Figure 9.5 The Avici TSR is a scalable Internet router that uses a 3-D torus interconnection network as its switching fabric. Each cabinet holds 40 line cards, each of which provides up to 5 Gbits/s of interface capacity, arranged as a $2 \times 4 \times 5$ sub-torus. Additional cabinets can be added to scale the machine to an $8 \times 8 \times 5$ torus.

$-y, -x$). Each character of this string specifies one hop of the route from source to destination. For example, $+x$ specifies that the first hop is in the positive x direction. A software process finds routes between each source s and each destination d through the possibly irregular topology and stores these routes in a table. Each packet queries the table in its source node s to find a route to its destination d. The route is appended to the packet header and used at each step along the way to select the next direction.

Because there is no backpressure over the input SONET links, the TSR must be able to route all traffic patterns without overloading any of its internal fabric

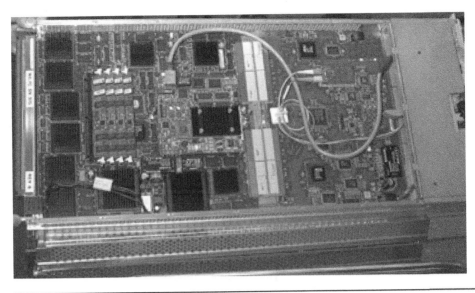

Figure 9.6 A single OC-48 (2.488 Gbits/s) SONET line card is divided into two parts. The upper sub-card is interface-specific. The lower main card is interface-independent. The six chips with heat sinks along the bottom of the board comprise a 24-bit-wide torus router bit sliced four bits to each chip.

channels — even in the presence of adversarial traffic. To satisfy this requirement, the TSR balances the load on network channels under worst-case traffic by using oblivious routing. Each packet that traverses the network from a source line card s to a destination line card d randomly selects from one of 24 routes to d stored in a routing table. Packets from the same *flow* are kept in order by making the random selection of routes based on a *flow identifier* so that all packets in the flow choose the same route.

To balance load on the 3 links out of s in the minimal quadrant for d, the 24 routes in the table for d include 8 routes for each of the 3 initial directions. For example, if $s = (0, 0, 0)$ and $d = (3, 2, 4)$, the minimal directions in a fully populated $8 \times 8 \times 5$ system are $+x$, $+y$, and $-z$. The routing table will include 8 routes starting with each of these 3 initial directions.

The details of the routing process are shown in Figure 9.7. A packet enters line card s and is input to a route lookup and classification process. This process selects a destination line card d for the packet and computes a flow identifier f for the packet. All packets belonging to the same flow will have an identical flow identifier.[6] The flow ID is hashed to give a route selector $r = \text{hash}(f) \in [0, 23]$. The route table is then indexed by $24d + r$ to select one of the 24 routes from s to d. The selected route is appended to the packet header and used for routing.

---

6. In IP, a flow consists of packets with the same source address and port and destination address and port.

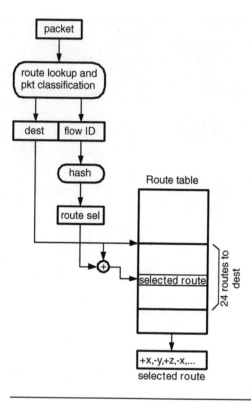

Figure 9.7 The TSR routing table on each source node s stores 24 routes for each destination d, 8 routes for each of 3 initial directions. To maintain packet order within a flow, the route to use for a particular packet is selected by hashing the flow identifier for that packet.

The key to load balancing with a table-driven oblivious routing algorithm is to construct a good table. If only a small number N_R of routes are stored for each (s, d) pair, these routes must be selected with care. One approach is to incrementally construct the routing table by generating one route at a time. Each time a route is generated it adds one unit of load to each channel that it uses. Each route is generated by selecting from among the minimal paths the path with the lowest load — summed across the channels in the path. We explore such route generation in Exercises 9.3 and 9.4.

9.6 **Bibliographic Notes**

Randomized routing was first described by Valiant [187]. The minimal oblivious routing algorithm described in Section 9.2.2 is due to Nesson and Johnsson [135]. The load-balanced oblivious routing algorithm of Section 9.3 was introduced by Singh et al. [168].Towles and Dally describe the worst-case analysis method for oblivious routing [185].

9.7 **Exercises**

9.1 *Load-balanced routing under uniform traffic.* Find a general expression for the channel load induced by the load-balanced routing algorithm (Section 9.3) on a k-ary n-cube under uniform traffic.

9.2 *Worst-case optimally of dimension-order routing.* Construct an argument as to why dimension-order routing has optimal worst-case throughput (optimal within a constant, additive term) for a minimal routing algorithm in 2-D tori. Does this argument hold for $n > 2$?

9.3 *Route generation in the Avici TSR.* Write a program to build a source routing table, like that of the Avici TSR, for an arbitrary topology network. Use the method discussed at the end of Section 9.5 to build the table. Assume you have $N_R = 8$ entries for each (s, d) pair. Test your table by generating a number of random permutations and determining the largest link load for these permutations. Now vary N_R and determine how load imbalance varies with the number of routes per (s, d) pair.

9.4 *Iterative route generation in the Avici TSR.* Perform the same analysis as in Exercise 9.3, but this time write an iterative route generation program. After determining an initial set of routes, use the channel loads from this initial set to generate a second set of routes. Repeat this process for N_I iterations. How does the performance of this iterative algorithm compare to that of the single pass algorithm?

9.5 *Instantaneous misbalance in the fat tree.* Suppose we use Valiant's algorithm on a 1,024-node folded 4-ary 5-fly (fat tree) in a batch mode. For each batch, each source terminal chooses a random *intermediate* destination terminal. Let $f(x)$ be the number of source terminals that have chosen intermediate terminal x. What is the expected value of $\max(f(x))$ over all intermediate terminals x?

CHAPTER 10

Adaptive Routing

An adaptive routing algorithm uses information about the network state, typically queue occupancies, to select among alternative paths to deliver a packet. Because routing depends on network state, an adaptive routing algorithm is intimately coupled with the flow-control mechanism. This is in contrast to deterministic and oblivious routing in which the routing algorithm and the flow control mechanisms are largely orthogonal.

A good adaptive routing algorithm theoretically should outperform an oblivious routing algorithm, since it is using network state information not available to the oblivious algorithm. In practice, however, many adaptive routing algorithms give poor worst-case performance. This is largely due to the local nature of most practical adaptive routing algorithms. Because they use only local network state information (e.g., local queue lengths) in making routing decisions, they route in a manner that balances local load but often results in global imbalance.

The local nature of practical adaptive routing also leads to delay in responding to a change in traffic patterns. The queues between a decision point and the point of congestion must become completely filled before the decision point can sense the congestion. A flow control method that gives stiff backpressure (e.g., by using shallow queues) is preferred with adaptive routing because it leads to more rapid adaptation to remote congestion.

10.1 Adaptive Routing Basics

Many of the issues involved with adaptive routing can be illustrated by considering the case of a simple 8-node ring, as shown in Figure 10.1. Node 5 is sending a continuous stream of packets to node 6, using all available bandwidth on channel (5,6).

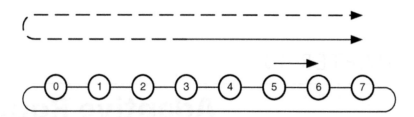

Figure 10.1   Adaptive routing on an 8-node ring. Node 3 wishes to send a packet to node 7 and can choose between the solid path via nodes 4 through 6 or the dotted path via nodes 2 through 0. Node 3 has no way to know of the simultaneous traffic between nodes 5 and 6.

At the same time, node 3 wishes to send a packet to node 7. It can choose either the clockwise route, denoted by a solid arrow, or the counterclockwise route, denoted by a dotted arrow.

To you, the reader (with global, instantaneous knowledge of the network state), it is obvious that the router at node 3 should choose the counterclockwise route to avoid the contention on channel (5,6). However, a typical router at node 3 has no knowledge of the contention on channel (5,6). That contention affects the queue on node 5, and in the absence of other traffic does not affect the queue on node 3.

How does the adaptive routing algorithm sense the state of the network? This is the key question in understanding adaptive routing. We can divide this question into subquestions involving space and time: Does the algorithm use *local* or *global* information? Does the algorithm use *current* or *historical* information? Of course, the answers to these questions are not binary. Rather, there is a continuum between local and global information and the currency of this information.

Almost all adaptive routers use flit-based or packet-based flow control (see Chapter 12) and use the state of the flit or packet queues at the present node to estimate the congestion on local links. They have no direct information on the state of links elsewhere in the network. Thus, in Figure 10.1, if node 3 is sending an isolated packet to node 7, its queues will not reflect the congestion on channel (5,6), and hence node 3 can do no better than choose randomly which route to take.

Routers are able to indirectly sense congestion elsewhere in the network through *backpressure*. When the queues on one node fill up, a backpressure signal stops transmission from the preceding node and hence causes the queues on that node to fill as well. Backpressure propagates backward through the network in the direction opposite traffic flow. However, backpressure propagates only in the presence of traffic routing into the congestion. In the absence of traffic, there is no propagation of backpressure and hence no information on remote congestion.

For example, in the case of Figure 10.1, the input queues on nodes 4 and 5 must be completely filled before node 3 senses the congestion on channel (5,6). The situation is illustrated in Figure 10.2, where each dot over a channel denotes some number of packets in the input buffer on the destination node of that channel.

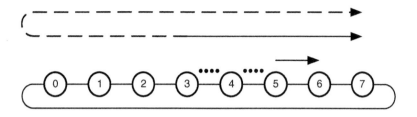

Figure 10.2   Node 3 senses the congestion on channel (5,6) only after the input buffers on nodes 4 and 5 are completely filled with packets, denoted by the dots in the figure.

This example demonstrates why adaptive routing performs better with *stiff* flow control.[1] Suppose each input queue can hold only $F = 4$ packets. Then node 3 will sense the congestion and begin routing in the other direction after sending just 8 packets. The network will be load-balanced relatively quickly, and only the first 8 packets will incur the higher latency of taking the congested path. If the input queues have a capacity of $F = 64$ packets, on the other hand, then node 3 would take 16 times as long to detect the congestion, and 16 times more packets would suffer the higher latency of the congested path.

With a mild load imbalance, it takes even longer for congestion information to propagate back to the source. If a channel is overloaded only by 10%, for example, 10 packets have to be sent over the congested path to back up one packet in the input buffer before the channel. In such a case, it can take an extremely long time, and many sub-optimally routed packets, for the source node to sense the congestion.

The example of Figure 10.1 also illustrates the problem of information currency for adaptive routing. Suppose that just at the moment node 3 senses congestion on channel (5,6) the traffic from 5 to 6 stops and is replaced by a flow of packets from 1 to 0. Node 3 would then mistakenly start routing packets into the newly congested channel (1,0). The problem is that node 3 is acting on historical information about the state of the network. It senses the state of channel (5,6) HF packets ago where $H = 2$ is the hop count to the source node of the congested channel and F is the capacity of the input buffer.

In topologies more complex than a ring, the adaptive routing decisions are made at every step, not just at the source. However, the local nature of congestion information still leads to sub-optimal routing, as illustrated in Figure 10.3. The figure shows how a good local decision can lead to a bad global route. In this case, a packet is traveling from $s = 00$ to $d = 23$. The path taken is highlighted in gray. The initial hop is made north to node 01. At node 01, the link to the north is slightly congested (congestion is denoted by the boldness of the line), so the packet next moves east to 11. After this move to the east, all paths north are highly congested, so the packet

---

1.   In Chapter 12, we shall see how virtual channels can be used to provide high performance with shallow queue depths — each queue holding just enough flits to cover the round-trip latency of the credit loop.

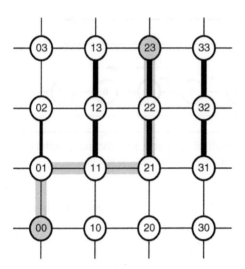

Figure 10.3   A locally optimum decision leads to a globally sub-optimal route. A packet is routed from $s = 00$ to $d = 23$ along the route highlighted in gray. To avoid the slight congestion on (01,02) the packet is routed from 01 to 11, after which it is forced to traverse two highly congested links.

is forced to traverse two highly congested links. The globally optimal route would have been to go north at 01, traversing only a single slightly congested link.

10.2 **Minimal Adaptive Routing**

A minimal adaptive routing algorithm chooses among the minimum (shortest) routes from source s to destination d, using information about the network state in making the routing decision at each hop. At each hop, a routing function generates a *productive* output vector that identifies which output channels of the current node will move the current packet closer to its destination. Network state, typically queue length, is then used to select one of the productive output channels for the next hop.

At node 01 in Figure 10.3, for example, two of the four output channels $(+x, -x, +y, -y)$ will move the packet closer to the destination, so the productive output vector is $(1, 0, 1, 0)$. Of the two productive outputs, $+x$ has lower congestion than $+y$, so the packet is routed over $+x$ to node 11.

Minimal adaptive routing is good at locally balancing channel load, but poor at global load balance. The route from 00 to 12 in Figure 10.4 illustrates how local congestion is avoided by adaptive routing. At node 01, both $+x$ and $+y$ are productive directions. The router chooses channel (01,11) to avoid the congested channel (01,02), locally balancing the load out of node 01. Figure 10.3 illustrates how locally adaptive routing is not able to avoid global congestion. As described above, the local

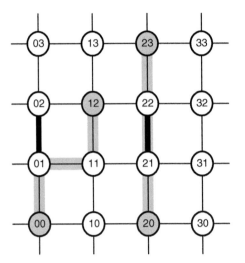

Figure 10.4 The route from 00 to 12 illustrates how adaptive routing avoids local congestion. The congestion on channel (01,02) is avoided by adaptively routing to 11 from node 01. The route from 20 to 23 illustrates how congestion cannot be avoided when minimal routing results in only a single productive output channel. The only minimal route includes the congested channel (21,22).

decision to traverse channel (01,11) rather than slightly congested channel (01,02) leads to much greater congestion downstream.

As with any minimal routing algorithm, minimal adaptive routing algorithms are unable to avoid congestion for source-destination pairs with no minimal path diversity ($|R_{sd}| = 1$). This situation is illustrated in the route from 20 to 23 in Figure 10.4. There is only a single productive direction at each hop, $+y$, so the packet cannot avoid congested channel (21,22). We shall see below how non-minimal adaptive routing avoids such bottlenecks.

While all of our examples in this section have involved torus networks, minimal adaptive routing can be applied to any topology. For example, the folded Clos of Figure 9.2 can be routed by *adaptively* routing to the right until a common ancestor of s and d is encountered, and then deterministically routing to the left to reach d. In this case, all outputs are productive during the right-going phase of the route, but only a single output is productive during the left-going phase. This is, in fact, exactly the routing method used by the data network of the Thinking Machines CM-5 (Section 10.6).

10.3 **Fully Adaptive Routing**

With non-minimal, or *fully* adaptive, routing, we no longer restrict packets to travel along a shortest path to the destination. Packets may be directed over channels that increase the distance from the destination to avoid a congested or failed channel.

For example, Figure 10.5 shows how adaptive routing can avoid congestion on the route from 20 to 23 from Figure 10.4. At node 21, the packet is directed to node 31, increasing the distance to the destination from 2 to 3 hops, to avoid congested channel (21,22). Directing a packet along such a non-productive channel is often called *misrouting*.

A typical fully adaptive routing algorithm gives priority to the productive outputs, so packets are routed toward the destination in the absence of congestion, but allows routing on unproductive outputs to increase path diversity. One possible algorithm is as follows: For a given packet, if there is a productive output with a queue length less than some threshold, the packet is routed to the productive output with the shortest queue length. Otherwise, the packet is routed to the output with the shortest queue length, productive or unproductive. Some algorithms limit the second step to avoid selecting the channel that would send the packet back to the node from which it just arrived (no U-turns) under the assumption that traversing a channel and then the reverse channel is clearly counterproductive.

While fully adaptive routing provides additional path diversity that can be used to avoid congestion, it can lead to livelock (see Section 14.5) unless measures are taken to guarantee progress. Livelock occurs when a packet travels indefinitely in the network without ever reaching its destination. With fully adaptive routing, this can happen if a packet is misrouted on an unproductive channel at least half the time. Figure 10.6 illustrates such an example of livelock. A packet from 00 to 03 encounters congestion at 02 and is misrouted to 12, where it encounters more congestion and is misrouted again to 11. This starts a cycle where the packet takes two steps forward, from 11 to 02, followed by two steps back, from 02 to 11.

To avoid livelock, a fully adaptive routing algorithm must include some mechanism to guarantee *progress* over time. One approach is to allow misrouting only a

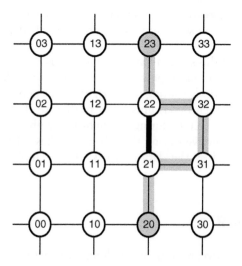

Figure 10.5   Fully adaptive routing may misroute packets to avoid congestion. At node 21 a packet destined for 23 is misrouted to 31 to avoid congestion on link (21,22).

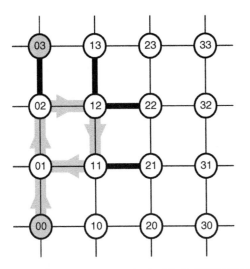

Figure 10.6 Fully adaptive routing can lead to livelock unless measures are taken to guarantee progress. Here, a packet being sent from 00 to 03 encounters congestion at 02 that sends it in a loop.

fixed number of times. After misrouting M times, for example, the algorithm reverts to minimum adaptive routing. This guarantees that if the packet starts H hops from the destination, it will be delivered after traversing at most $H+2M$ channels. Another alternative is to allow the packet to misroute one hop for every $H' > 1$ productive hops. Because this approach takes $H' + 1$ hops to reduce the distance to the destination by $H' - 1$, it is guaranteed to deliver the packet after $H\left(\frac{H'+1}{H'-1}\right)$ hops. Another approach, adopted by chaotic routing (Exercise 10.3), does not place a bound on the number of hops needed to deliver the packet, but instead makes a probabilistic argument that a packet will eventually be delivered.

In addition to potentially causing livelock, fully adaptive routing algorithms also raise new possibilities for causing deadlock. We defer a discussion of these issues until Chapter 14.

10.4 **Load-Balanced Adaptive Routing**

Adaptive routing algorithms have a difficult time achieving global load balance across the channels of an interconnection network because they typically make routing decisions based on entirely local information. One approach to overcome this problem is to employ a hybrid routing algorithm in which the *quadrant* to route in is selected obliviously using the method of Section 9.3. Then adaptive routing without backtracking is used within this quadrant to deliver the packet to its destination. The oblivious selection of quadrant balances the load globally, whereas the adaptive routing within the selected quadrant performs local balancing.

This hybrid algorithm results in very good load balance, and hence very good worst-case performance. Unfortunately, its performance on local traffic patterns is not as good as a pure adaptive algorithm (minimal or fully adaptive) because, like the oblivious algorithm of Section 9.3, it routes some of the packets the long way around the network. Although, this routing algorithm is not minimal, and some packets take the long way around, packets always make progress to their destinations. Once the routing quadrant is selected, the number of hops required to reach the destination H is determined and the packet is always delivered in exactly H hops. Hence, livelock is not an issue with load-balanced adaptive routing.

10.5 Search-Based Routing

So far, we have restricted our attention to routing strategies that are both greedy and conservative. They are greedy in the sense that they do not backtrack. Once they have taken a channel, they keep it. They are conservative in that they send a packet along just one path, rather than simultaneously broadcasting it over several paths.

One approach to non-greedy routing is to treat the routing problem as a search problem. The packet is instructed to search for the best path to the destination. This may involve the packet either backtracking once it finds a path blocked or congested, or alternatively broadcasting headers along multiple paths and then transmitting the data over the best of these paths.

Because they are both slow and make heavy use of resources, such search-based routing algorithms are seldom used in practice to actually route packets. They are useful, however, off line for finding paths in networks to build routing tables.

10.6 Case Study: Adaptive Routing in the Thinking Machines CM-5

Figure 10.7 shows a photograph of the Thinking Machines Connection-Machine CM-5 [114, 199]. This machine was the last connection machine built and was the first (and only) multiple instruction, multiple data (MIMD) machine built by Thinking Machines. The earlier CM-1 and CM-2 were bit-serial, single-instruction multiple-data (SIMD) parallel computers. The CM-5 consisted of up to 16K processing nodes, each of which contained a 32-MHz SPARC processor and a 4-wide vector unit. The machine included three separate interconnection networks: a data network, a control network, and a diagnostic network. The CM-5 is an interesting machine from many perspectives, including a cameo appearance in the movie *Jurassic Park*; however, we will focus on its data network.

As shown in Figure 10.8, the CM-5 data network uses a folded Clos topology with duplex connections to the processors and 2:1 concentration in the first two stages of switches. Each channel in the figure is 20 Mbytes/s (4 bits wide at 40 MHz) in each

Figure 10.7   A Thinking Machine CM-5 included up to 16K processing nodes. Each node incorporated a 32-MHz SPARC processor and a vector floating-point unit. The nodes were connected by a folded Clos (fat tree) network.

direction and employs differential signaling.[2] Each switch in the figure is a single chip 8×8 byte-wide router implemented in a 1 μm CMOS standard-cell technology.[3] The channels connecting the first two levels of the network are implemented using

2.   In addition to the 4 data bits in each direction, there is one flow control bit in each direction to apply backpressure when packets are blocked.
3.   With 4 connections on each side, at first glance the switches in Figure 10.8 look like 4×4 switches. However, it is important to remember that, unlike a butterfly network where each link is unidirectional, each connection to the switches in Figure 10.8 is bidirectional. Hence, each of these switches is an 8×8 crossbar with 8 inputs and 8 outputs.

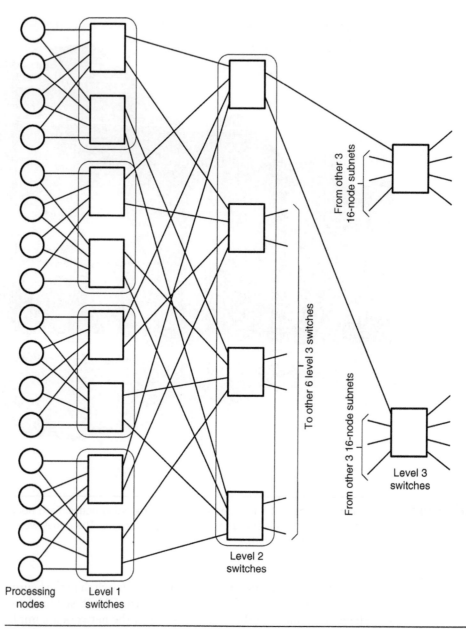

Figure 10.8   The CM-5 topology is a folded Clos (fat tree) network built from 8 × 8 switches, each with 4 upstream (right) ports and 4 downstream (left) ports. Each processing node is connected to 2 first-level switches. To provide a net 4:1 concentration, only 2 of the upstream ports are used on the first 2 ranks of switches. Starting with the third level of switches, all 4 upstream ports are used.

backplanes. Higher level channels are realized with cables that are either 9 feet or 26 feet in length.

Each processing node is connected to two separate switches via a pair of channels, giving an aggregate per-node interface bandwidth of 40 Mbytes/s. This duplex connection makes the network single-point fault tolerant. If a single router connected to a processing node fails, the processing node can continue to send and receive messages via its second channel. Each processor injects messages into the network via a memory-mapped interface. Messages can contain up to 5 32-bit words of data.[4] The two level-1 switches attached to a group of 4 processing elements logically act as a single node and collectively connect to each of 4 different level-2 switches. Similarly, the 4 level-2 switches in the figure collectively connect to each of 8 different level-3 switches (only 2 of which are shown). The topology is arranged so that a switch at level i can access 4^i nodes by sending messages only downstream (to the left).

The CM-5 routes messages in the manner described in Section 9.2.1, except that the upstream routing is adaptive rather than oblivious.[5] A message from node s to node d is routed in two phases. The message is first routed upstream (right) until it reaches a switch that is a common ancestor of s and d. This upstream routing is adaptive with the message choosing randomly among the idle upstream links. After reaching a common ancestor, the message is routed deterministically downstream (left) along the unique path to d using destination-tag routing.

A message in the CM-5 has the format shown in Figure 10.9. The message is divided into 4-bit flits. One flit is delivered each cycle over the 4-bit data channel as long as the sending node has a credit. The first flit of the message is a *height* flit that specifies the height h for this message: how high (how far right) in the network the message must travel to reach a common ancestor of s and d. Following the height flit are $\lceil h/2 \rceil$ route flits, each containing two two-bit route fields. Each route field specifies one step of the downstream route. The remainder of the flits in the message have to do with the payload and are not relevant for routing.

The upstream routing phase is controlled by the height field h of the message header. As an upstream message enters each router, h is compared to level l of the router. If $l < h$, the upstream phase continues by randomly selecting an idle upstream link to forward the message. If all links are busy, the message is blocked until a link becomes idle. When a message reaches a router where $l = h$, the common ancestor of s and d has been found and the downstream phase of routing begins.

At each step of the downstream route, one route field of the leading route flit r is consumed and then the height h is decremented. Decrementing h at each hop serves two useful purposes. First, this maintains the invariant that there are always $\lceil h/2 \rceil$ route flits following the head flit, since we decrement h as we consume route flits. Second, the LSB of h serves to select which route field in r to use for output port selection. If h is even, the left route field of r is used to select the downstream

---

4. This 5-word maximum message length was recognized as a limitation, and later versions of the machine allowed up to 18-word messages.
5. We explore the performance differences of the two approaches in Exercises 10.4 and 10.5.

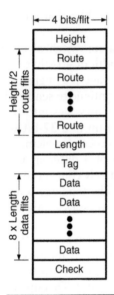

Figure 10.9   Format of a CM-5 message. Routing is controlled by a height field and a down route field. The 4-bit height field indicates how far upstream the message should propagate to reach an ancestor of the destination. The down route field is a suffix of the destination address and specifies the path from this ancestor to the destination node — two bits per level.

output port and h is decremented. If h is odd, the right route field of r is used to select the downstream port, h is decremented, and r is discarded. At the next hop h is again even and routing continues with the left route field of the next route flit. When the message arrives at the destination, $h = 0$ and all of the routing flits have been consumed.

The adaptivity of the upstream routing is governed by the flit-level blocking flow control (Section 13.2) employed by the CM-5. To regulate flow over channels, the CM-5 router employs a variant of on/off flow control (Section 13.3). When there is space in an input port buffer, the receiving router sends the sending router a *token*. The sender can use this token to send a flit immediately but cannot bank the tokens (as in credit-based flow control). If there is no flit to send, the token expires. When the buffer is full, no tokens are sent and traffic is blocked. Each CM-5 output port has a buffer large enough to hold one 5-word message (18-word message in later versions). Recall that during the upstream routing phase, the packet is randomly assigned to an *idle* upstream output port. An upstream port is considered *idle*, and eligible to be assigned to a new message, if its output buffer is empty. If no upstream port is idle, the packet is blocked in place — holding its input buffer busy, and, in turn, blocking packets further downstream. Because the router waits until an output buffer can accept an entire message before assigning the message to that output port, a message can never be blocked across the router's switch.

10.7 **Bibliographic Notes**

Much of the development of adaptive routing is closely tied to the flow control mechanisms needed to avoid deadlock and livelock. Early adaptive routing algorithms include those of Linder and Harden [118], Chien [36], Aoki and Dally [48], and Allen et al. [8]. Duato's protocol [61] enabled a family of adaptive routing algorithms including the one used in the Cray T3E [162]. Chaos routing (Exercise 10.3) was introduced by Konstantinidou and Snyder [104] and further explored by Bolding [26]. Minimal adaptive routing on a fat tree was used in the CM-5 and is described by Leiserson [114]. Boppana and Chalasani [27] present a comparison of several routing approaches and show that pratical, adaptive algorithms can be beaten by deterministic algorithms on some metrics.

10.8 **Exercises**

10.1 *Advantage of minimal adaptive routing in a mesh.* Find a permutation traffic pattern where minimal adaptive routing (Section 10.2) outperforms minimal oblivious routing (Section 9.2) in a 4 × 4 mesh network. Compute γ_{max} for both algorithms in the steady state. (Assume enough time has passed for backpressure information to propagate through the network.)

10.2 *Comparing minimal and load-balanced adaptive routing.* Find a permutation traffic pattern for which load-balanced adaptive routing (Section 10.4) outperforms minimal adaptive routing (Section 10.2) and a second traffic pattern for which minimal outperforms load-balanced.

10.3 *Livelock freedom of chaotic routing.* Chaotic routing is a *deflection routing* scheme. If multiple packets are contending for the same channel, the routers *randomly* grant that channel to one of the contenting packets. Any packets that lose this allocation are *misrouted* to any free output port. (This port may be non-productive.) Because the routers have the same number of input and output ports, it is always possible to match incoming packets to some output. Explain why this approach is probabilistically livelock-free by showing the probability that a packet does not reach its destination in T cycles approaches zero as T increases.

10.4 *Adaptive and oblivious routing in a fat tree.* Consider a 256-node folded Clos (fat tree) network constructed from 8 × 8 crossbar switches that uses dropping flow control. Which algorithm has a lower dropping probability? How do both dropping probabilities change with the traffic pattern?

10.5 *Worst-case traffic in CM-5 network.* Find a worst-case traffic pattern for the randomized oblivious routing algorithm (Section 9.2.1) for the CM-5 network. Compare the throughput of adaptive and oblivious routing on this traffic pattern.

10.6 **Simulation.** Explore tradeoff between buffer depth and response time of adaptive routing in an 8-ary 2-cube network. Alternate between two traffic permutations every T cycles and plot the average packet latency as a function of time. How does the amount of buffering per node affect the shape of this plot?

CHAPTER 11

Routing Mechanics

The term *routing mechanics* refers to the mechanism used to implement any routing algorithm: deterministic, oblivious, or adaptive. Many routers use *routing tables* either at the source or at each hop along the route to implement the routing algorithm. With a single entry per destination, a table is restricted to deterministic routing, but oblivious and adaptive routing can be implemented by providing multiple table entries for each destination. An alternative to tables is *algorithmic routing*, in which specialized hardware computes the route or next hop of a packet at runtime. However, algorithmic routing is usually restricted to simple routing algorithms and regular topologies.

11.1 Table-Based Routing

Recall Relations 8.1 through 8.3 from Section 8.3, which describe any routing algorithm

$$R : N \times N \mapsto \mathcal{P}(P)$$
$$R : N \times N \mapsto \mathcal{P}(C)$$
$$R : C \times N \mapsto \mathcal{P}(C).$$

Any of the three forms of the routing relation may be implemented using a table. The value of the relation for each pair of inputs is stored in the table and the table is indexed by the inputs. For example, for the first form of the routing relation, the set of paths for each pair of nodes is stored in the table, and the table is indexed by the source and destination node. Only that portion of the table that is needed on a particular node need be stored on that node. For example, at node x, only the part of the table for the source node (or current node) x needs to be stored.

The major advantage of table-based routing is generality. Subject to capacity constraints, a routing table can support any routing relation on any topology. A routing chip that uses table-based routing can be used in different topologies by simply reprogramming the contents of the table.

In the remainder of this section we look at the two extremes of table-based routing. With source-table routing, we directly implement the *all-at-once* routing relation by looking up the entire route at the source. With node-table routing, we perform incremental routing by looking up the hop-by-hop routing relation at each node along the route. There exists, of course, a continuum of points in between, with each table lookup returning a portion of the path.

11.1.1   Source Routing

With source routing, all routing decisions for a packet are made entirely in the source terminal by table lookup of a precomputed route. Each source node contains a table of routes, at least one per destination. To route a packet, the table is indexed using the packet destination to look up the appropriate route or set of routes. If a set is returned, additional state is used to select one route from the set in an oblivious or adaptive manner. This route is then prepended to the packet and used to rapidly steer the packet through the network along the selected path with no further computation. Because of its speed, simplicity, and scalability, source routing is one of the most widely used routing methods for deterministic and oblivious routing. Source table routing is not often used to implement adaptive routing because it does not provide a mechanism to take advantage of network state information at intermediate hops along the route.

Table 11.1 shows a source routing table for node 00 of the 4×2 torus network of Figure 11.1. The table contains two routes for each destination in the network. Each route in the table specifies a path from this source to a particular destination as a string of port selectors, one for each channel along the path. The port selectors in this case come from a five-symbol alphabet (NEWSX) encoded as 3-bit binary

Table 11.1   Source routing table for node 00 of 4×2 torus network of Figure 11.1.

Destination	Route 0	Route 1
00	X	X
10	EX	WWWX
20	EEX	WWX
30	WX	EEEX
01	NX	SX
11	NEX	ENX
21	NEEX	WWNX
31	NWX	WNX

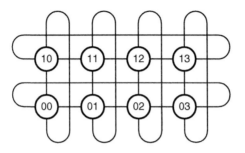

Figure 11.1   Network for source routing example of Table 11.1.

numbers. The first four symbols select one of the four output ports of a router (by compass direction) and the final symbol (X) specifies the exit port of the router.

Consider, for example, routing a packet from node 00 to node 21 in the network of Figure 11.1. At the source, the routing table, Table 11.1, is indexed with the packet's destination, 21, to determine the precomputed routes. Two routes, NEEX and WWNX, are returned from this lookup, and the source arbitrarily selects the first route, NEEX. This routing string is then digested one symbol at a time to steer the packet through the network. The first character, N, selects the north output port of the router at node 00 and is stripped off, passing the remainder of the string, EEX, on to node 01. At node 01, the east port is selected and EX is passed on to node 11. The east port is again selected at node 11 and the final character, X, is passed to node 21. At node 21, the X symbol selects the exit port of the router, steering the packet into the input queue of node 21.

Source routing has several advantages compared to many of the alternatives described in the rest of this chapter. The foremost advantage is speed. After the initial table lookup and route selection in the source, no further time is spent on routing. As each packet arrives at a router, it can immediately select its output port without any computation or memory reference. The routing delay (see Chapter 16) component of per-hop latency is zero. In addition to being fast, source routing results in a simple router design. There is no need for any routing logic or tables in each router.

Source routing is topology independent. It can route packets in any strongly connected topology subject only to the limitations of the number of router ports, the size of the source table, and the maximum length of a route. Using the four-port routers from the network of Figure 11.1, for example, one can wire an arbitrary degree-four network and route packets in it using source routing.

Routers that use source routing can be used in arbitrary-sized networks because the only limitations on network size, table size, and route length are determined by the source. This is advantageous, as it permits a single router design to be used in networks of arbitrary size without burdening a small network with a cost proportional to the largest network. The expense of supporting a large network is moved to the terminals where it need not be provided in the small network.

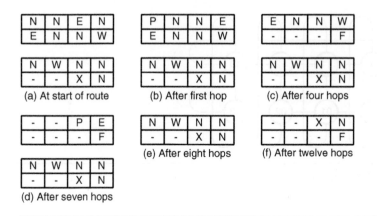

Figure 11.2   Arbitrary length encoding of source routes.

For a router to provide arbitrary scalability, it must be able to handle arbitrary-length routes. While routes are by definition arbitrary-length strings terminated by the exit port selector symbol, X, an encoding is required to pack these arbitrary-length strings into fixed-length routing phits and flits.

One method of encoding arbitrary-length source routes is illustrated in Figure 11.2. This example assumes a router with 16-bit phits, 32-bit flits, and up to 13 router ports. There are four 4-bit port selectors packed in each 16-bit phit and 8 selectors in each flit. To facilitate packing, we introduce two new port selection symbols, P and F, to denote phit continuation and flit continuation, respectively.

Figure 11.2(a) shows a 13-hop route (NENNWNNENNWNN) encoded into two flits. In each phit, the port selectors are interpreted from right to left. There are no continuation selectors in the original route. These are shifted in during subsequent steps of the route, as shown in Figure 11.2(b–f). After taking the first hop to the north, the lead flit is shifted right four bits to discard the first port selector, and a phit continuation selector, P, is shifted into the most significant position as illustrated in Figure 11.2(b). After four hops, the router encounters a phit with a P in the rightmost port selector. This P instructs the router to strip off the leading phit and append a flit continuation phit (F) to the end of the current flit (Figure 11.2[c]). After eight hops, the router encounters the flit continuation port selector, which instructs it to discard the current flit and continue processing port selectors starting with the next flit (Figure 11.2[e]). Processing continues in this manner across multiple phits and flits until an exit port selector, X, is encountered. A route of arbitrary length can be encoded in this manner by continuing over an arbitrary number of flits and phits.

The use of continuation port selectors enables us to process an arbitrary length route while dealing with only a single phit at a time. Because the leading phit output at each stage depends only on the leading phit input to that stage, it can be generated and output before the second phit arrives.

To handle arbitrary-length routes, the routing table in the source terminal must be designed to efficiently store arbitrary-length strings. This can be accomplished using

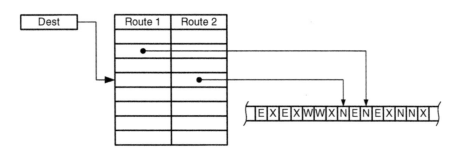

Figure 11.3 A routing table organized with one level of indirection to permit efficient encoding of variable-length source routes. The destination indexes a table of pointers into a string of concatenated source routes. The upper pointer points at a source route that is a suffix of the source route indexed by the lower pointer.

a single level of indirection, as illustrated in Figure 11.3. Indexing the table with the destination returns one or more pointers, each of which point to a variable-length route string. For the example shown in the figure, the destination selects the fourth row of the table, which contains two pointers, one for each of two routes. The second pointer locates the start of the route NENEX. Note that all routes are terminated by an X symbol. A route that is a suffix of another route can be represented by a pointer into the first route's string, requiring no additional storage. For example, the second row of the table's first route locates the string NEX, which is a suffix of the second route from row 4. This organization requires an extra memory access to look up a route. However, it is more storage efficient for all but the shortest routes because it avoids the storage waste associated with rounding up all routes to a maximum length. A compromise is to store the first few hops of the route in the table itself along with a pointer to the rest of the route. This organization avoids the extra memory reference for short routes while retaining the efficiency of indirection for long routes.

A source routing table may associate several routes with each destination. Table 11.1, for example, stores two routes to each destination. As described above with respect to oblivious and adaptive routing, providing multiple routes between a given source and destination has several advantages:

Fault tolerance: If some of the routes are edge disjoint, connection between the two nodes is maintained even if one or more channels between the nodes fail.

Load balance: The traffic between this pair of nodes is distributed over several network paths, balancing the load in the network across a broad range of traffic patterns.

Distribution: When the destination used to index the table is a logical destination, such as a service rather than a particular server, the multiple routes may in fact lead to multiple destinations. In this case, the routing table serves to distribute traffic among the servers providing a particular service.

When a routing table stores multiple routes for each destination, a route selection method is required to choose the route to be followed by each packet. When packet ordering is not a requirement, using a pseudo-random number generator works well to randomly distribute traffic over the routes. If there is a one-to-one correspondence between routes and table entries, and the random numbers are uniformly distributed, each route is equally likely to be selected and traffic will be balanced across the routes. In general, different destinations have different numbers of routes, while the table stores a fixed number of routes, so it is not usually possible to arrange for a one-to-one correspondence. For arbitrary numbers of routes, load balance can be approximated as closely as desired by using a very large number of table entries. If there are M routes, and N table entries, then each route is stored in k or $k + 1$ entries, where $k = \lfloor N/M \rfloor$ and the difference in load between two routes is bounded by $\frac{1}{N}$.

Another approach is to directly encode route probabilities as a binary field stored with each route. A logic circuit can then combine these probabilities with a random number to select among the routes. This approach allows arbitrary probability distributions to be encoded and is considerably more storage efficient than replicating table entries.

In some cases, we are required to maintain ordering among certain packets traveling through the network. Preserving the order of the packets in a message may be needed if the protocol is not able to reorder packets during assembly. Also, some protocols — for example, cache coherence protocols — depend on message order being preserved for correctness. A set of packets that must have its order preserved is called a *flow* and each packet of the flow is usually labeled by a flow identifier. This may be a combination of fields in the packet header that identify the protocol and type of packet. By hashing the flow identifier to generate the route selector rather than using a random number, we can ensure that all of the packets of a flow follow the same route and thus (assuming FIFO queueing) remain in order. However, if flows are not distributed uniformly, this approach may skew the probability of taking a particular route. Section 9.5 describes how flows are ordered in the randomized, table-based routing of the Avici TSR.

11.1.2 Node-Table Routing

Table-based routing can also be performed by storing the routing table in the routing nodes rather than in the terminals. Node-table routing is more appropriate for adaptive routing algorithms because it enables the use of per-hop network state information to select among several possible next-hops at each stage of the route. Using node rather than source tables significantly reduces the total storage required to hold routing tables because each node needs to hold only the next hop of the route to each destination rather than the entire route. However, this economy of storage comes at the expense of a significant performance penalty and restrictions on routing. With this arrangement, when a packet arrives at a router, a table lookup must be performed before the output port for the packet can be selected. As with source routers, the input port on which a packet arrives may be used in addition to the destination to select the table entry during the lookup. Performing a lookup at every step of the route, rather than just at the source, significantly increases the latency for a packet to pass through a router. Scalability is also sacrificed, since every node must contain a table large enough to hold the next hop to all of the destinations in the largest possible network. Also, with this approach, it is impossible to give packets from two different source nodes, x and y, destined for the same node, z, arriving at an intermediate node, w, over the same channel two different routes from w to z. This is easy to do with source routing.

Table 11.2 shows a possible set of node routing tables for the network of Figure 11.1. The table for each node is shown in two adjacent columns under the node number. There is a row for each destination. For each node and destination, two possible next hops are listed to allow for some adaptivity in routing. For example, if a packet is at node 01 and its destination is node 13, it may route in either the N or E direction. Note that some of the next hops actually move a packet further from its destination. These misroutes are shown in bold.

Livelock (see Section 14.5) may be a problem with node-table routing. It is possible to load the node tables with entries that direct packets in never-ending cycles. Consider, for example, a packet passing through node 00 destined for node 11. If the entry for $(00 \rightarrow 11)$ is N, directing the packet to 10 and the entry for $(10 \rightarrow 11)$ is S, the packet will wind up in a cycle between 00 and 10, never making progress to its destination. This is not a problem with source routing, since the fixed-length source route implies a finite number of hops before a packet reaches its destination. It can be avoided with node-table routing by careful construction of the routing tables.

Node table routing facilitates local adaptation of routes. A node can locally redirect traffic without coordinating with other nodes if one output link from a node becomes congested or fails. If a routing table includes multiple entries for each destination, as in the case of Table 11.2, this can be accomplished by simply biasing the selection of output ports based on congestion or availability. For example, if the N

Table 11.2 Node-routing table for the 4 × 2 torus network of Figure 11.1.

	From															
To	00		01		02		03		10		11		12		13	
00	X	X	W	**N**	W	E	E	**N**	S	N	S	W	S	W	S	E
01	E	**N**	X	X	W	**S**	E	W	S	**W**	S	N	S	W	S	W
02	E	W	E	**N**	X	X	W	**N**	S	W	S	E	S	N	S	W
03	W	**N**	E	W	E	**N**	X	X	S	**E**	S	E	S	E	S	N
10	N	S	N	W	N	W	N	E	X	X	W	**N**	W	E	E	**N**
11	N	E	N	S	N	W	N	W	E	**S**	X	X	W	**N**	E	W
12	N	E	N	E	N	S	N	W	E	W	E	**N**	X	X	W	**N**
13	N	W	N	E	N	E	N	S	W	**S**	E	W	E	**N**	X	X

Table 11.3   Hierarchical node-routing table for node 444_8 of an 8-ary 3-cube.

Destination	Next Hop	Remarks
100 100 100	X	This node
100 100 0XX	W	Nodes directly to west
100 100 11X	E	Nodes directly east
100 100 101	E	Neighbor east
100 0XX XXX	S	Nodes south
100 11X XXX	N	Nodes north
100 101 XXX	N	Node one row north
0XX 0XX 0XX	W	Octant down, south, and west
0XX 0XX 1XX	S	Octant down, south, and east (or equal)
0XX 1XX XXX	D	Quadrant down and north (or equal)
11X XXX XXX	U	Half-space two or more planes above
101 XXX XXX	U	Plane immediately above

port out of node 00 in Table 11.2 becomes congested, a packet destined to node 10 would choose S for its next hop, rather than N.

In very large networks, node-tables may be organized in a hierarchical manner to further reduce storage requirements.[1] In such a table, each nearby node gets its own next hop entry, small groups of remote nodes share entries, larger groups of distant nodes share entries, and so on. One approach to hierarchical routing is to use a content addressable memory (CAM) rather than a RAM to hold the routing table in each node. For example, a CAM-based routing table for node 444_8 in a 8-ary 3-cube network is shown in Table 11.3. Each row of this table consists of a destination group and the next hop for this group. The destination group is encoded as a 9-bit address (3 bits each of x, y, and z) that may include "don't care" symbols (X) in one or more bits. A CAM that allows "don't care" symbols is often referred to as ternary CAM (TCAM).

Depending on the number of Xs in a destination, it may encode a single node or a one-, two-, or 3-D group of contiguous nodes. For example, the first entry in the table encodes a single node, node 100 100 100 (444), the node containing this table. The next row encodes a one-dimensional array of nodes to the west of this node (440 to 443). A 2-D array of nodes (from 500 to 577) is encoded by the last entry of the table, and several entries in the table encode 3-D groups of nodes.

In effect, a CAM-based node routing table is a programmable logic array that computes the logical function for selecting the output port as a function of the destination address. Each entry that selects a given output port — W, for example — acts as a single product term (or implicant) in the logic equation that selects that

---

1.   This is done, for example, in the SGI Spider chip [69]. See Exercise 11.4.

output. For the W output, there are two implicants in the equation corresponding to the second and eighth rows of the table. Thus, by using a CAM, one can compress an arbitrary node routing table by replacing all of the entries for a given output with a smaller number of rows corresponding to a minimal sum-of-products realization of that output's function.

Node-table routing is the approach taken in most wide-area packet routers. This is because it lends itself to the use of distributed algorithms to compute network connectivity and routing. In interconnection networks, however, where nodes are co-located, node table routing is usually inferior to source routing because of the increased delay, unless the adaptivity is needed.

Source routing and node table routing represent two extremes in the space of table-based routers. A node table router stores just the next hop of the route in the table, whereas the source router stores the entire route. Of course, we can construct an arbitrary number of points between these two extremes, where the table stores the next $M > 1$ hops of the route. These intermediate solutions give local adaptivity and hop latency between the two extremes. To date they have not been used in practice, however, because they lack symmetry — a packet must perform different actions at different nodes with a hybrid approach.

11.2 Algorithmic Routing

Instead of storing the routing relation in a table, we can compute it using an algorithm. For speed, this algorithm is usually implemented as a combinational logic circuit. Such algorithmic routing sacrifices the generality of table-based routing. The algorithm is specific to one topology and to one routing strategy on that topology. However, algorithmic routing is often more efficient in terms of both area and speed than table-based routing. An algorithmic routing circuit can be as simple as a digit selector that pulls out a digit of the destination address for destination-tag routing (Section 8.4.1). Such a circuit is employed, for example, in the simple router of Section 2.5.

An algorithmic routing circuit for routing in a 2-D torus is shown in Figure 11.4. The circuit accepts a routing header including a sign bit for each dimension (sx and sy) and a relative address (x and y) that specify the number of hops to the destination in each dimension in the direction specified by the sign bit. The relative addresses are input to zero checkers that generate the signals xdone and ydone. If the relative address in a particular dimension is zero, the packet is done routing in that dimension. The sign bits and the done signals are then input to an array of five AND gates that determine which directions are *productive* in the sense that they will move the packet closer to the destination. For example, the gate second from the left determines that if xdone $= 0$, indicating that we are not done routing in the x dimension, and sx $= 0$, indicating that we are traveling in the $+x$ direction, then the $+x$ direction is productive. The 5-bit productive direction vector has a bit set for each direction that advances the packet toward the destination. This vector is input to a route selection block that selects one of the directions for routing and outputs a selected direction vector.

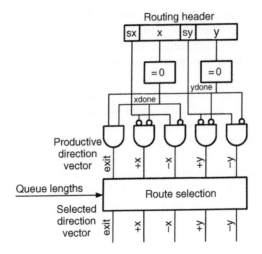

Figure 11.4   An example of algorithmic routing logic. This circuit accepts direction bits sx and sy and distance fields x and y from a packet routing header and generates a productive direction vector that indicates which channels advance the packet toward the destination, and a selected direction vector, a one-hot vector that selects one of the productive directions for routing.

The route selection block determines the type of routing performed by this circuit. A minimal oblivious router can be implemented by randomly selecting one of the active bits of the productive direction vector as the selected direction. Replacing this random selection with one based on the length of the respective output queues results in a minimal adaptive router. Allowing the selection to pick an unproductive direction if all productive directions have queue lengths over some threshold results in a fully adaptive router.

11.3 Case Study: Oblivious Source Routing in the IBM Vulcan Network

The IBM SP1 and SP2 series of message-passing parallel computers is based on a multistage interconnection network that employs oblivious source routing. Figure 11.5 shows a photo of a 512-processor IBM SP2. Each cabinet or *frame* in the figure holds 16 IBM RS6000 workstations (processing nodes) and one 32-port switch module.

Both the SP1 and SP2 use an interconnection network based on 8×8 switch chips controlled via source routing. Each of the eight bidirectional ports on the switch is 8-bits wide in each direction and operates at 50 MHz. In addition to the eight data lines in each direction there are two control lines, *tag* and *token*. The tag

Figure 11.5   A 512-processor IBM SP2 system. Each cabinet, or frame, holds 16 RS6000 workstations and one 32-port switch board.

line identifies valid flits, while the token line runs in the opposite direction to return credits (for flow control) to the sending node. Latency through the switch is five 20 ns cycles.

Eight switch chips are arranged on a 32-port switch module in a two-stage bidirectional network, as shown in Figure 11.6.[2] The module provides 16 bidirectional channels out each of two sides. All channels transmitted off the switch module are carried as differential ECL signals. At the bottom level of an SP machine, 16 processor modules are connected to the 16 ports on 1 side of a switch module. These 17 modules are packaged together in a cabinet, or *frame*.

SP machines larger than 16 processors are composed of frames and switch modules. Small SP machines, up to 80 processors (5 frames), are constructed by directly connecting frames to one another as illustrated in Figure 11.7 for 3- and 5-frame arrangements. Larger machines use the frames as the first 2 stages of a folded Clos network, and use switch modules to provide 16 × 16 intermediate stages and/or a 32-port final stage. For example, Figure 11.8 shows how a 512 processor machine is constructed from 32 frames and 16 additional switch modules. Up to 8K-processor machines can be constructed using two levels of switch modules.

---

2.   For fault tolerance, each switch module contains 2 copies of this 32-port switching arrangement (16 switch chips total). One copy performs the switching, while the other copy *shadows* it to check for errors.

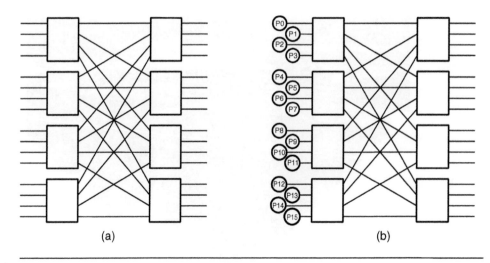

Figure 11.6   (a) A 32-port switch module is composed of eight 8 × 8 switches arranged in two ranks. All links are bidirectional. (b) Each frame consists of 16 processing nodes and 1 switch module.

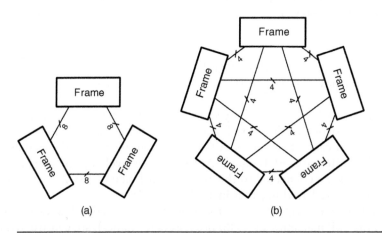

Figure 11.7   Small SP machines are realized by directly connecting frames. (a) A 48-processor (3-frame) configuration. (b) An 80-processor (5-frame) arrangement.

Each node includes a source routing table (as described in Section 11.1.1). The table includes four routes to each destination in the network. As shown in Figure 11.9, when a packet arrives to be routed, it indexes the table using the packet destination and the current value of a 2-bit packet counter. The packet counter is incremented before use by the next packet. An 8-byte route descriptor is read from the table. The first byte contains the number of valid routing flits N in the descriptor (0-7).

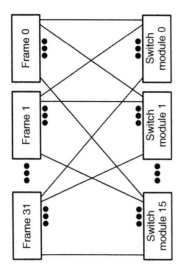

Figure 11.8 Larger SP machines are realized by connecting processor frames to switch modules that realize a folded Clos interconnection network. Here, a 512-node (32-frame) machine is connected via 16 switch modules (32-port).

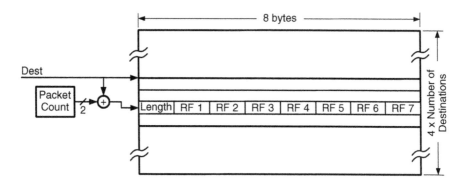

Figure 11.9 Source routing table for an IBM SP1/SP2. There are 4 routes for each destination. Each route occupies 8 bytes in the table. A length byte indicates how many routing flit (RF) bytes should be included in the route. Each RF byte encodes 2 hops of the route, as described in Figure 11.10.

The following N bytes contain the routing flits, each of which encodes 1 or 2 hops of the route. A maximal length route descriptor contains 7 routing flits encoding up to 14 hops. These N routing flits are prepended to the packet to be used during routing.

An SP packet is shown in Figure 11.10. After an initial length flit, the packet contains one or more routing flits followed by data flits. Each router examines the first routing flit. If the selector bit, S, is a zero, the router uses the upper hop field of the flit (bits 4–6) to select the output port and sets S. If S is a one, the second hop field (bits 0–2) is used, the route flit is discarded, and the length is decremented. Normally, the

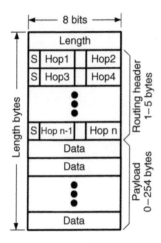

Figure 11.10   Packet format for an IBM SP1/SP2. Packets are divided into 8-bit flits. Each packet starts with a length flit followed by one or more routing flits. The remainder of the packet contains payload data. The length flit gives the overall length of the packet in flits, including the length and routing flits. Each routing flit specifies the ouput port (1 of 8) to be used for two hops. A selector bit S selects between the two hops within the flit.

S bit is initially 0. However, to encode routes that have an odd number of hops, the S bit of one of the routing flits — typically the last — must be set initially to 1. An SP packet proceeds along the source-determined route toward its destination with routing flits being consumed every two hops. When it arrives at its destination, it contains only the length flit and the data flits — all routing flits have been discarded. The length flit has been decremented with each routing flit dropped so at all points in time it reflects the current length of the packet.

The routing table is constructed in a manner that balances load over the network links and gives edge-disjoint paths [1]. For each node, the first route to each destination is determined by finding a minimum spanning tree for the network, assuming that each edge initially has unit cost. The edge weights are then adjusted by adding a small amount, ϵ, to the weight of each edge for each time it is used in a route. A new minimum spanning tree is then computed and used to generate the second route to each destination. The process is repeated for the third and fourth routes.

It is interesting to compare the IBM SP network to the networks of the CM-5 (Section 10.6) and the Avici TSR (Section 9.5). The SP network, in large configurations, has a topology that is nearly identical to that of the CM-5. Both are built from 8 × 8 port switches using a folded Clos topology. The main difference in topology is that there is no concentration in the SP networks, whereas CM-5 networks have a factor of four concentration. The two networks take very different approaches to routing in this common topology. The CM-5 uses adaptive routing on the upward links to balance load dynamically. In contrast, the SP network routes obliviously using multiple source routes to statically balance load. We examine this tradeoff in Exercise 11.9.[3]

While the CM-5 and the SP have a common topology and differ in routing, the TSR and the SP have very similar routing but apply this routing method to very different topologies. Both the TSR and the SP use oblivious source routing in which the source tables maintain a number of alternative routes to each destination: 4 in the SP and 24 in the TSR. They differ in that the SP selects the source route by using a packet counter, whereas the TSR selects the source route by hashing a flow identifier — this keeps related packets on the same route and hence in order. Both construct the alternative routes in the routing tables in a manner that statically balances load over the network channels. It is interesting that this technique of oblivious source routing is applicable to topologies as different as an irregular 3-D torus and a folded Clos network. A major strength of source routing is its topology independence.

11.4 **Bibliographic Notes**

Table-based routing is popular in implementation due to its flexibility and simplicity. As mentioned in Section 11.3, the IBM SP1 and SP2 networks [178, 176] use source routing tables, as does the Avici TSR (Section 9.5). However, for larger networks, source tables can become prohibitive in size. Many networks, such as the SGI Spider [69], InfiniBand [150], and the Internet, adopt node-tables for this reason. In addition, many node-table implementations use hierarchical routing, as introduced by McQuillan [127] and analyzed in detail by Kleinrock and Kamoun [100], to further reduce table size. The hierarchical routing mechanism used in the SGI Spider is considered in Exercise 11.4 . Another approach for reducing table size is interval routing (see Exercise 11.5), which was developed by Santoaro and Khatib [158], extended by van Leeuwen and Tan [188], and has been implemented in the INMOS Transputer [86]. CAMs are common in the implementation of routing tables as described by Pei and Zukowski [147] and McAuley and Francis [122]. Tries, introduced by Fredkin [68], are also popular for routing tables especially in IP routers. Pei and Zukowski [147] describe the design of tries in hardware and Doeringer et al. [59] give an extension of tries for efficient use in IP lookup.

11.5 **Exercises**

11.1 *Compression of source routes.* In the source routes presented in the chapter, each port selector symbol (N, S, E, W, and X) was encoded with three bits. Suggest an alternative encoding to reduce the average length (in bits) required to represent a source route. Justify your encoding in terms of "typical" routes that might occur on a torus. Also,

---

3.   Later versions of the SP machines used adaptive source routing on the uplinks.

compare the original three bits per symbol with your encoding on the following routes:

   (a) NNNNNEEEX

   (b) WNEENWWWWWNX

   (c) SSEESSX

11.2 *CAM-based routing tables.* Consider a node-table at node 34 of an 8-ary 2-cube topology. How many table entries are required if it is stored in a RAM? How much reduction is possible by storing the table in a CAM instead? List the entries for the CAM.

11.3 *Minimal adaptive route selection logic.* Sketch the logic required for the route selection block (Figure 11.4) with minimal adaptive routing. Design the logic to choose the productive path with the shortest downstream queues.

11.4 *Hierarchical routing in the SGI Spider.* The SGI Spider uses a hierarchical node-table with two levels: a *meta-table* and a *local table*. The ID of each node in the network is split into both a meta address (upper bits) and a local address (lower bits). To compute its next hop, a packet first checks if its meta-address matches the meta-address of the current node. If not, the next hop is retrieved from the meta-table, indexed by the meta-address of the packet's destination. If the meta-addresses match, the next hop is retrieved from the local table, indexed by the local address of the packet. Conceptually, the meta-address is used to move a packet to the correct "region" of the network, while the local address moves the packet to its exact destination. How much smaller could the routing tables for a 2-D torus network be if two bits of the node IDs are used in the meta-address of this approach?

11.5 *Linear interval routing in a mesh network.* Interval routing seeks to reduce the size of routing tables by labeling the outgoing edges of a node with non-overlapping intervals. Figure 11.11 shows an example of a *linear* interval labeling: each channel is labeled with an interval $[x, y]$, indicating all packets destined to a node d, such that $x \leq d \leq y$, should follow this channel. So, for example, to route from node 3 to node 1, a packet first follows the west channel from node 3 because the destination node falls in west channel's interval. Similarly, the next hop would take the packet

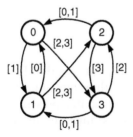

Figure 11.11   A topology and a linear interval labeling. A packet is routed hop-by-hop by choosing outgoing channels that contain the packet's destination node in their interval.

from node 1 to its destination, node 0. Find a linear interval labeling for a 4×4 mesh network. (Assign the numbers 0 through 15 to the destination nodes.)

11.6 *Route generation for the IBM SP2.* Write a program to compute spanning trees in a 512-node IBM SP2 network. Use these trees to generate a routing table that has four nodes per destination.

11.7 *Load balance of spanning tree routes.* Using the routing table you generated in Exercise 11.6 to compute the channel loads for a number of random permutations traversing the machine. How balanced is the load?

11.8 *Adaptive routing on the IBM SP network.* Later versions of the SP network employed *adaptive source routing*, which allowed CM-5-like adaptation on the uplinks and then source routing on the downlinks. Explain how this might work. Develop a route encoding that supports such adaptive source routing and that fits into the general framework of the SP network.

11.9 **Simulation.** Compare the performance of CM-5-like adaptive routing to IBM SP-like oblivious routing on a 512-node SP network.

CHAPTER 12

Flow Control Basics

Flow control determines how a network's resources, such as channel bandwidth, buffer capacity, and control state, are allocated to packets traversing the network. A good flow-control method allocates these resources in an efficient manner so the network achieves a high fraction of its ideal bandwidth and delivers packets with low, predictable latency.[1] A poor flow-control method, on the other hand, wastes bandwidth by leaving resources idle and doing unproductive work with other resources. This results in a network, like the one we examined in Chapter 2, in which only a tiny fraction of the ideal bandwidth is realized and that has high and variable latency.

One can view flow control as either a problem of resource allocation or one of contention resolution. From the resource allocation perspective, resources in the form of channels, buffers, and state must be allocated to each packet as it advances from the source to the destination. The same process can be viewed as one of resolving contention. For example, two packets arriving on different inputs of a router at the same time may both desire the same output. In this situation, the flow-control mechanism resolves this contention, allocating the channel to one packet and somehow dealing with the other, *blocked* packet.

The simplest flow-control mechanisms are *bufferless* and, rather than temporarily storing blocked packets, they either *drop* or *misroute* these packets. The next step up in complexity and efficiency is *circuit switching*, where only packet headers are buffered. In circuit switching, the header of a packet traverses the network ahead of any packet payload, reserving the appropriate resources along the

---

1. This assumes that the routing method does a good job load-balancing traffic and routes packets over nearly minimal distance paths.

path. If the header cannot immediately allocate a resource at a particular node, it simply waits at that node until the resource becomes free. Once the entire path, or *circuit*, has been reserved, data may be sent over the circuit until it is torn down *by deallocating the channels*. All of these flow-control mechanisms are rather inefficient because they waste costly channel bandwidth to avoid using relatively inexpensive storage space.

More efficient flow control can be achieved by buffering data while it waits to acquire network resources. Buffering *decouples* the allocation of adjacent channels in time. This decoupling reduces the constraints on allocation and results in more efficient operation. This buffering can be done either in units of packets, as with store-and-forward and cut-through flow control, or at the finer granularity of *flits*, as in the case of wormhole or virtual-channel flow control. By breaking large, variable-length packets into smaller, fixed-sized flits, the amount of storage needed at any particular node can be greatly reduced. Allocating resources in units of flits also facilitates creating multiple *virtual channels* per physical channel in the network, which can alleviate blocking and increase throughput.

We start our discussion of flow control in this chapter by discussing the flow control problem (Section 12.1), bufferless flow control (Section 12.2), and circuit switching (Section 12.3). In Chapter 13, we continue our exploration of flow control by discussing buffered flow-control methods. In these two chapters we deal only with allocation of resources in the network. As we shall see Chapter 14, there are additional constraints on allocation to ensure that the network remains deadlock-free. There is also a related problem to manage resources, in particular buffer memory, at the endpoints. Such end-to-end flow control employs similar principles but is not discussed here.

12.1   Resources and Allocation Units

To traverse an interconnection network, a message must be allocated resources: channel bandwidth, buffer capacity, and control state. Figure 12.1 illustrates these resources in a single node of a network. When a packet arrives at a node, it must first be allocated some control state. Depending on the flow control method, there may be a single control state per channel or, if an input can be shared between multiple packets simultaneously, there may be many sets of state. The control state tracks the resources allocated to the packet within the node and the state of the packet's traversal across the node. To advance to the next node, the packet must be allocated bandwidth on an output channel of the node. In some networks, allocating bandwidth on the output, or *forward*, channel also allocates bandwidth on a *reverse* channel traveling in the opposite direction. The reverse channel is typically used to carry acknowledgments and communicate flow control information from the receiving node. Finally, as the packet arrives at a node, it is temporarily held in buffer while awaiting channel bandwidth. All flow control methods include allocation of control state and channel bandwidth. However, some methods, which we will discuss in Section 12.2, do not allocate buffers.

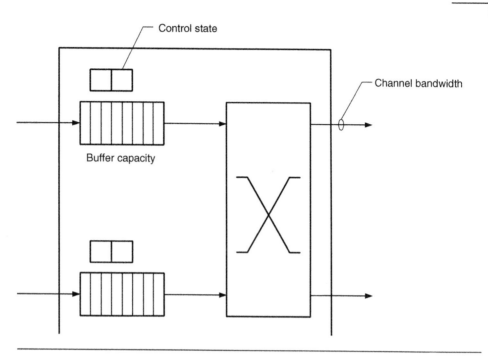

Figure 12.1   Resources within one network node allocated by a flow control method: control state records the allocation of channels and buffers to a packet and the current state of the packet in traversing the node. Channel bandwidth advances flits of the packet from this node to the next. Buffers hold flits of the packet while they are waiting for channel bandwidth.

To improve the efficiency of this resource allocation, we divide a message into packets for the allocation of control state and into flow control digits (flits) for the allocation of channel bandwidth and buffer capacity.

Figure 12.2 shows the units in which network resources are allocated. At the top level, a *message* is a logically contiguous group of bits that are delivered from a source terminal to a destination terminal. Because messages may be arbitrarily long, resources are not directly allocated to messages. Instead, messages are divided into one or more *packets* that have a restricted maximum length. By restricting the length of a packet, the size and time duration of a resource allocation is also restricted, which is often important for the performance and functionality of a flow control mechanism.

A *packet* is the basic unit of routing and sequencing. Control state is allocated to packets. As illustrated in Figure 12.2, a packet consists of a segment of a message to which a packet header is prepended. The packet header includes routing information (RI) and, if needed, a sequence number (SN). The routing information is used to determine the route taken by the packet from source to destination. As described in Chapter 8, the routing information may consist of a destination field or a source route, for example. The sequence number is needed to reorder the packets of a message if they may get out of order in transit. This may occur, for example, if different

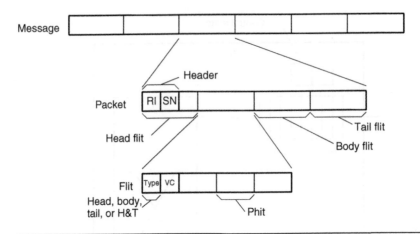

Figure 12.2   Units of resource allocation. Messages are divided into packets for allocation of control state. Each packet includes routing information (RI) and a sequence number (SN). Packets are further divided into flits for allocation of buffer capacity and channel bandwidth. Flits include no routing or sequencing information beyond that carried in the packet, but may include a virtual-channel identifier (VCID) to record the assignment of packets to control state.

packets follow different paths between the source and destination. If packets can be guaranteed to remain in order, the sequence number is not needed.

A *packet* may be further divided into flow control digits or *flits*. A flit is the basic unit of bandwidth and storage allocation used by most flow control methods. Flits carry no routing and sequencing information and thus must follow the same path and remain in order. However, flits may contain a virtual-channel identifier (VCID) to identify which packet the flit belongs to in systems where multiple packets may be in transit over a single physical channel at the same time.

The position of a flit in a packet determines whether it is a head flit, body flit, tail flit, or a combination of these. A head flit is the first flit of a packet and carries the packet's routing information. A head flit is followed by zero or more body flits and a tail flit. In a very short packet, there may be no body flits, and in the extreme case where a packet is a single flit, the head flit may also be a tail flit. As a packet traverses a network, the head flit allocates channel state for the packet and the tail flit deallocates it. Body and tail flits have no routing or sequencing information and thus must follow the head flit along its route and remain in order.

A flit is itself subdivided into one or more physical transfer digits or *phits*. A phit is the unit of information that is transferred across a channel in a single clock cycle. Although no resources are allocated in units of phits, a link level protocol must interpret the phits on the channel to find the boundaries between flits.

Why bother to break packets into flits? One could do all allocation: channel state, buffer capacity, and channel bandwidth in units of packets. In fact, we will examine several flow control policies that do just this. These policies, however, suffer from conflicting constraints on the choice of packet size. On one hand, we would like to make packets very large to amortize the overhead of routing and sequencing. On the other hand, we would like to make packets very small to permit efficient, fine-grained resource allocation and minimize blocking latency. Introducing flits eliminates this conflict. We can achieve low overhead by making packets relatively large and also achieve efficient resource utilization by making flits very small.

There are no hard and fast rules about sizes. However, phits are typically between 1 bit and 64 bits in size, with 8 bits being typical. Flits usually range from 16 bits (2 bytes) to 512 bits (64 bytes), with 64 bits (8 bytes) being typical. Finally, packets usually range from 128 bits (16 bytes) to 512 Kbits (64 Kbytes), with 1 Kbit (128 bytes) being typical. With these typical sizes, there are eight 8-bit phits to a 64-bit flit, and sixteen 64-bit flits to a 1-Kbit packet.

12.2 **Bufferless Flow Control**

The simplest forms of flow control use no buffering and simply act to allocate channel state and bandwidth to competing packets. In these cases, the flow-control method must perform an arbitration to decide which packet gets the channel it has requested. After the arbitration, the winning packet advances over this channel. The arbitration method must also decide how to dispose of any packets that did not get their requested destination. Since there are no buffers, we cannot hold the losing packets until their channels become free. Instead, we must either drop them or misroute them.

For example, consider the situation in Figure 12.3(a). Two packets, A and B, arrive at a bufferless network node and both request output channel zero. Figure 12.3(b) shows how a dropping flow control method, similar to that used in Chapter 2, handles this conflict. In this case, A wins the arbitration and advances over the output link. Packet B, on the other hand, loses the arbitration and is discarded. Any resources, such as channel bandwidth, that are expended advancing packet B to this point are wasted. Packet B must be retransmitted from its source, which we assume has a buffered copy, repeating the effort already expended getting the packet to this point in the network. Also, some acknowledgment mechanism is required to inform the source when B has been received successfully or when it needs to be retransmitted. Alternatively, packet B may be misrouted to the other output, as shown in Figure 12.3(c). In this case, there must be sufficient path diversity and an appropriate routing mechanism to route packet B to its destination from this point.

A time-space diagram for the dropping flow control policy of Figure 12.3(b), using explicit negative acknowledgments or *nacks* (N), is shown in Figure 12.4. This diagram is similar to a Gantt chart in that it shows the utilization of resources (channels) on the vertical axis plotted against time on the horizontal axis. The figure shows a five-flit packet being sent along a four-hop route. The vertical axis shows the forward (F) and reverse (R) directions of the four channels (0–3). The horizontal axis shows the flit cycle (0–17). In the figure, the first transmission of the packet is unable to allocate channel 3 and is dropped. A nack triggers a retransmission of the packet, which succeeds.

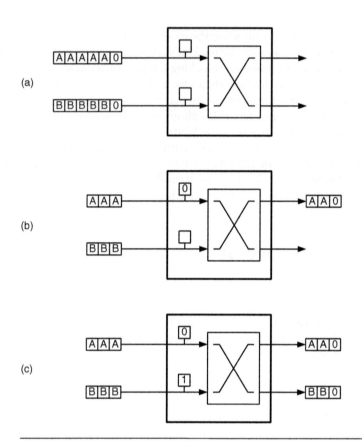

Figure 12.3   Bufferless flow control: (a) Two packets, A and B, arrive at a network node. Both request output channel 0. (b) Dropping flow control: A acquires channel 0 and B is dropped. B must be retransmitted from the source. (c) Misrouting: A acquires channel 0 and B is misrouted onto channel 1. Later in the network, B must be rerouted for it to reach its correct destination.

A packet delivery begins in flit cycle 0 with the head flit (H) of the packet traversing channel 0. The body flits (B) follow on cycles 1 through 3 and the tail flit (T) follows on cycle 4. In this case, the tail flit does not deallocate channel 0, as the packet must retain ownership of the channel until it receives an acknowledgment (positive or negative). During cycles 1 and 2, the head flit traverses channels 1 and 2, respectively. After traversing channel 2, however, the head flit encounters contention, is unable to acquire channel 3, and is dropped. To signal this failure, the router at the far end of channel 2 sends a nack along the reverse direction of channel 2 during cycle 3. The nack traverses the reverse direction of channels 1 and 0 on cycles 4 and 5, respectively, arriving at the source at the end of cycle 5. As the nack arrives at each node, it releases the resources held by that node, making them available for other packets. For example, in cycle 4, the nack arrives at the near end of channel 1. At that point, the packet releases the forward and reverse directions of channel 1.

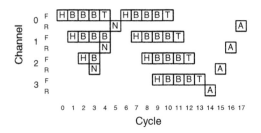

Figure 12.4   Time-space diagram showing dropping flow control with explicit negative acknowledgment. Time is shown on the horizontal axis in cycles. Space is shown on the vertical axis in channels. Forward and reverse channels for each link are shown on alternating lines. A five-flit packet is transmitted across channel 0 in cycle 0 and proceeds across channels 1 and 2. It is unable to acquire channel 3 in cycle 3 and thus is dropped. A negative acknowledgment or *nack* (N) is transmitted across reverse channel 2 in cycle 3. The arrival of this nack triggers a retransmission of the packet, starting in cycle 6. The last flit of the packet is received at the destination in cycle 13 and an acknowledgment is sent along the reverse channels in cycle 14.

After the nack is received by the source, the packet is retransmitted starting in cycle 6. The retransmitted packet is able to acquire all four of the channels needed to reach the destination. The head flit reaches the destination during cycle 9 and the tail arrives during cycle 13. After the tail is received, an acknowledgment (A) is sent along the reverse channel in cycle 14 and arrives at the source in cycle 17. As the acknowledgment arrives at each node, it frees the resources held by that node.

The dropping flow control we implemented in Chapter 2 does not use explicit negative acknowledgments. Rather, it uses a timeout to detect when a packet is dropped, as illustrated in Figure 12.5. As before, the packet fails to acquire channel 3 on the first transmission. In this case, however, a nack is not sent. Instead, the packet continues its transmission across channels 0 through 2. On each of these channels, the tail flit deallocates the resources held by the packet as it leaves the node. Thus, channels 0, 1, and 2 become free during cycles 5, 6, and 7, respectively.

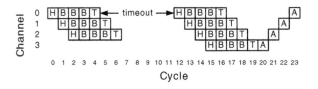

Figure 12.5   Time-space diagram showing dropping flow control without explicit nack: A 5-flit packet is transmitted starting in cycle 0 and proceeds across channels 0, 1, and 2. The packet is unable to acquire channel 3 in cycle 3 and is dropped at this point. However, the preceding channels continue to transmit the packet until the tail flit is received. A timeout triggers a retransmission of the packet in cycle 12. The tail flit of the packet is received during cycle 19 and an acknowledgment is sent starting in cycle 20.

After a timeout has elapsed without the source receiving an acknowledgment, it retransmits the packet starting in cycle 12. This time the packet is successfully received during cycles 15 through 19. An acknowledgment is sent in cycle 20, arriving in cycle 23. Since no resources are held after the tail flit passes, the acknowledgment must compete for the reverse channels and may itself be dropped. In this case, the packet will be retransmitted even though it was correctly received the first time. Some mechanism, which would typically employs sequence numbers, is needed at the receiver to delete such duplicate packets, ensuring that every packet is received exactly once.

Although simple, dropping flow control is very inefficient because it uses valuable bandwidth transmitting packets that are later dropped. A method of calculating the throughput of dropping flow control (without explicit nacks) is given in Chapter 2.

While misrouting, as in Figure 12.3(c), does not drop packets, it wastes bandwidth by sending packets in the wrong direction. In some cases, this leads to instability, where the throughput of the network drops after the offered traffic reaches a point. Misrouting also applies only to networks that have sufficient path diversity for a packet to be able to reach the destination after being misrouted. A butterfly network, for example, cannot use misrouting, since one incorrect hop will prevent a packet from ever reaching its destination. In networks like tori that do have sufficient path diversity, livelock is an issue when misrouting is used. If a packet misroutes too often, it may never get closer to its destination. Any flow control policy that involves misrouting should include some provable guarantee of forward progress to ensure that every packet eventually gets delivered.

12.3 Circuit Switching

Circuit switching is a form of bufferless flow control that operates by first allocating channels to form a *circuit* from source to destination and then sending one or more packets along this circuit. When no further packets need to be sent, the circuit is deallocated. As illustrated in Figure 12.6, the process involves four phases. During the first phase (cycles 0–4), a request (R) propagates from the source to the destination and allocates channels. In this example, no contention is encountered and the request reaches the destination without delay. After the circuit is allocated, an acknowledgment (A) is transmitted back to the source during the second phase (cycles 6–10). Once the acknowledgment is received, the circuit is established and can handle an arbitrary number and size of data packets with no further control. In the example, two four-flit packets are sent, each followed by three idle cycles. Finally, when no further data needs to be sent, a tail flit (T) is sent to deallocate the channels (cycles 26–30), freeing these channels for use in other circuits.

Circuit switching differs from dropping flow control in that if the request flit is blocked, it is held in place rather than dropped. This situation is illustrated in Figure 12.7, which is identical to Figure 12.6 except that the request is delayed four cycles before it is able to allocate channel 3. During this period, the head flit is blocked. It is held in the router at the near end of channel 3 and it rearbitrates

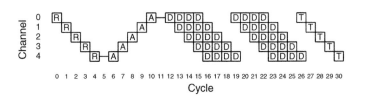

Figure 12.6   Time-space diagram showing transmission of two 4-flit packets over a 5-hop route using circuit switching with no contention. For this example, $t_r = 1$ cycle, and $D = 0$. The transmission proceeds in 4 phases. First, a request (R) is sent to the destination to acquire the channel state at each hop of the route. Second, when the request reaches the destination, an acknowledgment (A) is returned along a reverse channel to the source. Third, when the acknowledgment reaches the source, the data flits (D) are sent over the reserved channel. As long as the channel (circuit) is open, additional packets may be sent. Finally, a tail flit (T) deallocates the reserved channels as it passes.

for access to channel 3 each cycle. Eventually (in this case, in cycle 7), the head flit acquires channel 3 and proceeds with allocation of the circuit. So, compared to dropping flow control, circuit switching has the advantage that it never wastes resources by dropping a packet. Because it buffers the header at each hop, it always makes forward progress. However, circuit switching does have two weaknesses that make it less attractive than buffered flow control methods: high latency and low throughput.

From the time-space diagram of Figures 12.6 and 12.7 one can see that the zero-load latency of a single packet using circuit switching is

$$T_0 = 3Ht_r + \frac{L}{b},$$

ignoring wire latency. The first term reflects the time required to set up the channel (not including contention) and deliver the head flit, the second term is serialization latency, the third term is the time of flight, and the final term is contention time. This equation has three times the header latency given in Equation 3.11 because the path from source to destination must be traversed three times to deliver the packet: once in each direction to set up the circuit and then again to deliver the first flit. These

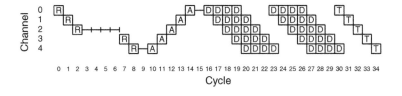

Figure 12.7   Time-space diagram showing circuit switching with contention. For this example, $t_r = 1$ cycle, and $D = 0$. The case is identical to that of Figure 12.6 except that the request is blocked for 4 cycles (cycles 3–6) before it is able to allocate channel 3.

three traversals represent a significant increase in latency in the case of a single short packet.[2]

Throughput also suffers in circuit switching because the period of time that a channel is reserved is longer than the time it is active. In the case where a single packet is sent, each channel is busy (held by the circuit) for

$$T_{b0} = 2Ht_r + \frac{L}{b}.$$

During the $2Ht_r$ setup time the channel is idle. It cannot be allocated to another circuit and the current circuit is not yet ready to send data over it. The channel bandwidth represented by this time is thus lost. For short-duration circuits, this is a significant overhead.[3]

Circuit switching has the advantage of being very simple to implement. The logic for a router differs only slightly from that of the dropping flow control router described in Chapter 2. A register is added to each input to hold a request in the event of contention and a reverse path is added.

12.4 Bibliographic Notes

Circuit switching has its origins in telephony, but is not commonly used in modern interconnection networks. Dropping flow control, while inefficient, is simple and was used in the BBN Butterfly (Section 4.5) and its follow-on, the BBN Monarch (Section 23.4). Misrouting packets, also refered to as *deflection* or *hot-potato* routing, was introduced by Baran [12] and was used in both the HEP Multiprocessor [174] and the Tera Computer System [9].

12.5 Exercises

12.1 *Dropping flow control with explicit nacks.* Consider the dropping flow control technique with explicit nacks discussed in Section 12.2 and shown in Figure 12.4. Compute an upper-bound on the throughput (as a fraction of capacity) of a network using this flow control method. Assume a maximum packet length of F flits, an average hop count of H_{avg}, uniform traffic, and a symmetric topology.

12.2 *Timeout interval for dropping flow control.* Consider the dropping flow control technique with timeout discussed in Section 12.2 and shown in Figure 12.5. Assuming that the maximum packet length is F flits and each hop of the network requires one

---

2. In the absence of contention, this disadvantage can be eliminated by using optimistic circuit switching. See Exercise 12.4.
3. Some of this overhead can be eliminated, in the absence of contention, by having the circuit setup request reserve the channel for future use rather than claiming it immediately. See Exercise 12.5.

flit time (cycle), give an expression for the minimum timeout interval in terms of these parameters and the network's diameter.

12.3 *Livelock with dropping flow control and timeout.* In dropping flow control with a time-out mechanism, the reverse channels for acknowledgments are not reserved and therefore the acknowledgments themselves may be dropped because of contention. Explain the livelock issues associated with this and suggest a simple solution.

12.4 *Optimistic circuit switching.* An optimistic circuit switching technique could lower zero-load latency by speculatively sending the data along with the header as the circuit is being set up. If the header becomes blocked, the data is dropped and a nack is send back along the partially reserved circuit. Otherwise, the data can be deliver to the destination immediately after the circuit is established. Draw two time-space diagrams of optimistic circuit switching for when the speculative data is and is not dropped. Assume $t_r = 1$ cycle and $D = 0$. Can optimistic circuit switching reduce the number of cycles that channels are reserved but idle? If so, by how much?

12.5 *Reservation circuit switching.* Consider a flow control method similar to circuit switching but where the request message reserves each channel for a fixed period of time in the future (for example, for 10 cycles starting in 15 cycles). At each router along the path, if the request can be accommodated a reservation is made. If the request cannot be accommodated a nack is sent that cancels all previous recommendations for the connection, and the request is retried. If a request reaches the destination, an acknowledgement is sent back to the source, confirming all reservations. Draw a time-space diagram of a situation that demonstrates the advantage of reservation circuit switching over conventional circuit switching.

CHAPTER 13

Buffered Flow Control

Adding buffers to our networks results in significantly more efficient flow control. This is because a buffer decouples the allocation of adjacent channels. Without a buffer, the two channels must be allocated to a packet (or flit) during consecutive cycles, or the packet must be dropped or misrouted. There is nowhere else for the packet to go. Adding a buffer gives us a place to store the packet (or flit) while waiting for the second channel, allowing the allocation of the second channel to be delayed without complications.

Once we add buffers to an interconnection network, our flow control mechanism must allocate buffers as well as channel bandwidth. Moreover, we have a choice as to the granularity at which we allocate each of these resources. As depicted in Figure 13.1, we can allocate either buffers or channel bandwidth in units of flits or packets. As shown in the figure, most flow control mechanisms allocate both resources at the same granularity. If we allocate both channel bandwidth and buffers in units of packets, we have packet-buffer flow control, either store-and-forward flow control or cut-through flow control, which will be described further in Section 13.1.

If we allocate both bandwidth and buffers in units of flits, we have flit-buffer flow control, which we will describe in Section 13.2. Allocating storage in units of flits rather than packets has three major advantages. It reduces the storage required for correct operation of a router, it provides stiffer backpressure from a point of congestion to the source, and it enables more efficient use of storage.

The off-diagonal entries in Figure 13.1 are not commonly used. Consider first allocating channels to packets but buffers to flits. This will not work. Sufficient buffering to hold the *entire* packet must be allocated before transmitting a packet across a channel. The case where we allocate buffers to packets and channel bandwidth to flits is possible, but not common. Exercise 13.1 explores this type of flow control in more detail.

		Channel allocated in units of	
		Packets	Flits
Buffer allocated in units of	Packets	Packet-buffer flow control	See Exercise 13.1
	Flits	Not possible	Flit-buffer flow control

Figure 13.1 Buffered flow control methods can be classified based on their granularity of channel bandwidth allocation and of buffer allocation.

This chapter explores buffered flow control in detail. We start by describing packet-buffer flow control (Section 13.1) and flit-buffer flow control (Section 13.2). Next, we present methods for managing buffers and signaling backpressure in Section 13.3. Finally, we present flit-reservation flow control (Section 13.4), which separates the control and data portions of packets to improve allocation efficiency.

13.1 Packet-Buffer Flow Control

We can allocate our channel bandwidth much more efficiently if we add buffers to our routing nodes. Storing a flit (or a packet) in a buffer allows us to decouple allocation of the input channel to a flit from the allocation of the output channel to a flit. For example, a flit can be transferred over the input channel on cycle i and stored in a buffer for a number of cycles j until the output channel is successfully allocated on cycle $i + j$. Without a buffer, the flit arriving on cycle i would have to be transmitted on cycle $i + 1$, misrouted, or dropped. Adding a buffer prevents the waste of the channel bandwidth caused by dropping or misrouting packets or the idle time inherent in circuit switching. As a result, we can approach 100% channel utilization with buffered flow control.

Buffers and channel bandwidth can be allocated to either flits or packets. We start by considering two flow control methods, *store-and-forward* and *cut-through*, that allocate both of these resources to packets. In Section 13.2, we see how allocating buffers and channel bandwidth to flits rather than packets results in more efficient buffer usage and can reduce contention latency.

With store-and-forward flow control, each node along a route waits until a packet has been completely received (stored) and then forwards the packet to the next node. The packet must be allocated two resources before it can be forwarded: a packet-sized buffer on the far side of the channel and exclusive use of the channel. Once the entire packet has arrived at a node and these two resources are acquired, the packet is forwarded to the next node. While waiting to acquire resources, if they are not immediately available, no channels are being held idle and only a single packet buffer on the current node is occupied.

A time-space diagram showing store-and-forward flow control is shown in Figure 13.2. The figure shows a 5-flit[1] packet being forwarded over a 4-hop route with no contention. At each step of the route the entire packet is forwarded over one channel before proceeding to the next channel.

The major drawback of store-and-forward flow control is its very high latency. Since the packet is completely received at one node before it can begin moving to the next node, serialization latency is experienced at each hop. Therefore, the overall latency of a packet is

$$T_0 = H \left(t_r + \frac{L}{b} \right).$$

Cut-through flow control[2] overcomes the latency penalty of store-and-forward flow control by forwarding a packet as soon as the header is received and resources (buffer and channel) are acquired, without waiting for the entire packet to be received. As with store-and-forward flow control, cut-through flow control allocates both buffers and channel bandwidth in units of packets. It differs only in that transmission over each hop is started as soon as possible without waiting for the entire packet to be received.

Figure 13.3 shows a time-space diagram of cut-through flow control forwarding a 5-flit[3] packet over a 4-hop route without contention (Figure 13.3[a]) and with contention (Figure 13.3[b]). The figure shows how transmission over each hop is started as soon as the header is received as long as a buffer and a channel are available.

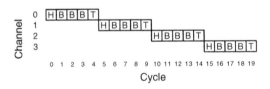

Figure 13.2 Time-space diagram showing store-and-forward flow control used to send a 5-flit packet over 4 channels. For this example, $t_r = 1$ cycle and $D = 0$. At each channel, a buffer is allocated to the packet and the entire packet is transmitted over the channel before proceeding to the next channel.

1. With store-and-forward flow control, packets are not divided into flits. We show the division here for consistency with the presentation of other flow control methods.
2. This method is often called *virtual cut-through*. We shorten the name here to avoid confusion with virtual-channel flow control.
3. As with store-and-forward, cut-through does not divide packets into flits. We show the division here for consistency.

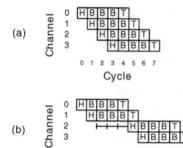

Figure 13.3   Time-space diagram showing cut-through flow control sending a 5-flit packet over 4 channels. For this example, $t_r = 1$ cycle and $D = 0$. (a) The packet proceeds without contention. An entire packet buffer is allocated to the packet at each hop; however, each flit of the packet is forwarded as soon as it is received, resulting in decreased latency. (b) The packet encounters contention for three cycles before it is able to allocate channel 2. As the buffer is large enough to hold the entire packet, transmission over channel 1 proceeds regardless of this contention.

In the first case, the resources are immediately available and the header is forwarded as soon as its received. In the second case, the packet must wait three cycles to acquire channel 2. During this time transmission continues over channel 1 and the data is buffered.

By transmitting packets as soon as possible, cut-through flow control reduces the latency from the product of the hop count and the serialization latency to the sum of these terms:

$$T_0 = Ht_r + \frac{L}{b}.$$

At this point, cut-through flow control may seem like an ideal method. It gives very high channel utilization by using buffers to decouple channel allocation. It also achieves very low latency by forwarding packets as soon as possible. However, the cut-through method, or any other packet-based method, has two serious shortcomings. First, by allocating buffers in units of packets, it makes very inefficient use of buffer storage. As we shall see, we can make much more effective use of storage by allocating buffers in units of flits. This is particularly important when we need multiple, independent buffer sets to reduce blocking or provide deadlock avoidance. Second, by allocating channels in units of packets, contention latency is increased. For example, a high-priority packet colliding with a low-priority packet must wait for the entire low-priority packet to be transmitted before it can acquire the channel. In the next section, we will see how allocating resources in units of flits rather than packets results in more efficient buffer use (and hence higher throughput) and reduced contention latency.

13.2 **Flit-Buffer Flow Control**

13.2.1 **Wormhole Flow Control**

Wormhole flow control[4] operates like cut-through, but with channel and buffers allocated to flits rather than packets. When the head flit of a packet arrives at a node, it must acquire three resources before it can be forwarded to the next node along a route: a virtual channel (channel state) for the packet, one flit buffer, and one flit of channel bandwidth. Body flits of a packet use the virtual channel acquired by the head flit and hence need only acquire a flit buffer and a flit of channel bandwidth to advance. The tail flit of a packet is handled like a body flit, but also releases the virtual channel as it passes.

A virtual channel holds the state needed to coordinate the handling of the flits of a packet over a channel. At a minimum, this state identifies the output channel of the current node for the next hop of the route and the state of the virtual channel (idle, waiting for resources, or active). The virtual channel may also include pointers to the flits of the packet that are buffered on the current node and the number of flit buffers available on the next node.

An example of a 4-flit packet being transported through a node using wormhole flow control is shown in Figure 13.4. Figure 13.4(a–g) illustrates the forwarding process cycle by cycle, and Figure 13.4(h) shows a time-space diagram summarizing the entire process. The input virtual channel is originally in the idle state (I) when the head flit arrives (a). The upper output channel is busy, allocated to the lower input (L), so the input virtual channel enters the waiting state (W) and the head flit is buffered while the first body flit arrives (b). The packet waits two more cycles for the output virtual channel (c). During this time the second body flit is blocked and cannot be forwarded across the input channel because no buffers are available to hold this flit. When the output channel becomes available, the virtual channel enters the active state (A), the header flit is forwarded and the second body flit is accepted. The remaining flits of the packet proceed over the channel in subsequent cycles (e, f, and g). As the tail flit of the packet is forwarded, it deallocates the input virtual channel, returning it to the idle state (g).

Compared to cut-through flow control, wormhole flow control makes far more efficient use of buffer space, as only a small number of flit buffers are required per virtual channel.[5] In contrast, cut-through flow control requires several packets of buffer space, which is typically at least an order of magnitude more storage than wormhole flow control. This savings in buffer space, however, comes at the expense of some throughput, since wormhole flow control may block a channel mid-packet.

---

4. Historically, this method has been called "wormhole routing." However, it is a flow control method and has nothing to do with routing. Hence, we refer to this method as "wormhole flow control."
5. Enough buffers should be provided to cover the round-trip latency and pipeline delay between the two nodes. This is discussed in more detail in Sections 13.3 and 16.3.

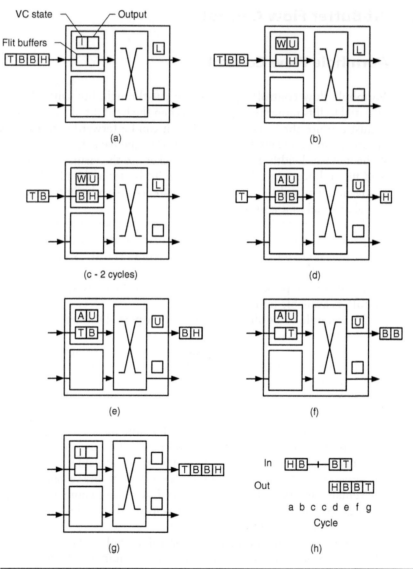

Figure 13.4   Wormhole flow control: (a–g) a 4-flit packet routing through a node from the upper input port to the upper output port, (h) a time-space diagram showing this process. In (a) the header arrives at the node, while the virtual channel is in the idle state (I) and the desired upper (U) output channel is busy — allocated to the lower (L) input. In (b) the header is buffered and the virtual channel is in the waiting state (W), while the first body flit arrives. In (c) the header and first body flit are buffered, while the virtual channel is still in the waiting state. In this state, which persists for two cycles, the input channel is blocked. The second body flit cannot be transmitted, since it cannot acquire a flit buffer. In (d) the output virtual channel becomes available and allocated to this packet. The state moves to active (A) and the head is transmitted to the next node. The body flits follow in (e) and (f). In (g) the tail flit is transmitted and frees the virtual channel, returning it to the idle state.

Blocking may occur with wormhole flow control because the channel is owned by a packet, but buffers are allocated on a flit-by-flit basis. When a flit cannot acquire a buffer, as occurs in Figure 13.4(c), the channel goes idle. Even if there is another packet that could potentially use the idle channel bandwidth, it cannot use it because the idled packet owns the single virtual channel associated with this link. Even though we are allocating channel bandwidth on a flit-by-flit basis, only the flits of one packet can use this bandwidth.

13.2.2 **Virtual-Channel Flow Control**

Virtual-channel flow control, which associates several virtual channels (channel state and flit buffers) with a single physical channel, overcomes the blocking problems of wormhole flow control by allowing other packets to use the channel bandwidth that would otherwise be left idle when a packet blocks. As in wormhole flow control, an arriving head flit must allocate a virtual channel, a downstream flit buffer, and channel bandwidth to advance. Subsequent body flits from the packet use the virtual channel allocated by the header and still must allocate a flit buffer and channel bandwidth. However, unlike wormhole flow control, these flits are not guaranteed access to channel bandwidth because other virtual channels may be competing to transmit flits of their packets across the same link.

Virtual channels allow packets to pass blocked packets, making use of otherwise idle channel bandwidth, as illustrated in Figure 13.5. The figure shows 3 nodes of a 2-D, unidirectional torus network in a state where a packet B has entered node 1 from the north, acquired channel p from node 1 to node 2 and blocked. A second packet A has entered node 1 from the west and needs to be routed east to node 3. Figure 13.5(a) shows the situation using wormhole routing with just a single virtual channel per physical channel. In this case, packet A is blocked at node 1 because it is unable to acquire channel p. Physical channels p and q are both idle because packet B is blocked and not using bandwidth. However, packet A is unable to use these idle channels because it is unable to acquire the channel state held by B on node 2.

Figure 13.5(b) shows the same configuration with two virtual channels per physical channel. Each small square represents a complete virtual channel state: the channel state (idle, waiting, or active), the output virtual channel, and a flit buffer. In this case, packet A is able to acquire the second virtual channel on node 2 and thus proceed to node 3, making use of channels p and q that were left idle with just a single virtual channel per node.

The situation illustrated in Figure 13.5 is analogous to a stopped car (packet B) on a single-lane road waiting to make a left turn onto a congested side road (south exit of node 2). A second car (packet A) is blocked behind the first car and unable to proceed even though the road ahead (channels p and q) is clear. Adding a virtual channel is nearly analogous to adding a left turn lane to the road. With the left turn lane, the arriving car (packet A) is able to pass the car waiting to turn (packet B) and continue along the road. Also, just as adding a virtual channel does not add bandwidth to the physical channel, a left turn lane does not increase the width of the roads between intersections.

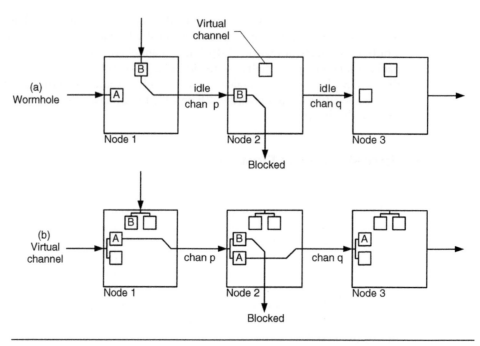

Figure 13.5   (a) With wormhole flow control when a packet, B, blocks while holding the sole virtual channel associated with channel p, channels p and q are idled even though packet A requires use of these idle channels. (b) With virtual channel flow control, packet A is able to proceed over channels p and q by using a second virtual channel associated with channel p on node 2.

As this simple example demonstrates, virtual-channel flow control decouples the allocation of channel state from channel bandwidth. This decoupling prevents a packet that acquires channel state and then blocks from holding channel bandwidth idle. This permits virtual-channel flow control to achieve substantially higher throughput than wormhole flow control. In fact, given the same total amount of buffer space, virtual-channel flow control also outperforms cut-through flow control because it is more efficient to allocate buffer space as multiple short virtual-channel flit buffers than as a single large cut-through packet buffer.

An example in which packets on two virtual channels must share the bandwidth of a single physical channel is illustrated in Figure 13.6. Here, packets *A* and *B* arrive on inputs 1 and 2, respectively, and both acquire virtual channels associated with the same output physical channel. The flits of the two packets are interleaved on the output channel, each packet getting half of the flit cycles on the channel. Because the packets are arriving on the inputs at full rate and leaving at half rate, the flit buffers (each with a capacity of three flits) fill up, forcing the inputs to also throttle to half rate. The gaps in the packets on the input do not imply an idle physical channel, however, since another virtual channel can use these cycles.

When packets traveling on two virtual channels interleave flits on a single physical channel, as shown in Figure 13.7, this physical channel becomes a bottleneck that

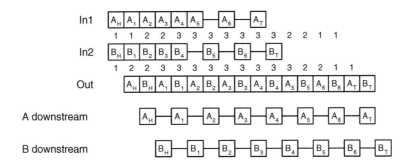

Figure 13.6 Two virtual channels interleave their flits on a single physical channel. Packet *A* arrives on input 1 at the same time that packet *B* arrives on input 2, and both request the same output. Both packets acquire virtual channels associated with the output channel but must compete for bandwidth on a flit-by-flit basis. With fair bandwidth allocation, the flits of the two packets are interleaved. The numbers under the arriving flits show the number of flits in the corresponding virtual channel's input buffer, which has a capacity of 3 flits. When the input buffer is full, arriving flits are blocked until room is made available by departing flits. On downstream links, the flits of each packets are available only every other cycle.

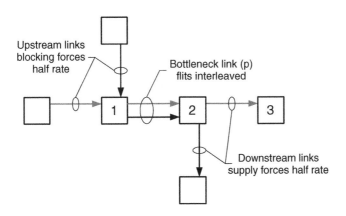

Figure 13.7 Sharing bandwidth of a bottleneck link by two virtual channels reduces both upstream and downstream bandwidth. Packet *A* (gray) and packet *B* (black) share bandwidth on bottleneck link *p*. Each packet gets 50% of the link bandwidth. The topology and routes are the same as in Figure 13.5. The supply of flits out of the bottleneck link limits these packets to use no more than 50% of the bandwidth of links downstream of the bottleneck. Also, once the flit buffers of the virtual channels on node 1 are filled, bandwidth upstream of the bottleneck channels is reduced to 50% by blocking. Packets traveling on other virtual channels may use the bandwidth left idle by the bottleneck.

affects the bandwidth of these packets both upstream and downstream. On the downstream channels (south and east of node 2), the head flit propagates at full rate, allocating virtual channels, but body flits follow only every other cycle. The idle cycles are available for use by other packets. Similarly, once the flit buffers on node 1 fill, blocking limits each packet to transmit flits only every other cycle on the channels immediately upstream. This reduced bandwidth may cause the buffers on the upstream nodes to fill, propagating the blocking and bandwidth reduction further upstream. For long packets, this blocking can reach all the way back to the source node. As with the downstream case, the idle cycles on the upstream channels are available for use by other packets.

Fair bandwidth arbitration, which interleaves the flits of competing packets over a channel (Figure 13.6) results in a higher average latency than *winner-take-all* bandwidth allocation, which allocates all of the bandwidth to one packet until it is finished or blocked before serving the other packets (Figure 13.8). With interleaving in our 2-packet example, both packets see a contention latency that effectively doubles their serialization latency. With winner-take-all bandwidth arbitration, however, first one packet gets all of the bandwidth and then the other packet gets all of the bandwidth. The result is that one packet has no contention latency, while the other packet sees a contention latency equal to the serialization latency of the first packet — the same as with interleaving. Of course, if packet *A* blocks before it has finished transmission, it would relinquish the channel and packet *B* would use the idle cycles. Making bandwidth arbitration unfair reduces the average latency of virtual-channel flow control with no throughput penalty.

The block diagram of Figure 13.9 illustrates the state that is replicated to implement virtual channels.[6] The figure shows a router with two input ports, two output

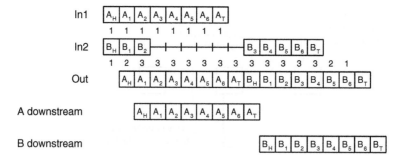

Figure 13.8   Packets *A* and *B* arrive at the same time on two virtual channels, sharing a single physical channel, just as in Figure 13.6. With *winner-take-all* arbitration, however, packet *A* transmits all of its flits before relinquishing the physical channel to packet *B*. As a result, the latency of packet *A* is reduced by 7 cycles without affecting the latency of packet *B*.

---

6.   The internal organization of virtual-channel routers is described in more detail in Chapter 16.

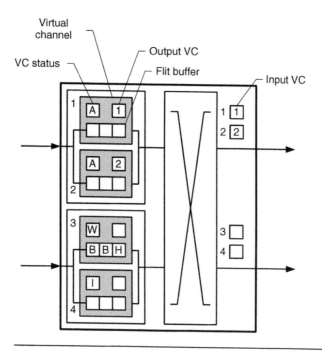

Figure 13.9 Block diagram of a virtual channel router showing the state associated with each virtual channel. Each input virtual channel includes the channel status, the output virtual channel allocated to the current packet, and a flit buffer. Each output virtual channel identifies the input virtual channel, if any, to which it is allocated. In this configuration, input virtual channel 1 is in the active state forwarding a packet to output virtual channel 1. Similarly, input virtual channel 2 is forwarding a packet to output virtual channel 2. Input virtual channel 3, associated with the lower physical channel, is waiting to be allocated an output virtual channel. Input virtual channel 4 is idle, waiting for the arrival of a packet.

ports, and two virtual channels per physical channel. Each input virtual channel includes a complete copy of all channel state including a status register (idle, waiting, or active), a register that indicates the output virtual channel allocated to the current packet, and a flit buffer.[7] For each output virtual channel, a single status register records whether the output virtual channel is assigned and, if assigned, the input virtual channel to which it is assigned. The figure shows a configuration in which input virtual channels 1 and 2, associated with the upper input port, are active, forwarding packets to output virtual channels 1 and 2, associated with the upper output port. Input virtual channel 3 is waiting to be allocated an output virtual channel, and input virtual channel 4 is idle.

---

7. As we shall see in Section 17.1, the flit buffers of the virtual channels associated with a single input are often combined into a single storage array, and in some cases have storage dynamically allocated from the same pool.

Virtual-channel flow control organizes buffer storage in two dimensions (virtual channels and flits per virtual channel), giving us a degree of freedom in allocating buffers. Given a fixed number of flit buffers per router input, we can decide how to allocate these buffers across the two dimensions. Figure 13.10 illustrates the case where there are 16 flits of buffer storage per input. This storage can be organized as a single virtual channel with 16 flits of storage, two 8-flit virtual channels, four 4-flit virtual channels, eight 2-flit virtual channels, or sixteen 1-flit virtual channels. In general, there is little performance gained by making a virtual channel *deeper* than the number of flits needed to cover the round-trip credit latency so that full throughput can be achieved with a single virtual channel operating. (See Section 16.3.) Thus, when increasing the buffer storage available, it is usually better to add more virtual channels than to add flits to each virtual channel.[8]

Virtual channels are the Swiss-Army™ Knife of interconnection networks. They are a utility tool that can be used to solve a wide variety of problems. As discussed above, they can improve network performance by allowing active packets to pass blocked packets. As we shall see in Chapter 14, they are widely used to avoid deadlock in interconnection networks. They can be applied to combat both deadlock in the network itself and deadlock due to dependencies in higher-level protocols. They can be used to provide multiple levels of priority or service in an interconnection network by separating different classes of traffic onto different virtual channels and allocating bandwidth preferentially to the higher-priority traffic classes. They can even be used to make a network non-interfering by segregating traffic with different destinations.

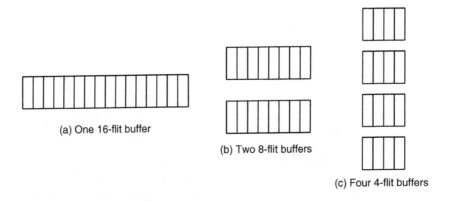

(a) One 16-flit buffer

(b) Two 8-flit buffers

(c) Four 4-flit buffers

Figure 13.10   If 16 flits of buffer storage are available at an input port, the storage can be arranged as a single virtual channel with a 16-flit queue as 2 virtual channels with 8-flit queues or as 4 virtual channels with 4-flit queues.

---

8. This is not a hard and fast rule. The performance analysis of virtual-channel flow control is discussed in more detail in Chapter 23

13.3 **Buffer Management and Backpressure**

All of the flow control methods that use buffering need a means to communicate the availability of buffers at the downstream nodes. Then the upstream nodes can determine when a buffer is available to hold the next flit (or packet for store-and-forward or cut-through) to be transmitted. This type of buffer management provides *backpressure* by informing the upstream nodes when they must stop transmitting flits because all of the downstream flit buffers are full. Three types of low-level flow control mechanisms are in common use today to provide such backpressure: credit-based, on/off, and ack/nack. We examine each of these in turn.

13.3.1 **Credit-Based Flow Control**

With *credit-based* flow control, the upstream router keeps a count of the number of free flit buffers in each virtual channel downstream. Then, each time the upstream router forwards a flit, thus consuming a downstream buffer, it decrements the appropriate count. If the count reaches zero, all of the downstream buffers are full and no further flits can be forwarded until a buffer becomes available. Once the downstream router forwards a flit and frees the associated buffer, it sends a *credit* to the upstream router, causing a buffer count to be incremented.

A timeline illustrating credit-based flow control is shown in Figure 13.11. Just before time t_1, all buffers on the downstream end of the channel (at node 2) are full. At time t_1, node 2 sends a flit, freeing a buffer. A credit is sent to node 1 to signal that this buffer is available. The credit arrives at time t_2, and after a short processing interval results in a flit being sent at time t_3 and received at time t_4. After a short interval, this flit departs node 2 at time t_5, again freeing a buffer and sending a credit back to node 1.

The minimum time between the credit being sent at time t_1 and a credit being sent for the *same* buffer at time t_5 is the credit round-trip delay t_{crt}. This delay, which includes a round-trip wire delay and additional processing time at both ends, is a critical parameter of any router because it determines the maximum throughput that can be supported by the flow control mechanism. If there were only a single flit buffer on this virtual channel, each flit would have to wait for a credit for this single buffer before being transmitted. This would restrict the maximum throughput of the channel to be no more than one flit each t_{crt}. This corresponds to a bit rate of $\frac{L_f}{t_{crt}}$, where L_f is the length of a flit in bits. If there are F flit buffers on this virtual channel, F flits could be sent before waiting for the credit for the original buffer, giving a throughput of F flits per t_{crt} or $\frac{FL_f}{t_{crt}}$ bits per second. Thus, we see that to prevent low-level flow control from limiting throughput over a channel with bandwidth b, we require:

$$F \geq \frac{t_{crt}b}{L_f}. \tag{13.1}$$

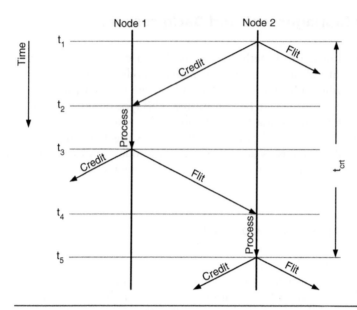

Figure 13.11   Timeline of credit-based flow control. At time t_1 all buffers at the node 2 end of the channel are full and node 2 sends a flit, freeing a buffer. At the same time, a credit for this buffer is sent to node 1 and is received at t_2. After a short processing delay, node 1 sends a flit to occupy the buffer granted by the credit at t_3, and this flit is received at t_4. A short time after arrival, this flit departs node 2 at t_5, freeing the buffer again and hence generating another credit for node 1. The minimum time between successive credits for the same buffer is the credit round-trip time t_{crt}.

Figure 13.12 illustrates the mechanics of credit-based flow control. The sequence begins in Figure 13.12(a). Upstream node 1 has two credits for the virtual channel shown on downstream node 2. In Figure 13.12(b) node 1 transmits the head flit and decrements its credit count to 1. The head arrives at node 2 in Figure 13.12(c), and, at the same time, node 1 sends the first body flit of the packet, decrementing its credit count to zero. With no credits, node 1 can send no flits during Figure 13.12(d) when the body flit arrives and (e) when node 2 transmits the head flit. Transmitting the head flit frees a flit buffer, and node 2 sends a credit back to node 1 to grant it use of this free buffer. The arrival of this credit in Figure 13.12(f) increments the credit count on node 1 at the same time the body flit is transmitted upstream, and a second credit is transmitted downstream by node 2. With a non-zero credit count, node 1 is able to forward the tail flit and free the virtual channel by clearing the input virtual channel field (Figure 13.12[g]). The tail flit is forwarded by node 2 in Figure 13.12(h), returning a final credit that brings the count on node 1 back to 2 in Figure 13.12(i).

A potential drawback of credit-based flow control is a one-to-one correspondence between flits and credits. For each flit sent downstream, a corresponding credit is eventually sent upstream. This requires a significant amount of upstream signaling and, especially for small flits, can represent a large overhead.

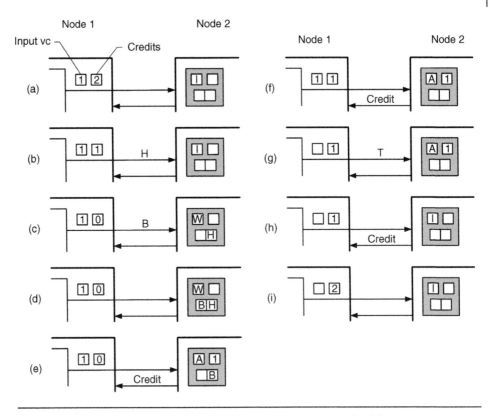

Figure 13.12 A sequence showing the operation of credit-based flow control. Each subfigure represents one flit time. In (a) node 1 has two credits for the two downstream flit buffers of the virtual channel on node 2. In (b) and (c) node 1 transmits two flits, a head and body, using both of these credits. These flits arrive in (c) and (d), respectively, and are buffered. At this point node 1 has no credits, making it unable to send additional flits. A credit for the head flit is returned in (e), incrementing the credit count in (f) and causing the tail flit to be transmitted in (g). The body credit is returned in (f) and the tail credit in (h), leaving node 1 in (i) again with two credits (corresponding to the two empty flit buffers on node 2).

13.3.2 **On/Off Flow Control**

On/off flow control can greatly reduce the amount of upstream signaling in certain cases. With this method the upstream state is a single control bit that represents whether the upstream node is permitted to send (*on*) or not (*off*). A signal is sent upstream only when it is necessary to change this state. An *off* signal is sent when the control bit is on and the number of free buffers falls below the threshold F_{off}. If the control bit is off and the number of free buffers rises above the threshold F_{on}, an *on* signal is sent.

A timeline illustrating on/off flow control is illustrated in Figure 13.13. At time to t_1, node 1 sends a flit that causes the number of free buffers on node 2 to fall

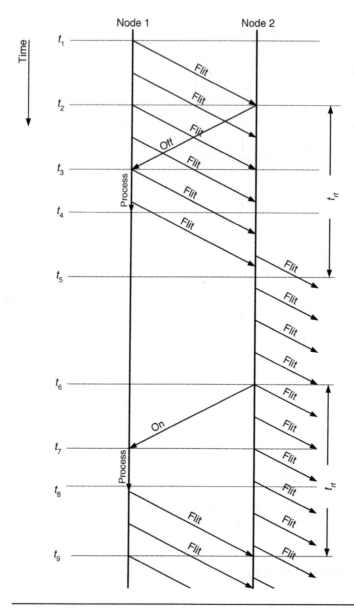

Figure 13.13   Timeline of on/off flow control. The flit transmitted at t_1 causes the free buffer count on node 2 to fall below its limit, F_{off}, which in turn causes node 2 to send an *off* signal back to node 1. Before receiving this signal, node 1 transmits additional flits that node 2 buffers in the F_{off} buffers that were free when the off signal was transmitted. At t_6 the free buffer count on node 2 rises above F_{on}, causing node 2 to send an *on* signal to node 1. During the interval between sending this *on* signal at t_6 and the receipt of the next flit from node 1 at t_9, node 2 sends flits out of the $F - F_{on}$ buffers that were full at t_6.

below F_{off} and thus triggers the transmission of an *off* signal at t_2. At t_3 the *off* signal is received at node 1, and after a small processing delay, stops transmission of flits at t_4. In the time t_{rt} between t_1 and t_4, additional flits are sent. The lower limit must be set so that

$$F_{\text{off}} \geq \frac{t_{rt}b}{L_f} \tag{13.2}$$

to prevent these additional flits from overflowing the remaining flit buffers.

After sending a number of flits, node 2 clears sufficient buffers to bring the free buffer count above F_{on} at t_6 and sends an *on* signal to node 1. Node 1 receives this signal at t_7 and resumes sending flits at t_8. The first flit triggered by the *on* signal arrives at t_9. To prevent node 2 from running out of flits to send between sending the *on* signal at t_6 and receiving a flit at t_9, the *on* limit must be set so that

$$F - F_{\text{on}} \geq \frac{t_{rt}b}{L_f}. \tag{13.3}$$

On/off flow control requires that the number of buffers be at least $\frac{t_{rt}b}{L_f}$ to work at all or Equation 13.2 cannot be satisfied. To operate at full speed, twice this number of buffers is required so that Equation 13.3 can also be satisfied ($F_{\text{on}} \geq F_{\text{off}}$):

$$F \geq F_{\text{on}} + \frac{t_{rt}b}{L_f} \geq F_{\text{off}} + \frac{t_{rt}b}{L_f} \geq \frac{2t_{rt}b}{L_f}.$$

With an adequate number of buffers, on/off flow control systems can operate with very little upstream signaling.

13.3.3 **Ack/Nack Flow Control**

Both credit-based and on/off flow control require a round-trip delay t_{rt} between the time a buffer becomes empty, triggering a credit or an *on* signal, and when a flit arrives to occupy that buffer. Ack/nack flow control reduces the minimum of this buffer vacancy time to zero and the average vacancy time to $t_{rt}/2$. Unfortunately there is no net gain because buffers are held for an additional t_{rt} waiting for an acknowledgment, making ack/nack flow control less efficient in its use of buffers than credit-based flow control. It is also inefficient in its use of bandwidth which it uses to send flits only to drop them when no buffer is available.

With ack/nack flow control, there is no state kept in the upstream node to indicate buffer availability. The upstream node optimistically sends flits whenever they become available.[9] If the downstream node has a buffer available, it accepts the flit and sends an acknowledge (ack) to the upstream node. If no buffers are available when the flit arrives, the downstream node drops the flit and sends a negative

---

9. Because of this behavior, ack/nack flow control is often referred to as optimistic flow control.

acknowledgment (nack). The upstream node holds onto each flit until it receives an ack. If it receives a nack, it retransmits the flit.

In systems where multiple flits may be in flight at the same time, the correspondence between acks or nacks and flits is maintained by ordering — acks and nacks are received in the same order as the flits to which they correspond. In such systems, however, nacking a flit leads to flits being received out of order at the downstream node. The downstream node must then reorder the flits by holding the later flits until the earlier flits are successfully retransmitted.

The timeline of Figure 13.14 illustrates ack/nack flow control. At time t_1, node 1 sends flit 1 to node 2, not knowing that no buffers are available. This flit arrives at time t_2, at which time node 2 sends a nack back to node 1. At the same time, node 1 begins sending flit 2 to node 2. At time t_3, node 1 receives the nack, but before it can be processed starts sending flit 3 to node 2. It has to wait until the transmission of flit 3 is completed, at t_4, to retransmit flit 1. Flit 1 is finally received at t_5 after flits 2 and 3. Node 2 must buffer flits 2 and 3 until flit 1 is received to avoid transmitting the flits out of order.

Because of its buffer and bandwidth inefficiency, ack/nack flow control is rarely used. Rather, credit-based flow control is typically used in systems with small numbers of buffers, and on/off flow control is employed in most systems that have large numbers of flit buffers.

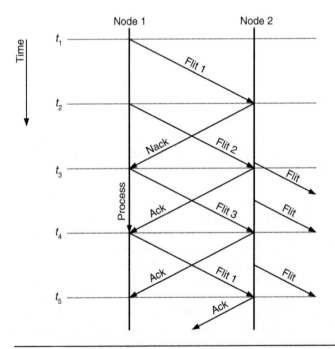

Figure 13.14   Timeline of ack/nack flow control. Flit 1 is not accepted by node 2, which returns a nack to sending node 1. Node 2 then accepts flits 2 and 3 from node 1 and responds to each by sending an ack to node 1. After flit 3 has been completed, node 1 responds to the nack by retransmitting flit 1. Node 2 cannot send flits 2 or 3 onward until flit 1 is received. It must transmit flits in order.

13.4 **Flit-Reservation Flow Control**

While traditional wormhole networks greatly reduce the latency of sending packets through an interconnection network, the idealized view of router behavior can differ significantly from a pipelined hardware implementation. Pipelining breaks the stages of flit routing into several smaller steps, which increases the hop time. Accounting for these pipelining delays and propagation latencies gives an accurate view of buffer utilization.

An example of buffer utilization for a wormhole network with credit-based flow control is shown in Figure 13.15. Initially, a flit is sent from the current node to the next node in a packet's route. The flit occupies the buffer until it is forwarded to its next hop; then a credit is sent along the reverse channel to the current node, which takes a wire propagation time $T_{w,credit}$. The credit is processed by the current node, experiencing a credit pipeline delay, and is then ready for use by another flit. Once this flit receives the credit, it must traverse the flit pipeline before accessing the channel. Finally, the flit propagates across the channel in $T_{w,data}$ and occupies the buffer. As shown in the figure, the actual buffer usage represents a small fraction of the overall time. The remaining time required to recycle the credit and issue another flit to occupy the buffer is called the *turnaround time*. Lower buffer utilization reduces network throughput because fewer buffers are available for bypassing blocked messages and absorbing traffic variations. Flit-reservation flow control can reduce turnaround time to zero and hide the flit pipeline delay in a practical implementation.

Flit-reservation hides the overhead associated with a pipelined router implementation by separating the control and data networks. Control flits race ahead of the data flits to reserve network resources. As the data flits arrive, they have already been allocated an outgoing virtual channel and can proceed with little overhead. Reservation

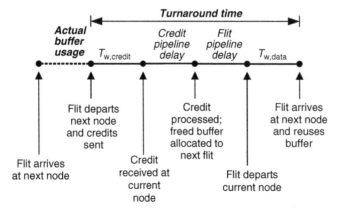

Figure 13.15 An illustration of realistic delays on buffer utilization in a typical wormhole router. As the buffer turnaround time increases, the fraction of time a buffer is actually in use decreases.

also streamlines the delivery of credits, allowing zero turnaround time for buffers. Of course, it is not always possible to reserve resources in advance, especially in the case of heavy congestion. In these situations, data flits simply wait at the router until resources have been reserved — the same behavior of a standard wormhole router.

A flit-reservation packet is shown in Figure 13.16. As shown, the control portion of the flit is separated from the data portion. The *control head flit* holds the same information as a typical virtual channel flow control head flit: a type field (control head), a virtual channel, and routing information. An additional field t_{d0} shows the time offset to the first data flit. This data offset couples control flits to their associated data flits. So, for example, when a control flit arrives at a node at time $t = 3$ with $t_{d0} = 2$, the router knows that the associated data flit will be arriving on the data network at time $t = 3 + t_{d0} = 5$. By knowing the future arrival time of the data, the router can start reserving resources to the data before its actual arrival. Since the router knows when a particular data flit is arriving, no control fields are required in the data flits themselves.

Since packets can contain an arbitrary number of data flits and the control head flit has only a limited number of data offset fields (one, in this case), additional *control body flits* can follow the head flit. The control body flits are analogous to body flits in a typical wormhole router, and they contain additional data offset fields. In Figure 13.16, for example, the control body flits have two additional data offset fields, allowing two data flits per control body flit.

13.4.1 **A Flit-Reservation Router**

A flit-reservation router is shown in Figure 13.17. Before we describe the detailed operation of the modules in the router, the basic operation of the router is discussed. Control flits arrive in the router and are immediately passed to the *routing logic*. When a control head flit arrives at the routing logic, its output virtual channel (next hop) is computed and associated with the incoming virtual channel. As subsequent control body flits arrive for that virtual channel, they are marked with the same output channel as the control head flit. All control flits then proceed through a switch to

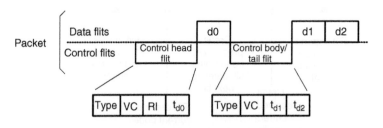

Figure 13.16 A flit-reservation packet showing the separation between control flits and data flit. Control flits are associated with particular data flits by an offset in their arrival times (e.g., for a control head flit arriving a time t, its associated data flit d0 arrives at time $t + t_{d0}$).

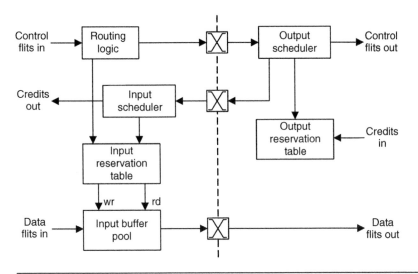

Figure 13.17 A flit-reservation flow control architecture. Only a single input and single output are shown for simplicity (per input and per output structures separated by the dashed line).

their output port. Up to this point, control flits are processed identically to flits in a standard virtual channel router.

After a control flit reaches its output port, it passes through the *output scheduler* before being forwarded to its next hop. The output scheduler is responsible for reserving buffer space at the next hop for each of a control flit's associated data flits, which are given by the data offset fields. A schedule of the next router's buffer usage is kept in the *output reservation table*, which is continually updated by incoming credits. Once the output scheduler has allocated all of a control flit's associated data flits and marked those reservations in the output reservation table, the control flit is forwarded to its next hop.

The *input reservation table* provides the connection between control flits and their data flits within the router. As a control flit passes through the routing logic, its destination is also marked in the input reservation table. Additionally, each of the data offset fields of the control flit is marked in this table. Once a flit has been scheduled for departure by the output scheduler, this information is passed back to the input scheduler associated with the input the flit arrived on. This allows the input scheduler to know on which cycles flits have been scheduled to depart, indicating when their buffers will become free.

13.4.2 **Output Scheduling**

The output scheduler is responsible for determining the future departure time of each data flit associated with each control flit. For a data flit to depart, the physical output channel must be reserved and the output scheduler must also ensure that

Output channel	Time	8	9	10	11	12	13	14	15	16	17
East channel	Channel busy			✕							
	Free buffers on next node	2	1	1	0	1	2	3	4	4	4

(a)

Output channel	Time	8	9	10	11	12	13	14	15	16	17
East channel	Channel busy			✕		✕					
	Free buffers on next node	2	1	1	0	0	1	2	3	3	3

(b)

Figure 13.18 The output reservation table. (a) Example state. (b) Updated state of the table after scheduling a new data flit arriving on cycle 9. Cycle 12 is the first available departure time, since the channel is busy on cycle 10 and no buffers are free on cycle 11.

there will be sufficient buffer space in the next router to store the data flit. To track the future output channel utilization and availability of buffers at the next hop, the output scheduler maintains the output reservation table (Figure 13.18). This table maintains both channel usage and buffer availability counts for a number of cycles in the future.

An example state of the output reservation table is shown in Figure 13.18(a). A new control flit arrives at time $t = 0$ and the arrival time of its data flit(s) are determined. In this example, a single new data flit is indicated with $t_{d0} = 9$.[10] Therefore, the data flit will arrive at cycle $t + t_{d0} = 9$ destined to the east output. This information is forwarded to the output scheduler and the input reservation table. (The input reservation table is discussed in the next section.)

The output scheduler attempts to find an available departure time for the new data flit. Cycle 10 is unavailable because the output channel is already marked as busy. Cycle 11 is also unusable, but in this case, it is because there will be no free buffers on the next hop. The first available cycle is 12, so the output scheduler updates the output reservation table to indicate the channel as busy during cycle 12 (Figure 13.18[b]). Also, the flit occupies a buffer at the next node, so the number of free buffers is decremented from cycle 12 onwards.[11] This buffer will eventually be added back to output reservation table count once the credit from the next router returns, which is also discussed in the next section.

---

10.   For simplicity, we consider the wire propagation times to be zero in this example.
11.   Since the output scheduler does not know the departure time of the flit from the next router's buffer, the buffer is initially assumed to be occupied until further notice. For this same reason, the data flit in the example could not depart on cycle 10 even if the output channel was free because there must be a free buffer for *every* time step after that point.

Now that the output reservation is complete, the reservation information is passed to an input scheduler. Once all the data flits associated with a control flit are reserved, the control flit is forwarded to the next hop. However, the relative time of its data flits may have changed in the output scheduling process, so its data offset field(s) must be updated. If the control flit in our example departs the router and then arrives at the next router at $t = 2$, then its t_{d0} field is changed to 10. This is because the data flit has been delayed an additional cycle relative to the control flit by the output scheduler and its departure time is cycle 12 — therefore, $t_{d0} = 12 - 2 = 10$.

13.4.3 Input Scheduling

The input scheduler and the input reservation table organize the movement of data flits through the flit-reservation router. By accessing the data offset fields of incoming control flits, the routing logic marks arrival times of data flits and their destination port in the input reservation table (Figure 13.19). Once the data flit has been reserved an output time by the output scheduler, this reservation information is passed back to the input scheduler. The input scheduler updates the input reservation table with the departure time and output port of the corresponding data flit. Also, the input scheduler sends a credit to the previous router to indicate when the data flit's buffer is available again.

Continuing the example from the previous section, when the arrival time of the new data flit (cycle 9) is decoded in the routing logic, this information is passed to the input reservation table (Figure 13.19). The data flit arriving during cycle 9 is latched at the beginning of cycle 10, so the arrival is marked in the table along with its output channel. The remaining entries of the table remain undetermined at this point.

When the reservation information for the new data flit returns from the output scheduler, the departure time of the flit is stored in the input reservation table. At this time, a credit is sent to the previous node, indicating this departure on cycle 12. Although flit reservation guarantees the arriving data flit can be buffered, the actual buffer location is not allocated until one cycle before the data flit arrives. After allocation, the buffer in and buffer out fields in the table are written.

Input channel / Time		8	9	10	11	12	13	14	15	16	17
West channel	Flit arriving?			☒							
	Buffer in			5							
	Departure time			+2							
	Buffer out					5					
	Output channel					E					

Figure 13.19 The input reservation table. A flit arrives from the west channel and is stored in buffer 5 of the input buffer pool in cycle 10. Two cycles later, the flit departs on the east channel of the router.

13.5 **Bibliographic Notes**

Cut-through flow control was introduced by Kermani and Kleinrock [94]. Cut-through is an attractive solution for networks with small packets and has been used in recent networks such as the Alpha 21364 [131]. Dally and Seitz introduced wormhole flow control, which was first implemented on the Torus Routing Chip [56]. These ideas were further refined by Dally, who also introduced virtual-channel flow control [44, 47]. Both wormhole and virtual-channel flow control have appeared in numerous networks — the Cray T3D [95] and T3E [162], the SGI Spider [69], and the IBM SP networks [178, 176], for example. Flit-reservation flow control was developed by Peh and Dally [145].

13.6 **Exercises**

13.1 *Packet-buffer, flit-channel flow control.* Consider a flow control method that allocates buffers to packets, like cut-through flow control, but allocates channel bandwidth to flits, like virtual-channel flow control. What are the advantages and disadvantages of this approach? Construct an example and draw a time-space diagram for a case in which this flow control scheme gives a different ordering of events than pure cut-through flow control.

13.2 *Overhead of credit-based flow control.* Compute the overhead of credit-based flow control as a fraction of the packet length L. Assume a flit size of L_f and V virtual channels.

13.3 *Flow control for a shared buffer pool.* Consider a router where the flit buffers are shared between all the inputs of the switch. That is, at any point in time, flits stored at each input may occupy a portion of this shared buffer. To ensure fairness, some fraction of this shared buffer is statically assigned to particular inputs, but the remaining fraction is partitioned dynamically. Describe a buffer management technique to communicate the state of this buffer to upstream nodes. Does the introduction of a shared buffer increase the signaling overhead?

13.4 *A single-ported input buffer pool.* Consider a flit-reservation router whose input buffer pool is single-ported (one read *or* write per cycle). Describe how to modify the router described in Section 13.4 and what additional state, if any, is required in the router.

13.5 *Synchronization issues in flit-reservation.* The flit data in a flit-reservation router is indentified only by its arrival relative to its control information. However, in a *plesiochronous* system, the clocks between routers may occasionally "slip," causing the insertion of an extra cycle of transmission delay. If ignored, these extra cycles would change what data was associated with a particular control flit. Suggest a simple way to solve this problem.

CHAPTER 14

Deadlock and Livelock

Deadlock occurs in an interconnection network when a group of *agents*, usually packets, are unable to make progress because they are waiting on one another to release resources, usually buffers or channels. If a sequence of waiting agents forms a cycle, the network is deadlocked. As a simple example, consider the situation shown in Figure 14.1. Connections A and B traversing a circuit-switched network each hold two channels, but cannot proceed further until they acquire a third channel, currently held by the other connection. However, neither connection can release the channel needed by the other until it completes its transmission. The connections are deadlocked and will remain in this state until some intervention. Deadlock can occur over various resources. In this example, the resource is a physical channel. It can also be a virtual channel or a shared packet buffer.

Deadlock is catastrophic to a network. After a few resources are occupied by deadlocked packets, other packets block on these resources, paralyzing network operation. To prevent this situation, networks must either use deadlock avoidance (methods that guarantee that a network cannot deadlock) or deadlock recovery (in which deadlock is detected and corrected). Almost all modern networks use deadlock avoidance, usually by imposing an order on the resources in question and insisting that packets acquire these resources in order.

A closely related network pathology is livelock. In livelock, packets continue to move through the network, but they do not make progress toward their destinations. This becomes a concern, for example, when packets are allowed to take non-minimal routes through the network — either a deterministic or probabilistic guarantee must ensure that the number of misroutes of a packet away from its destination is limited.

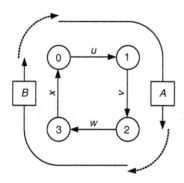

Figure 14.1 Deadlock in a circuit-switched network. Connection A holds channels u and v but cannot make progress until it acquires channel w. At the same time, connection B holds channels w and x but cannot make progress until it acquires channel u. Neither connection will release the channel needed by the other. Hence, they are deadlocked.

14.1 **Deadlock**

14.1.1 **Agents and Resources**

The agents and resources that are involved in deadlock differ depending on the type of flow control employed, as shown in Table 14.1. For circuit switching, as shown in Figure 14.1, the agents are connections and the resources are physical channels. As a connection is set up, it acquires physical channels and will not release any of them until after the connection is completed. Each connection may indefinitely hold *multiple* channels, all of the channels along the path from source to destination. With a packet-buffer flow control method (like store-and-forward or virtual cut-through), the agents are packets and the resources are packet buffers. As the head of the packet propagates through the network, it must acquire a packet buffer at each node. At any given point in time, a packet may indefinitely hold only a single packet buffer. Each time the packet acquires a new packet buffer, it releases the old packet buffer a short, bounded time later. With a flit-buffer flow control method, the agents are again packets, but the resources are *virtual channels*. As the head of

Table 14.1 Agents and resources causing deadlock for different flow control methods.

Flow Control	Agent	Resource	Cardinality
Circuit switching	Connection	Physical channel	Multiple
Packet-buffer	Packet	Packet buffer	Single
Flit-buffer	Packet	Virtual channel	Multiple

the packet advances, it allocates a virtual channel (control state and a number of flit buffers) at each node. It may hold several virtual channels indefinitely, since if the packet blocks, the buffer space in each virtual channel is not sufficient to hold the entire packet.

14.1.2 **Wait-For and Holds Relations**

The agents and resources are related by wait-for and holds relations. Consider, for example, the case of Figure 14.1. The wait-for and holds relationships for this case are illustrated in Figure 14.2(a). Connection A holds (dotted arrows) channels u and v and waits for (solid arrow) w. Similarly connection B also holds two channels and waits for a third. If an agent holds a resource, then that resource is waiting on the agent to release it. Thus, each *holds* relation induces a *wait-for* relation in the opposite direction: holds$(a, b) \Rightarrow$ waitfor(b, a). Redrawing the holds edges as wait-for edges in the opposite direction gives the wait-for graph of Figure 14.2(b). The cycle in this graph (shaded) shows that the configuration is deadlocked.

The cycle of Figure 14.2 consists of alternating edges between agents and resources. The edges from an agent to a resource indicate that the agent is waiting on that resource. The edges in the opposite direction indicate that the resource is held by the indicated agent (and, hence, is waiting on that agent to be released).

Such a cycle will exist, and hence deadlock will occur, when:

1. Agents hold and do not release a resource while waiting for access to another.

2. A cycle exists between waiting agents, such that there exists a set of agents $A_0, \ldots, A_{n-1}$, where agent A_i holds resource R_i while waiting on resource $R_{(i+1 \bmod n)}$ for $i = 0, \ldots, n-1$.

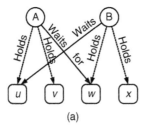 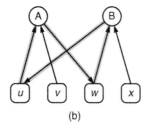

Figure 14.2 Wait-for and holds relationships for the deadlock example of Figure 14.1. (a) Connections A and B each hold two channels (dotted arrows) and wait for a third (solid arrows). (b) Each holds relation implies a wait-for relation in the opposite direction. Redrawing the graph using only wait-for relations reveals the wait-for cycle causing deadlock (shaded).

14.1.3 **Resource Dependences**

For two resources R_i and R_{i+1} to be two edges apart in the wait-for graph, it must be possible for the agent A_i holding resource R_i to wait indefinitely on resource R_{i+1}. Whenever it is possible for an agent holding R_i to wait on R_{i+1}, we say that a *resource dependence* exists from R_i to R_{i+1} and denote this as $R_i \succ R_{i+1}$. In the example of Figure 14.1, we have $u \succ v \succ w \succ x \succ u$. Note that resource dependence ($\succ$) is a transitive relation. If $a \succ b$ and $b \succ c$, then $a \succ c$.

This cycle of resource dependences is illustrated in the resource (channel) dependence graph of Figure 14.3. This graph has a vertex for each resource (in this case, each channel) and edges between the vertices denote dependences — for example, to denote that $u \succ v$ we draw an edge from u to v.

Because the example of Figure 14.1 deals with circuit switching, the resources are physical channels and our resource dependence graph is a physical channel dependence graph. With packet-buffer flow control, we would use a packet-buffer dependence graph. Similarly, with flit-buffer flow control, we would use a virtual-channel dependence graph.

A cycle of resource dependences in a resource dependence graph (as in Figure 14.3) indicates that it is possible for a deadlock to occur. For a deadlock to actually occur requires that agents (connections) actually acquire some resource and wait on others in a manner that generates a cycle in the wait-for graph. A cycle in a resource dependence graph is a necessary but not sufficient condition for deadlock. A common strategy to avoid deadlock is to remove all cycles from the resource dependence graph. This makes it impossible to form a cycle in the wait-for graph, and thus impossible to deadlock the network.

14.1.4 **Some Examples**

Consider the four-node ring network of Figure 14.1 but using packet-buffer flow control with a single packet buffer per node. In this case, the agents are packets and the resources are packet buffers. The packet-buffer dependence graph for this situation is shown in Figure 14.4 (a). A packet resident in the buffer on node 0 (B_0) will not release this buffer until it acquires B_1, so we have $B_0 \succ B_1$ and the dependence graph has an edge between these two buffers.

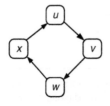

Figure 14.3   Resource (channel) dependence graph for the example of Figure 14.1

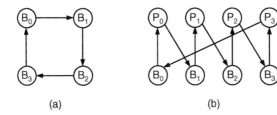

Figure 14.4 Dependence and wait-for graphs for packet-buffer flow control. (a) Resource (packet buffer) dependence graph for the network of Figure 14.1 using packet-buffer flow control with a single packet buffer per node. (b) Wait-for graph for a deadlocked configuration with four packets holding four packet buffers.

The cycle in the packet buffer dependence graph indicates the potential for deadlock in this network. To actually construct a deadlock situation in this case requires four packets, $P_0, \ldots, P_3$, each holding one buffer and waiting for the next. The wait-for graph for this deadlocked configuration is shown in Figure 14.4(b). Each buffer B_i waits on the packet that holds it, P_i, to release it. Each packet in turn waits on buffer B_{i+1} to advance around the ring. We cannot construct a cycle in this wait-for graph with fewer than four packets, since the cycle in the buffer dependence graph is of length four and, with packet-buffer flow control, each packet can hold only a single buffer at a time.

Now consider the same four-node ring network, but using flit-buffer flow control with a two virtual channels for each physical channel. We assume that a packet in either virtual channel of one physical channel can choose either virtual channel of the next physical channel to wait on. Once a packet has chosen one of the virtual channels, it will continue to wait on this virtual channel, even if the other becomes free.[1] The virtual channel dependence graph for this case is shown in Figure 14.5(a). Because a packet holding either of the virtual channels for one link can wait on either of the virtual channels for the next link, there are edges between all adjacent channels in this graph.

A wait-for graph showing a deadlocked configuration of this flit-buffer network is shown in Figure 14.5(b). The situation is analogous to that shown in Figure 14.2, but with packets and virtual channels instead of connections and physical channels. Packet P_0 holds virtual channels $u0$ and $v0$ and is waiting for $w0$. At the same time, P_1 holds $w0$ and $x0$ and is waiting for $v0$. The "1" virtual channels are not used at all. If packet P_0 were allowed to use either w_0 or w_1, this configuration would not represent a deadlock. A deadlocked configuration of this network when packets are allowed to use any unclaimed virtual channel at each hop requires four packets. Generating this configuration is left as Exercise 14.1.

---

1. We leave the construction of a deadlocked configuration for the case where the packet takes the first free virtual channel as Exercise 14.1 .

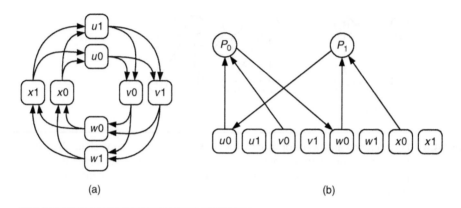

(a)   (b)

Figure 14.5   Dependence and wait-for graphs for flit-buffer flow control. (a) Resource (virtual channel) dependence graph for the network of Figure 14.1 using flit-buffer flow control with two virtual channels per physical channel. (b) Wait-for graph for a deadlocked configuration with two packets holding two virtual channels each.

14.1.5 **High-Level (Protocol) Deadlock**

Deadlock may be caused by dependences external to the network. For example, consider the case shown in Figure 14.6. The top network channel is waiting for the server to remove a request packet from the network. The server in turn has limited buffering and thus cannot accept the request packet until the lower channel accepts a reply packet from the server's output buffer. In effect, the upper channel is waiting on the lower channel due to the external sever. This edge of the wait-for graph is due not to the network itself, but to the server. Deadlock caused by wait-for loops that include such external edges are often called high-level deadlock or protocol deadlock.

In a shared-memory multiprocessor, for example, such an external wait-for edge may be caused by the *memory server* at each node, which accepts memory read and write request packets, reads or writes the local memory as requested, and sends a response packet back to the requesting node. If the server has limited internal

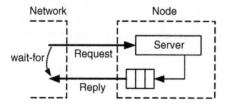

Figure 14.6   Implicit resource dependence in a request-reply system.

buffering, the situation is exactly as depicted in Figure 14.6 and the channel into the server may have to wait on the channel out of the server. The effect of these external wait-for edges can be eliminated by using different logical networks (employing disjoint resource sets — for example, separate virtual channels or packet buffers) to handle requests and replies. The situation can become even more complex in cache-coherent, shared memory machines where a single transaction may traverse two or three servers (directory, current owner, directory) in sequence before returning the final reply. Here separate logical networks are often employed at each step of the transaction.Using these separate logical networks to avoid protocol deadlock is a special case of resource ordering, as described in Section 14.2.

14.2 **Deadlock Avoidance**

Deadlock can be avoided by eliminating cycles in the resource dependence graph. This can be accomplished by imposing a partial order on the resources and then insisting that an agent allocate resources in ascending order. Deadlocks are therefore avoided because any cycle must contain at least one agent holding a higher-numbered resource waiting for a lower-numbered resource, and this is not allowed by the ordered allocation. While a partial order suffices to eliminate cycles, and hence deadlocks, for simplicity we often impose a total order on the resources by numbering them.

While all deadlock avoidance techniques use some form of resource ordering, they differ in how the restrictions imposed by this resource ordering affect routing. With some approaches, resources can be allocated in order with no restrictions on routing. In other approaches, the number of required resources is reduced at the expense of disallowing some routes that would otherwise violate resource ordering.

With packet-buffer flow control, we have the advantage that there are typically many packet buffers associated with each node. Similarly, there are typically many virtual channels associated with each physical channel in systems using flit-buffer flow control. With multiple resources per physical unit, we can achieve our ordering by assigning different resources on the same physical unit (for example, different packet buffers on a node) to different positions in the order. With circuit switching, the resources are the physical units (channels) and thus each channel can appear only at a single point in the ordering. Thus, to order resources with circuit switching, we have no alternative but to restrict routing.

14.2.1 **Resource Classes**

Distance Classes: One approach to ordering resources (virtual channels or packet buffers) is to group the resources into numbered classes and restrict allocation of resources so that packets acquire resources from classes in ascending order. One method of enforcing ascending resource allocation is to require a packet at distance i from its source to allocate a resource from class i. At the source, we inject packets into resource class 0. At each hop, the packet acquires a resource of the next highest class. With this system, a packet holding a packet-buffer from class i can wait on a buffer only in class $i + 1$ (Figure 14.7).[2] Similarly, a packet holding a virtual channel in class i can only wait on virtual channels in higher numbered classes. Packets only travel *uphill* in terms of resource classes as they travel through the network. Because a packet holding a resource (packet-buffer or virtual channel) from class i can never wait, directly or indirectly, on a resource in the same or lower numbered class, no cycle in the resource dependence graph exists and hence deadlock cannot occur.

As a concrete example of distance classes, Figure 14.8 shows a four-node ring network using buffer classes based on distance. Each node i has four buffers B_{ji}, each of which holds packets that have traveled j hops so far. Packets in buffer B_{ji} are either delivered to the local node i or forwarded to buffer $B_{j+1,i+1}$ on node i. This buffer structure leads to an acyclic buffer dependence graph that consists of four spirals, and hence avoids deadlock.

To enforce the uphill-only resource allocation rule, each packet needs to remember its previous resource class when it allocates its next resource. Thus, for packet-buffer flow control with distance classes, the routing relation takes the form:

$$R : Q \times N \rightarrow Q$$

where Q is the set of all buffer classes in the network. A similar relation is used to allocate virtual channels in a network with flit-buffer flow control. This hop-by-hop routing relation allows us to express the uphill use of buffer classes.

Distance classes provide a very general way to order resources in any topology. However, they do so at considerable expense — they require a number of packet buffers (or virtual channels) proportional to the diameter of the network. For some

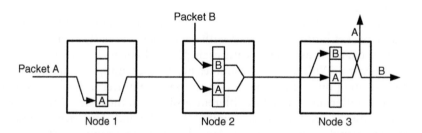

Figure 14.7   An example of routing packets through several buffer classes. Each node contains five buffer classes with the lowest class at the bottom of the buffers and the highest class at the top. As packets A and B progress through the network, their buffer classes increase.

2.   We could allow a packet to wait for any buffer class greater than the one it currently holds. If we do this, however, we cannot guarantee that it will not run out of classes by skipping too many on the way up the hill.

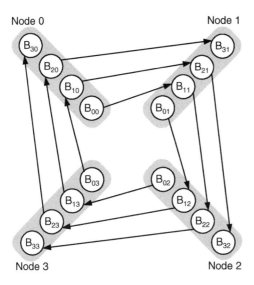

Figure 14.8   Distance classes applied to a four-node ring network. Each node i has four classes, with buffer B_{ji} handling traffic at node i that has taken j hops toward its destination.

networks, we can take advantage of the topology to reduce the number of buffer classes significantly. For example, in a ring network, we can order resources by providing just two classes of resources.[3]

Figure 14.9 shows how buffer dependences can be made acyclic in a ring by using dateline buffer classes. Each node i has two buffers, a "0" buffer B_{0i} and a "1" buffer B_{1i}. A packet injected at source node s is initially placed in buffer B_{0s} and remains in the "0" buffers until it reaches the *dateline* between nodes 3 and 0. After crossing the dateline, the packet is placed in "1" buffer B_{10} and remains in the "1" buffer until it reaches its destination. Dividing the use of the two buffer classes based on whether or not a packet has passed the dateline in effect converts the cycle of buffer dependences into an acyclic spiral. Hence deadlock is avoided.

Dateline classes can also be applied to flit-buffer flow control. Figure 14.10 shows the virtual channel dependence graph for an application of dateline classes to a four-node ring with two virtual channels per physical channel. Each physical channel c has two virtual channels $c0$ and $c1$. All packets start by using the "0" channels and switch to the "1" channels only when they cross the dateline at node 3. To restrict the selection of output virtual channel based on input virtual channel, this approach requires that the routing function be of the form

$$R : C \times N \rightarrow C$$

---

3.   We will see in Section 14.2.3 how to extend this method to handle arbitrary torus networks.

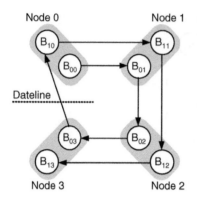

Figure 14.9   With dateline buffer classes, each node *i* in a ring has two buffers B_{1i} and B_{0i}. A packet injected on node *s* starts in buffer B_{s0} and remains in the "0" buffers until it reaches the *dateline* between nodes 3 and 0. After crossing the dateline, the packet is placed in buffer B_{10} and remains in the "1" buffers until it reaches its destination.

where C here represents the set of *virtual* channels. Restricting the selection of virtual channels here takes the cyclic channel dependence graph of Figure 14.5 and makes it acyclic by removing a number of edges.

Overlapping Resource Classes: Restricting the use of resource classes, either according to distance or datelines, while making the resource dependence graph acyclic, can result in significant load imbalance. In Figure 14.10, for example, under uniform traffic, more packets will use virtual channel $v0$ (5 routes) than will use $v1$ (1 route). This load imbalance can adversely affect performance because some resources may be left idle, while others are oversubscribed. Similarly, with distance classes, not every

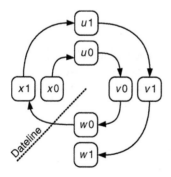

Figure 14.10   Virtual channels divided into dateline classes. Each physical channel *c* on a four-node ring is divided into two virtual channels *c0* and *c1*. All packets start routing on the "0" virtual channels and switch to the "1" virtual channels when they cross the dateline at node 3.

route will use the maximum number of hops, so the higher numbered classes will tend to have lower utilization.

One approach to reducing load imbalance is to overlap buffer classes. For example, with dateline classes, suppose we have 32 packet buffers. We could assign 16 buffers each to classes "0" and "1" as illustrated in Figure 14.11(a). However, a better approach is to assign one buffer each for exclusive use by each class, and allow the remaining 30 buffers to be used by either class as shown in Figure 14.11(b). This approach reduces load imbalance by allowing most of the buffers to be used by packets requiring either class.

It is important when overlapping classes, however, to never allow a packet to wait on a busy resource in the overlap region. That is, the packet cannot select a busy buffer that belongs to both classes — say, B_{11} — and then wait on B_{11}. If it does so, it might be waiting on a packet of the other class and hence cause a deadlock. To avoid deadlock with overlapped classes, a packet must not select a particular buffer to wait on until an idle buffer of the appropriate class is available. By waiting on the class, the packet is waiting for *any* buffer in the class to become idle and thus will eventually be satisfied by the one *exclusive* buffer in the class. If a *non-exclusive* buffer becomes available sooner, that can boost performance, but it doesn't alter the correctness of waiting for the exclusive buffer.

14.2.2 Restricted Physical Routes

Although structuring the resources of a network into classes allows us to create a deadlock-free network, this can, in some cases, require a large number of resources

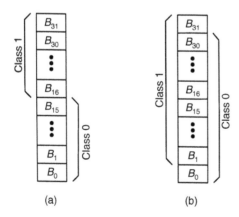

(a)   (b)

Figure 14.11 Two methods to partition 32 buffers into 2 classes. (a) Sixteen buffers are assigned to each class with no overlap. (b) 31 buffers are assigned to each class with an overlap of 30 buffers. As long as the overlap is not complete, the classes are still independent for purposes of deadlock avoidance.

to ensure no cyclic resource dependences. An alternative to structuring the resources to accommodate all possible routes is to restrict the routing function. Placing appropriate restrictions on routing can remove enough dependences between resources so that the resulting dependence graph is acyclic without requiring a large number of resource classes.

Dimension Order (e-cube) Routing: One of the simplest restrictions on routing to guarantee deadlock freedom is to employ dimension-order routing in k-ary n-meshes. (See Section 8.4.2.) For example, consider a 2-D mesh. Within the first dimension x, a packet traveling in the $+x$ direction can only wait on a channel in the $+x$, $+y$, and $-y$ directions. Similarly, an $-x$ packet waits only on the $-x$, $+y$, and $-y$ directions. In the second dimension, a $+y$ packet can only wait on other $+y$ channels and a $-y$ packet waits only on $-y$. These relationships can be used to enumerate the channels of the network, guaranteeing freedom from deadlock.

An example enumeration for dimension-order routing is shown in Figure 14.12 for a 3×3 mesh. Right-going channels are numbered first, so that their values increase to the right. Then the left, up, and down channels are numbered, respectively. Now, any dimension-order route through the network follows increasingly numbered channels. Similar enumerations work for an arbitrary number of dimensions once a fixed dimension order is chosen.

The Turn Model: While dimension-order routing provides a way of restricting the routing algorithm to prevent cyclic dependences in k-ary n-mesh networks, a more general framework for restricting routing algorithms in mesh networks is the turn model. In the turn model, possible deadlock cycles are defined in terms of the particular turns needed to create them. We will consider this model in two dimensions, although it can be extended to an arbitrary number of dimensions. As shown in

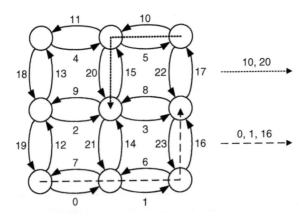

Figure 14.12   Enumeration of a 3×3 mesh in dimension-order routing. The channel order for two routes is also shown.

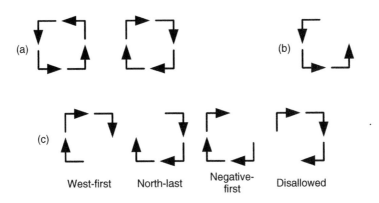

Figure 14.13 (a)

(b)

(c)

West-first   North-last   Negative-first   Disallowed

Figure 14.13 The turn model for a two-dimension mesh network. (a) The two abstract cycles. (b) If the North to West is eliminated (c) three possible routing algorithms can be created by eliminating another turn in the other abstract cycle.

Figure 14.13(a), there are eight turns in a 2-D routing algorithm ($+x$ to $+y$, $+x$ to $-y$, $-x$ to $+y$, and so on), which can be combined to create two abstract cycles. By inspection, at least one turn from each of these two cycles must be eliminated to avoid deadlock. Dimension-order routing, for example, eliminates two turns in each of the cycles — those from any y dimension to any x dimension.

We can explore a set of routing functions that is less restrictive than dimension-order routing by first eliminating one of the turns from the first abstract cycle. Figure 14.13(b) shows the elimination of the North to West turn (that is, the turn from $+y$ to $-x$). Combining this with a turn elimination in the second cycle yields three deadlock-free routing algorithms, as illustrated in Figure 14.13(c). The fourth possible elimination is *not* deadlock-free, as explored in Exercise 14.3.

Each of the three choices yields a different routing algorithm. When the south-to-west turn is eliminated, the west-first algorithm is generated. In west-first routing, a packet must make all of its west hops before moving in any other direction. After it turns from the west direction, it may route in any other direction except west. Removing the north-to-east turn results in north-last routing. In north-last, a packet may move freely between the directions except north. Once the packet turns north, it must continue in the north direction until its destination. Finally, eliminating the east-to-south turn gives negative-first routing. In negative-first routing, a packet must move completely in the negative directions (south and west) before changing to the positive directions (north and east). Once in the positive directions, the packet stays there until it reaches its destination.

At first, it may appear that the turn model is doing something different — breaking deadlock by restricting turns rather than by imposing an ordering of resources. This is not the case. Instead, by restricting turns, the turn model imposes a total order on the network channels, but a different one than that induced by dimension-order routing. Figure 14.14 shows the order imposed by the west-first turn model on the channels of a 3×3 mesh network. The west-going channels are numbered first.

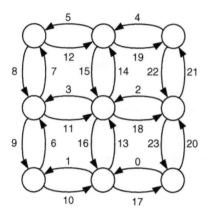

Figure 14.14   Channel ordering induced by the west-first turn model.

The remaining channels are then numbered from west to east, with the north/south channels in each column numbered in direction order.

The turn-model and dimension-order routing give two possibilities for restricting a routing algorithm so that it is deadlock-free. However, there are still drawbacks to these techniques. Restricting the routing function reduces the path diversity of the network, which can diminish a network's performance and fault tolerance. In the case of dimension-order routing, the diversity is reduced to zero. Additionally, these techniques cannot remove the channel cycles inherent in topologies such as the torus.

14.2.3   Hybrid Deadlock Avoidance

In the previous sections, we saw that deadlock-free networks can be designed by either splitting network resources and enumerating these resources or appropriately restricting the paths packets could take from source to destination. However, both of these approaches have several drawbacks. The buffer-class approach required a large amount of buffering per node, and restricting the routing algorithm led to reduced path diversity and could not be employed on all topologies. As with many design problems, the most practical solutions to deadlock avoidance combine features of both approaches.

Torus Routing: We can use dimension-order routing to route deadlock free in a torus by applying dateline classes to each dimension. In effect, the dateline classes turn the torus into a mesh — from the point of view of resource dependence — and dimension-order routing routes deadlock-free in the resulting mesh. For example, with flit-buffer flow control, we provision two classes of virtual channels for each physical channel. As a packet is injected into the network, it uses virtual channel 0.

If the packet crosses a predefined *dateline* for each dimension, it is moved to virtual channel 1. When a packet is done routing in a particular dimension, it always enters the next dimension using virtual channel 0.[4] This continues until the packet is ejected.

The technique of breaking dependences and then reconnecting paths with virtual channels generalizes to many situations. Revisiting the protocol deadlock described in Section 14.1.5, the dependence between requests and replies may lead to potential deadlocks. A common technique used to work around this problem is to simply assign requests and replies to distinct virtual channels. Now, as long as the underlying algorithm used to route requests and replies is deadlock-free, the request-reply protocol cannot introduce deadlock.

Planar-Adaptive Routing: Another routing technique that combines virtual channels with restricted physical routes is *planar-adaptive routing*. In planar-adaptive routing, a limited amount of adaptivity is allowed in the routing function while still avoiding cycles in the channel dependence graph. The algorithm starts by defining adaptive planes in a k-ary n-mesh. An adaptive plane consists of two adjacent dimensions, i and $i + 1$. Within an adaptive plane, any minimal, adaptive routing algorithm can be used. By limiting the size of a plane to two-dimensions, the number of virtual channels required for deadlock avoidance is independent of the size of the network and the number of dimensions.

This is illustrated in Figure 14.15 for a k-ary 3-mesh. A minimal, adaptive algorithm could use any path within the minimal sub-cube for routing. However, as n grows, this requires an exponential number of virtual channels to avoid deadlock. Planar-adaptive routing allows adaptive routing within the plane A_0 (defined by dimensions 0 and 1) followed by adaptive routing within the plane A_1 (defined by dimensions 1 and 2). This restriction still allows for large path-diversity, but uses only a constant number of virtual channels.

For general planar-adaptive routing, each channel is divided into three virtual channels, denoted $d_{i,v}$, where i is the dimension the virtual channel is in and v is the

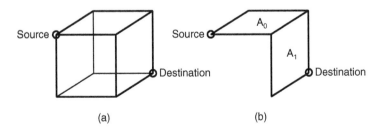

(a)   (b)

Figure 14.15   (a) The set of all minimal paths between a source and destination node in a k-ary 3-mesh (b) and the subset of paths allowed in planar-adaptive routing.

---

4.   This naive assignment creates imbalance in the utilization of the the VCs. More efficient assignments are explored in Exercise 14.6.

virtual channel ID. The adaptive plane A_i contains the virtual channels $d_{i,2}$, $d_{i+1,0}$, and $d_{i+1,1}$ for $i = 0, \ldots, n - 2$. Routing begins in A_i with $i = 0$. Any minimal, adaptive routing algorithm is used to route the packet until the ith dimension of the packet's location matches the ith dimension of its destination. Then i is incremented and routing in the next adaptive plane begins. These steps continue until routing in A_{n-2} is complete. It may be necessary to finish routing minimally in the nth dimension before the packet reaches its destination.

To ensure deadlock-free routing within the ith adaptive plane, the virtual channels within that plane are divided into increasing and decreasing subsets. The increasing subset is $d_{i,2+}$ and $d_{i+1,0}$ and the decreasing subset is $d_{i,2-}$ and $d_{i+1,1}$, where "+" and "−" denotes the channels that increase or decrease an address in a dimension, respectively. Packets that need to increase their address in the ith dimension route exclusively in the increasing network and vice versa. This approach is verified to be deadlock-free in Exercise 14.7.

14.3 Adaptive Routing

In the previous sections, we modified the network and routing algorithms to eliminate cyclic dependences in the resource dependence graph. Some of these techniques can naturally be expressed as adaptive routing algorithms. For example, a packet using west-first routing can move an arbitrary number of hops to the west before routing in other directions — this decision could be made adaptively based on network conditions. Some buffer-class approaches can also easily incorporate adaptively.

However, the focus of this section is on an important difference between adaptive and oblivious routing algorithms:[5] adaptive routing algorithms can have cycles in their resource dependence graphs while remaining deadlock-free. This result allows the design of deadlock-free networks without the significant limitations on path diversity necessary in the oblivious case.

14.3.1 Routing Subfunctions and Extended Dependences

The key idea behind maintaining deadlock freedom despite a cyclic channel dependence graph is to provide an escape path for every packet in a potential cycle. As long as the escape path is deadlock-free, packets can move more freely throughout the network, possibly creating cyclic channel dependences. However, the existence of the escape route ensures that if packets ever get into *trouble*, there still exists a deadlock-free path to their destination.

---

5.   Actually, some oblivious routing algorithms can have cyclic dependences and remain deadlock-free, but this is not true in general [160].

An Example: Consider, for example, a 2-D mesh network that uses flit-buffer flow control with two virtual channels per physical channel. We denote a channel by *xydv* — that is, its node, direction, and virtual channel class (for example, 10e0 is virtual channel class 0 in the east direction on node 10). Routing among the virtual channels is restricted by the following rules.

1. All routing is minimal.

2. A packet in virtual channel xyd1 is allowed to route to any virtual channel on the other end of the link — any direction, any virtual channel. For example, 00e1 may route to any of the eight virtual channels on node 10.

3. A packet in virtual channel xyd0 is allowed to route to virtual channel 1 of any direction on the other end of the link. For example, 00e0 may route to 10d1 for any of the four directions *d*.

4. A packet in virtual channel xyd0 is allowed to route in dimension order (*x* first, then *y*) to virtual channel 0 at the other end of the link. For example, 00e0 may route to 10e0 or 10n0 as well as the four "1" channels on node 10. Channel 00n0 may route only to 01n0 and the four "1" channels on node 01.

In short, routing is unrestricted as long as the source or destination virtual channel is from class "1". However, when both source and destination virtual channels are from class "0," routing must be in dimension order.

This set of routing rules guarantees deadlock freedom even though cycles exist in the virtual channel dependence graph. An example cycle is shown in Figure 14.16, which shows four nodes' worth of the virtual channel dependence graph for a network employing these routing rules. Only a portion of the dependence edges are shown to avoid cluttering the figure. A dependence cycle is shown that includes the virtual channels 00e0, 10n0, 11w1, and 01s1. Each of the four edges of this cycle are legal routes according to the four routing rules above.

Despite this dependence cycle, this routing arrangement is deadlock-free because escape paths exist. Suppose, for example, that packet *A* holds virtual channels 00e0 and 10n0 and packet *B* holds virtual channels 11w1 and 01s1. A deadlocked configuration would occur if *A* waits for 11w1 and *B* waits for 00e0. However, these packets need not wait on the cyclic resource because they have other options. Packet *A* can choose 11n0 instead of 11w1. As long as (1) at least one packet in the potential cycle has an acyclic option and (2) packets do not commit to waiting on a busy resource, a deadlocked configuration will not occur. In this example, we are guaranteed that an acyclic option exists for every packet in the cycle, because they can always revert to dimension-order routing on the "0" channels, which is guaranteed to be acyclic, as described in Section 14.2.2.

Indirect Dependences: Now consider our 2-D mesh example, but without routing restriction 1 — that is, where non-minimal routing is allowed. Without this restriction, the mesh is no longer guaranteed to be deadlock-free because indirect dependences may result in cycles along the escape channels.

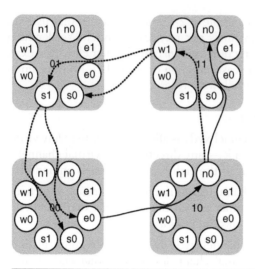

Figure 14.16 Example of cyclic virtual channel dependences with escape paths. The figure shows four nodes of the virtual channel dependence graph for a 2-D mesh network that must follow dimension order routing within the "0" channels. A cycle exists in the virtual-channel dependence graph including virtual-channels 00e0, 10n0, 11w1, and 01s1. No deadlock is possible, however, because at three points the packet is free to choose another virtual channel not on the cycle — the escape path.

For example, in Figure 14.16 suppose packet A is routing to node 12 (not shown) and currently holds channel 00e0. Packet B, also routing to node 12 has routed non-minimally so that it holds channels 10n0, 11w1, and 01s1. Even though the misrouting was performed on the "1" channels (11w1 and 01s1), it creates an *indirect dependence* on the "0" channels — from 10n0 to 00e0. This indirect dependence creates a cycle on the "0" channels. If all of the "1" channels become blocked due to cyclic routing, the "0" channels are no longer able to drain the network. Packet A is waiting on packet B to release 10n0 before it can make progress along its dimension-order route. At the same time, packet B is waiting for packet A to release 00e0 to it to route in dimension order. Hence the "0" channels, which are supposed to be a deadlock-free escape path, have become deadlocked due to an indirect dependence.

We can avoid indirect dependences by insisting that a packet that visits escape channel ("0" channel) a is not allowed to route to escape channel b via any channels, escape or otherwise, if $b \succ a$. One easy way to enforce this ordering for our example 2-D network is to insist that routing on the non-escape channels ("1" channels) is minimal.

Indirect dependence is a concern only for networks that use flit-buffer flow control. With packet-buffer flow control, a packet can hold only a single buffer at a time. Hence, it is impossible for a packet holding an escape buffer to wait on another escape buffer via some non-escape buffers. The initial escape buffer would be released as soon as the non-escape buffer is acquired.

Formal Development: Now that we've introduced the concepts of escape channels and indirect dependence, we can more formally describe deadlock avoidance for adaptive routing. If we have an adaptive routing *relation* $R : C \times N \rightarrow \mathcal{P}(C)$ over a set of virtual channels C, we can define a *routing subrelation* $R_1 \subseteq R$ over a subset of the virtual channels $C_1 \subseteq C$ so that R_1 is *connected* — that is, any source s can route a packet to any destination d using R_1. The entire routing relation R is deadlock-free if routing subrelation R_1 has no cycles in its *extended channel dependence graph*. The routes in R, but not in R_1, are referred to as the *complement subrelation* $R_C = R - R_1$. When the arguments of the routing relations are included $R(c, d)$, they indicate the packet's current channel c and the packet's destination d.

Returning to our example, R is the entire set of routing rules (rules 1 through 4, above), C is the set of all virtual channels (both the "0" and "1" channels, above), R_1 is the set of rules for the escape channels (rule 4, above), and C_1 is the set of escape virtual channels (the "0" channels, above). Also, R_C is the set of routes permitted by R but not by R_1 (rules 1 through 3, above).

The extended resource dependence graph for R_1 has as vertices the virtual channels in C_1 and as edges both the direct dependences and the indirect dependences between channels in C_1.[6] In short, this is our standard dependence graph extended with indirect dependences. We define the two types of edges in the extended dependence graph more precisely as:

direct dependence — This is the same channel dependence considered for deadlock avoidance with oblivious routing. If there exists a node x such that $c_j \in R_1(c_i, x)$, there is a channel dependence from c_i to c_j. That is, if a route in R_1 uses two channels c_i and c_j in sequence, then there is a direct dependence (or just a dependence) between the two channels, which we denote as $c_i \succ c_j$. For example, in Figure 14.16 there is a dependence from 00e0 to 10n0, since any packet routing in row 0 to column 1 will use these two virtual channels in sequence.

indirect dependence — An indirect dependence is created because our assumptions about flit-buffer flow control allow packets to occupy an arbitrary number of channels concurrently. In this situation, the dependence is created when a path to node x uses a channel $c_i \in R_1$, followed immediately by some number of channels $c_1, \ldots, c_m$ through R_C, and finally routing through a channel $c_j \in R_1$. For example, in Figure 14.16 with non-minimal routing there is an indirect dependence from 10n0 to 00e0, since a packet holding 10n0 can misroute in R_C via 11w1 and 01s1 to node 00 where R_1 dimension order routing requires channel 00e0. With non-minimal routing, this dependence would not exist because a packet that wants to use 00e0 would not be allowed to route in the other direction on 11w1. Implementations sometimes remove indirect dependences by simply disallowing a transition from R_1 to R_C — once a packet uses the routing sub-function R_1, it must continue to use R_1 for the rest of its route.

---

6. In some routing algorithms where R_1 and R_C share channels, *cross dependences* must also be added to the extended dependence graph. This sharing of channels almost never happens in practice, and we will not discuss cross dependences here.

A key result, proved by Duato [61], is that an acyclic, extended channel dependence graph implies a deadlock-free network.

THEOREM 14.1	An adaptive routing relation R for an interconnection network is deadlock-free if there exists a routing subrelation R_1 that is connected and has no cycles in its extended channel dependence graph.

While our discussion of extended dependences has focused on wormhole flow control, these ideas can be applied to store-and-forward and cut-through networks as well. The key differences are that the extended dependences are simpler and dependences are formed between buffers instead of channels. The simplification is because a packet cannot occupy an arbitrary number of channels when blocked, which eliminates the indirect dependences leaving only direct dependences. If the *buffer* dependence graph for R_1 of store-and-forward network is acyclic, the network is deadlock-free. The routing subrelation can be examined in isolation from R, allowing a very flexible definition of R as long as R_1 is deadlock-free.

14.3.2 Duato's Protocol for Deadlock-Free Adaptive Algorithms

While Duato's result provides the theoretical groundwork for designing deadlock-free adaptive routing functions, it does not immediately reveal how one might use the ideas to create a practical network design. In this section, we describe the most common technique for applying Duato's result, which allows fully adaptive, minimal routing.

Duato's protocol has three steps and can be used to help design deadlock-free, adaptive routing functions for both wormhole and store-and-forward networks:

1. The underlying network is designed to be deadlock-free. This can include the addition of virtual resources to break any cyclic dependences such as those in the torus.

2. Create a new virtual resource for each physical resource in the network. For wormhole networks, these resources are the (virtual) channels and for store-and-forward networks, buffers are the resources. Then, the original set of virtual resources uses the routing relation from Step 1. This is the escape relation, R_1. The new virtual resources use a routing relation R_C.

3. For packet-buffer flow control, there are no restrictions on R_C. For flit-buffer flow control, R_C must be constructed so that the extended dependence graph of R_1 remains acyclic.

The most common use of Duato's protocol creates a minimally adaptive routing algorithm with dimension-order routing as the routing subfunction R_1. For example, in a 3-ary 2-mesh with wormhole flow control, dimension-order routing is used for

R_1. Then R_C can contain all minimal paths and $R = R_1 \cup R_C$. The resulting extended dependence graph is acyclic (Figure 14.17).

14.4 **Deadlock Recovery**

Until now, we have focused on techniques to eliminate the conditions that cause deadlock. These methods required restricted routing functions or additional resources to break cyclic dependences. However, another approach to dealing with deadlock is not to avoid it, but rather to recover from it. Such an approach is attractive when the design cannot accommodate the additional resources or the performance degradation necessary to avoid deadlock. Of course, such techniques rely on the fact that deadlocks will be infrequent and that the average-case performance, rather than the worst-case performance, is considered important.

There are two key phases to any deadlock recovery algorithm: *detection* and *recovery*. In the detection phase, the network must realize it has reached a deadlocked configuration. Determining exactly whether the network is deadlocked requires finding a cycle in the resource wait-for graph. Solving this problem is difficult and costly, so most practical detection mechanisms are *conservative*. A conservative detection always correctly identifies a deadlock, but may also flag network conditions that are not deadlocks (false positives). The introduction of false positives decreases performance, but makes implementation more feasible while still correctly detecting all deadlocks. This type of detection is usually accomplished with timeout counters: a counter is associated with each network resource, which is reset when the resource makes progress (that is, data is sent through the resource). However, if the counter reaches a predetermined threshold, the resource is considered deadlocked and the recovery phase is started.

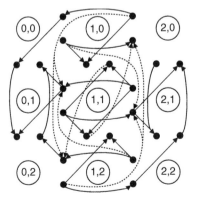

Figure 14.17 Extended channel dependence graph for a 3-ary 2-mesh. Direct (indirect) dependences shown as solid (dotted) lines. Network channels are represented by the filled circles and dependences that can be expressed as a combination of other dependences are not shown.

14.4.1 **Regressive Recovery**

In regressive recovery, packets or connections that are deadlocked are removed from the network. This technique can be applied to circuit switching, for example. If a group of partially constructed circuits has formed a wait-for cycle, they are deadlocked. This state persists until a timeout counter for one of the channel reaches a threshold. Then any circuit waiting on that channel is torn down (removed from the network) and retried. This breaks the wait-for dependence, removing the deadlock. To ensure that the source retries the circuit, a Nack can be sent to the source, or the source can retry any circuit that has been pending for more than a threshold time. Here, as with many regressive recovery techniques, special attention must be given to livelock to ensure a retried packet eventually succeeds.

Compressionless routing uses a similar recovery idea for wormhole-based networks. Extra, empty flits are appended to the packets so that the source node can ensure that the destination has received at least one flit by the time it has transmitted the final flit. This is feasible because of the small amount of buffering common to wormhole networks — the minimum length of any packet in compressionless routing is roughly the number of buffers along its path from the source to destination.

When a packet is started, the source node begins a timeout count. This count is reset after each new flit is accepted into the network. If this timeout reaches a threshold value before the last flit has been injected into the network, the entire packet is removed from the network and retried. Once the final flit has been injected, the source is guaranteed that the packet's head has already reached the destination. This implies that the packet has already allocated a path of (virtual) channels all the way to its destination and can no longer be blocked. The primary cost of this approach is the extra padding that must be added to packets. If many short packets are being sent, the maximum throughput can be significantly reduced.

14.4.2 **Progressive Recovery**

Progressive recovery resolves deadlock conditions without removing deadlocked packets from the network. Since network resources are not wasted by sending and then removing packets, progressive recovery techniques have potentially higher performance. Also, livelock issues associated with resending packets are eliminated.

The prevalent progressive recovery approaches implement the ideas of Duato's protocol in hardware. For example, the DISHA [186] architecture provides a shared escape buffer at each node of the network. When a suspected deadlock condition is detected, the shared escape buffer is used as a drain for the packets in the deadlock. Like the routing sub-function, routing using the escape buffer is designed to be deadlock-free.

Whether or not a hardware implementation of Duato's algorithm is appropriate depends on the relative costs of resource in a particular design. DISHA was designed under the assumption that virtual channels and buffers are expensive network resources, which is certainly true in some applications. However, in other applications the introduction of a centralized resource, in this case the shared escape buffer, may outweigh the impact of additional virtual channels.

14.5 Livelock

Unlike deadlock, livelocked packets continue to move through the network, but never reach their destination. This is primarily a concern for non-minimal routing algorithms that can *misroute* packets. If there is no guarantee on the maximum number of times a packet may be misrouted, the packet may remain in the network indefinitely. Dropping flow control techniques can also cause livelock. If a packet is dropped every time it re-enters the network, it may never reach its destination.

There are two primary techniques for avoiding livelock, *deterministic* and *probabilistic* avoidance. In deterministic avoidance, a small amount of state is added to each packet to ensure its progress. The state can be a *misroute count*, which holds the number of times a packet has been misrouted. Once the count reaches a threshold, no more misrouting is allowed. This approach is common in non-minimal, adaptive routing. A similar approach is to store an *age-based priority* in each packet. When a conflict between packets occurs, the highest priority (oldest) packet wins. When used in deflection routing or dropping flow control, a packet will become the highest-priority packet in the network after a finite amount of time. This prevents any more deflections or drops and the packet will proceed directly to its destination.

Probabilistic avoidance prevents livelock by guaranteeing the probability that a packet remains in the network for T cycles approaches zero as T tends to infinity. For example, we might want to avoid livelock in a 2-ary k-mesh with deflection routing and single flit packets.[7] The maximum number of hops a packet can ever be from its destination is $H_{\max} = 2(k-1)$. We then write a string for the history of a packet, where t denotes a routing decision toward the destination and d represents a deflection (such as $tddtdtt\ldots$). If the number of t's in the string minus the number of d's ever exceeds H_{max}, then we know the packet must have reached its destination. As long as the probability of a packet routing toward destination is always non-zero, the probability of this occurring approaches one. Therefore, our network is livelock-free as long as we can always guarantee a non-zero chance of a packet moving toward its destination at each hop.

14.6 Case Study: Deadlock Avoidance in the Cray T3E

The Cray T3E [162] is the follow-on to the T3D (Section 8.5). It is built around the Alpha 21164 processor and is scalable up to 272 nodes for a single cabinet machine. Up to 8 cabinets can be interconnected to create liquid-cooled configurations as large as 2,176 nodes. Like the T3D, the T3E's topology is a 3-D torus network. For the

---

7. This problem becomes more complex with multi-flit packets. See Exercise 14.11 .

2,176-node machines, the base topology is an 8,32,8-ary 3-cube, which accounts for 2,048 of the nodes. Additional redundant/operating system nodes are then added in half of the z-dimensional rings, expanding their radix to 9, to bring the total number of nodes to 2,176.

Additional latency tolerance in the T3E allowed a step back from the fast ECL gate arrays used to implement the T3D routers and each of T3E routers is implemented as a CMOS ASIC. The extra latency tolerance also allowed the T3E designers to use adaptive routing for improved throughput over the dimension-order routing of the T3D as well as increased fault-tolerance from the routing algorithm's ability to route around faulty links or nodes.

Adaptivity is incorporated into the T3E's routing using the same approach as Duato's protocol (Section 14.3.2). The T3E network uses cut-through flow control and each node has enough buffering to always store the largest packet. This ensures that no indirect dependences will ever be created in the extended channel dependence graph. Therefore, it is sufficient for the routing sub-function to be deadlock-free to guarantee that the entire network will be deadlock-free.

The routing sub-function used in the T3E is called *direction-order routing*: packets are routed first in the increasing x dimension ($+x$), then in $+y$, $+z$, $-x$, $-y$, and finally $-z$. Three bits in the packet header indicate whether each packet traverses a dimension in the increasing or decreasing direction. For example, a packet increasing in all dimensions would route in $+x$, $+y$, then $+z$, while a packet increasing in all but the y dimension would route $+x$, $+z$, then $-y$. Momentarily ignoring the intra-dimension cycles caused by the torus topology, direction-order routing is easily shown to be deadlock-free. Conceptually, any cycle in the channel dependence graph would have to contain routes from increasing channels ($+$) to decreasing channels ($-$) and vice versa. Although routes are allowed from increasing to decreasing channels, the converse is not true. Therefore, direction-order routing is deadlock-free for the mesh.

The T3E also allows a slight variation of direction-order routing through the use of initial and final hops. The initial hop a packet takes in the network can be in any of the increasing directions, and the final hop can be in $-z$ once all other hops are complete. The addition of initial and final hops improves the fault tolerance of the routing algorithm, as illustrated in Figure 14.18. As shown, the faulty channel between node 21 and node 22 blocks the default direction-order route. However, by taking an initial hop in the $-y$ direction, the packet successfully routes from node 20 to node 03.

Now considering the wrap-around channels of the torus, the T3E adopts a *dateline* approach as described in Section 14.2.3: any packet traveling across a predetermined dateline within a dimension must start on virtual channel (VC) 0 and transition to virtual channel 1 after the dateline. Packets that do not cross the dateline can use either virtual channel 0 or virtual channel 1, but any particular packet must choose one. In the T3E, the choice of virtual channels for these packets is made with the goal of balancing the load between virtual channels. That is, an optimal assignment of virtual channels exactly balances the average number of packets that traverse each of the virtual channels of a particular physical channel. Because the space of possible VC assignments is so large, a simulated annealing algorithm is

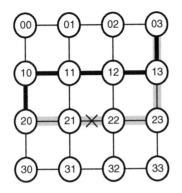

Figure 14.18 The fault-tolerance benefits of allowing an initial hop before direction-order routing begins (only two dimensions are used for simplicity). The default direction-order route from node 20 to node 03 is shown in gray and crosses the faulty channel (marked with X). By taking an initial hop in the $-y$ direction, the fault is avoided along the bold route.

used in the T3E to find an approximate solution to this virtual channel balancing problem.

In the dateline approach, two virtual channels are required to break deadlocks within the dimensions of the tori and another two are used to create separate virtual networks for request and reply messages. This eliminates the high-level deadlock illustrated in Figure 14.6. Finally, one virtual channel is used for minimal adaptive routing, which accounts for the five VCs used in the T3E network.

14.7 **Bibliographic Notes**

Early work on deadlock-free interconnection networks identified the technique of enumerating network resources and traversing these resources an increasing order [129, 70, 78, 57]. Linder and Harden [118] developed a method that makes arbitrary adaptive routing deadlock-free but at the cost of a number of virtual channels that increases exponentially with the number of dimensions. Glass and Ni developed the turn model, which allowed limited adaptivity while still remaining deadlock-free [73]. In different approach, Dally and Aoki allowed some non-minimal adaptivity with a small number of dimension reversals [48].

Duato introduced the extended channel dependence graph [60] and refined and extended these results in [61] and [62]. Duato's approach for designing deadlock-free adaptive routing algorithms has since been used in several networks such as the Reliable Router [52], the Cray T3E, and the Alpha 21364 [131].

Other, more specific, deadlock-free routing algorithms include planar adaptive routing [36] as described in Section 14.2.3. In irregular network topologies, deadlock can be avoided with the up*/down* algorithm [70] which is employed in the DEC AutoNet [159], for example. Gravano et al. introduced the *-channels and other algorithms [76], which use similar ideas as Duato to incorporate adaptivity into the torus. Kim et al. describe compressionless routing [98], which employs circuit switching ideas rather than the mainly packet switching approaches we have focused on in this chapter. While most systems choose deadlock avoidance, Anjan and Pinkston describe the DISHA deadlock recovery scheme [186].

The problem of VC misbalance caused by many deadlock avoidance schemes was studied specifically in the torus by Bolding [25]. In addition to the simulated annealing approach by Scott and Thorson described in the Cray T3E case study, the U.S. Patent held by Cray covers some additional VC load-balancing approaches [30].

Schwiebert [160] shows that it is possible to have oblivious routing algorithms with cyclic channel dependences while still being deadlock-free. However, current examples of this behavior require unusual routing function and topologies.

14.8 **Exercises**

14.1 *Deadlocked configuration.* Consider the example of Figure 14.5, but with the criteria that a packet may use either virtual channel at each hop, whichever becomes free first. Draw a wait-for graph for a deadlocked configuration of this network.

14.2 *Deadlock freedom of simple routing algorithms.* Determine whether the following oblivious routing algorithms are deadlock-free for the 2-D mesh. (Show that the channel dependence graph is either acyclic or contains a cycle.)

   (a) Randomized dimension-order: All packets are routed minimally. Half of the packets are routed completely in the X dimension before the Y dimension and the other packets are routed Y before X.

   (b) Less randomized dimension-order: All packets are routed minimally. Packets whose minimal direction is increasing in both X and Y always route X before Y. Packets whose minimal direction is decreasing in both X and Y always route Y before X. All other packets randomly choose between X before Y and vice versa.

   (c) Limited-turns routing for the 2-D mesh: All routes are restricted to contain at most three right turns and no left turns.

14.3 *Compound cycles in the turn model.* Explain why the fourth turn elimination option in Figure 14.13(c) does not result in a deadlock-free routing algorithm.

14.4 *Necessary number of disallowed turns.* Use the ideas of the turn model to find a lower bound on the number of turns that must be eliminated when routing in a k-ary n-mesh to ensure deadlock freedom.

14.5 *Enumerating turn model channels.* Number the channels of a 3×3 mesh network in a manner similar to that shown in Figure 14.14, but for the north-last routing restriction.

14.6 *Balancing virtual-channel utilization.* In Section 14.2.3, virtual channels were used to remove the cyclic dependences in a ring network. However, the simple dateline approach discussed has poor balancing between the virtual channels. For example, virtual channel 1 from node 4 to 1 is used only by packets routed from node 2 to node 1 and virtual channel 1 from 1 from node 1 to 2 is not used at all. Describe a new routing algorithm of the form $C \times N \mapsto C$ that better balances the virtual channels and is also deadlock-free.

14.7 *Deadlock freedom of planar adaptive routing.* Show that planar adaptive routing (Sec tion 14.2.3) is deadlock-free in k-ary n-meshes.

14.8 *Fault-tolerant planar adaptive routing.* Fault-tolerant planar adaptive routing (FPAR) extends the basic planar adaptive routing algorithm to avoid faults (non-functioning nodes or channels).

(a) To allow for faulty nodes in the final dimension, FPAR adds the adaptive plane A_{n-1} containing the previously unused channels $d_{n-1,2}$, $d_{0,0}$, and $d_{0,1}$. Then routing proceeds from adaptive plane A_0 to A_{n-1}, as in planar-adaptive routing, but with the addition of the extra plane. Show that there are no cyclic dependences *between* adaptive planes.

(b) When routing in the plane A_i, if d_i or d_{i+1} is blocked by a fault, a non-faulty channel is chosen as long as that channel is also minimal. It may happen that the packet has finished routing in the $(i + 1)^{\text{th}}$ dimension and a fault exists in d_i. In this case, misrouting is required in d_{i+1} — if the packet was routing in d_{i+1}, it continues in the same direction, otherwise an arbitrary direction in d_{i+1} is chosen for misrouting. This misrouting continues until the first non-faulty

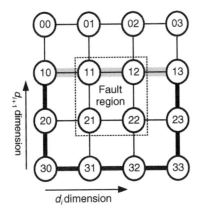

Figure 14.19 An example of misrouting in fault-tolerant planar-adaptive routing. A faulty region of nodes and channels (denoted by the dotted box) blocks the minimal path from node 10 to 13 in the A_i plane and the minimal distance in the $(i+1)^{th}$ dimension is zero. So, misrouting is required in d_{i+1} to route around the faulty region.

channel in d_i is found, which is taken. Then the packet continues in d_i until its first opportunity to "correct" the misrouting in d_{i+1}. After the distance in d_{i+1} had again been reduced to zero, normal routing is resumed. This procedure's details are complex, but in essence it attempts to simply steer around faulty regions (Figure 14.19). The converse case of being finished in d_i with a fault in d_{i+1} does not occur because once routing is done in d_i, i is incremented. Show that FPAR is deadlock-free *within* a single adaptive plane A_i in a k-ary n-mesh. Does it follow that the entire routing algorithm is also deadlock-free?

14.9 *Deadlock analysis of routing in the Cray T3E.* Show that direction-order routing (Section 14.6) is deadlock-free in a k-ary 3-mesh by enumerating the channels. Does this enumeration also prove that the initial and final hop variation used in the Cray T3E is also deadlock free? If not, create a new channel enumeration that does. Now consider a further extension of the T3E's routing algorithm that allows an arbitrary number of initial hops in any of the positive (+) dimensions, in any order. Is this extension also deadlock free?

14.10 *Virtual channel cost of randomized dimension traversal order.* You have designed a randomized variant of dimension-order routing for the k-ary n-mesh in which the order dimensions are traversed is completely random. This gives $n!$ possible dimension traversal orders. Of course, if each traversal order is given its own virtual channel, the routing algorithm is deadlock-free. However, $n!$ virtual channels can be costly for high-dimensional networks. Can you find a deadlock-free virtual channel assignment that uses fewer virtual channels?

14.11 *Probabilistic livelock avoidance with variable size packets.* Explain why the probabilistic argument for livelock freedom in a deflection routing network does not necessarily apply to networks with multi-flit packets. Assume that packets are pipelined across the network as in wormhole flow control and each input port has one flit of buffering. Construct a scenario with multi-flit packets and deflection routing that does livelock.

CHAPTER 15

Quality of Service

Previously, we focused on increasing the *efficiency* of resource allocation, such as achieving higher bandwidth and lower latency. However, even with perfect routing and flow control, situations remain in which the requests for a particular resource will exceed its capacity. In this regime, our attention shifts from efficiently allocating the resource to *fairly* allocating the resource according to to some service policies. In this chapter, we focus on both the typical services requested by network clients and the mechanisms for providing these services. Broadly speaking, we refer to these topics as providing QoS.

15.1 Service Classes and Service Contracts

In some applications of interconnection networks, it is useful to divide network traffic into a number of *classes* to more efficiently manage the allocation of resources to packets. Different classes of packets may have different requirements — some classes are latency-sensitive, while others are not. Some classes can tolerate latency but not jitter. Some classes can tolerate packet loss, while others cannot. Also, different classes of packets may have different levels of importance. The packets that keep the life-support systems going will take priority over the packets that carry digital audio for the entertainment system.

Allocating resources based on classes allows us to prioritize services so more important classes get a higher level of services and to tailor services so resource allocation can account for the particular requirements of each packet class. With prioritized services, we may give packets of one class strict priority in allocation of buffers and channels over packets of a lower class, or we may provide one class with a larger fraction of the overall resource pool. With tailored resource allocation, packets that belong to a class that needs low latency can advance ahead of packets that are latency-insensitive. Packets of another class that can tolerate latency but not jitter may be scheduled to make their delay large but predictable. Packets of a class that can tolerate loss will be dropped before packets of a class that cannot tolerate loss. Knowing the priority and requirements of each class allows us to allocate resources more efficiently than if all packets received exactly the same service.

Traffic classes fall into two broad categories: *guaranteed service* classes and *best efforts* classes. Guaranteed service classes are guaranteed a certain level of performance as long as the traffic they inject complies with a set of restrictions. There is a *service contract* between the network and the client. As long as the client complies with the restrictions, the network will deliver the performance. The client side of the agreement usually restricts the volume of traffic that the client can inject — that is, the maximum offered throughput. In exchange for keeping the offered traffic below a certain level, the network side of the contract may specify a guaranteed loss rate, latency, and jitter. For example, we may guarantee that 99.999% of the packets of class A will be delivered without loss and have a latency no larger than 1 μs as long as A injects no more than 1 Kbits during any 100 ns period.

In contrast, the network makes no strong guarantees about *best efforts* packets. Depending on the network, these packets may have arbitrary delay or even be dropped.[1] The network will simply make its best effort to deliver the packet to its destination.

Service classes in interconnection networks are analogous to service classes in package delivery. While guaranteed service packets are like Federal Express® (they guarantee that your package will get there overnight as long as it fits in their envelope and is not more than a specified weight), best efforts packets are like mail in the U.S. Postal Service (your package will probably get there in a few days, but there are no guarantees). There may be several classes of best efforts traffic in a network, just as the mail has several classes of package delivery (such as first-class, third-class, and so on).

Within a given class of service and, in particular, a best-efforts class, there is often a presumption of *fairness* between *flows*. Flows are simply the smallest level of distinction made between the packets that comprise a class. A flow might be all the packets generated by a particular source, those traveling toward a common destination or all the packets sent by an application running on the network clients. One expects that two flows of the same class will see roughly the same level of service: similar levels of packet loss, similar delay, etc. If one flow has all of its packets delivered with low latency and the other flow has all of its packets dropped, the network is being unfair. We shall define this notion of fairness more precisely below.

---

1.   Some networks guarantee that no packet will be dropped, even best-effort packets.

15.2 **Burstiness and Network Delays**

As mentioned, service guarantees may contain delay and jitter constraints in addition to rate requests. To implement these guarantees, the *burstiness* of particular traffic flows must be considered. Conceptually, a non-bursty flow sends its data in a regular pattern. A non-bursty flow is shown in Figure 15.1(a), sending at a rate of two-third packets/cycle. In contrast, a bursty flow tends to send its data in larger clumps rather than smoothly over time. For example, Figure 15.1(b) shows a bursty flow with a rate of one-third packets/cycle.

The result of these two flows sharing a 1-packet/cycle-channel is shown in Figure 15.1(c). First, the jitter of the non-bursty flow has been increased from 1 to 2 cycles because of the interaction with the bursty flow. Also, the bursty flow was delayed up to 4 cycles while traversing the channel. It can be easily verified that reducing the burstiness of the second flow would reduce both its delay and the jitter of the first flow. This simple example shows that the delay and jitter guarantees of flow can be greatly affected by the flows with which it shares resources. To quantify these affects, we first introduce a technique for characterizing bursty flows. The delay analysis of a simple network element subjected to two bursty flows is presented.

15.2.1 (σ, ρ) **Regulated Flows**

A common and powerful way to model flows when making QoS guarantees is in terms of two parameters: σ and ρ. The ρ parameter simply represents the average rate of the flow, while the σ term captures burstiness. For any time interval of length T, the number of bits injected in a (σ, ρ) *regulated* flow is less than or equal to

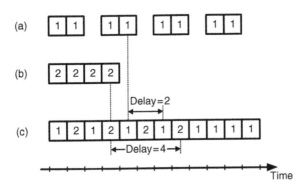

Figure 15.1   An example of a (a) non-bursty and (b) bursty flow (c) sharing the same channel. The jitter of the non-bursty flow is increased to 2 cycles and the delay of the bursty flow is 4 cycles.

$\sigma + \rho T$. That is, the number of bits can only exceed the average number ρT by the maximum burst size σ. We have already seen a (σ, ρ) description of a flow in Section 15.1: packets of class A are delivered in under 1 μs as long as A injects no more than 1 Kbits during any 100-ns period. The rate of this flow is $\rho = 1$ Kbits/100 ns $=$ 10 Gbps and σ is at most 1 Kbits.

Not only is it useful to express the nature of a flow in terms of σ and ρ, but it is also possible to control these parameters of any particular flow. This control can be achieved using a (σ, ρ) regulator, as shown in Figure 15.2. The regulator consists of an input queue that buffers the unregulated input flow. This queue is served only when tokens are available in the *token queue*. For each byte served from the input queue, a single token is removed from the token queue. To control the rate ρ of the outgoing flow, tokens are inserted into the token queue at a rate of ρ. Then the amount of burstiness in the output flow is set by the depth of the token queue σ. Controlling flows with a (σ, ρ) regulator can reduce the impact of bursty flows on other flows in the network.[2]

15.2.2 Calculating Delays

Assuming (σ, ρ) characterized flows, *deterministic* bounds on network delays can be computed. As an example, we focus on the delay of a simple network element: a two-input multiplexer with queueing (top of Figure 15.3). The multiplexer accepts two regulated input flows, denoted (σ_1, ρ_1) and (σ_2, ρ_2), which are multiplexed onto a single output. Both inputs and the output channel are assumed to have bandwidth b.

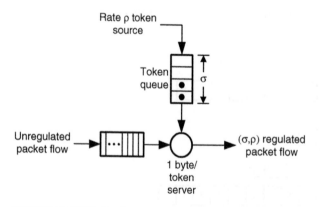

Figure 15.2 A (σ, ρ) regulator.

---

2. In practice, the token queue of the regulator can be realized by a saturating credit counter that increments at rate ρ and saturates when it reaches a count of σ. A packet can be transmitted only if the credit counter is non-zero and the counter is decremented each time a packet is transmitted.

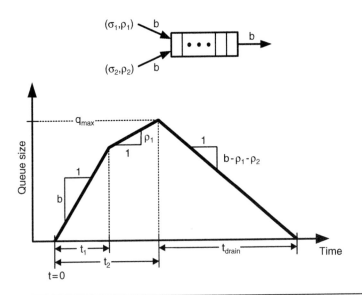

Figure 15.3 Queue size at a two-input multiplexer (shown at top) under adversarial inputs designed to maximize the interval for which the queue is non-empty. The increasing slopes represent the portions of the input where either both or one input is sending at a peak rate, limited only by the injection channel bandwidth. After both input bursts are exhausted, the queue slowly drains.

Finally, our only assumption about the way in which packets are selected for the output is that the multiplexer is *work-conserving*. That is, the output channel is never idle if the multiplexer contains any packets.

For the system to be stable and the maximum delay to be defined, it is sufficient for $\rho_1 + \rho_2 < b$. This condition also implies that the multiplexer queue will be empty at times, which leads to a simple observation about packet delay: the longest delay of any packet through the multiplexer is at most the maximum time the queue can be non-empty. So, our strategy for determining maximum packet delay will be to find the adversarial behavior of the two input flows that will keep the multiplexer queue non-empty for the longest interval possible.

Our adversary strategy is summarized in the graph of queue time versus time shown in Figure 15.3. The strategy has three phases. Initially, the input queue is assumed to be empty. The first phase begins at time $t = 0$ with both input flows simultaneously sending packets at rate b, the maximum allowed by the constraints of the channels. This fills the multiplexer's queue at rate b because the input channels are injecting packets at a total rate of $2b$, and since the multiplexer is work-conserving, it is draining packets from the queue at a rate of b. Therefore, the net rate of increase in the queue size is b, which is reflected in the slope of the first line segment of Figure 15.3. This phase continues until one of the inputs can no longer sustain a rate of b without violating its (σ, ρ) constraints. Without loss of generality, we assume the first flow reaches this point first at time t_1. By our definition of (σ, ρ) flows, this occurs when $bt_1 = \sigma_1 + \rho_1 t_1$. Rewriting,

$$t_1 = \frac{\sigma_1}{b - \rho_1}.$$

During the second phase, the first flow can send only at a rate of ρ_1 so that its (σ, ρ) constraint is not violated. The second flow continues to send at b, giving a net injection rate of $\rho_1 + b$ and a drain rate of b. Therefore, the queue still grows during the second phase, but with a smaller rate of ρ_1, as shown in the figure. Similarly, this phase ends at t_2 when the second flow can no longer send at rate b:

$$t_2 = \frac{\sigma_2}{b - \rho_2}.$$

At the beginning of the third phase, both input flows have exhausted their bursts and can send only at their steady-state rates of ρ_1 and ρ_2, respectively. This yields a decreasing rate of $b - \rho_1 - \rho_2$ in the queue size. At this rate, the queue becomes empty after $t_{drain} = q_{max}/(b - \rho_1 - \rho_2)$, where q_{max} is the queue size at the beginning of the phase. The queue size is simply the sum of the net amount after the first and second phases:

$$q_{max} = bt_1 + \rho_1(t_2 - t_1) = \sigma_1 + \frac{\rho_1 \sigma_2}{b - \rho_2}.$$

By our previous argument, we know the delay D must be bounded by the length of this non-empty interval.

$$D_{max} = t_2 + t_{drain} = \frac{\sigma_1 + \sigma_2}{b - \rho_1 - \rho_2}.$$

While we have not made a rigorous argument that our choice of input behavior gives the largest possible value of D_{max}, it can be shown that it is in fact the case [42]. Intuitively, any adversary that exhausts the entire bursts of both input streams before the queue re-empties will give the largest possible non-empty interval. Additionally, our strategy of starting both bursts immediately maximizes the size of the queue, which also bounds the total amount of buffering needed at the multiplexer to q_{max}. Similar techniques can be applied to a wide variety of basic network elements to determine their delays. We analyze additional network elements in the exercises.

15.3   Implementation of Guaranteed Services

There are a wide range of possibilities for implementing guaranteed services. We begin with *aggregate resource allocation*, where no specific resources are allocated to any flow; rather, the network accepts requests from its clients based on their aggregate resource usage. This aggregate approach is inexpensive in terms of hardware cost, but does not provide the tightest delay bounds. Lower delays can be obtained by *reserving* specific resources in either space or time and space together. The additional costs of these methods is the hardware required to store the resource reservations.

15.3.1 **Aggregate Resource Allocation**

The simplest way to implement a service guarantee is to require that the aggregate demand Λ_C of a traffic class C is less than a bound. Traffic conforming to this bound then is guaranteed not to saturate the network, and hence can be guaranteed lossless delivery with certain delay characteristics. This is the simplest method of providing guaranteed service. Because no specific resources are reserved for individual flows, little if any additional hardware is needed to support aggregate allocation.[3] However, because all of the (possibly bursty) input flows in class C are mixed together, the resulting output flows become even more bursty. As a result, aggregate resource allocation gives the loosest delay bounds of the methods we shall describe.

With aggregate allocation, requests can be explicitly supplied by the network clients or can be implicit in nature. In a packet switching network, for example, the network might be able to accept any set of resource allocations that did not *oversubscribe* any input or output port of the network. A port is oversubscribed if the total amount of traffic it is required to source or sink exceeds its bandwidth — this corresponds to a row or column sum of the request matrix Λ_C.

Now, to see how burstiness affects aggregate resource allocation, consider the 2-ary 2-fly with an extra stage shown in Figure 15.4. To balance load, this network uses a randomized routing algorithm in which all traffic routes from the source to a random switch node in the middle stage of the network before being routed to its destination. The figure also shows two flows: a bursty flow from node 0 to node 1 (solid lines) and a non-bursty flow from 2 to 0 (dotted lines). Because aggregate resource allocation does not reserve particular resources to flows, there is no way to prevent coupling between these two flows. This makes low-jitter requirements on the non-bursty flow more difficult to achieve in this example. Additionally, the use of

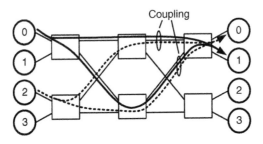

Figure 15.4 Two flows under aggregate resource allocation. Because the flows share channel resources, there is coupling between the bursty (solid lines) and non-bursty (dotted lines) flows, which affects their delay and jitter.

3. Typically, a network employing aggregate allocation will employ hardware at each network input to perform admissions control. This hardware admits traffic to the network only if it is in compliance with the aggregate demand bound. No additional hardware is required internal to the network.

randomized routing introduces more burstiness. Previously, we considered the bursti-ness in time of the traffic flows, but randomized routing also introduces burstiness in space — the routing balances load on average, but instantaneous loads may be unbalanced. This further complicates a guarantee of low jitter or delay.

Taking these factors into account and using the delay result from Section 15.2.2, we can compute a delay bound for this aggregate resource allocation. Both flows in this example are (σ, ρ) regulated, and for the bursty flow $\rho_1 = 0.75$ Gbps and $\sigma_1 = 1,024$ bits and for the non-bursty flow $\rho_2 = 0.75$ Gbps and $\sigma_2 = 64$ bits. Channel bandwidth is $b = 1$ Gbps and the maximum packet length is $L = 128$ bits.

Following the flow from 0 to 1 through the system, it is split in the first stage by the randomized routing algorithm. We assume that the routing algorithm splits the flow into two sub-flows, both with rate $\rho_1/2$. Although we cannot assume that the burstiness of the sub-flows is also halved because routing occurs at packet granularity, this burstiness can be upper-bounded by $(\sigma_1 + L)/2$. (See Exercise 15.2.)

Using these results, we know the sub-flows from the allocation are $(\sigma_1', \rho_1/2)$ regulated, where

$$\sigma_1' \le (\sigma_1 + L)/2 = (1024 + 128)/2 = 576 \text{ bits}.$$

Similarly, for the second sub-flow

$$\sigma_2' \le (\sigma_2 + L)/2 = (64 + 128)/2 = 96 \text{ bits}.$$

So, in splitting the second flow, its burstiness is actually increased because of the packet granularity limit.

The first delay incurred by either of these flows comes as they are multiplexed onto the output channel of the second stage. Using the result from Section 15.2.2, we know this delay is at most

$$D_{\max} = \frac{\sigma_0' + \sigma_1'}{b - (\rho_0 + \rho_1)/2} = \frac{\sigma_1 + \sigma_2 + L}{2b - \rho_1 - \rho_2}.$$

as long as $\rho_0 + \rho_1 < 2b$. Substituting the values from the example,

$$D_{\max} = \frac{576 + 96}{1 - 0.375 - 0.375} = \frac{672 \text{ bits}}{0.25 \text{Gbps}} = 2.688 \,\mu\text{s}.$$

Without any additional information, the jitter can be as large as $D_{\max}$ because packets could conceivably pass through the multiplexer with no delay. A similar calculation gives the delay incurred by the sub-flows as they are merged in the final stage of the network before reaching their destinations.

15.3.2 Resource Reservation

In situations where stronger guarantees on delay and jitter are required, it may be necessary to reserve specific resources rather than rely on aggregate allocation. Of course, this comes at greater hardware overhead because these reservations must also be stored in the network. We present two reservation approaches: virtual circuits, where resources are reserved in space, and time-division multiplexing, where resources are reserved in both space and time.

With virtual circuits, each flow is assigned a specific route through the network. This reservation technique is used in Asynchronous Transfer Mode (ATM), for example (Section 15.6). The use of virtual circuits addresses several sources of delay and jitter. First, because resources are allocated in space, any variations in resource usage due to factors such as randomized routing are completely eliminated. The second advantage is that flows can be routed to avoid coupling with other flows. Consider the previous example of the 2-ary 2-fly with an extra stage from Section 15.3.1. By controlling their routes, the non-bursty flow (dotted line) can be routed around the bursty flow (solid line), improving its jitter (Figure 15.5).

When extremely tight guarantees are required, time-division multiplexing (TDM) provides the strictest controls. To avoid the variability introduced by flows sharing a resource over time, TDM "locks-down" all the resources needed by a particular flow in both time and space. Because no other flows are allowed to access these resources, they are always available to the allocated flow, making guarantees easy to maintain. A TDM implementation divides time into a fixed-number of small slots. The size and number of slots then govern the granularity at which a resource can be allocated. So, for example, time might be broken into 32 slots, with each slot equal to the transmission time of a single flit. If the channel bandwidth is 1 Gbyte/s, flows could allocate bandwidth in multiples of 32 Mbytes/s. If a flow required 256 Mbytes/s, it would request 8 of the 32 time slots for each resource it needed.

Figure 15.6 revisits the 2-ary 2-fly example, in which flows have been allocated using TDM. Although some channels carry both flows, the flows are isolated because each of the four time slots is assigned to a unique flow. With time-slot allocation, a flow can share a resource with a bursty flow without increasing its own burstiness.

Time-slot allocations such as this can either be computed off-line, in the case where the required connections are known in advance, or on-line, when connections will be both added and removed over time. In either situation, finding "optimal"

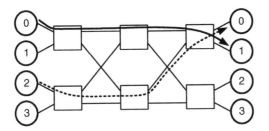

Figure 15.5   Two flows under virtual circuit resource reservation. Coupling between the bursty (solid lines) and non-bursty (dotted lines) flows is avoided by choosing independent routes through the network.

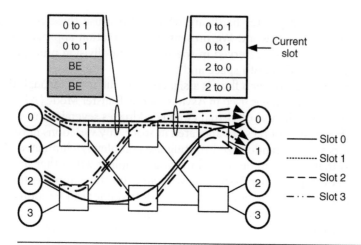

Figure 15.6   An allocation of flows in a 4-slot TDM network along with timewheels from two channels of the network.

allocations is generally NP-hard, so most practical implementations resort to heuristic approaches. One heuristic is explored in Exercise 15.5.

As shown in Figure 15.6, a *timewheel* can be associated with each resource in the network to store its allocation schedule. Then, a pointer into the timewheel table indicates the current time slot and the owner of the resource during that time slot. For this example, unused slots are marked as "BE" to indicate the resource's availability for best-effort traffic. Used slots match the resource allocation for the channel. As time progresses, the pointer is incremented to the next table entry, wrapping to the top once it reaches the bottom of the table.

15.4   Implementation of Best-Effort Services

The key quality of service concern in implementing best-effort services is providing *fairness* among all the best-effort flows. Similar concerns may also arise within the flows of a guaranteed service when resources are not completely reserved in advance. We present two alternative definitions of fairness, latency and throughput fairness, and discuss their implementation issues.

15.4.1   Latency Fairness

The goal of latency-based fairness is to provide equal delays to flows competing for the same resources. To see the utility of such an approach, consider an example of cars leaving a crowded parking lot, as shown in Figure 15.7. Each column of the parking lot is labeled with a letter and contains a line of cars waiting to turn onto the exit row, which leads to the exit of the parking lot. Cars are labeled with

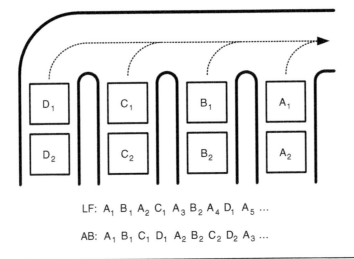

LF: A_1 B_1 A_2 C_1 A_3 B_2 A_4 D_1 A_5 ...

AB: A_1 B_1 C_1 D_1 A_2 B_2 C_2 D_2 A_3 ...

Figure 15.7   A parking lot with 4 columns of cars waiting for access to a single, shared exit. Cars are labeled with their column and a relative entrance time into the parking lot. The sequence of exiting cars under locally fair (LF) and age-based (AB) arbitrations are also shown.

their column along with a relative time that they started leaving the parking lot. So, D_2 started leaving before A_4, for example. This is analogous to packets queued in the vertical channels of a mesh network waiting for access to a shared horizontal channel of that network. We will assume a car can leave the parking lot every 5 seconds.

Standard driving courtesy dictates that at each merge point, cars from either entrance to a merge alternate access to that merge. We call this the *locally fair* arbitration policy. As shown by the dotted lines, our parking lot example has three merge points. Now consider the sequence of cars leaving the lot under the locally fair policy. The first car from the rightmost column A_1 leaves first, followed by B_1. Because of the locally fair policy at the right merge point, A_2 leaves next, followed by C_1, and so on. Although D_1 was one of the first cars waiting to leave the parking lot, it must wait 8 cars, or 40 seconds, before finally leaving. By this time, 4 cars from column A have left. Obviously, the delays under locally fair arbitration are distributed unfairly. In fact, if this example was extended to contain 24 columns, the first car of the last column X_1 would have to wait over a year to leave the parking lot! Of course, this assumes a relatively large number of cars in the parking lot.

To remedy this problem, we can replace our arbitration policy with one that is *latency fair*. An arbitration is latency fair if the oldest requester for a resource is always served first. For our parking lot example, we can simply use the relative starting times of each car to make decisions at the merge points — the oldest of two cars at a merge point goes first. This gives a much better exit sequence with one car leaving from each column before any column has two cars that have left. For networks, we refer to this policy as *age-based arbitration*. When multiple packets are competing for a resource, the oldest, measured as the time since its injection into the network, gets access first.

While age-based arbitration greatly improves the latency fairness of networks, it is generally used only as a local approximation to a truly latency fair network. This caveat arises because high-priority (older) packets can become blocked behind low-priority (younger) packets, which is known as *priority inversion*. As we will see, a similar problem arises in throughput fairness and both can be solved by constructing a *non-interfering* network, which we address in Section 15.5.

15.4.2 **Throughput Fairness**

An alternative to latency-based fairness, throughput fairness, seeks to provide equal bandwidth to flows competing for the same resource. Figure 15.8(a) illustrates this idea with three flows of packets crossing a single, shared channel in the network. As shown, each flow requests at rate of 0.5 packets per cycle, but the channel can support only a total 0.75 packets per cycle. Naturally, a throughput-fair arbitration would be to simply divide the available bandwidth between the three flows so that each received 0.25 packets per cycle across the channel.

This example becomes more complex when the rates of each of the flows are no longer equal. Figure 15.8(b) shows the case in which the rates have been changed to 0.15, 0.5, and 0.5 packets per cycle. Many reasonable definitions of fairness could lead to different allocations in this situation, but the most common definition of fairness used for throughput is *max-min fairness*. An allocation is max-min fair if the allocation to any flow cannot be increased without decreasing the allocation to a flow that has an equal or lesser allocation. The resulting allocation, shown in Figure 15.8(b), is max-min fair.

Algorithmically, max-min fairness can be achieved by the following procedure. For the N flows, let b_i be the bandwidth requested by the ith flow, where $0 \leq i < N$. The bandwidth requests are also sorted such that $b_{i-1} \leq b_i$ for $0 < i < N$. Then the

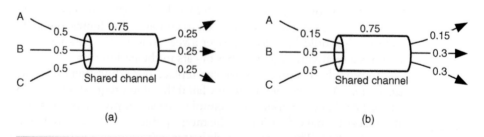

(a)   (b)

Figure 15.8   Throughput-fair assignment of bandwidth to three flows sharing a single channel: (a) an allocation when the flows have equal requests, (b) a max-min fair allocation for unequal requests.

bandwidths are allocated using the following recurrence:

$$R_0 = b,$$

$$a_i = \min\left[b_i, \frac{R_i}{N - i}\right],$$

$$R_{i+1} = R_i - a_i,$$

where b is the total bandwidth of the resource, R_i is the amount of bandwidth available after scheduling i requests, and a_i is the amount of bandwidth assigned to request i. This algorithm satisfies the smallest requests first and any excess bandwidth for each request is distributed evenly among the remaining larger requests.

Max-min fairness can be achieved in hardware by separating each flow requesting a resource into a separate queue. Then, the queues are served in a round-robin fashion. Any empty queues are simply skipped over. This implementation is often referred to as *fair queueing*. While not conceptually difficult, several practical issues can complicate the implementation. For example, additional work is required if the packets have unequal lengths and *weighted fair queueing* adds the ability to weight some flows to receive a higher priority than others. As we saw in latency fairness, true throughput fairness also requires per-flow structures to be maintained at each resource in the network.

15.5 **Separation of Resources**

To meet service and fairness guarantees, we often need to isolate different classes of traffic. Sometimes, we also need to distinguish between the flows within a class. For brevity, we will collectively refer to classes and flows simply as classes throughout this section. With reservation techniques such as TDM, the cost of achieving this comes with the tables required to store the resource reservations. When resources are allocated dynamically, the problem is more complicated. Ideally, an algorithm could globally schedule resources so that classes did not affect one another. However, interconnection networks are distributed systems, so a global approach is not practical. Rather, local algorithms, such as fair queueing and age-based arbitration, must generate resource allocations, and hardware resources must separate classes to prevent the behavior of one class from affecting another. Before introducing *non-interfering networks* for isolating traffic classes, we discuss *tree-saturation*, an important network pathology that can result from poor isolation.

15.5.1 **Tree Saturation**

When a resource receives a disproportionally high amount of traffic, one or more *hot-spots* in the network can occur. A hot-spot is simply any resource that is being loaded more heavily than the average resource. This phenomenon was first observed in shared memory interconnects: a common synchronization construct is a lock, where multiple processors continuously poll a single memory location in order to obtain the lock and gain access to a shared data structure. If a particular lock is located at one node in the network and many other nodes simultaneously access this lock, it is easy to see how the destination node can become overwhelmed with requests and become a hot-spot. In an IP router application, random fluctuations in traffic to a particular output or a momentary misconfiguration of routing tables can both cause one or more output ports to become overloaded, causing a hot-spot.

Of course, facilitating resource sharing is an important function of any interconnection network and in the example of a shared lock, the network's flow control will eventually exert backpressure on the nodes requesting the lock. This is expected behavior of the network. However, a possibly unexpected impact of hot-spots is their affect on the network traffic not requesting a hot-spot resource. *Tree-saturation* occurs as packets are blocked at a hot-spot resource (Figure 15.9). Initially, channels adjacent to the hot-spot resource become blocked as requests overwhelm the resource, forming the first level of the tree. This effect continues as channels two hops from the resource wait on the blocked channels in the first level, and so on. The resulting pattern of blocked resources forms a tree-like structure.

As also shown in Figure 15.9, it is quite possible for a packet to request a channel in the saturation tree, but never request the hot-spot resource. If there is not an adequate separation of resources, these packets can become blocked waiting for channels in the saturation tree even though they do not require access to the overloaded resource. Tree-saturation is a universal problem in interconnection networks and is not limited to destination nodes being overwhelmed. For example, the same effect

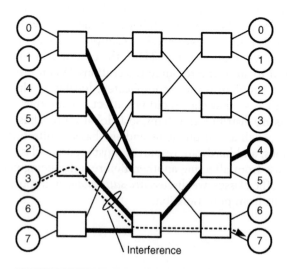

Figure 15.9　Tree-saturation in a 2-ary 3-fly. Destination 4 is overloaded, causing channels leading toward it to become blocked. These blocked channels in turn block more channels, forming a tree pattern (bold lines). Interference then occurs for a message destined to node 7 because of the tree-saturated channels.

occurs when a network channel becomes loaded beyond its capacity. The quantitative effects of tree-saturation in shared memory systems are explored in Exercise 15.4.

15.5.2 Non-interfering Networks

To achieve isolation between two classes *A* and *B*, there cannot be any resource shared between *A* and *B* that can be held for an indefinite amount of time by *A* (*B*) such that *B* (*A*) cannot interrupt the usage of that resource. A network that meets this definition is referred to as *non-interfering*. For example, with virtual-channel flow control, a non-interfering network would have a virtual channel for each class in the network. However, physical channels do not have to be replicated because they are reallocated each cycle and cannot be held indefinitely by a single class. The partitioning also applies to buffers at the inputs of the network, where each client needs a separate injection buffer for each class.

While non-interfering networks provide the separation between classes necessary to meet service and fairness guarantees, their implementation can be expensive. Consider, for example, the use of an interconnection network as a switching fabric for an Internet router where we require that traffic to one output not *interfere* with traffic destined for a different output. In this case, we provide a separate traffic *class* for each output and provide completely separate virtual channels and injection buffers for each class. Even in moderate-sized routers, providing non-interference in this manner requires that, potentially, hundreds of virtual channels be used, which corresponds to a significant level of complexity in the network's routers (Chapters 16 and 17).[4] To this end, the number of classes that need true isolation should be carefully chosen by the designer. In many situations, it may be possible to combine classes without a significant degradation in service to gain a reduction in hardware complexity.

15.6 Case Study: ATM Service Classes

Asynchronous transfer mode (ATM) is a networking technology designed to support a wide variety of traffic types, with particular emphasis on multimedia traffic such as voice and video traffic, but with enough flexibility to efficiently accommodate best-effort traffic [154]. Typical applications of ATM include Internet and campus network backbones and as well as combined voice, video, and data transports within businesses.

ATM is connection-based, so before any data can be sent between a source-destination pair, a virtual circuit (Section 15.3.2) must be established to reserve network resources along a path connecting the source and destination. While the connections are circuit-based, data transfer and switching in ATM networks is

---

4. The torus network used in the Avici TSR [49] provides non-interference between up to 1,024 classes in this manner.

packet-based. An unusual feature of ATM is that all packets, called *cells*, are fixed-length: 53 bytes. This reduces packetization latency and simplifies router design.

Each ATM connection is characterized under one of five basic service classes:

- Constant bit rate (CBR) — a constant bit rate connection, such as real-time, uncompressed voice.

- Variable bit rate, real-time (VBR-rt) — a bursty connection in which both low delay and jitter are critical, such as a compressed video stream for tele-conferencing.

- Variable bit rate (VBR) — like VBR-rt, but without tight delay and jitter constraints.

- Available bit rate (ABR) — bandwidth demands are approximately known with the possibility of being adjusted dynamically.

- Unspecified bit rate (UBR) — best-effort traffic.

The service classes, excluding UBR, also require additional parameters, along with the class type itself, to specify the parameters of the service required. For example, a (σ, ρ) regulated flow, which is required for VBR-rt, is specified by a sustained cell rate (SCR) and burst tolerance (BT) parameter. ATM also provides many other parameters to further describe the nature of flows, such as the minimum and peak cell rates (MCR and PCR) and cell loss ratio (CLR), for example.

While ATM switches can use TDM-like mechanisms for delivering CBR traffic, efficient support of the other traffic types requires a dynamic allocation of resources. Most switches provide throughput fairness and must also separate resources to ensure isolation between virtual circuits.

15.7 Case Study: Virtual Networks in the Avici TSR

In Section 9.5, we introduced the Avici TSR and its network and examined its use of oblivious source routing to balance load. In this section, we take a further look at this machine and study how it uses virtual channels to realize a non-interfering network.

The Avici TSR uses a separate virtual channel for each pair of destination nodes to make traffic in the network completely non-interfering [50]. That is, traffic to one destination A shares no buffers in the network with traffic destined to any other destination $B \neq A$. Thus, if some destination B becomes overloaded and backs up traffic into the network, filling buffers and causing tree saturation, this overload cannot affect packets destined for A. Because packets destined for A and B share no buffers, A's packets are isolated from the overload. Packets destined for A and B do share physical channel bandwidth. However, this is not an issue, since traffic is spread to balance load on the fabric channels and the network load to B cannot in the steady state exceed the output bandwidth at node B.

The problem is illustrated in Figure 15.10. Overloaded destination node B backs up traffic on all links leading to B, shown as dark arrows. This is a form of tree

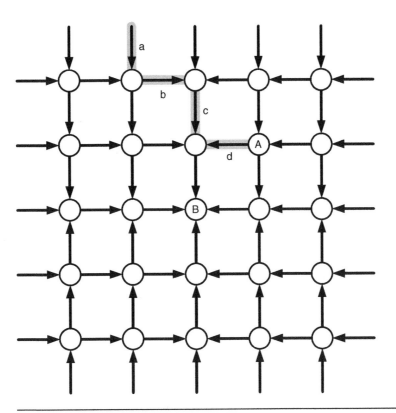

Figure 15.10 A fragment of a 2-D mesh network showing links blocked by tree-saturation from overloaded destination node *B*. Packets routing to node *A* over the gray path (a, b, c, d) will encounter interference from packets destined for *B* on all but link *d* unless resources are kept separate.

saturation (Section 15.5.1). All of the virtual channels and flit buffers usable by *B* on these links are full, blocking further traffic. Now consider a packet routing to node *A* over the path shown in gray. If this packet shares resources with packets destined for *B*, it will be blocked waiting on a virtual channel or flit buffer at links a, b, and c and delayed indefinitely. The packet destined for *A* will not be blocked at link *d* because it is proceeding in the opposite direction from packets destined for *B* and hence using a different set of virtual channels and flit buffers.

The bandwidth consumed by packets destined to *B* is not an issue. Assuming that node *B* can consume at most one unit of bandwidth and that load is distributed evenly over incoming paths, only $\frac{1}{12}$ unit of bandwidth on link c ($\frac{1}{20}$ on link a) is consumed by traffic destined for *B*. This leaves ample bandwidth to handle the packet destined for *A*. The problem is that packets destined for *B* hold all of the virtual channels and flit buffers and release these resources very slowly because of the backup. The problem is analogous to trying to drive to a grocery store (node *A*) near the stadium (node *B*) just before a football game. You, like the packet to node *A*, are blocked waiting for all of the cars going to the game to clear the road.

The solution is to provide separate virtual networks for each destination in the machine. The Avici TSR accomplishes this, and also provides differentiated service for two classes of traffic by providing 512 virtual channels for each physical channel. For each destination d, the TSR reserves two virtual channels on every link leading *toward* the destination. One virtual channel is reserved for *normal* traffic and a second virtual channel is reserved for *premium* traffic. Separate source queues are also provided for each destination and class of service so that no interference occurs in the source queues. With this arrangement, traffic destined for A does not compete for virtual channels or buffers with traffic destined to B. Traffic to A only shares physical channel bandwidth with B, and nothing else. Hence, packets to A are able to advance without interference. Returning to our driving analogy, it is as if a separate lane were provided on the road for each destination. Cars waiting to go to the stadium back up in the stadium lane but do not interfere with cars going to the grocery store, which are advancing in their own lane. The great thing about interconnection networks is that we are able to provide this isolation by duplicating an inexpensive resource (a small amount of buffering) while sharing the expensive resource (bandwidth). Unfortunately this can't be done with roads and cars.

To support up to 512 nodes with 2 classes of traffic using only 512 virtual channels, the TSR takes advantage of the fact that minimal routes to 2 nodes that are maximally distant from one another in all dimensions share no physical channels. Hence, these nodes can use the same virtual channel number on each physical channel without danger of interference. This situation is illustrated for the case of an 8-node ring in Figure 15.11. Packets heading to node X along a minimal route use only the link directions shown with arrows. Packets heading to node Y along a minimal route use only links in the opposite direction (not shown). Because minimal routes to the two nodes share no channels, they can safely use the same virtual channel number without danger of actually sharing a virtual channel, and, hence, without danger of interference.

15.8 **Bibliographic Notes**

More detailed discussions of general issues related to QoS along with implementation issues, especially those related to large-scale networks such as the Internet, are covered by Peterson and Davie [148]. The general impact of QoS on router design is address by Kumar et al. [107]. Cruz provides a detailed coverage of (σ, ρ) flows and their utility in calculating network delays [42, 43]. Early definitions of

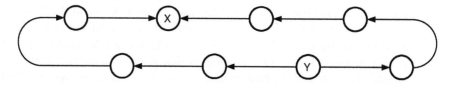

Figure 15.11 Two destinations X and Y maximally distant from one another on this 8-node ring can share a set of virtual channels without interference on minimal routes. All links that lead to X lead away from Y and vice versa.

fairness in networks is covered by Jaffe [88]. Throughput fairness is introduced by Nagle [132] and extended to account for several practical considerations by Demers et al. [58]. True max-min fairness can be expensive to implement and several notable approximations exist, such as Golestani's stop-and-go queueing [74] and the rotating combined queueing [97] algorithm presented by Kim and Chien, the latter of which is specifically designed in the context of interconnection networks. Yum et al. describe the MediaWorm router [198], which provides rate-based services using an extension of the virtual clock algorithm [200]. Several other commercial and academic routers incorporate various levels of QoS, such as the SGI SPIDER [69], the Tandem ServerNet [84], and the MMR [63]. Tree saturation was identified by Pfister and Norton, who proposed combining buffers to consolidate requests to the same memory address in shared-memory systems [151]. One well-known technique that separates resources of different flows is virtual output queueing, covered by Tamir and Frazier [181].

15.9 **Exercises**

15.1 *Burstiness example.* Verify that the combining example shown in Figure 15.1 falls within the delay bound given by the general multiplexer model of Section 15.2.2.

15.2 *Burstiness of an equally split flow.* Describe how a single (σ, ρ) characterized flow can be split into two flows, each with rate $\sigma/2$. Show that the bound on the burstiness of either of the two new sub-flows is $(\sigma + L)/2$.

15.3 *First-come, first-served multiplexer.* Apply the approach of Section 15.2.2 to a multiplexer with a first-come, first-served (FIFO) service discipline. Assuming the two inputs to the multiplexer are a (σ_1, ρ_1) characterized flow and a (σ_2, ρ_2) characterized flow, respectively, what is the maximum delay of this multiplexer?

15.4 *Impact of tree-saturation.* Consider a system that has p nodes, each of which contains a processor and a memory module. Every node generates a fixed-length memory request at a rate r, with a fraction h of the requests destined to a "hot" memory location and the remaining fraction $1 - h$ uniformly distributed across all other memory locations. What is the total rate of requests into the hot memory module? Assuming that tree saturation blocks all requests once the hot module is saturated, $p = 100$, and $h = 0.01$, what fraction of the total memory bandwidth in the system can be utilized?

15.5 **Simulation.** Consider the reservation of TDM flows in a 4×4 mesh network. Assume time is divided into T slots and individual flows require one of these slots. Also, flows are scheduled incrementally in a greedy fashion: given a flow request between a particular source-destination pair, the flow is scheduled along the path with the smallest cost. The cost of a path is defined as the maximum number of TDM slots used along any of its channels. Then the quality of the schedule is determined by the number of slots required to support it (the maximally congested channel). Use this heuristic to schedule a set of random connections and compare it to the lower-bound on congestion using the optimization problem from Equation 3.9.

CHAPTER 16

Router Architecture

A router is composed of registers, switches, function units, and control logic that collectively implement the routing and flow control functions required to buffer and forward flits en route to their destinations. Although many router organizations exist, in this chapter, we examine the architecture of a typical virtual-channel router and look at the issues and tradeoffs involved in router design.

Modern routers are pipelined at the flit level. Head flits proceed through pipeline stages that perform routing and virtual channel allocation and all flits pass through switch allocation and switch traversal stages. Pipeline stalls occur if a given pipeline stage cannot be completed in the current cycles. These stalls halt operation of all pipeline stages before the stall, since flits must remain in order.

Most routers use credits to allocate buffer space. As flits travel downstream on a channel, credits flow upstream to grant access to the buffers just vacated. In routers that have a limited number of buffers, the latency of credit handling can have a large impact on performance. Credits also affect deadlock properties. If a virtual channel is allocated before all credits are returned, a dependency is created between the packet being allocated the channel and any packets still in the downstream buffer.

16.1 Basic Router Architecture

16.1.1 Block Diagram

Figure 16.1 shows a block diagram of a typical virtual-channel router. These blocks can be partitioned broadly into two groups based on functionality: the *datapath* and *control plane*. The datapath of the router handles the storage and movement of a packet's payload and consists of a set of input buffers, a switch, and a set of output

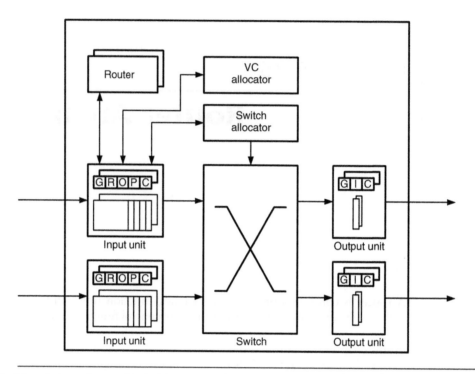

Figure 16.1   Virtual-channel router block diagram. The datapath consists of an input unit for each input port, an output unit for each output port, and a switch connecting the input units to the output units. Three modules make decisions and perform allocation: packets are assigned an output port by a router and an output virtual channel by a virtual-channel allocator. Each flit is then assigned a switch time slot by a switch allocator.

buffers. The remaining blocks implement the control plane of the router and are responsible for coordinating the movement of packets through the resources of the datapath. For our virtual-channel router, the control blocks perform route computation, virtual-channel allocation, and switch allocation. We also associate input control state with the input buffers, forming an input unit, and similarly for the outputs.

Each flit of a packet arrives at an *input unit* of the router. The input unit contains a set of flit buffers to hold arriving flits until they can be forwarded and also maintains the state of each virtual channel associated with that input link. Typically, five state fields are maintained to track the status of each virtual channel, as described in Table 16.1.

To begin advancing a packet, route computation must first be performed to determine the output port (or ports) to which the packet can be forwarded.[1] Given

---

1.   In some situations, when the routing is simple or bandwidth demands, the router may be duplicated with each input unit having its own router. Such duplication is not possible with the two allocators as they are allocating a shared resource.

Table 16.1 Virtual channel state fields, represented by a 5-vector: GROPC.

Field	Name	Description
G	Global state	Either idle (I), routing (R), waiting for an output VC (V), active (A), or waiting for credits (C).
R	Route	After routing is completed for a packet, this field holds the output port selected for the packet.
O	Output VC	After virtual-channel allocation is completed for a packet, this field holds the output virtual channel of port R assigned to the packet.
P	Pointers	Flit head and tail pointers into the input buffer. From these pointers, we can also get an implicit count on the number of flits in the buffer for this virtual channel.
C	Credit count	The number of credits (available downstream flit buffers) for output virtual channel O on output port R.

its output port, the packet requests an output virtual channel from the *virtual-channel allocator*. Once a route has been determined and a virtual channel allocated, each flit of the packet is forwarded over this virtual channel by allocating a time slot on the switch and output channel using the *switch allocator* and forwarding the flit to the appropriate output unit during this time slot.[2] Finally, the output unit forwards the flit to the next router in the packet's path. As with the input unit, several state fields contain the status of each output virtual channel (Table 16.2).

The control of the router operates at two distinct frequencies: packet rate and flit rate. Route computation and virtual-channel allocation are performed once per packet. Therefore, a virtual channel's R, O, and I state fields are updated once per packet. Switch allocation, on the other hand, is performed on a per-flit basis. Similarly, the P and C state fields are updated at flit frequency.

Table 16.2 Output virtual channel state fields, represented by a 3-vector: GIC.

Field	Name	Description
G	Global state	Either idle (I), active (A), or waiting for credits (C).
I	Input VC	Input port and virtual channel that are forwarding flits to this output virtual channel.
C	Credit count	Number of free buffers available to hold flits from this virtual channel at the downstream node.

---

2. When the switch has an output speedup of one (switch bandwidth equals output bandwidth) the switch and output channels can be scheduled together with no loss of performance. When the switch has output speedup, the output unit typically incorporates a FIFO to decouple switch scheduling from output scheduling.

16.1.2 **The Router Pipeline**

Figure 16.2 shows a Gantt chart illustrating the pipelining of a typical virtual channel router. To advance, each head flit must proceed through the steps of routing computation (RC), virtual-channel allocation (VA), switch allocation (SA), and switch traversal (ST). The figure reflects a router in which each of these steps takes one clock cycle — each has its own pipeline stage.[3] Figure 16.2 shows a situation in which the packet advances through the pipeline stages without any stalls. As we will discuss shortly, stalls may occur at any stage.

The routing process begins when the head flit of a packet arrives at the router during cycle 0. The packet is directed to a particular virtual channel of the input port, at which it arrives as indicated by a field of the head flit. At this point, the global state (G) of the target virtual channel is in the idle state. The arrival of the head flit causes the virtual channel to transition to the routing (G = R) state at the start of cycle 1.

During cycle 1, information from the head flit is used by the router block to select an output port (or ports). The result of this computation updates the route field (R) of the virtual channel state vector and advances the virtual channel to the waiting for output virtual channel (G = V) state, both at the start of cycle 2. In parallel with RC, the first body flit arrives at the router.

During cycle 2, the head flit enters the VA stage, the first body flit passes through the RC stage, and the second body flit arrives at the router. During the VA stage, the result of the routing computation, held in the R field of the virtual-channel state vector, is input to the virtual-channel allocator. If successful, the allocator assigns a single output virtual channel on the (or on one of the) output channel(s) specified by the R field. The result of the VA updates the output virtual channel field (O) of

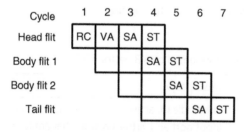

Figure 16.2   Pipelined routing of a packet. Each flit of a packet proceeds through the stages of routing computation (RC), virtual channel allocation (VA), switch allocation (SA), and switch traversal (ST). The RC and VA stages perform computation for the head flit only (once per packet). Body flits pass through these control stages with no computation. The SA and ST stages operate on every flit of the packet. In the absence of stalls, each flit of the packet enters the pipeline one cycle behind the preceding flit and proceeds through the stages one per cycle.

---

3. This is not always the case. In some routers, two or more of these steps may be combined into a single pipeline stage, whereas in other routers a single step may itself be pipelined across many stages.

the input virtual-channel state vector and transitions the virtual channel to the active
(G = A) state. The result of the allocation also updates the global field of the selected
output virtual-channel state vector to the active (G = A) state and sets the input (I)
field of the output virtual channel state vector to identify the successful input virtual
channel. From this point until the release of the channel by the tail flit, the C (credit)
field of the output virtual channel is also reflected in the C field of the input virtual
channel.

For purposes of deadlock analysis, the VA stage is the point at which a dependency is created from the input virtual channel to the output virtual channel. After
a single output virtual channel is allocated to the packet, the input virtual channel
will not be freed until the packet is able to move its entire contents into the output
virtual channel's buffers on the next node.

At the start of cycle 3, all of the per-packet processing is complete and all
remaining control is the flit-by-flit switch allocation of the SA stage. The head flit
starts this process, but is handled no differently than any other flit. In this stage,
any active virtual channel (G = A) that contains buffered flits (indicated by P) and
has downstream buffers available (C > 0) bids for a single-flit time slot through the
switch from its input virtual channel to the output unit containing its output virtual
channel. Depending on the configuration of the switch, this allocation may involve
competition not only for the output port of the switch, but also competition with
other virtual channels in the same input unit for the input port of the switch. If the
switch allocation bid is successful, the pointer field is updated to reflect the departure
of the head flit in the virtual channel's input buffer, the credit field is decremented to
reflect the fact that a buffer at the far side has been allocated to the departing head
flit, and a credit is generated to signal the preceding router that a buffer is available
on this input virtual channel.

Successful switch allocation is shown in Figure 16.2 during cycle 3 and the head
flit traverses the switch (the ST stage) during cycle 4. Since we are assuming an
otherwise empty switch, the head flit can start traversing the channel to the next
router in cycle 5 without competition.

While the head flit is being allocated a virtual channel, the first body flit is in
the RC stage, and while the head flit is in the SA stage, the first body flit is in the
VA stage and the second body flit is in the RC stage. However, since routing and
virtual-channel allocation are per-packet functions, there is nothing for a body flit to
do in these stages. Body flits cannot bypass these stages and advance directly to the
SA stage because they must remain in order and behind the head flit. Thus, body flits
are simply stored in the virtual channel's input buffer from the time they arrive to
the time they reach the SA stage. In the absence of stalls, three input buffer locations
are needed to hold the three body flits in the pipeline at any one time. We will see
below that a larger buffer is needed to prevent the pipeline from bubbling due to
credit stalls.

As each body flit reaches the SA stage, it bids for the switch and output channel
in the same manner as the head flit. The processing here is per-flit and the body flits
are treated no differently than the head flit. As a switch time slot is allocated to each
flit, the virtual channel updates the P and C fields and generates a credit to reflect its
departure.

Finally, during cycle 6, the tail flit reaches the SA stage. Allocation for the tail flit is performed in the same manner as for head and body flits. When the tail flit is successfully scheduled in cycle 6, the packet releases the virtual channel at the start of cycle 7 by setting the virtual channel state for the output virtual channel to idle (G = I) and the input virtual channel state to idle (G = I) if the input buffer is empty. If the input buffer is not empty, the head flit for the next packet is already waiting in the buffer. In this case, the state transitions directly to routing (G = R).

16.2 Stalls

In the previous discussion, we assumed the ideal case in which packets and flits proceeded down the pipeline without delays: allocations were always successful, and channels, flits, and credits were available when needed. In this section, we examine six cases, listed in Table 16.3, where this ideal behavior is interrupted by stalls. The first three types of stalls are *packet stalls* — stalls in the packet processing functions of the pipeline. Packet stalls prevent the virtual channel from advancing into its R, V, or A states. Once a virtual channel has a packet in the active (A) state, packet processing is complete and only *flit stalls* can occur. A flit stall occurs during a cycle in which a flit does not successfully complete switch allocation. This can be due to lack of a flit, lack of a credit, or simply losing the competition for the switch time slot.

Table 16.3 Types of router pipeline stalls. For each stall, the table gives the type of stall, whether it is a packet (P) or flit (F) stall, the packet state during which the stall occurs (all flit stalls occur during the active (A) state), and a description of the stall.

Type of Stall	P/F	State	Description
VC Busy	P	¬I	Back-to-back packet stall. The head flit for one packet arrives before the tail flit of the previous packet has completed switch allocation. The head flit of the new packet is queued in the buffer. It will be dequeued and the new packet will enter the routing state once the tail flit of the preceding packet completes switch allocation.
Route	P	R	Routing not completed, input virtual channel remains in R state and reattempts routing next cycle.
VC Allocation	P	V	VA not successful, input virtual channel remains in the V state and reattempts allocation next cycle.
Credit	F	A	No credit available. Flit is unable to attempt switch allocation until a credit is received from the downstream virtual channel.
Buffer Empty	F	A	No flit available. Input buffer is empty, perhaps due to an upstream stall. Virtual channel is unable to attempt switch allocation until a flit arrives.
Switch	F	A	Switch allocation attempted but unsuccessful. Flit remains in switch-allocation pipe stage and reattempts switch allocation on the next cycle.

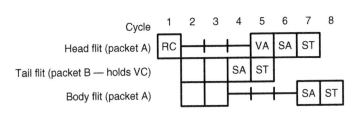

Figure 16.3   Virtual-channel allocation stall — an example of a packet stall. Packet *A* arrives during cycle 1, but is not able to allocate a virtual channel *V* until cycle 5, after *V* is freed by packet *B*. While awaiting the virtual channel, the pipeline is stalled for three cycles.

As an example of a packet stall, Figure 16.3 shows a virtual-channel allocation stall. The head flit of packet *A* completes the routing computation in cycle 1 and bids for an output virtual channel starting in cycle 2. Unfortunately, all of the output virtual channels compatible with its route are in use, so the packet and its head flit fail to gain an output virtual channel and are unable to make progress. The bid is repeated without success each cycle until the tail flit of packet *B*, which holds an output virtual channel that *A* can use, completes switch allocation and releases the output virtual channel in cycle 4. During cycle 5, packet *A* successfully completes virtual-channel allocation and acquires the channel that *B* just freed. It proceeds on to switch allocation in cycle 6 and switch traversal in cycle 7. Note that since the head of *A* cannot perform vitual-channel allocation until the tail of *B* completes switch allocation, there is an idle cycle on the output virtual channel between forwarding the tail flit of *B* and the head flit of *A*. The body flits of *A* stall as well, since the first body flit is not able to enter switch allocation until the head flit has completed this stage. However, this virtual-channel stall need not slow transmission over the input channel as long as there is sufficient buffer space (in this case, six flits) to hold the arriving head and body flits until they are able to begin switch traversal.

A switch allocation stall, an example of a flit stall, is shown in Figure 16.4. The head flit and first body flit proceed through the pipeline without delay. The second body flit, however, fails to allocate the switch during cycle 5. The flit successfully reattempts allocation during cycle 6 after stalling for one cycle. All body flits following this flit are delayed as well. As above, this stall need not slow transmission over the input channel if sufficient buffer space is provided to hold arriving flits. A credit stall has an identical timing diagram, but instead of the switch allocation failing, no switch allocation is attempted due to lack of credits. A buffer empty stall is shown in Figure 16.5. The head flit and first body flit proceed without delay, but then the second body flit does not arrive at the router until cycle 6. The second body flit bypasses the RC and VA stages, which it would normally spend waiting in the input buffer, and proceeds directly to the SA stage in cycle 7, the cycle after it arrives. Body flits following the second flit can proceed directly to the SA stage as well.

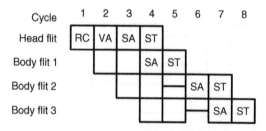

Figure 16.4 Switch allocation stall — an example of a flit stall. Here, the second body flit of a packet fails to allocate the requested connection through the switch in cycle 5, and hence is unable to proceed. The request is retried successfully in cycle 6, resulting in a one-cycle stall that delays body flit 2 and all subsequent flits.

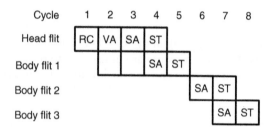

Figure 16.5 Buffer empty stall — another type of flit stall. Here, body flit 2 is delayed by three cycles, arriving in cycle 6 rather than 3. Its departure from the router is only delayed by one cycle; however, the other two cycles are absorbed by bypassing the empty RC and VA pipeline stages.

16.3 Closing the Loop with Credits

A buffer is allocated to a flit as it leaves the SA stage on the *transmitting* or *upstream* side of a link. To reuse this buffer, a credit is returned over a reverse channel after the same flit departs the SA stage on the *receiving* or *downstream* side of the link. When this credit reaches the input unit on the upstream side of the link, the buffer is available for reuse. Thus, the token representing a buffer can be thought of as traveling in a loop: starting at the SA stage on the upstream side of the link, traveling downstream with the flit, reaching the SA stage on the downstream side of the link, and returning upstream as a credit. The buffer is only occupied by the flit during the time that the flit is in the first three pipeline stages (RC, VA, and SA) on the upstream side. The remainder of the time around the loop represents an overhead period during which the buffer is idle.

The credit loop latency t_{crt}, expressed in flit times, gives a lower bound on the number of flit buffers needed on the upstream side for the channel to operate at full bandwidth, without credit stalls. Since each flit needs one buffer and the buffers cannot be recycled until their tokens traverse the credit loop, if there are fewer than t_{crt} buffers, the supply of buffers will be exhausted before the first credit is returned. The credit loop latency in flit times is given by

$$t_{crt} = t_f + t_c + 2T_w + 1, \qquad (16.1)$$

where t_f is the flit pipeline delay, t_c is the credit pipeline delay, and T_w is the one-way wire delay.[4]

If the number of buffers available per virtual channel is F, the duty factor on the channel will be

$$d = \min\left(1, \frac{F}{t_{crt}}\right). \qquad (16.2)$$

The duty factor will be unity (full bandwidth achieved) as long as there are sufficient flit buffers to cover the round trip latency, as previously described in Equation 13.1. As the number of flit buffers falls below this number, credit stalls will cause the duty factor for one virtual channel to be reduced proportionally. Of course, other virtual channels can use the physical channel bandwidth left idle due to credit stalls.

Figure 16.6 illustrates the timing of the *credit loop* and the credit stall that can result from insufficient buffers. The figure shows the timing of a 6-flit packet: 1 head flit followed by 5 body flits in crossing a channel that has 4 flit buffers available at the far end. Upstream pipeline stages are shown in white and downstream pipeline stages are shaded grey. For simplicity, the flit time is equal to one clock cycle. It is assumed that traversing the physical link in each direction takes two clock cycles, labeled W1 and W2 — hence $T_w = 2$ cycles. Credit transmission takes place in pipeline stage CT on the upstream node and updating the credit count in response to a credit takes place in pipeline stage CU on the downstream node.

The pipeline in Figure 16.6 has a flit latency of $t_f = 4$ cycles and a credit latency of $t_c = 2$ cycles (the CT and CU stages). With $T_w = 2$ cycles, Equation 16.1 gives us a total round-trip credit delay of $t_{crt} = 11$ cycles. Hence, there are 11 cycles from when the head flit is in the SA stage in cycle 1 to when body flit 4 reuses the head flit's credit to enter the SA stage. The last term (the one) in Equation 16.1 is due to the fact that the SA stage appears twice in the forward path of the flit. With only $F = 4$ buffers, Equation 16.2 predicts that the duty factor will be 4/11, which is correct.

16.4 **Reallocating a Channel**

Just as a credit allows a flit buffer to be recycled, the passing of a tail flit enables an output virtual channel to be recycled. When we recycle an output virtual channel, however, it can have a great impact on the deadlock properties of a network.

---

4.   Although we normally express these delays in seconds, or sometimes cycles, for convenience we assume it is expressed in units of flit times (which may take several cycles each) here.

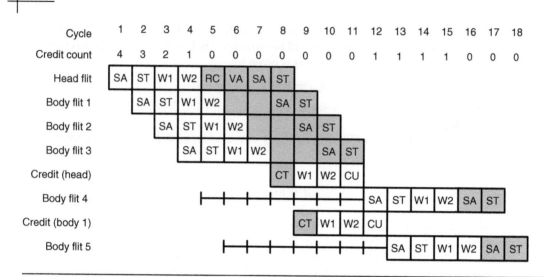

Cycle	1	2	3	4	5	6	7	8	9	10	11	12	13	14	15	16	17	18
Credit count	4	3	2	1	0	0	0	0	0	0	0	1	1	1	1	0	0	0

Figure 16.6   Credit stall: a buffer is allocated to the head flit when it is in the upstream SA stage in cycle 1. This buffer cannot be reassigned to another flit until after the head flit leaves the downstream SA stage, freeing the buffer, *and* a credit reflecting the free buffer propagates back to the upstream credit update (CU) stage in cycle 11. Body flit 4 uses this credit to enter the SA stage in cycle 12.

When the tail flit of a packet, *A*, enters the ST stage, the output virtual channel, *x*, allocated to *A* can be allocated to a different packet, *B* by the virtual-channel allocator, as illustrated in Figure 16.3. Packet *B* may come from any input virtual channel: the same input virtual channel as *A*, a different virtual channel on the same input, or a virtual channel on a different input.

Allocating *x* to *B* at this point, however, creates a dependency from *B* to *A*. Because *B* will be in the input buffer of virtual channel *x* behind *A*, it will have to wait on *A* if *A* becomes blocked. If, instead of reallocating output virtual channel *x* when the tail flit enters the ST stage of the upstream router, we wait until the tail flit enters the ST stage of the *downstream* router, we eliminate this dependence. By waiting until *A* vacates the input buffer of the downstream router, packet *B* ensures that it will not allocate an output virtual channel holding a blocked packet ahead of it. This wait is critical to the deadlock properties of some routing algorithms, particularly adaptive routing algorithms, that assume when a virtual channel is allocated, it is empty. In practice, the upstream router does not know when *A* enters the ST stage of the downstream router, but instead waits until the credit from the tail flit returns, bringing the credit count to its maximum, before reallocating the virtual channel. This wait can result in considerable lost bandwidth when there are many back-to-back short packets traversing a single virtual channel while the other virtual channels sharing the same physical channel are idle.

Three options for reallocating an output virtual channel are shown in Figure 16.7. All three panels, (a) to (c), show a two-flit packet *A* followed by the head flit of a packet *B*. Panel (a) also shows the credit from the tail flit of *A*. No credits are shown in the other panels.

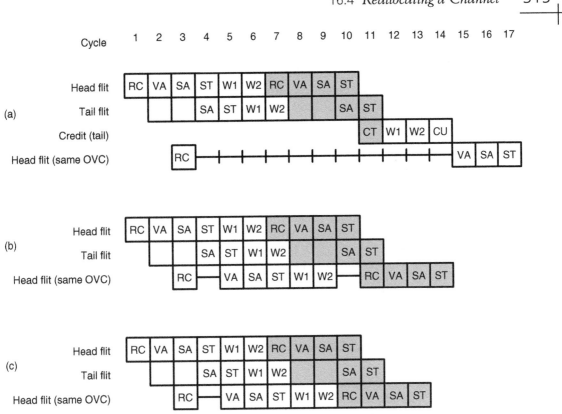

Figure 16.7 Reallocating an output virtual channel (OVC) to a new packet. (a) The most conservative approach is to wait until the downstream flit buffer for the virtual channel is completely empty, as indicated by the arrival of the credit from the tail flit. This avoids creating a dependency between the current packet and a packet occupying the downstream buffer. (b) If we can afford to create a dependency, we can reallocate the virtual channel as soon as the tail flit of the previous packet completes the SA stage. We don't need to wait for its credit to return. However, unless we duplicate state, the new packet is not able to enter the RC stage on the downstream router until the tail flit completes the SA stage on this router. This is because the global state for each virtual channel can only handle a single packet at a time. (c) If we extend the global state to allow two packets to be on the virtual channel at the same time, the new one routing while the old one is still active, we can avoid this second stall at the cost of some control complexity.

Panel (a) shows the case in which we wait for the credit from the tail flit before reallocating the virtual channel. In this case, the head flit of B arrives during cycle 2, completes the RC stage in cycle 3, and then enters a virtual channel stall for 12 cycles until cycle 15 when it finally completes the VA stage. The delay here is $t_{crt} + 1$ since the delay of the VA stage is added to the normal credit loop.

Panels (b) and (c) of Figure 16.7 illustrate the case in which we reallocate the output virtual channel as soon as the tail flit is in the ST stage, as in Figure 16.3. In both panels, packet B is able to complete the VA stage in cycle 5 after only a single stall cycle.

Panels (b) and (c) differ in the timing of the reallocation of the input virtual channel at the downstream side of the link. In panel (b), packet *B* is not allowed to complete the RC stage until cycle 11, when the tail flit of *A* has entered the downstream ST stage, completely freeing the input virtual channel. This results in a one-cycle virtual-channel busy stall but simplifies the router logic slightly, since the virtual channel makes a clean transition from handling *A* to handling *B*. The VC state reflects *A* from cycles 7 to 10 and packet *B* from cycle 11 onward.

Panel (c) shows a more aggressive approach that eliminates the virtual-channel busy stall at the expense of more complicated virtual-channel control logic that can handle two packets, in different stages, at the same time. Here, we allow *B* to enter the RC stage before the tail flit has completed the SA stage. In this case, the input virtual-channel is handling packets *A* and *B* simultaneously, and hence must track the state of both (for example, with two G fields in the state vector). Also, if *B* completes the RC stage, but the tail of *A* fails switch allocation, the R state field will belong to B, whereas the O, P, and C fields belong to *A*. This may persist for several cycles. If this situation occurs, *B* encounters a virtual-channel busy stall, but at the VA stage rather than at the RC stage.

16.5 Speculation and Lookahead

Latency of an interconnection network is directly related to pipeline depth. Thus, it is advantageous to reduce the pipeline to as few stages as possible, as long as this is accomplished without lengthening the delay of each stage. We can reduce the number of stages by using speculation to enable us to perform operations in parallel that otherwise would be performed sequentially. We can also reduce the number of stages by performing computation ahead of time to remove it from the critical path.

We can reduce the number of pipeline stages in a router without packing more logic into each stage by operating our router *speculatively*. For example, we can speculate that we will be granted a virtual channel and perform switch allocation in parallel with virtual-channel allocation, as illustrated in Figure 16.8. This reduces the pipeline for head flits to three stages and results in body flits idling for only one cycle during the RC stage. Once we have completed the routing computation for a packet, we can take the resulting set of virtual channels and simultaneously bid for one of these virtual channels and for a switch time slot on the corresponding physical channel.[5] If the packet fails to allocate a virtual channel, the pipeline stalls and both the SA and VA are repeated on the next cycle. If the packet allocates a virtual channel, but fails switch allocation, the pipeline stalls again. However, only switch allocation is required on the next cycle and since the packet has an output virtual channel allocated, the allocation is no longer speculative.

---

5.   If the set of routes possible produced by the RC stage spans more than one physical channel, we can speculate again by choosing one.

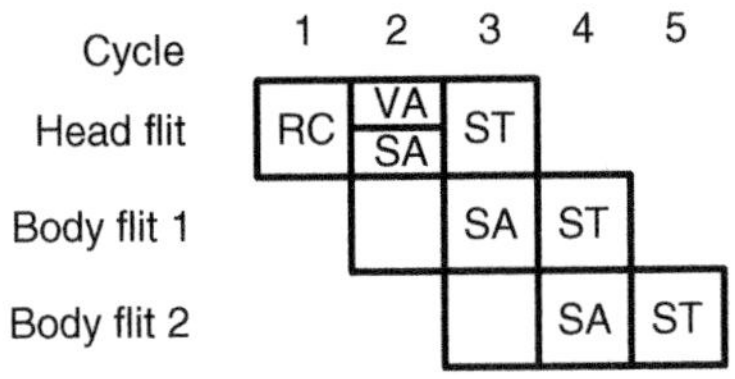

Cycle	1	2	3	4	5
Head flit	RC	VA / SA	ST		
Body flit 1			SA	ST	
Body flit 2				SA	ST

Figure 16.8 Speculative router pipeline with VA and SA performed in parallel. The switch allocation for the head flit is speculative, since it depends on the success of the virtual channel allocation being performed at the same time. If the VA fails, the switch allocation will be ignored even if it succeeds.

To guarantee that speculation does not hurt performance, we can speculate conservatively. We do this by giving priority in switch allocation to non-speculative requests. This guarantees that a switch cycle that could be used by a non-speculative request will never go idle because it was granted to a speculative request that later failed to allocate a virtual channel.

In practice, speculation works extremely well in that it almost always succeeds in the cases where pipeline latency matters. Pipeline latency matters when the router is lightly loaded. In this situation, almost all of the virtual channels are free and the speculation is almost always successful. When the router becomes heavily loaded, queueing latency dominates pipeline latency, so the success of the speculation does not greatly influence performance.

We can take speculation one step further by performing switch traversal speculatively in parallel with the two allocation steps, as illustrated in Figure 16.9. In this case, we forward the flit to the requested output without waiting for switch allocation. To do this, we require that the switch have sufficient *speedup*, or at least *input speedup* so it can accept speculative requests from all of the inputs. (See Section 17.2.) While the speculative flits are traversing the switch, the switch allocation is performed in parallel.[6] The allocation is completed at the same time that the flits reach the far side of the switch and the results of the allocation select the winning flit. Whether or not this *double speculation* can be done without affecting cycle time depends on the details of the switch and allocators. In many routers, however, the virtual channel allocator is the critical path and the switch allocation and switch output selection can be performed in series without lengthening the clock cycle.

We have gotten the pipeline down to two cycles. Could it be shortened to a single cycle? We could try to apply speculation a third time and perform the routing computation in parallel with the other three functions. However, this would be very inefficient, since we really need to know where a flit is going before allocating channels, allocating switch cycles, or forwarding it across a switch.

---

6.   With complete input speedup, this allocation reduces to an arbitration. (See Chapter 19.)

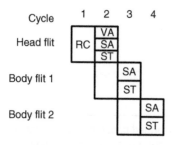

Figure 16.9   Speculative router pipeline with VA, SA, and ST all performed in parallel.

Rather than speculate on the result of the routing stage, we can perform this work ahead of time, so that the result is available when we need it. To do this, while we are traversing a given router — say, for hop *i* of a route — we perform the routing computation not for this router, but for the *next* router, the router at hop $i + 1$ of the route, and pass the result of the computation along with the head flit. Thus, when the packet reaches each router, it has already selected the channel (or set of channels) on which it will leave the router and the VA and SA operations can begin immediately using this information. In parallel with any of the pipeline stages of the router, the routing computation for the next hop is performed, denoted NRC, so this router will also have its output channel(s) selected when the flit arrives.

Figure 16.10 shows two pipelines employing this type of one-hop route lookahead. On the left side, route lookahead is applied to our non-speculative four-stage pipeline to shorten it to three stages. Here, we have put the NRC (next hop routing computation) into the VA stage. We could have easily put it in the SA stage, or even had it span the two stages without loss of performance. On the right side, route

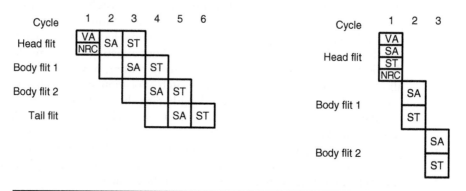

Figure 16.10   Route lookahead. By performing the routing computation one hop ahead, we can overlap it with any of the other pipeline stages. On the left, we show this applied to our standard four-stage pipeline, and on the right, applied to the pipeline of Figure 16.9, shortening it to a single stage.

lookahead is applied to the doubly speculative pipeline of Figure 16.9, shortening it to a single cycle.[7] For applications in which latency matters, we cannot do much better than this.

16.6 **Flit and Credit Encoding**

Up to now we have discussed flits and credits traversing the channel between two routers without describing the details of how this is done. How do we separate flits from credits? How do we know where each flit begins? How is the information within a flit or a credit encoded. In this section, we will briefly examine each of these issues.

First, we address the issue of separating flits from credits. The simplest approach is to send flits and credits over completely separate channels, as shown in Figure 16.11(a). The flit channel is usually considerably wider (as denoted by the thicker line) than the credit channel, as it carries substantially more information. While simple, having a separate credit channel is usually quite inefficient. There is often excess bandwidth on the credit channel that is wasted and serializing the credits to get them over a narrow channel increases credit latency. It is usually more efficient to carry the flits and credits on a single physical channel, as illustrated in Figure 16.11(b). One approach is to include one credit (for the reverse channel) within each flit (for the forward channel). We call this *piggybacking* the credit on the flit as if the flit were giving the credit a piggyback ride. This approach is simple and takes advantage of the fact that, although they are different sizes, flits and credits have the same *rates* — on average, one credit is needed each flit time. Alternatively, flits and credits can be multiplexed at the phit level on the wires. This has the advantage that during transients when the flit and credit rates are different, all of the bandwidth can be devoted to flits or all of the bandwidth can be devoted to credits

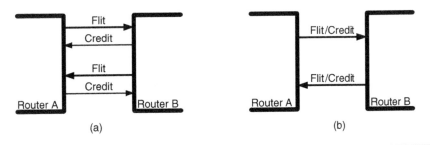

Figure 16.11   Flits and credits may (a) be sent over separate physical channels or (b) share a channel. If they share a channel, the credits be *piggybacked* within each flit or may be multiplexed with the flits using a phit-level protocol.

---

7. This does not necessarily mean one-cycle-per-hop latency. While it is only a single cycle through the router, many systems will require an additional cycle (or more) to traverse the channel.

rather than having a fixed allocation. However, there can be some fragmentation loss with this approach if a credit does not exactly fit into an integer number of phits.

To multiplex flits and credits at the phit level, we need to address the issue of how we know where each flit and credit begins and how we know whether it is a flit or a credit. This is accomplished by a phit-level protocol, which may use an encoding like the one illustrated in Figure 16.12. Here, each phit indicates whether it is starting a credit, starting a flit, or idle. Because we know how long credits and flits are, we do not need to encode *middle of flit* in the type field; it can be inferred, and we may use the phit type field of continuation phits for additional payload information. If credits are piggybacked in the flit, we can dispense with the start-of-credit encoding and reduce the phit-type field to a single bit. Other link-level protocols, such as those based on framing, can completely eliminate this field at the expense of some flexibility.

Now that we know where each flit and credit begins, let's examine their contents. As shown in Figure 16.13, all flits include a virtual channel identifier that indicates which virtual channel this flit is associated with, a type field that identifies the type of flit, either head or not (we might also encode the tail or not), and a check field that is used to check for errors in the flit in addition to the payload. Head flits add a route information field that is used by the routing function to select an output virtual channel for the packet. If credits are piggybacked on flits, a credit field is added to both head and non-head flits. A credit, whether it is piggybacked or sent separately, includes a virtual-channel identifier that indicates the virtual channel that is to receive the credit and a check field. Credits can optionally include a type field that could, for example, encode on/off flow control information or other back-channel information for the virtual channel.

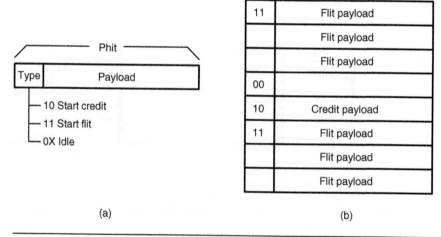

(a)   (b)

Figure 16.12 (a) Each phit consists of a type field that encodes the start of flits and credits and a payload field. (b) An example of phit-level multiplexing, the type field encodes a three-phit flit followed by an idle phit, a one-phit credit, and a second three-phit flit.

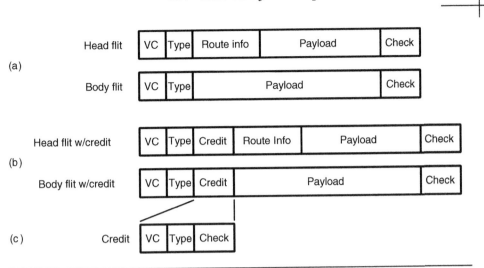

Figure 16.13   (a) Head and body flit formats. All flits contain a virtual channel identifier (VC), a type field, payload, and a check field. Head flits also include routing information. (b) Head and body flit formats with piggybacked credit. (c) Credit format. Credits include a virtual channel identifier and check bits. They may also include a type field.

A cursory examination of Figures 16.12 and 16.13 might suggest that we are doing the same work twice. After all, the phit encoding identifies a virtual channel (credit or flit) and a type (start or idle). Why then do we need to repeat this information at the flit level? The answer is overhead. Consider, for example, multicomputer router with 16-bit phits, 16-byte flits, 16 virtual channels, and piggybacked credits. If we were to put a 4-bit virtual-channel field and a 2-bit type field in every 16-bit phit, our overhead would be 37.5%. Adding an 8-bit check field would leave only 2 bits left for information. With a single-bit type field on the phits and a total of 14 bits for virtual channel, type, and check bits on the flit, our total overhead is just 17%.

16.7 Case Study: The Alpha 21364 Router

The Alpha 21364 is the latest in the line of the Alpha microprocessor architectures [89, 131]. The 152 million transistor, 1.2-GHz 21364 integrates an Alpha 21264 processor core along with cache coherence hardware, two memory controllers, and a network router, onto a single die. This allows the 21364 to be used as a single-chip building block for large, shared-memory systems. The router has four external ports with a total 22.4 Gbytes/s of network bandwidth and up to 128 processors can be connected in a 2-D torus network. The architecture of the 21364's router closely matches the basic router architecture introduced in this chapter: it employs deep pipelining of packets through the router to meet its 1.2-GHz operating frequency and also uses speculation to recover some of the latency incurred from this pipelining.

Figure 16.14 shows the high-level aspects of the 21364's router architecture. The router uses cut-through flow control, not wormhole flow control. Although this

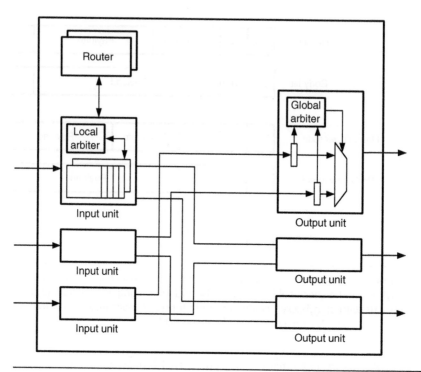

Figure 16.14   A simplified view of the Alpha 21364 router architecture. Although the actual implementation has 8 input ports and 7 output ports, only a 3 × 3 router is shown for simplicity. The 21364's crossbar is implemented as a mux for each output port, as shown. Also, the switch allocator is divided into two distinct units: a local arbiter and a global arbiter.

design decision requires each input unit to be able to store at least one entire packet, this is practical for the 21364 because the largest packets used in its coherence protocol are 76 bytes long. Each input unit contains enough buffering for several maximum-length packets and the router has a total of 316 packet buffers. This is a sufficient amount of packet storage for the router to have several virtual channels, which are used to both avoid deadlock and implement adaptive routing.

As shown in the router's pipeline diagram (Figure 16.15), routing and buffering occupies the first four pipeline cycles of the first phit of a packet — we consider the phit pipeline because this is a cut-through router. Routing occurs during the RC stage, followed by a transport/wire delay (T) cycle. Header decode and update of the input unit state occupies another cycle (DW) before the payload of the packet is finally written to the input queues (WrQ). In parallel with the storage of the first phit in the input buffers, switch allocation can begin.

Because the Alpha 21364 uses cut-through flow control, the normally separate virtual-channel and switch allocation steps can be merged into a single allocation. Although multiple virtual channels can be active per output in a general virtual-channel router, a cut-through router can simplify its design by allowing only one

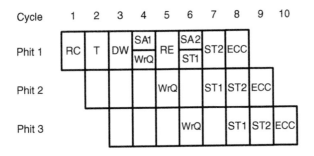

Figure 16.15 The pipeline stages of the Alpha 21364 router. The entire pipeline latency is 13 cycles, or 10.8ns at 1.2GHz, including several cycles for synchronization, driver delay, receiver delay, and transport delay to and from the pins. Because the flit size equals the packet size in cut-through flow control, pipeline delays at the phit level are shown.

active virtual channel per output. This ensures that any virtual-channel allocation is also a valid switch allocation and, for the 21364 design, we will simply refer to this merged VA and SA step as SA. Once the packet using a virtual channel is completely transferred downstream, the corresponding output becomes available to other waiting packets. Of course, this approach could defeat the purpose of virtual channels under wormhole flow control because the single virtual channel active at each output may become blocked. However, cut-through ensures that the downstream router will always have enough buffering to store an entire packet, therefore avoiding the possibility that a packet becomes blocked while also holding an output.

To meet its aggressive timing requirements, the switch allocation of the 21364 is split into two stages, referred to as the local and global arbiters. (In Section 19.3, we discuss these *separable allocators* in greater detail.) First, the local arbiters in each input unit choose a packet among all waiting packets that are ready. A packet is considered ready when its output port is available and its downstream router has a free packet buffer. Local arbitration takes one pipeline cycle (SA1) and, during the following cycle, the header information for each packet selected by a local arbiter is read and transported to the correct output (RE). This header information is enough to begin the global arbitration cycle (SA2). Since the local arbitration decisions are not coordinated, its quite possible that multiple packets have been selected for a single output. The global arbiters resolve this conflict by choosing one packet for each output. In parallel with this global arbitration decision, the data corresponding to each packet that was selected by the local arbiters is speculatively read from the input buffers and transferred to the outputs (ST1). This transfer is speculative because these packets may still lose at the global arbiters, but if they win, the cycle required to read their data and transfer it to the outputs is hidden. Switch transfer is completed in stage ST2 once the global arbitration results are known. The final stage (ECC) appends a 7-bit error correcting code to each phit before it is transferred to the downstream router. Although not shown in Figure 16.15, backchannels within the router convey the results of the speculative switch transfer to the input units. If the speculation succeeds, subsequent phits of the packet can continue to be sent. If it fails, the speculative phits are dropped from within the switch and must be resent by the input unit. The result of the 21364's deeply pipelined design and speculative switch transfer is an extremely low per-hop latency of 10.8 ns combined with sustained throughputs of 70% to 90% of peak [131].

16.8   **Bibliographic Notes**

The basic virtual-channel router architecture presented in this chapter dates back to the Torus Routing Chip [56]. Refinements over the years can be found in the Reliable Router [52], the routers in the Cray T3D [95] and T3E [162], the IBM SP series [176], and the SGI SPIDER chip [69]. The SPIDER chip was the first router to compute the route one step ahead. The use of speculation to reduce router latency was first described by Peh and Dally [146].

16.9   **Exercises**

16.1 *Simplifications with one virtual channel.* How do the pipeline and global state change if there is only a single virtual channel per physical channel? Describe each pipeline stage and each field of the global state.

16.2 *Architecture of a circuit switching router.* Sketch the architecture of a router for circuit switching (as in Figure 12.6). What is the global state? What is the pipeline? What are the steps in handling each circuit and each flit?

16.3 *Impact of ack/nack flow control on the pipeline.* Suppose ack/nack flow control is used instead of credit-based flow control. How does this change the router pipeline? What happens to credit stalls?

16.4 *Dependencies from aggressive virtual-channel reallocation.* Consider a simple minimal adaptive routing algorithm for 2-D mesh networks based on Duato's protocol: VC 0 is reserved for dimension order routing (the routing subfunction) and VC 1 is used for adaptive routing. Which of the reallocation strategies shown in Figure 16.7 are deadlock-free for this routing algorithm?

16.5 *Flit formatting for narrow channels.* For the flit fields shown in Figure 16.13, let the VC field require 4 bits, the type information require 2 bits, the credit field require 8 bits, and the route information require 6 bits. If the input channel receives 8 bits per clock cycle, how many cycles are required before routing can begin? If so, give the corresponding flit format. If route lookahead is employed, can virtual-channel allocation begin on the first cycle?

CHAPTER 17

Router Datapath Components

In this chapter, we explore the design of the components that comprise the datapath of a router: the input buffer, switch, and output unit. Input buffers hold flits while they are waiting for virtual channels, switch bandwidth, and channel bandwidth. Input buffers may be centralized, partitioned across physical channels, or partitioned across virtual channels. A central buffer offers the most flexibility in allocating storage across physical channels and virtual channels, but is often impractical due to a lack of bandwidth. Within each buffer partition, storage can be statically allocated to each virtual channel buffer using a circular buffer, or dynamically allocated using a linked list.

The switch is the core of the router and where the actual switching of flits from input ports to output ports takes place. A centralized buffer itself serves as a switch with the input multiplexer and output demultiplexers of the buffer performing the function. A shared bus can also serve as a switch and has the advantage that it does not require complex switch allocation — each input broadcasts to all outputs when it acquires the bus. However, a bus makes inefficient use of internal bandwidth. Most high-performance routers use crosspoint switches. A crosspoint switch is usually provided with some excess bandwidth, or speedup, to simplify the allocation problem and to prevent switch allocation from becoming a performance bottleneck and hence idling expensive channel bandwidth.

17.1 Input Buffer Organization

The input buffers are one of the most important structures in a modern router. The flow control protocol allocates space in these buffers to hold flits awaiting channel bandwidth to depart these routers. The buffers provide space for arriving flits so that the incoming channel need not be slowed when a packet is momentarily delayed due to a pipeline stall or contention for a virtual channel or physical channel.

17.1.1 Buffer Partitioning

How the input buffers are partitioned is tightly tied to the design of the switch. Input buffers can be combined across the entire router, partitioned by physical input channel, or partitioned by virtual channel, as illustrated in Figure 17.1. If a single memory is used to hold all of the input buffers on a router, as shown in Figure 17.1(a), there is no need for a switch. However, no real savings exist in such a case. Instead, there are really two switches: one that multiplexes inputs into the memory and one that distributes outputs from the memory.[1] The major advantage of the central memory organization is flexibility in dynamically allocating memory across the input ports. Unfortunately, there are two serious shortcomings with this approach. First, the bandwidth of the single memory can become a serious bottleneck in high-performance routers. For example, in a router with $\delta_i = 8$ input ports, the central memory bandwidth must be at least $2\delta_i = 16$ times the port bandwidth so it can write 8 flits and read 8 flits each flit time. To get high-bandwidth requires a very wide memory, often many flits wide, which leads to the second problem. Router latency is increased by the need to deserialize flits to make a wide enough word to write to the shared memory and then reserialize the flits at the output.

The bandwidth and latency problems of the central buffer memory can be avoided by providing a separate memory per input port,[2] as illustrated in Figure 17.1(b). While the total memory bandwidth is still $2\delta_i$ times the port bandwidth in this arrangement, each individual memory needs only twice the bandwidth of the input port, making implementation more feasible and avoiding the latency penalties of a wide memory. This arrangement enables buffers to be shared across the virtual channels associated with a particular physical channel, but not across the physical channels.

One can subdivide the flit buffers further, providing a separate memory for each virtual channel, as shown in Figure 17.1(c). This has the advantage of enabling input speedup in the switch by allowing the switch to access more than one virtual channel from each physical channel in a single cycle. For example, suppose a router has $\delta_i = 8$ input ports, each of which has $V = 4$ virtual channels. A per-virtual-channel buffer organization would then provide $\delta_i V = 32$ inputs to a 32×8 switch. As shall be discussed in Section 17.2 providing input speedup on the switch in this manner increases the throughput of the router. Dividing the input buffers this finely, however, can lead to poor memory utilization because memory space in the buffers of

---

1. Switching by demultiplexing into the shared memory and multiplexing out of the shared memory is similar to the switching performed by a bus switch (Section 17.2.1) except that the memory decouples the timing of the input and output sides of the switch.
2. Of course, there are possibilities between these two points — for example, sharing one memory between each pair of input ports.

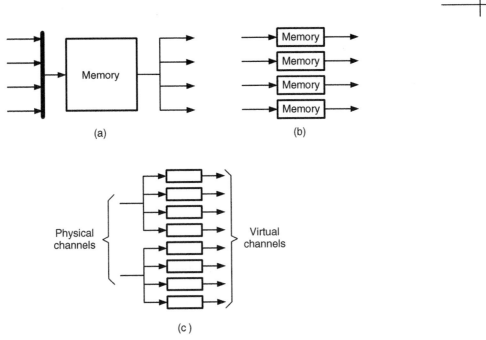

Figure 17.1 Flit buffer organization. (a) A central memory is shared across all input ports of the router. This allows flexibility in allocating buffer space across channels, and the memory serves as a switch as well. However, the bandwidth of the single central memory can become a bottleneck, which makes this organization unsuitable for high-performance routers. (b) A separate buffer memory is provided for each input physical channel. This organization provides bandwidth that scales with the number of ports and requires a switch with a number of input ports equal to the number of physical channels. (c) A separate buffer memory is provided for each virtual channel. This enables the switch to source flits from multiple virtual channels on the same physical channel in a single cycle. However, it often results in significant buffer fragmentation, as the memory of idle virtual channels cannot be shared by busy virtual channels.

idle virtual channels cannot be allocated to busy virtual channels. Also, the overhead associated with so many small memories is considerable. It is much more efficient to have a few larger memories.

A good compromise between a per-physical-channel buffer, as in Figure 17.1(b), and a per-virtual-channel buffer, Figure 17.1(c), is to provide a buffer on each physical channel with a small number of output ports, as shown in Figure 17.2(a). This approach gives us the buffer-sharing and economy-of-scale advantages of per-physical-channel buffers while still permitting switch speedup. The expense of multiport memory in Figure 17.2(a) can be avoided, with a very small penalty in performance, by using a partitioned physical channel buffer rather than a multiport physical channel buffer, as shown in Figure 17.2(b). With this approach, the V virtual channels associated with a physical channel are partitioned across a small number $B < V$ of per physical channel buffers. For example, in the case shown in the figure

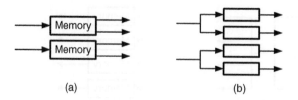

Figure 17.2   Multiple buffer ports per physical channel. (a) Switch input speedup is enabled without the expense of a per-virtual-channel buffer by providing multiple output ports on a per-physical-channel buffer. (b) A less expensive approach that gives almost the same performance is to divide the virtual channels across multiple single-port buffers. For example, all even virtual channels are stored in one buffer and all odd channels in the other.

where $B = 2$, the even virtual channels would be stored in one buffer and the odd virtual channels in the other. This organization sacrifices some flexibility in both buffer allocation (storage cannot be shared across partitions) and switch scheduling (the switch can select only one virtual channel per partition). In practice, however, the performance of a partitioned buffer is nearly identical to a multiport buffer and the cost (in chip area) is significantly less.

17.1.2 Input Buffer Data Structures

In any buffer organization, we need to maintain a data structure to keep track of where flits and packets are located in memory and to manage free memory. Also, in those buffer organizations in which multiple virtual channel buffers share a single memory we need a mechanism to allocate space fairly between the virtual channels.

Two data structures are commonly used to represent flit buffers: circular buffers and linked lists. Circular buffers are simpler to manage, but require a fixed partition of buffer space. Linked lists have greater overhead, but allow space to be freely allocated between different buffers.

In our description of both types of buffers, we use the term *buffer* or *flit buffer* to refer to the entire structure used to hold all of the flits associated with one virtual channel, for example, and the term *cell* or *flit cell* to refer to the storage used to hold a single flit.

A circular buffer is illustrated in Figure 17.3. Fixed FIRST and LAST pointers mark the boundaries of the buffer within the memory. These pointers may be made configurable by placing them in registers. However, they cannot be changed once the buffer is in use. The contents of the buffer lie at and below the Head pointer and above the Tail pointer, possibly wrapping around the end of the buffer, as shown in Figure 17.3(b).

An element is added to the buffer by storing it in the cell indicated by the tail pointer and incrementing the tail pointer in a circular manner. In "C" code, this operation is expressed as:

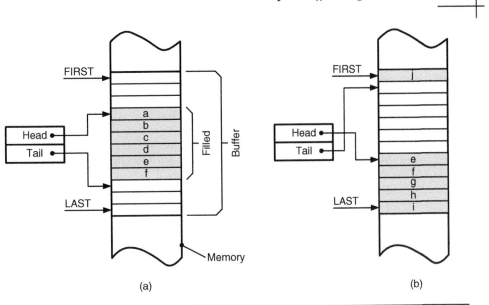

(a)   (b)

Figure 17.3   (a) A buffer containing six flits *a* through *f*. (b) The same buffer after flits *a* through *d* are removed and flits *g* through *j* are added, causing the tail to wrap around the end of the buffer. A circular buffer is a fixed partition of a memory, delimited by constant pointers FIRST and LAST. Variable Head and Tail pointers indicate the current contents of the buffer. Data is inserted into the buffer at the head pointer, which is then *circularly* incremented. Data is removed from the buffer at the tail pointer, which is then circularly incremented.

```
mem[Tail] = new_flit ;
Tail = Tail + 1 ;
if(Tail > LAST) Tail = FIRST ;
if(Tail == Head) Full = 1 ;
```

The third line of this code makes the buffer circular by wrapping the tail pointer back to the start of the buffer when it reaches the end.

Similarly, an element is removed from the buffer by reading the value at the head pointer and circularly incrementing the head pointer:

```
flit = mem[Head] ;
Head = Head + 1 ;
if(Head > LAST) Head = FIRST;
if(Head == Tail) Empty = 1 ;
```

The last line of both code sequences detects when the buffer is full and empty, respectively. These conditions occur when the head and tail pointers are equal. The difference is that a full condition occurs when an insert causes the pointers to be equal and an empty condition occurs when a remove causes the pointers to be equal. The full and empty indications are used by logic external to the flit buffer to avoid making insert requests when the buffer is full and remove requests when the buffer is empty.

In typical implementations, circular buffer data structures are implemented in logic that carries out all four lines of the above code sequences in a single clock cycle. In fact, many implementations are structured so that a buffer can simultaneously insert an element and remove an element in a single clock cycle.

If the buffers are a power of two in size and aligned to a multiple of their size (for example, 16-flit buffers starting at a multiple of 16) then the wrapping of pointers can be performed without a comparison by just blocking the carry from the bits that indicate a location within the buffer to the bits that indicate a buffer within the memory. For example, if flit buffers are 16 cells in size, bits 0 to 3 of the pointer select the cell within the buffer and bits 4 and up identify the buffer. Thus, if we block the carry from bit 3 to bit 4, incrementing the pointer from the last position of a buffer, BF for buffer B, will result in a pointer that wraps to the first position of the buffer B0.

A linked list data structure, illustrated in Figure 17.4, gives greater flexibility in allocating free storage to buffers, but incurs additional overhead and complexity. To manage a memory that has linked lists, each cell is augmented with a pointer field to indicate the next cell in a list. To facilitate the multiple pointer operations required to manage the list, the pointers are usually kept in a separate memory than the flit cells themselves, and this pointer memory is often multiported. For each buffer stored in the memory, a head and tail pointer indicate the first and last cells of the linked list of buffer contents. All cells that do not belong to some buffer are linked together on a free list. Initially, all of the cells in the memory are on the free list and the head and tail pointers for all flit buffers are NULL. For purposes of memory allocation, we also keep a count register for each buffer that indicates the number of cells in that buffer and a count register holding the number of cells on the free list.

Inserting a flit into the buffer requires the following operations:

```
if(Free != NULL) {
mem[Free] = new_flit ;// store flit into head of free list
if(Tail != NULL)
 ptr[Tail] = Free ;    // link onto buffer list - if
Tail = Free ;          // it exists
Free = ptr[Free] ;     // unlink free cell
ptr[Tail] = NULL ;     // optional - clear link to free list
if(Head == NULL)       // if buffer was empty
 Head = Tail ;         // head and tail both point at
                       // this cell
} else ERROR ;         // no cells left on free list
```

Inserting a flit into a linked list like this can be done in a single cycle, but it requires a dual-ported flit buffer. We must write the pointer of the old tail cell and read the pointer of the head of the free list (the new tail cell). However, there are no dependencies between these two operations, so they can go on at the same time if we have enough ports. If we want to clear the pointer of the new tail cell, so it no longer points to the free list (a good idea for consistency checking — see below — but not absolutely required), we need a read-modify-write operation on at least one of the two ports.

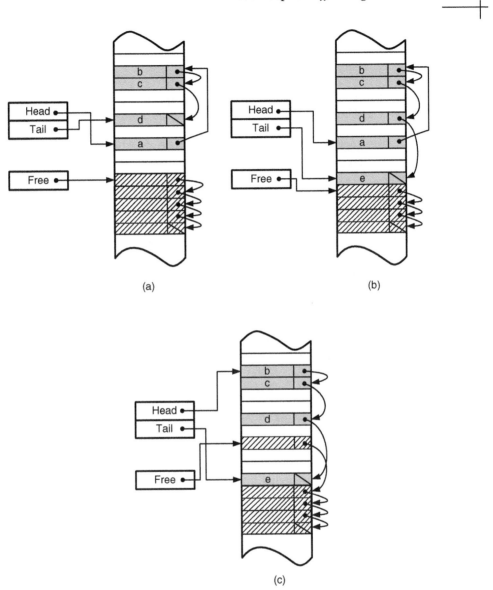

Figure 17.4 A linked list data structure maintains each buffer as a linked list of cells. Free cells are linked together on a free list. (a) A flit buffer contains four cells in a linked list. Head points to the first cell in the list and Tail points to the last cell in the list. Free points to a linked list of five free cells. (b) Flit e is added to the list by removing a cell from the free list, filling it, and linking it into the buffer list. (c) Flit a is removed from the list and its cell is returned to the free list.

Removing a flit from the buffer goes as follows:

```
if(Head != NULL) {
  flit = mem[Head] ;  // get the flit
  Next = ptr[Head] ;  // save pointer to next flit in buffer
  ptr[Head] = Free ;  // put old head cell on the free list
  Free = Head ;       //
  if(Head == Tail) {  // if this was the last cell - empty list
    Head = NULL ;
    Tail = NULL ;
  } else {            // otherwise, follow the link to the
    Head = Next ;     // next cell
  }
} else ERROR ;        // attempt to remove from empty buffer
```

As this code shows, removing a flit is considerably easier. It can be done in a single cycle with a single-port pointer memory that supports a read-modify-write operation to both read the link in the head cell and link it into the free list.

The main cost of using a linked-list structure is the complexity of manipulating the pointer memories, which require multiple pointer-memory operations per insert or remove, and the increased complexity of error handling. The actual storage overhead of the pointers is small. For example, to associate an 8-bit pointer with a typical 8-byte flit requires an overhead of only 12.5%.

Handling bit errors that occur within linked list data structures is much more involved than handling such errors in circular buffers. In a circular buffer, a bit error in the memory affects only the contents of the buffer, not its structure. A bit error in the head or tail pointer of a circular buffer at most results in inserting or deleting a bounded number of flits from the buffer. With circular buffers, errors can always be contained to a single buffer by resetting the state of the buffer (setting Head = Tail = FIRST).

With a linked-list buffer, a single bit error in the pointer field of a flit can have much greater consequences. A bit error in a pointer in the free list disconnects many free flits of storage, resulting in allocation failures, while the buffer counter indicates that many free cells are still available. If the error causes this pointer to point at a cell allocated to a flit buffer, allocating this cell to another buffer will both break the chain of the original buffer and cause the two buffers to share a sequence of the chain. A bit error in a cell allocated to a buffer has similar consequences.

The important point is that bit errors in linked-list structures have effects that are global across all buffers in a memory. Thus, it is important to detect and contain these errors when they occur before the effects of the error have corrupted all buffers in the memory.

Error control methods for linked-list buffers combine defensive coding of entries and background monitoring. Defensive coding starts by adding parity to the pointers themselves so that single bit errors will be detected before a pointer is used. Each flit cell is tagged with a type field that indicates whether it is free or allocated, and this field is used to check consistency. Storing the buffer number, or at least a hash of it, in allocated cells enables us to check that cells have not become swapped between buffers. Finally, storing a sequence number, or at least the low few bits of it, into each cell enables the *input* unit to check that cells have not become reordered. The insert and remove operations check the validity of their pointers, that the type and buffer numbers of the cells read match expectations, and that the sequence numbers of cells removed are in order. If exceptions are detected, the buffers affected are disabled and an error handling routine, which is usually implemented in software, is triggered.

A background monitoring process complements defensive coding by checking the consistency of cells that are not being accessed. This detects errors in the middle of linked lists that would otherwise lie dormant until much later. The monitoring process walks each buffer list, from head to tail, checking that the number of cells in the buffer matches the count, that each cell in the buffer is allocated, that each cell belongs to this buffer, and that there are no parity errors. A similar check is made of the free list. Such list walking is often complemented by a *mark-and-sweep* garbage collection pass to detect cells that somehow have become disconnected from any list. In most implementations the background monitor itself is a hardware process that uses idle cycles of the buffer memory to perform these consistency checks. When an error is detected, the buffers affected are disabled and a software-error handling routine is invoked.

17.1.3 Input Buffer Allocation

If we build our input buffers using a linked-list data structure, we can dynamically allocate storage between the different buffers sharing a memory. This dynamic memory allocation offers the potential of making more efficient use of memory by allocating more memory to busy virtual channels and less to idle virtual channels. Many studies have shown that when there are large numbers of virtual channels, load is not balanced uniformly across them. Some are idle, while others are overloaded. Thus, it is advantageous to allocate more memory to the busy channels and less to the idle channels. However, if not done carefully, dynamic buffer allocation can negate the advantages of virtual channels. If one greedy virtual channel were allowed to allocate all of the storage and then block, no other virtual channel would be able to make progress. This, in turn, creates a dependency between the virtual channels and can potentially introduce deadlock.

To control input buffer allocation, we add a count register to the state of each virtual channel buffer that keeps track of the number of flits in that channel's buffer. An additional count register keeps track of the number of cells remaining on the free list. Using these counts, we can implement many different buffer allocation policies. The input to the policy is the current state of the counts. The output is an indication of whether or not each virtual channel is allowed to allocate an additional buffer.

For example, a simple policy is to reserve one flit of storage for each virtual channel buffer. This guarantees that we won't inadvertently cause deadlock by allocating all of the storage to a single virtual channel that then blocks. Each virtual channel is able to continue making progress with at least one flit buffer. To implement this policy, we allow a virtual channel to add a flit to its buffer if (a) its buffer is currently empty (every virtual channel gets at least one flit), or (b) the count of free flit buffers is greater than the number of virtual channel buffers with a count of zero.

To simplify administration of this policy, we divide our free count into a count of reserved cells and a count of floating cells. A reserved cell count is maintained per virtual channel and is incremented or decremented when a buffer becomes empty or the first flit is inserted into an empty buffer, respectively. We increment or decrement the shared floating count when inserting or removing cells into buffers with at least one additional cell in them. With this arrangement, condition (b) can be checked by checking the count of floating cells for zero.

Although this simple policy does preserve our deadlock properties, it can still lead to great imbalance in buffer allocation. One way to correct such imbalance is to use a sliding limit allocator, which is characterized by two parameters, r and f. Parameter r is the number of cells reserved for each buffer and must be at least one to avoid deadlock. Parameter f is the fraction of the remaining cells (after reserving r for each buffer) that can be allocated to any one buffer.

As with the fixed reservation scheme, we implement the sliding limit allocator by keeping two free cell counters, one for reserved cells N_r and one for floating cells N_f. We allow an insert into buffer b with count N_b if $N_b < r$, in which case the buffer comes from the reserved pool and we decrement N_r, or $r \leq N_b < fN_f + r$, in which case the buffer comes from the floating pool and we decrement N_f.

The aggressiveness of this policy is set by the parameter f. When $f = 1$, a single buffer can allocate the entire free pool and this policy degrades into the reservation policy above. When $f = \frac{1}{B}$, where B is the number of buffers in the memory, each buffer can allocate at most $\frac{1}{B}$ of the pool and the policy is equivalent to a fixed partition. More useful values of f exist between these two extremes. For example, with $f = 0.5$, any buffer can use at most half of the remaining floating pool and with $f = \frac{2}{B}$, no buffer can use more than twice the space as would be granted with a fixed partition.

17.2 Switches

The switch is the heart of a router. It is here that packets and flits are actually directed to their desired output port. The main design parameter with a switch is its *speedup*—the ratio of the switch bandwidth provided to the minimum switch bandwidth needed to support full throughput on all inputs and outputs. As we shall see, adding speedup to a switch simplifies the task of switch allocation and results in higher throughput and lower latency. We can set the speedup on the input side of a switch and on the output side of a switch independently.

A switch of a given speedup can be realized in a number of different ways. If a flit time covers many clock periods, we can perform switching in time (as with a bus) instead of or in addition to switching in space. Also, there are situations in which we compose a switch from a series of smaller switches and thus form a small network in its own right. However, except in the simplest cases, multistage switches raise a host of difficult allocation and flowcontrol problems.

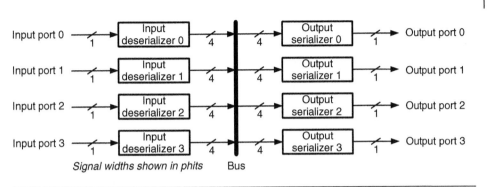

Figure 17.5 A 4 × 4 switch realized as a bus. Each input deserializer accumulates phits, acquires a bus cycle, and then transmits P phits in parallel over the bus to one or more output serializers. The output serializers then send the individual phits over the output ports.

17.2.1 **Bus Switches**

When a flit time consists of more internal clock cycles (phit times) than the number of ports, a bus may be used as a switch, as illustrated in Figure 17.5. The timing of this type of bus switch is shown in Figure 17.6. Each input port of the switch accepts phits of a flit and accumulates them until it has at least P phits, where P is the number of input switch ports. Once an input port has P phits, it arbitrates for the bus. When it acquires the bus, it transmits the P phits, possibly an entire flit, broadside to any combination of output units. The receiving output unit(s) then deserializes the P phits and transmits them one at a time to the downstream logic. For this to work without loss of bandwidth due to fragmentation, the number of phits per flit must be a multiple of P.

The figures show a case in which there are four input ports and four output ports on the bus switch, each flit is four phits in length, and flits are aligned on the inputs and outputs (that is, all flits have their first phit during the same clock cycle). In this case, each input port accumulates an entire flit and then transmits it over the bus. Because each input port has access to all output ports during its bus cycle, there is no need for a complex switch allocation. (See Chapter 19.) Each input port is guaranteed access to the output it desires when it acquires the bus because it gets access to all of the outputs. This also means that multicast and broadcast can be accomplished (ignoring for the moment the flowcontrol issues associated with multicast) with no additional effort. In Figure 17.6, a fixed bus arbitration scheme is used. Input port 0 gets the bus during phit cycle 0 (the first phit of each flit), input port 1 gets the bus during phit cycle 1, and so on. This type of arbitration is extremely simple.

There is no point in having a speedup of greater than one with a bus switch. Because an input port always gets its output port during its bus cycle, it can always dispose of the input flit that just arrived with a speedup of one. Additional speedup increases cost with no increase in performance.

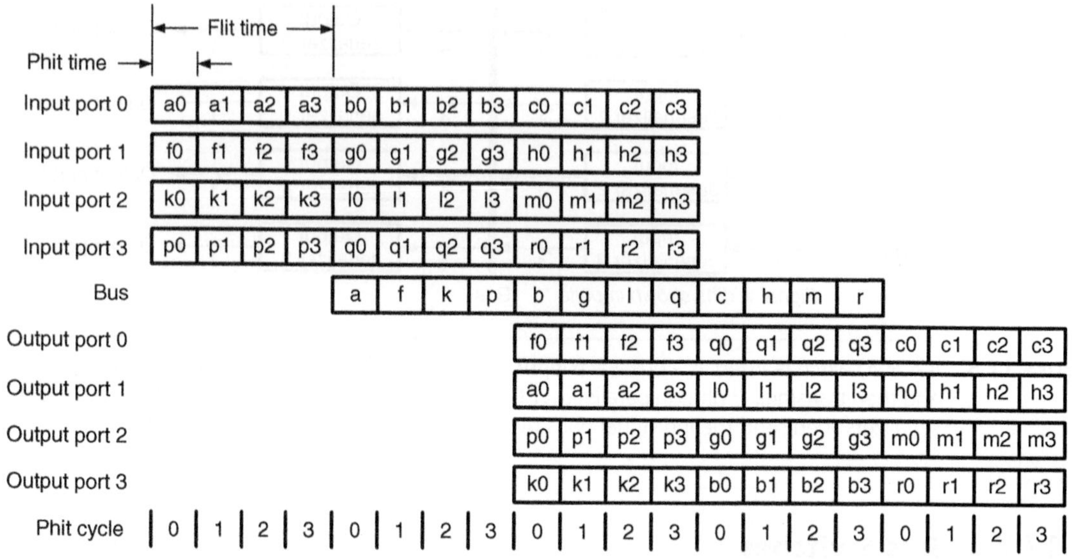

Figure 17.6   Timing diagram for a 4 × 4 bus switch in which each flit consists of four phits. Flits *a, f, k,* and *p* arrive on the four input ports, one phit at a time, during the first flit time; flits *b, g, l,* and *q* arrive during the second flit time; and flits *c, h, m,* and *r* arrive during the third flit time. While the ith flits are arriving, the $(i − 1)^{th}$ flits are transmitted broadside across the bus, one flit being transmitted each phit time, and the $(i − 2)^{th}$ flits are being reserialized on the output ports.

Bus switches have the advantage of being extremely simple. But this simplicity comes at the expense of wasted port bandwidth and increased latency. The bus bandwidth equals the aggregate bandwidth of the router, Pb, where P is the number of ports and b is the per-port bandwidth. To interface to this bus, however, each input deserializer must have an output bandwidth of Pb and each output serializer must have an input bandwidth of Pb. Thus, the aggregate bandwidth out of the input deserializers, and into the output serializers, is P^2b, because only $\frac{1}{P}$ of this bandwidth (on the input side, at least) is active during any given cycle.

If the router is implemented on a chip, this excess bandwidth has some cost but is not a serious limitation.[3] Figure 17.7 shows how the bus lines can be run directly over the input deserializers and output serializers so the cost of this excess bandwidth is low. In fact, this bus switch has a wiring complexity that is exactly double that of a crossbar switch with the same bandwidth, $2P \times P$ phit-wide buses for the bus switch versus $P \times P$ for the crossbar. Each serializer or deserializer in the bus switch also requires $2P$ phits of storage, while a crossbar switch is stateless.

---

3.   If one tried to build a bus switch with the serializers and deserializers on separate chips, however, the pin bandwidth would be prohibitively expensive.

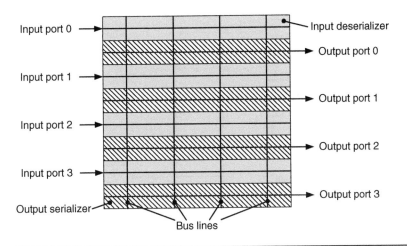

Figure 17.7 Floorplan and wiring plan of a 4 × 4 bus switch. The phit-wide input and output lines run horizontally. Each input line feeds an input deserializer (shaded), while each output line is driven by an output serializer (hatched). The 4-phit-wide bus runs vertically over the serializers and deserializers.

The latency and coarse granularity of a bus switch is a more serious limitation than its excess bandwidth. As shown in Figure 17.6, the bus switch must switch in units of at least P phits. If the flit size is smaller than this, a bus switch is not feasible. Also, the latency of the bus switch is $2P$ phit times because of the time required to deserialize the input and align the outputs.[4] This is in comparison to the delay of a single phit time for a crossbar.

An intermediate point between a straight bus switch and a complete crossbar switch is a multiple-bus switch. Figure 17.8 shows an output partitioned switch with two buses. Outputs 0 and 1 share bus 1 and outputs 2 and 3 share bus 2. By halving the number of outputs supplied by each bus, we halve the number of phits that must be transported over the bus each cycle to keep the outputs busy. Hence, serialization latency is halved. This reduction in latency comes at the expense of more complex allocation. We can no longer use a fixed allocation. Instead, we must solve a matching problem to allocate input ports and output ports for each cycle. (See Chapter 19.)

The multiple bus configuration shown in Figure 17.8 is equivalent to a 2-phit wide 4 × 2 crosspoint switch (four 2-phit-wide inputs switched onto two 2-phit-wide buses). Each output of this 4 × 2 switch drives two output serializers that receive 2-phit-wide data from a bus and output this data one phit at a time. In the extreme case, where we put each output on a separate bus, the multiple bus switch is equivalent to a $P \times P$ crosspoint.

---

4. This latency can be reduced to $P + 1 + a$ phit times, where a is the number of cycles lost during bus arbitration on a bus switch with unaligned output flits. The average value of a if all P ports are competing is $\frac{P}{2}$.

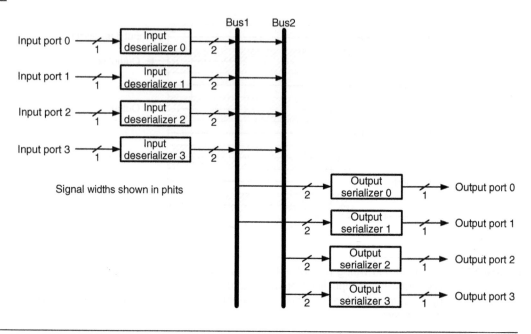

Figure 17.8   Output-partitioned multiple-bus switch: The four inputs can drive two phits at a time onto either bus. Outputs 0 and 1 receive two-phit parcels from bus 1, while outputs 2 and 3 receive from bus 2. This organization reduces latency and excess bandwidth at the expense of more complicated allocation.

We can build input-partitioned multiple-bus switches as well as output-partitioned multiple bus switches, as is explored further in Exercise 17.4. Also, as discussed in Exercise 17.5, we can partition a switch across both the inputs and the outputs at the same time.

Memory switches, such as the one shown in Figure 17.1(a), are bus switches with a memory separating the input and output sides of the bus. These memory switches can be partitioned across inputs or outputs or both as described in Exercise 17.6.

17.2.2   **Crossbar Switches**

Recall from Section 6.2 that an $n \times m$ crossbar or crosspoint switch can connect any of the n inputs to each of the m outputs. We represent crossbars with the symbol shown in Figure 17.9. The structure of crossbars is discussed in detail in Section 6.2 and will not be repeated here.

The primary issue when using a crossbar as the switch in a router datapath is speedup: the ratio of provided bandwidth to required bandwidth. We can provide speedup on the input of the crossbar, the output of the crossbar or on both sides. This speedup can be provided in space (additional inputs) or time (higher bandwidth

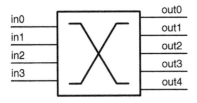

in0
in1
in2
in3

out0
out1
out2
out3
out4

Figure 17.9   Symbol for a 4 × 5 crossbar switch.

inputs).[5] Figures 17.10 and 17.11 show examples of switches with input speedup (Figure 17.10[b]), output speedup (Figure 17.11[a]), and both input and output speedup (Figure 17.11[b]).

Providing a crossbar switch with speedup simplifies the task of allocating the switch, or gives better performance with a simple allocator.[6] Especially with small flit sizes and low latency requirements, it is often cheaper to build a switch with more speedup than it is to build an allocator that achieves the required performance with less speedup. Switch allocation is the process of allocating crossbar input ports and crossbar output ports in a non-conflicting manner to a subset of the flits waiting in the input buffers. The topic of allocation is treated in detail in Chapter 19. Here we assume the use of a simple random separable allocator to compare the performance of switches with varying amounts of speedup.

A random separable allocator works by first having each input buffer randomly select a flit from among all waiting flits to be forwarded on each input port of the crossbar. If traffic is uniform, each output has equal probability of being the destination of each selected input flit. Each output port then randomly selects a flit from among the flits on the input ports bidding for that output.

Suppose our k input, k output router has input speedup of s and no output speedup; thus, the crossbar has sk inputs. With a random separable allocator, an output will forward a flit as long as some input has selected a flit for that output. The throughput Θ of the switch is the probability that at least one of the sk flits selected in the first phase of allocation is destined for a given output. This probability is given by

$$\Theta = 1 - \left(\frac{k-1}{k}\right)^{sk}. \tag{17.1}$$

---

5.   Speedup in space always has an integral value. (You cannot have half of a switch port). Speedup in time, however, may have non-integral values. For example, you can operate a crossbar that passes a flit in 3 cycles in a router that handles a flit every 4 cycles for a speedup of 1.33.

6.   In other contexts, speedup is also used to compensate for overhead that may be incurred by adding headers to packets or flits or due to fragmentation when segmenting packets into flits. In this case, we assume we are switching flits without adding additional overhead and, hence, use speedup strictly to simplify allocation.

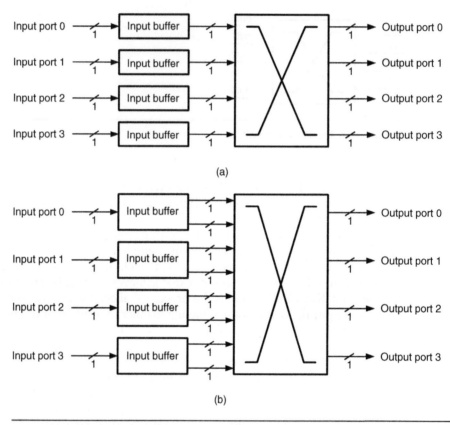

Figure 17.10 Crossbar input speedup. The 4 × 4 switch shown in (a) has an input speedup of one — the number of inputs matches the number of outputs and all are of unit bandwidth. The switch in (b) has an input speedup of two — there are twice as many inputs as outputs — resulting in a simpler allocation problem.

The throughput vs. input speedup curve is plotted for a 4 output switch in Figure 17.12. Using the separable allocator, a switch with no speedup achieves only 68% of capacity, while an input speedup of 2 yields 90% of capacity and 3 brings capacity up to 97%.

When the input speedup of the switch reaches 4 in this case (or k in the general case), there is no point in using an allocator. At this point, the k crossbar inputs out of each input buffer can each be dedicated to a single output port so that each crossbar input handles flits between a single input-output port pair. The situation is then identical to that of the bus switch described above in Section 17.2.1 and 100% throughput is achieved without allocation. Since 100% throughput can be achieved with trivial arbitration with a speedup of k, there is no point in ever making a switch with an input speedup larger than $s = k$.

A switch with output speedup can achieve the exact same speedup as a switch with an identical amount of input speedup by using a separable allocator operating in

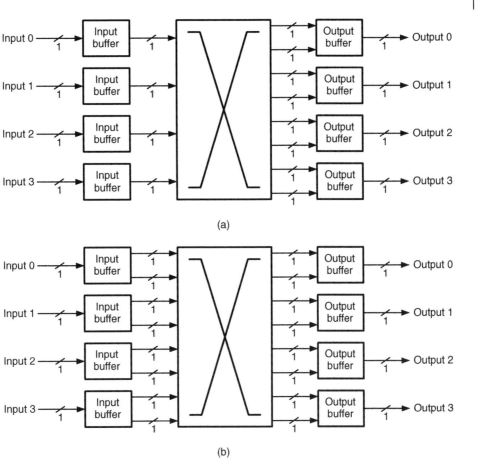

Figure 17.11 Crossbar output speedup. (a) A 4 × 4 switch with an output speedup of 2. (b) A switch that has a speedup of 2 — both an input speedup of 2 and an output speedup of 2.

the reverse direction. Each output port selects a waiting flit destined for that output port and then bids on the input port holding this flit. In practice, this is difficult to implement, since it requires getting information about waiting input flits to the output ports. For this reason, and also to eliminate the need for output buffering, input speedup is generally preferred over output speedup.

A switch that has both input and output speedup can achieve a throughput greater than one. Of course, it cannot sustain this, since the input and output buffers are finite. If the input and output speedups are s_i and s_o, respectively, and assuming $s_i \geq s_o$, we can view this as a switch with overall speedup s_o and with input speedup of s_i/s_o relative to the overall speedup. Thus, the throughput in this case is given by

$$\Theta = s_o \left(1 - \left(\frac{k-1}{k} \right)^{\frac{s_i k}{s_o}} \right).$$

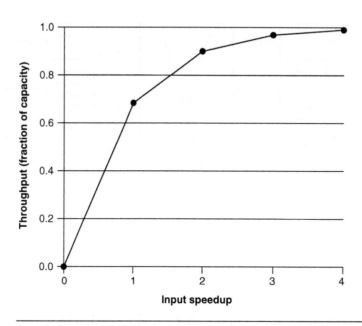

Figure 17.12   Throughput as a function of input speedup for a $k = 4$ output crossbar switch using a single iteration random separable allocator.

Most often, crossbars are operated with a small amount of overall speedup (between 1.2 and 1.5) so that 100% throughput can be achieved without heroic allocation and with low latency and then with the input ports doubled to further simplify allocation.

17.2.3 Network Switches

A switch can be realized as a network of smaller switches. For example, the 7-input × 7-output switch needed to implement a dimension-order router for a 3-D torus network can be realized with three 3 × 3 switches, as shown in Figure 17.13. This reduces the number of crosspoints needed to realize the switch from 49 to 27. Each dimension of this switch can be partitioned further by separating the *plus* and *minus* directions, as shown in Figure 17.14. However, this additional partitioning increases the crosspoint count to 33.

Partitioning the switch has the advantage of reducing the total number of cross-points and enhancing the locality of the logic, thus reducing wire length.[7] On the downside, however, the partitioned switch requires more complex control, if an entire

---

7.   This was the motivation for introducing this type of partitioning as *dimension-slicing* or *port-slicing* in Section 7.2.2

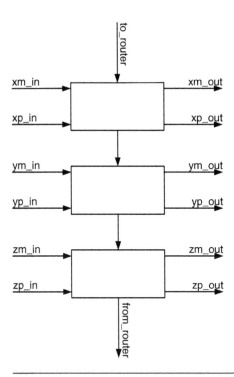

Figure 17.13 A dimension-order router can be divided into smaller routers by implementing the restricted 7 × 7 switching function with three 3 × 3 switches.

path is to be allocated at once, or flit buffers between the partitions to hold a flit while it arbitrates for the next switch. Also, depending on the network, traffic, and routing, the loading on the internal links between the sub-routers may become greater than the load on the channels themselves.

In general, partitioning a switch into a network of switches should be avoided unless it is necessary to fit each partition into a level of the packaging hierarchy.

17.3 **Output Organization**

The datapath of the output unit of Figure 16.1 consists primarily of a FIFO buffer to match the rate of the output channel to the rate of the switch. For a switch that has no output speedup, no datapath is required. As flits exit the switch, they can be placed directly onto the output channel. For a switch with output speedup (in time or space), however, a FIFO buffer is needed to stage the flits between the switch and the channel.

There is no need to partition the output buffer storage across virtual channels. A single FIFO buffer can be shared across all of the virtual channels multiplexed onto

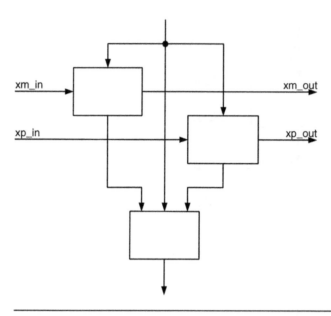

Figure 17.14 The dimension-order router can be divided further, reducing each 3 × 3 switch into two 2 × 2 switches and a 3:1 multiplexer.

the output physical channel without introducing dependencies between the virtual channels. No dependencies exist because the output FIFO never blocks. Every flit time it dequeues the flit at the head of the FIFO (if there is one) and places it on the channel regardless of the state of downstream flit buffers. Backpressure does not affect the output buffer, but rather throttles traffic on the input side of the switch during the SA stage of the router pipeline.

Although there is no backpressure from the channel to the output buffer, there must be backpressure from the output buffer to the switch allocator to prevent buffer overflow. When the buffer occupancy is over a threshold, the output buffer signals to the switch allocator to stop traffic to that output port until the channel can drain the buffer below the threshold.

Output buffers need not be large. Usually 2 to 4 flits of buffering is sufficient to match the speed between the switch and the channel and larger buffers consume area without improving performance.

17.4 Case Study: The Datapath of the IBM Colony Router

The IBM Colony router is the third generation router in the IBM SP family of interconnection networks [177]. The earlier generation Vulcan routers were discussed in the case study presented in Section 11.3 and the Colony router adopts a similar internal architecture and the same multistage network topology. The Colony also includes several enhancements, such as higher signaling speeds, a faster internal cycle time, and support for adaptive and multicast routing. The result is a router that can scale to implement massive interconnection networks. For example, Colony routers are used in ASCI White, an 8,192-processor IBM SP supercomputer, which was the world's fastest computer from 2000 to 2001.

An overview of the Colony's internal architecture is shown in Figure 17.15. The 8×8 switching element adopts a hybrid datapath organization, using both a shared, central memory for storing and switching packets and an input-queued bypass crossbar for switching packets. Any packet arriving at an input port of the switch requests access to the bypass crossbar. If the requested output port of the crossbar is available, the packet is immediately pipelined through the router, minimizing latency. As long as load on the network is low, packets have a good chance of winning access to the bypass crossbar, and their overall latency closely matches that of an input-queued router.

If an input port has failed to gain access to the bypass crossbar and a *chunk*, defined as 8 flits, of a packet has arrived at the input port, then that packet requests access to the central memory. This request is guaranteed to eventually be served. Since it is possible for all 8 input ports of the switch to be blocked and require access to the central memory, the bandwidth into the memory must be equal to the total input bandwidth of the router. While the central memory could conceptually be implemented with 16-port SRAM (one port for each input and output of the router), highly ported SRAMs are expensive and slow and thus impractical for a

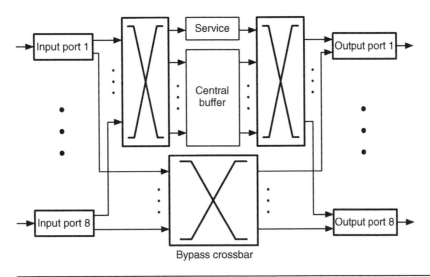

Figure 17.15 The IBM Colony router's architecture. The 8×8 switch is organized around a shared, central buffer that is time-multiplexed between the router ports. A bypass crossbar allows packets to cut through the router, reducing latency when contention is low. The service block initiates and executes network service commands such as diagnostics and error reporting.

high-bandwidth router design. Instead, the designers of the Colony router implemented the central memory as an 8-flit wide (one chunk) 2-port SRAM. Then, the read and write ports of the SRAM are multiplexed in time: each input (output) port transfers an entire 8-flit chunk of data to the SRAM. If multiple input (output) ports want access to the central memory on the same cycle, they are serviced in least recently used order. Time-multiplexing a wide, banked SRAM makes the implementation of the central memory feasible, but it also increases the granularity of the switch to 8-flit chunks. If packet lengths are not a multiple of 8 flits, a significant portion of the switch's bandwidth can be wasted.

As a further optimization, the wide SRAM is banked into four 2-flit wide SRAMs. Instead of each port accessing all the SRAM banks in the same cycle, access is staggered in time. For example, input one transfers the first flit of chunk in one cycle, the second flit in the next cycle, and so on.[8] This approach allows the connections from ports to the SRAM to be implemented with 1-flit wide channels switched through a crossbar. One such crossbar is required for the input and output ports to access the central memory, as shown in Figure 17.15. If bank accesses were not staggered in time, the amount of wiring required would be greatly increased — each port would need a channel 8 times as wide to connect to the central memory.

The usage of the central memory of the Colony router is tracked with several linked-list structures. Each chunk of data stored in the central memory has two additional fields for maintaining these linked-lists. The first field is a next packet pointer and each output maintains a head (`FirstP`) and tail pointer (`LastP`) to the list of packets destined to the output. Since packets may be comprised of several chunks, the second field contains a next chunk pointer. Similarly, a head (`FirstC`) and tail pointer (`LastC`) is also maintained for the list of chunks. One free list tracks all the available chunks in the central memory and the pointer to the head of the list is maintained (`Free`). By also tracking the second entry in the free list (`FreeList`), the chunk lists can support a read and write per cycle while requiring only a 2-port SRAM. (See Exercise 17.8.)

An example of these linked lists for one output of the router is shown in Figure 17.16. Initially, two packets are stored in the central buffer, as in Figure 17.16(a). In the next cycle, Figure 17.16(b), the first chunk of packet A is read and a chunk of packet B is written to the memory. The read begins by checking the head of the chunk list `FirstC`, which is NULL in this case. This indicates that the next chunk should be retrieved from a new packet, packet A in this case. So, by following `FirstP`, the first chunk to the new packet is found and read from the memory. `FirstC` is set to the next chunk pointer from the read chunk and `FirstP` is updated to point to the next packet in the memory. The freed chunk is then added to the free list. The write operation begins by taking a chunk from the free list and storing the chunk's data. Because this write is for packet B, the write operation simply follows `LastC` and adds a next chunk pointer from B2 to B3. (See Figure 17.16(b).) If the write had been to a new packet, `LastP` would have been followed and updated instead.

8.   Two-flit (de)serializing buffers match the 1-flit datapath width of the switch with the 2-flit width of the SRAMs.

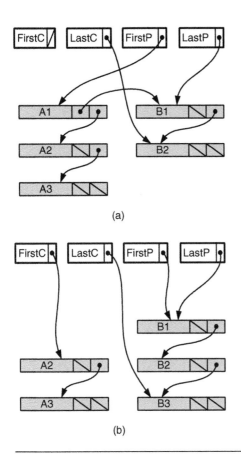

(a)

(b)

Figure 17.16 An example of the linked-list data structures used in the central memory of the Colony router. Every entry in the linked-lists has three fields, which are shown from left to right in the figure: chunk data, next packet, and next chunk. Only one output port's lists are shown for simplicity.

17.5 **Bibliographic Notes**

Tamir and Frazier [182] give a detailed implementation of a dynamic buffer management approach, as do Stunkel et al. [177]. Many earlier interconnection networks, such as the J-Machine [138] and the Cray T3D [95], partitioned their switches by dimension to simplify implementation. A maximally partitioned design is presented by Nojima et al. [137]. Other networks, such as the Torus Routing Chip [56], the SGI SPIDER [69], and the Cray T3E [162] opted for crossbar-based switches with varying amounts of speedup. Shared memory approaches include the design of Katevenis et al. [92] and the IBM Colony switch [177] detailed in the case study. The HIPIQS architecture, introduced by Sivaram et al. [170], combines an input-queued approach with speedup and dynamic buffer management. Finally, both Ahmadi and Denzel [5] and Tobagi [184] provide excellent surveys of different datapath organizations beyond those covered in this chapter.

17.6 **Exercises**

17.1 *Circular buffer management hardware.* Sketch a register-transfer-level diagram showing the logic required to insert and remove flits simultaneously from a circular buffer. Pay particular attention to the logic to determine the full and empty states. Optionally, code this in Verilog.

17.2 *Linked-list buffer management hardware.* Draw a schematic showing the logic required to insert or remove flits (one or the other) from a linked-list flit buffer in a single cycle. Optionally, code this in Verilog.

17.3 *Design of a bus-based switch.* Suppose you have a router with $P = 7$ input ports and $P = 7$ output ports and a flit size of 8 phits. Explain how to build an efficient bus switch for this router. How many phits does the bus need to carry each cycle?

17.4 *An input partitioned bus-based switch.* Draw a block diagram for a two-bus input partitioned switch, like Figure 17.8, but with two inputs driving each bus and each output receiving from both buses. Allow each output to read from only one of these two buses during any given cycle. Describe a method for allocating such a switch. Now consider the case where an output can receive simultaneously from both buses. How does this change your allocation method?

17.5 *A doubly partitioned bus-based switch.* Draw a diagram of an input and output partitioned switch that has eight ports — partition both the inputs and outputs into two groups and use four independent buses for switching. Explain the allocation of this switch.

17.6 *Partitioning a memory switch.* Generalize bus partitioning for the memory switch organization shown in Figure 17.1(a). How many separate memories are required if the outputs are partitioned into two groups? How wide are these memories?

17.7 *A buffered crossbar.* Recursively apply partitioning to the memory switch from Exercise 17.6. That is, for a P input switch, partition the inputs into P groups. Likewise for the outputs. How many separate memories are required in this switch? How wide are these memories?

17.8 *Advantage of tracking two free list entries.* Write pseudocode for insertion and deletion from the IBM Colony linked-list structure using two free list entries (store pointers to both the first and second entry of the free list). Use three separate memories (data, packet pointers, and chunk pointers) that each have a single read and a single write port. Hint: Consider three different cases: insertion only, deletion only, and simultanous insertion and deletion.

CHAPTER 18

Arbitration

While the datapaths of routers are made up of buffers and switches, the control paths of routers are largely composed of arbiters and allocators. As discussed in Chapter 16, we use allocators to allocate virtual channels to packets and to allocate switch cycles to flits. This chapter discusses arbiters, which resolve multiple requests for a single resource. In addition to being useful in their own right, arbiters form the fundamental building block for allocators that match multiple requesters with multiple resources. Allocators are discussed in Chapter 19.

Whenever a resource, such as a buffer, a channel, or a switch port is shared by many agents, an *arbiter* is required to assign access to the resource to one agent at a time. Figure 18.1(a) shows a symbol for an *n*-input arbiter that is used to arbitrate the use of a resource, such as the input port to a crossbar switch, among a set of agents, such as the virtual channels (VCs) connected to that input port. Each virtual channel that has a flit to send requests access to the input port by asserting its request line. Suppose, for example, that there are $n = 8$ VCs and VCs 1, 3, and 5 assert their request lines, r_1, r_3, and r_5. The arbiter then selects one of these VCs — say, VC $i = 3$ — and grants the input port to VC 3 by asserting grant line g_3. VCs 1 and 5 lose the arbitration and must try again later by reasserting r_1 and r_5. Most often, these lines will just be held high until the VC is successful and receives a grant.

18.1 Arbitration Timing

The duration of the grant depends on the application. In some applications, the requester may need uninterrupted access to the resource for a number of cycles. In other applications, it may be safe to rearbitrate every cycle. To suit these different

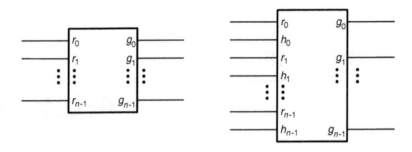

Figure 18.1   Arbiter symbols. (a) An arbiter accepts n request lines, $r_0, \ldots, r_{n-1}$, arbitrates among the asserted request lines, selecting one, r_i, for service, and asserting the corresponding grant line, g_i. (b) To allow grants to be held for an arbitrary amount of time without gaps between requests, a hold line is added to the arbiter.

applications, we may build arbiters that issue a grant for a single cycle, for a fixed number of cycles, or until the resource is released.

For example, an arbiter used in the allocation of a switch that handles flits in a single cycle will grant a virtual channel the right to a switch port for just one cycle. In this case, a new arbitration is performed every cycle, each granting the switch for one cycle — usually, but not always, the cycle immediately following the arbitration. Figure 18.2(a) illustrates this case. Here, VCs 0 and 1 both request the switch during cycle 1 by asserting r_0 and r_1. VC 0 wins this arbitration, is granted the requested switch port during cycle 2, and drops its request during the same cycle. VC 1 keeps its request high and since there is no competition is granted the switch port during cycle 3. If both VCs keep their request lines high, and if the arbiter is fair (see Section 18.2), the switch port will be alternated between them on a cycle-by-cycle basis, as shown in cycles 6 through 9 in the figure.

Figure 18.2(b) shows a situation in which each grant is for four cycles; for example this would occur in a router where the switch takes four cycles to transmit each flit and flit transmissions cannot be interrupted. Here, VC 0 wins the arbitration during cycle 1 and is granted the switch for cycles 2 through 5. VC 1, which asserts its request continuously, must wait until cycle 6 for access to the switch.

Finally, Figure 18.2(c) shows a situation in which the resource is granted until released. In this case, VC i interfaces to the arbiter via both a request line r_i and a hold line h_i, as in Figure 18.1(b). After winning an arbitration, a VC has exclusive use of the resource as long as the hold line is held high. For example, here VC 0 holds the switch for 3 cycles (cycles 2 through 4) and then VC 1 holds the switch for 2 cycles (cycles 5 and 6).

Some arbiters implement variable length grants by holding the resource until the request line is dropped. However, such an approach requires at least one idle cycle between two uses of the resource by the same agent. With a separate hold line, an agent can release the resource and request it again during the same cycle. VC 1 does exactly this during cycle 6, winning the switch again for a single cycle (cycle 7).

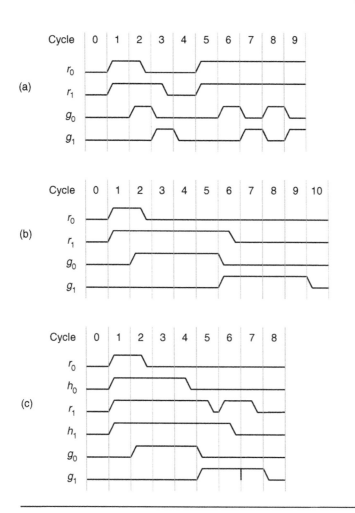

Figure 18.2 Arbiter timing. (a) An arbiter grants a resource for one cycle at a time. A new arbitration each cycle determines the owner for the next cycle. (b) An arbiter grants a resource for a fixed period of four cycles. (c) An arbiter grants a resource for as long as the requester asserts the hold line. There are three grants in this example: to requester 0 for 3 cycles, to requester 1 for 2 cycles, and to requester 1 for one cycle. The last two grants are back-to-back so that g_1 does not go low between them.

18.2 **Fairness**

A key property of an arbiter is its *fairness*. Intuitively, a fair arbiter is one that provides *equal* service to the different requesters. The exact meaning of *equal*, however, can vary from application to application. Three useful definitions of fairness are:

Weak fairness: Every request is **eventually** served.

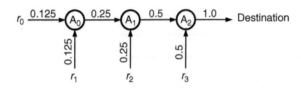

Figure 18.3   In a system with multiple arbiters, each arbiter can be locally fair, while at the system level the overall arbitration is unfair. In this example, requester r_3 receives four times the bandwidth to the destination as r_1 even though each of the three arbiters is individually fair in the strong sense.

Strong fairness: Requesters will be served **equally often**. This means that the number of times one requester is served will be within ϵ of the number of times some other requester is served when averaged over a sufficient number of arbitrations, N. Often, we modify strong fairness to be *weighted* so that the number of times requester i is served is proportional to its weight, w_i.

FIFO fairness: Requesters are served **in the order** they made their requests. This is like customers at a bakery who are served in the order they arrived by each taking a number.

Even if an arbiter is *locally* fair, a system employing that arbiter may not be fair, as illustrated in Figure 18.3. The figure illustrates the situation in which four sources, $r_0, \ldots, r_3$, are sending packets to a single destination. The sources are combined by three strongly fair 2:1 arbiters. Although each arbiter fairly allocates half of its outgoing bandwidth to each of its two inputs, bandwidth is not fairly allocated to the four sources. Source r_3 receives half of the total bandwidth, since it needs to participate in only one arbitration, while sources r_0 and r_1 receive only $\frac{1}{8}$ of the bandwidth — since they must participate in three rounds of arbitration. This same situation, and ways to address it, are discussed in Section 15.4.1 and in Section 18.6 of this chapter.

18.3 Fixed Priority Arbiter

If we assign priority in a linear order, we can construct an arbiter as an *iterative circuit*, as illustrated for the fixed-priority arbiter of Figure 18.4. We construct the arbiter as a linear array of *bit cells*. Each bit cell i, as shown in Figure 18.4(a), accepts one request input, r_i, and one carry input, c_i, and generates a grant output, g_i, and a carry output, c_{i+1}. The carry input c_i indicates that the resource has not been granted to a higher priority request and, hence, is available for this bit cell. If the current request is true and the carry is true, the grant line is asserted and the carry output is deasserted, signaling that the resource has been granted and is no longer available.

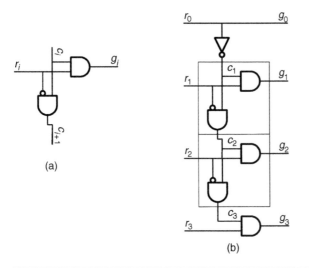

Figure 18.4   A fixed priority arbiter. (a) A bit cell for an iterative arbiter. (b) A four-bit fixed priority arbiter.

We can express this logic in equation form as:

$$g_i = r_i \wedge c_i$$
$$c_{i+1} = \neg r_i \wedge c_i$$

Figure 18.4(b) shows a four-bit fixed priority arbiter constructed in this manner.[1] The arbiter consists of four of the bit cells of Figure 18.4(a). The first and last bit cells, however, have been simplified. The first bit cell takes advantage of the fact that c_0 is always 1. While the last bit cell takes advantage of the fact that there is no need to generate c_4.

At first glance, it appears that the delay of an iterative arbiter must be linear in the number of inputs. Indeed, if they are constructed as shown in Figure 18.4, then that is the case, since in the worst case the carry must propagate from one end of the arbiter to the other. However, we can build these arbiters to operate in time that grows logarithmically with the number of inputs by employing carry-lookahead techniques — similar to those employed in adders. See Exercise 18.1.

Although useful for illustrating iterative construction, the arbiter of Figure 18.4 is not useful in practice because it is completely unfair. It is not even fair in the weak sense. If request r_0 is continuously asserted, none of the other requests will *ever* be served.

---

1.   This is the same type of arbiter used in the circuit of Figure 2.6.

18.4 **Variable Priority Iterative Arbiters**

We can make a fair iterative arbiter by changing the priority from cycle to cycle, as illustrated in Figure 18.5. A one-hot priority signal p is used to select the highest priority request. One bit of p is set. The corresponding bit of r has high priority and priority decreases from that point cyclically around the circular carry chain.[2]

The logic equations for this arbiter are:

$$g_i = r_i \wedge (c_i \vee p_i)$$
$$c_{i+1} = \neg r_i \wedge (c_i \vee p_i)$$
$$c_0 = c_n$$

18.4.1 **Oblivious Arbiters**

Different schemes can be used to generate the priority input into the circuit of Figure 18.5, resulting in different arbitration behavior. If p is generated without any information about r or g, an oblivious arbiter results. For example, generating p with

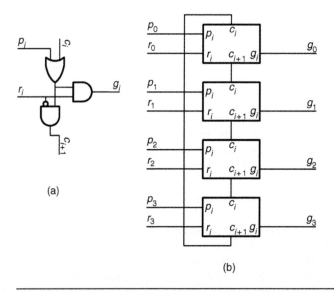

(a)

(b)

Figure 18.5   A variable priority iterative arbiter. The high-priority request input is selected by the one-hot p signal. (a) A one-bit slice of a variable priority arbiter. (b) A four-bit arbiter built from four such bit slices.

2.   Some timing verifiers have trouble with cyclic carry chains like this. This problem can be avoided by replicating the arbiter, connecting the carry-in of the first arbiter to "0" connecting the carry-out of the first arbiter into the carry-in of the second, and OR-ing the grants together.

a shift register rotates the priority by one position each cycle, and generating p by decoding the output of a random number generator gives an arbiter with a different random priority each cycle.

Both the rotating arbiter and the random arbiter have weak fairness. Eventually, each request will become high priority and get service. These arbiters, however, do not provide strong fairness. Consider the case where two adjacent inputs r_i and r_{i+1} repeatedly request service from an n-input arbiter, while all other request inputs remain low. Request r_{i+1} wins the arbitration only when p_{i+1} is true while r_i wins the arbitration for the other $n - 1$ possible priority inputs. Thus, r_i will win the arbitration $n - 1$ times as often as r_{i+1}. This unfairness can be overcome by using a *round-robin* arbiter.

18.4.2 **Round-Robin Arbiter**

A round-robin arbiter operates on the principle that a request that was just served should have the lowest priority on the next round of arbitration. This can be accomplished by generating the next priority vector p from the current grant vector g. In Verilog, this logic is given by:

```
assign next_p = |g ? {g[n-2:0],g[n-1]} : p ;
```

Figure 18.6 shows in schematic form, a four-bit round-robin arbiter. If a grant was issued on the current cycle, one of the g_i lines will be high, causing p_{i+1} to go high on the next cycle. This makes the request next to the one receiving the grant highest priority on the next cycle, and the request that receives the grant lowest priority. If no grant is asserted on the current cycle, any g is low and the priority generator holds its present state.

The round-robin arbiter exhibits strong fairness. After a request, r_i is served, it becomes the lowest priority. All other pending requests will be serviced before priority again rotates around so that r_i can be serviced again.

18.4.3 **Grant-Hold Circuit**

The arbiters we have discussed so far allocate the resource one cycle at a time. A client that receives a grant can use the resource for just one cycle and then must arbitrate again for the resource. This cycle-by-cycle arbitration is fine for many applications. However, as explained in Section 18.1, in some applications clients require uninterrupted use of the resource for several cycles. For example, a client may need uninterrupted access to a switch for three cycles to move a packet into an output buffer.

The duration of a grant can be extended in an arbiter by using a grant-hold circuit, as illustrated in Figure 18.7. The Verilog code for this circuit is shown below. If a particular bit had a grant on the last cycle (last is true), and the hold line for that bit (h) is asserted, that bit will continue to receive a grant (g stays high), and the

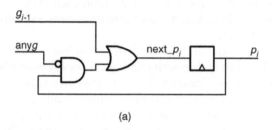

(a)

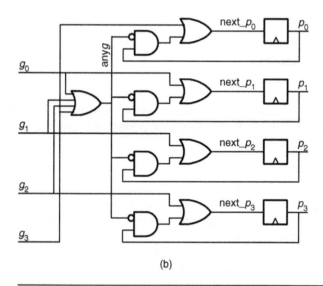

(b)

Figure 18.6　A round-robin arbiter makes the last winning request lowest priority for the next round of arbitration. When there are no requests, the priority is unchanged. (a) A one-bit slice of a round-robin priority generator. (b) A four-bit priority generator build from four of these slices.

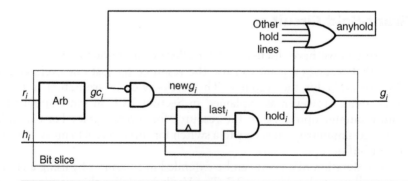

Figure 18.7　An arbiter with a grant-hold circuit holds a grant signal g_i and disables further arbitration until hold line h_i is released.

output of the arbiter (gc) is suppressed by signal anyhold. We have already seen an example of this type of grant-hold circuit, in a slightly different form, in Figure 2.6.

```
assign g = anyhold ? hold : gc ;    // n-bit grant vector
assign hold = last & h ;            // n-bit hold vector
assign anyhold = |hold ;            // hold not zero
always @(posedge clk) last = g ;    // last cycle's grant vector
```

18.4.4 **Weighted Round-Robin Arbiter**

Some applications require an arbiter that is unfair to a controlled degree, so that one requester receives a larger number of grants than another requester. A weighted round-robin arbiter exhibits such behavior. Each requester i is assigned a weight w_i that indicates the maximum fraction f_i of grants that requester i is to receive according to $f_i = \frac{w_i}{W}$ where $W = \sum_{j=0}^{n-1} w_j$. A requester with a large weight will receive a large fraction of the grants while a requester with a small weight receives a smaller fraction. For example, suppose a weighted round-robin arbiter has four inputs with weights of $1, 3, 5$, and 7, respectively. The inputs will receive $\frac{1}{16}, \frac{3}{16}, \frac{5}{16}$, and $\frac{7}{16}$ of the grants, respectively.

As shown in Figure 18.8, a weighted round-robin arbiter can be realized by preceding an arbiter with a circuit that disables a request from an input that has already used its quota. The figure shows a one-bit slice of a weighted round-robin arbiter which consists of a weight register, a counter, an AND-gate, and a conventional arbiter. When the preset line is asserted, the counter in each bit-slice is loaded with the weight for that slice. As long as the counter is non-zero, the AND-gate is enabled and requests are passed along to the arbiter. Each time the requester receives a grant, the counter is decremented. When the counter has reached zero (the requester has received its share of grants), the AND-gate is disabled and requests from this requester are disabled until the next preset.

The preset line is activated periodically each W cycles. If all inputs are requesting their quotas, the counters will have all reached zero between presets. If some inputs do not use their quota, the counters of the other inputs will reach zero before the

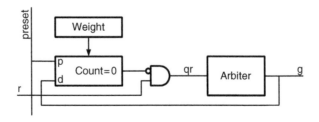

Figure 18.8   A one-bit slice of a weighted round-robin arbiter.

end of the preset interval, and the resource will remain idle until the preset interval is completed. In some cases, a multistage arbiter is used to allow requesters that have consumed their quota to compete for these otherwise idle cycles.

Choosing the number of weight bits presents a tradeoff between precision and burstiness. A large number of weight bits allows weights to be specified with very high precision. Unfortunately, large weights also result in burstiness. Requesters are only guaranteed their shares over the preset interval, W cycles, and increasing the size of weights makes this interval larger. Within this interval usage is uneven with the low weight requesters getting more than their share at the beginning of the interval and the high-weight requesters making it up at the end. (See Exercise 18.2.)

18.5 Matrix Arbiter

A matrix arbiter implements a *least recently served* priority scheme by maintaining a triangular array of state bits w_{ij} for all $i < j$. The bit w_{ij} in row i and column j indicates that request i takes priority over request j. Only the upper triangular portion of the matrix need be maintained since the diagonal elements are not needed and $w_{ji} = \neg w_{ij} \forall i \neq j$. Each time a request k is granted, it clears all bits in its row, w_{k*}, and sets all bits in its column, w_{*k} to give itself the lowest priority since it was the most recently served.

Figure 18.9 shows a four-input matrix arbiter. The state is maintained in the six flip-flops denoted by solid boxes in the upper triangular portion of the matrix. Each of the shaded boxes in the lower triangular portion of the matrix represents the complementary output of the diagonally symmetric solid box.

When a request is asserted, it is AND-ed with the state bits in its row to disable any lower priority requests. The outputs of all of the AND gates in a column are OR-ed together to generate a disable signal for the corresponding request. For example, if bit w_{02} is set and request r0 is asserted, signal dis2 will be asserted to disable lower priority request 2. If a request is not disabled, it propagates to the corresponding grant output via a single AND gate. When a grant is asserted, it clears the state bits in its row and sets the state bits in its column to make its request the lowest priority on subsequent arbitrations. The modification to the state bits takes place on the next rising edge of the clock. See Figure 18.10.

To operate properly, a matrix arbiter must be initialized to a legal state.[3] Not all states of the array are meaningful since for an n-input arbiter there are $2^{n(n-1)/2}$ states and only $n!$ permutations of inputs. States that result in cycles are particularly dangerous because they can result in deadlock. For example, if $w_{01} = w_{12} = 1$ and $w_{02} = 0$ and requests 0, 1, and 2 are all asserted, the requests will disable one another and no grants will be issued.

---

3.   Its also a good idea to have some mechanism for the arbiter to recover from an illegal state should it ever enter one due to a bit error.

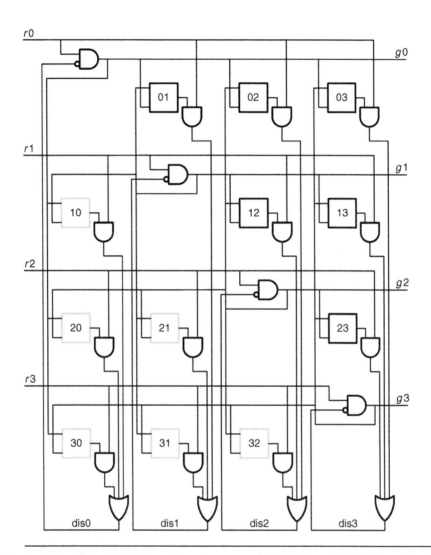

Figure 18.9   A matrix arbiter maintains a triangular array of state bits w_{ij} that implement a least recently served order of priority. If w_{ij} is true, then request i will take priority over request j. When a request k wins an arbitration, it clears bits w_{k*} and sets bits w_{*k} to make itself lowest priority.

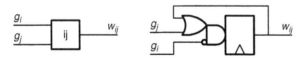

Figure 18.10   Each state bit of the matrix arbiter w_{ij} consists of a flip-flop that is synchronously set on column grant g_j and reset on row grant g_i.

The matrix arbiter is our favorite arbiter for small numbers of inputs because it is fast, inexpensive to implement, and provides strong fairness.

18.6 Queuing Arbiter

While a matrix arbiter provides least recently served priority, a queueing arbiter provides FIFO priority. It accomplishes this, as illustrated in Figure 18.11, by assigning a time stamp to each request when it is asserted. During each time step, the earliest time stamp is selected by a tree of comparators.

The `timer` block in Figure 18.11 is a counter that increments a time stamp during each *arrival interval*, typically each clock cycle. When a request arrives[4] the current time stamp is latched in the `stamp` block and associated with the index of the request; the index is 0 for request r_0, 1 for request r_1, and so on. At the output of the `stamp` block, a request is represented by an index-time stamp pair, i, t_i. Inputs with no request are represented by a unique time value that is considered to be larger than all possible time stamps, and an unused index that does not decode to a grant. A tree of comparators then selects the *winning* index, the one with the lowest (earliest) time stamp, i_w. Each node of the tree performs a pair-wise comparison between two requests: the request (index and time stamp) that has the lower time stamp *wins* the comparison and proceeds to the next level of the tree. Finally, the output index is decoded to give a radial grant signal.

The cost of the arbiter is largely determined by the number of bits used to represent the time stamp since this determines the size of the registers in the `stamp`

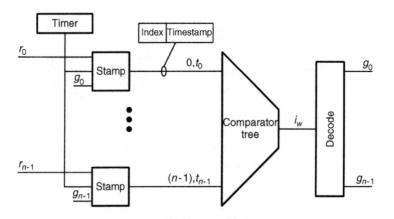

Figure 18.11 A queueing arbiter serves requests in order of arrival by assigning a time stamp to each request and selecting the earliest time stamp by using a tree of comparators.

---

4. A request *arrives* during a cycle when the request line toggles from 0 to 1 or during a cycle after a grant is received when the request line remains high.

blocks and the width of the comparators in the comparator tree. The ratio of two time parameters determines the required time stamp width w_t, the *time stamp range* Δt, and the *arrival interval* t_a.

$$w_t = \log_2 \frac{\Delta t}{t_a}$$

Both parameters are usually represented in units of clock cycles. Δt must be large enough to cover the anticipated maximum difference in arrival time of requests and also must be large enough to accommodate timer *rollover*:

$$\Delta t = 2n\,T_{\max} \tag{18.1}$$

where n is the number of inputs to the arbiter and $T_{\max}$ is the maximum service time. The factor of two is described below.

When a large Δt is required, w_t can be reduced by increasing t_a at the expense of less precision. The arbiter serves requests in the order of their arrival time rounded to the nearest t_a. The larger t_a becomes, the more requests may be served out of order.

When the timer *rolls over* from its maximum value to zero, care is required to prevent new requests from being assigned smaller time stamps than older requests and, hence, cutting to the head of the queue.

This unfair behavior can be prevented by XNOR-ing the most significant bit of all time stamps with the most significant bit of the timer before comparison. For example, consider the case in which $\Delta t = 16$, so we have a $w_t = 4$ bit timer, and $t_a = 1$. The maximum difference in request times is $nT_{\max} = 8$. When the timer rolls over from 15 to 0, all time stamps will be in the range of [8,15]. Thus, if we complement the MSB of each time stamp, we translate the old time stamps to the range [0,7] and new time stamps are assigned a value of 8. Similarly, when the counter rolls over from 7 to 8, the oldest time stamp has a value of 0. At this point, we uncomplement the MSB of the time stamps, so that new requests, which will be assigned a time stamp of 8, will appear younger than these existing requests. This approach does require that Δt be twice the maximum difference in request times, hence the factor of two in Equation 18.1. Exercise 18.7 investigates how to relax this requirement so that Δt need only be slightly larger than $n\,T_{\max}$.

In applications where service times are small, range requirements are very modest. For example, consider a four-input arbiter in which all service times are one cycle. In this case $nT_{\max} = 4$ and, with the MSB complement rollover method, $\Delta t = 8$. Thus, three-bit time stamps suffice. If we are willing to relax our precision so that $t_a = 2$, we can get by with two-bit time stamps. We could reduce the time stamps to a single bit by choosing $t_a = 4$. However, at that point the fairness achieved by the queueing arbiter is no better than would be achieved by a round-robin arbiter.

Queueing arbiters are particularly useful at ensuring global (or system-level) fairness through multiple stages of arbitration. This problem was discussed in Section 18.2 and illustrated in Figure 18.3. To achieve system-level fairness, each request (for example, each packet) is issued a time stamp when it enters the system and before it is queued to wait for any arbitration. This is accomplished with a time-stamp counter (synchronized across the system) and a `stamp` module at each entry point to the system. Once it has been stamped at entry, each request keeps this initial time stamp and uses it for all arbitrations. It is not assigned a new time stamp at each arbiter. Also, all queues in the system must be ordered by time stamp so that the oldest time stamps are served first. A system that uses arbiters that select the lowest global time stamp in this manner will provide strong system-level fairness. For example, in the case of Figure 18.3, all requesters will receive an identical fraction (0.25) of the total bandwidth.

18.7 Exercises

18.1 *Carry-lookahead arbiter.* Sketch a 64-input fixed-priority arbiter that operates in time proportional to the logarithm of the number of inputs. Hint: Partition the inputs into 4-bit groups and compute a *propagate* signal for each group that is true if a carry entering the group will result in a carry leaving the group. That is, the carry will propagate through the group. Then use these 4-bit propagate signals to generate corresponding propagate signals for 16-bit groups.

18.2 *Weighted round-robin.* Suppose you have a 4-input weighted round-robin arbiter with weights of 5, 5, 5, and 49, respectively, and that all inputs are making requests all of the time. Describe the time sequence of grants returned by the arbiter in this case and how it differs from the *ideal* time sequence. Optionally, sketch a weighted round-robin arbiter that achieves the ideal time sequence.

18.3 *Operation of a matrix arbiter.* Suppose you have a 4-input matrix arbiter like the one shown in Figure 18.9. The arbiter is initialized by clearing all 6 state bits. Then the following request vectors are asserted during 4 cycles, one per cycle: $r_3, \ldots, r_0 =$ 1111, 1111, 1010, 1001. Which grant signals are asserted each cycle, and what is the final state of the matrix?

18.4 *Design of a queueing arbiter.* You are to design a queueing arbiter that has 8 inputs. The maximum service time for any input is 8 cycles. However, the sum of any 8 service times is guaranteed to be less than 32 cycles. You can tolerate an arbiter timing precision of $t_a = 4$ cycles. How many bits of time stamp are required to construct this arbiter, assuming you use the MSB complement scheme to handle timer rollover?

18.5 *Timer rollover.* Describe how to deal with timer rollover in a queueing arbiter along the lines of the one in Figure 18.11 in which $\Delta t = nT_{max} + 1$.

18.6 *Errors in a matrix arbiter's state.* Is there any illegal state of a matrix arbiter (values of w_{ij} for $i > j$) where a set of requests can result in multiple grants? Also, suggest a simple way to detect and correct any errors in the state of the arbiter.

CHAPTER 19

Allocation

While an arbiter assigns a single resource to one of a group of requesters, an *allocator* performs a matching between a group of resources and a group of requesters, each of which may request one or more of the resources. Consider, for example, a set of router input units, each of which holds several flits destined for different output ports of a switch. We have already been introduced to allocators in Section 17.2.2, where we discussed their use to allocate crossbar switches. On each cycle, a switch allocator must perform a matching between the input units and the output ports so that at most one flit from each input port is selected and at most one flit destined to each output port is selected.

19.1 **Representations**

An $n \times m$ allocator is a unit that accepts n m-bit vectors as inputs and generates n m-bit vectors as outputs, as illustrated in Figure 19.1 for an $n = 4$ by $m = 3$ allocator. When request input r_{ij} is asserted, requester i wants access to resource j. Each requester can request any subset of the resources at the same time. For allocators used in router designs, the requesters often correspond to switch inputs and the resources correspond to switch outputs. So, we will usually refer to the requesters and resources as inputs and outputs, respectively.

The allocator considers the requests and generates the grant vectors subject to three rules:

1. $g_{ij} \Rightarrow r_{ij}$, a grant can be asserted only if the corresponding request is asserted.
2. $g_{ij} \Rightarrow \neg g_{ik} \forall k \neq j$, at most, one grant for each input (requester) may be asserted.

363

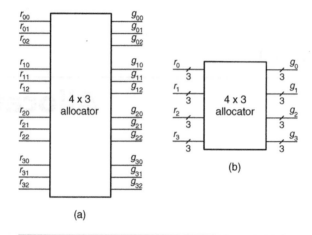

(a)

(b)

Figure 19.1   An $n \times m$ allocator accepts n m-bit request vectors and generates n m-bit grant vectors. (a) Symbol for an allocator with individual request and grant lines shown. (b) Symbol with bundled requests and grants.

3. $g_{ij} \Rightarrow \neg g_{kj} \forall k \neq i$, at most, one grant for each output (resource) can be asserted.

The allocator can be thought of as accepting an $n \times m$ request matrix R containing the individual requests, r_{ij} and generating a grant matrix G containing the individual grants, g_{ij}. R is an arbitrary binary-valued matrix.[1] G is also a binary-valued matrix that only contains ones in entries corresponding to non-zero entries in R (rule 1), has at most one one in each row (rule 2), and at most one one in each column (rule 3).

Consider, for example, the following request matrix R and two possible grant matrices, G_1 and G_2:

$$R = \begin{bmatrix} 1 & 1 & 1 \\ 1 & 1 & 0 \\ 1 & 0 & 0 \\ 0 & 1 & 0 \end{bmatrix} \quad G_1 = \begin{bmatrix} 1 & 0 & 0 \\ 0 & 1 & 0 \\ 0 & 0 & 0 \\ 0 & 0 & 0 \end{bmatrix} \quad G_2 = \begin{bmatrix} 0 & 0 & 1 \\ 0 & 1 & 0 \\ 1 & 0 & 0 \\ 0 & 0 & 0 \end{bmatrix}$$

Both G_1 and G_2 are valid grant matrices for request matrix R in that they comply with the three rules listed above. G_2, however is generally a more desirable grant matrix in that it has assigned all three resources to inputs.[2] A solution to the allocation problem such as G_2 containing the maximum possible number of assignments is called a *maximum matching*. A matrix such as G_1, in which no additional requests

1. In some applications, R can be integer-valued — where multiple requests can be made for the same output. However, we shall not consider those applications here.
2. Actually, a maximum-size match is not always best for non-uniform traffic [124].

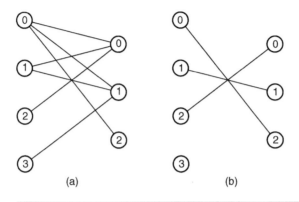

(a)   (b)

Figure 19.2   Bipartite graph representation of the allocation problem. (a) Bipartite graph corresponding to request matrix R. (b) Bipartite graph matching corresponding to grant matrix G_2.

can be serviced without removing one of the existing grants, is called a *maximal matching*.

Matrix G_1 illustrates the difficulty of computing a good matching. Once grant g_{00} is set in the first row of the matrix, a maximum matching is no longer possible. Because input 2 is requesting only resource 0, allocating this resource to any other input causes input 2 to go idle. Allocating resource 1 to input 0 will also prevent a maximum matching since inputs 1 and 2 cannot both be assigned to resource 0. For difficult request matrices, making one bad assignment can result in a sub-optimal matching.

An allocation problem can also be represented as a bipartite graph.[3] For example, Figure 19.2(a) shows a bipartite graph representation of request matrix R. Cast in this form, the allocation problem is identical to the bipartite matching problem: the problem of finding a maximum subset of graph edges so that each vertex is incident on at most one edge in the subset. Figure 19.2(b) shows a maximum matching of the graph of Figure 19.2(a) corresponding to grant matrix G_2.

A maximum matching for any bipartite graph can be computed in time that is $O(P^{2.5})$ for a P port switch [83]. Generally speaking, computing a maximum-size match requires backtracking — grants may be added temporarily, but later removed during the course of the algorithm. Unfortunately, a switch allocator or virtual channel allocator cannot afford the complexity of such an exact solution and the necessity of backtracking makes maximum-size algorithms difficult to pipeline. Since there is typically only a single cycle, or at most a small number of cycles, available to perform an allocation, maximal-size matches become a more realistic target for a high-speed implementation. However, as we shall see, maximal matches can also be quite expensive and, in practice, hardware allocators often introduce simple heuristics that give fast but approximate solutions.

3.   Recall that in a bipartite graph, the vertices can be partitioned into two sets A and B, and all edges connect a vertex in A to a vertex in B.

19.2 **Exact Algorithms**

Although an exact maximum matching is not feasible in the time budget available for most router applications, there are times when we can compute such matchings offline. Also, when evaluating a new heuristic, we would like to compute the maximum matching for comparison with the result of our heuristic. In these cases, we can compute a maximum matching by iteratively augmenting a sub-maximum matching.

The augmenting path algorithm is similar in spirit to the looping algorithm used to schedule Clos networks (Figure 6.11). First, we start with a sub-optimal matching M of the bipartite graph. From this matching, we construct a directed *residual graph*. The residual graph has the same edges and nodes as the bipartite graph representing the request matrix, but the edges are oriented: if an edge is in the current matching M, it points from its output to its input (from right to left in the Figure 19.3); otherwise the edge points from its input to its output. Now, an *augmenting path* is found, which is any directed path through the residual graph from an unmatched input to an unmatched output. Then the matching is updated using this path: any edge from an input to output in the path is added to the matching and the edge from the output to the corresponding input in path is removed from the matching. This gives a matching that includes one more edge, and this process is repeated until the residual graph no longer admits an augmenting path.

To see how the algorithm works, consider the following request matrix:

$$R = \begin{bmatrix} 1 & 1 & 1 & 1 & 0 & 0 \\ 0 & 1 & 0 & 1 & 0 & 0 \\ 0 & 1 & 0 & 0 & 0 & 0 \\ 0 & 1 & 0 & 1 & 1 & 1 \\ 0 & 0 & 0 & 0 & 1 & 0 \\ 0 & 0 & 0 & 1 & 1 & 0 \end{bmatrix}. \tag{19.1}$$

A bipartite graph representing this request matrix and the construction of a maximum matching on this graph are illustrated in Figure 19.3. Figure 19.3(a) shows the graph with an initial matching shown in bold. Matched inputs and outputs are shown in gray. This initial matching happens to be maximal. We construct the corresponding residual graph, as shown in Figure 19.3(b). Panel (c) of the figure shows how an augmenting path is constructed from an unassigned input (2) to an unassigned output (0). The path traverses edge (i2,o1), edge (o1,i0), and edge (i0,o0). After the path is found, the edges along the path change their assignment status in the matching, as shown in panel (d) of the figure: left to right edges, (i2,o1) and (i0,o0), are added to the matching, and the right to left edge, (o1,i0), is removed.

This process then repeats. Panel (e) shows how augmenting path from input 5 to output 5 is found on the configuration of panel (d). Updating the edge assignments along this second augmenting path gives the matching of panel (f). At this point, the algorithm is done because no more augmenting paths exist in the residual graph. By the theory of Ford and Fulkerson [67], we know that if no augmenting paths exist, the matching is maximum.

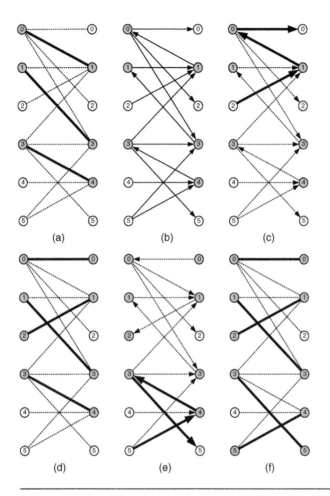

Figure 19.3   Example of the augmenting paths algorithm.

19.3 **Separable Allocators**

While the augmenting path method always finds the maximum matching, it is difficult to parallelize or pipeline and is too slow for applications in which latency is important. Latency-sensitive applications typically employ fast heuristics that find a good matching, but one that is not guaranteed to be maximum, or in some cases not even maximal. This lack of optimality can be compensated for by providing additional speedup in the resource being allocated (Section 19.7) or by performing multiple iterations (Sections 19.3.1 and 19.3.2), or both.

Most heuristic allocators are based on a basic *separable allocator*. In a *separable allocator*, we perform allocation as two sets of arbitration: one across the inputs and one across the outputs. This arbitration can be performed in either order. In an input-first separable allocator, an arbitration is first performed to select a single request at each input port. Then, the outputs of these input arbiters are input to a set of output arbiters to select a single request for each output port. The result is a legal matching, since there is at most one grant asserted for each input and for each output. However, the result may not even be maximal, let alone maximum. It is possible for an input request to win the input arbitration, locking out the only request for a different output, and then lose the output arbitration. This leaves an input and an output, which could have been trivially connected, both idle.

Figure 19.4 shows a 4×3 (4-input $\times$ 3-output) input-first separable allocator. Each input port has separate request lines for each output. For example, for a flit at input 2 to request output 1, request line r_{21} is asserted. The first rank of four three-input arbiters selects the winning request for each input port. Only one of the signals x_{ij} will be asserted for each input port i. The results of this input arbitration, the signals x_{ij}, are forwarded to a rank of three 4-input output arbiters, one for each output. The output arbiters select the winning request for each output port and assert the grant signals g_{ij}. The output arbiters ensure that only one grant is asserted for each output, and the input arbiters ensure that only one grant is asserted for each input. Thus, the result is a legal matching.

A separable allocator can also be realized by performing the output arbitration first and then the input arbitration. A 4×3 output-first separable allocator is shown in Figure 19.5. In this case, the first rank of three 4-input arbiters selects the winning request for each *output* port. Only one of the resulting signals y_{ij} will be asserted for each output port j. The four 3-input input arbiters then take y_{ij} as input, pick the winning request for each input, and output this result on grant signals, g_{ij} ensuring that at most one of g_{ij} is asserted for each input i.

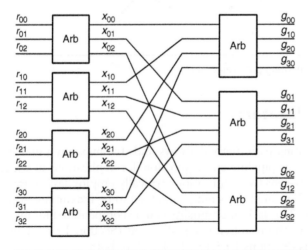

Figure 19.4   A 4×3 input-first separable allocator. A separable allocator performs allocation using two ranks of arbiters. With an input-first allocator, the first rank picks one request from each input. The second rank picks one of these selected input requests for each output.

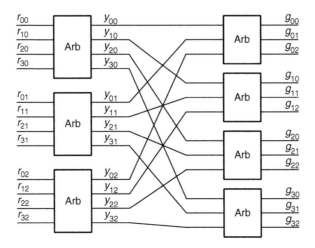

Figure 19.5 An *4 × 3* output-first separable allocator. The first rank of arbiters picks one request for each output while the second rank picks one of the surviving requests from each input.

An input-first separable allocator takes a request matrix and performs arbitration across the rows first and then down the columns. In contrast, an output-first separable allocator performs arbitration down the columns first and then across the rows. For square request matrices, both work equally well. For rectangular matrices, there is an advantage to performing arbitration across the shorter dimension first (such as performing input arbitration first for our 4×3 arbiter), since this tends to propagate more requests to the output stage. Thus, for switches that have more input speedup than output speedup, it is usually more efficient to perform allocation input-first.

For example, consider the request matrix

$$R = \begin{bmatrix} 1 & 1 & 1 \\ 1 & 1 & 0 \\ 0 & 1 & 0 \\ 0 & 1 & 1 \end{bmatrix}. \tag{19.2}$$

One possible input-first separable allocation for this matrix is shown in Figure 19.6. In this example, each arbiter selects the first asserted input. Thus, the intermediate request matrix X after the input arbiters is

$$X = \begin{bmatrix} 1 & 0 & 0 \\ 1 & 0 & 0 \\ 0 & 1 & 0 \\ 0 & 1 & 0 \end{bmatrix}.$$

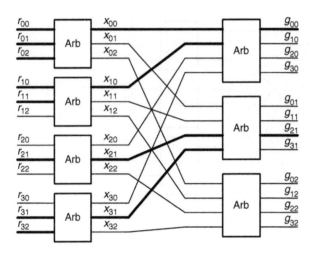

Figure 19.6   An example separable allocation. In this example, each arbiter picks the first asserted input. The result is a non-maximal matching.

Note that X has eliminated input conflicts and thus has at most one non-zero entry in each row. The output arbiters then eliminate output conflicts, giving a final grant matrix G with at most one non-zero in each column as well:

$$G = \begin{bmatrix} 1 & 0 & 0 \\ 0 & 0 & 0 \\ 0 & 1 & 0 \\ 0 & 0 & 0 \end{bmatrix}.$$

We could have made an assignment to all three outputs if either the first or the last input arbiter had selected its last request rather than its first.

Applying the same request matrix (Equation 19.2) to an output-first allocator where each arbiter is also initialized to pick the first asserted input gives an intermediate matrix and a grant matrix, as shown below.

$$Y = \begin{bmatrix} 1 & 1 & 1 \\ 0 & 0 & 0 \\ 0 & 0 & 0 \\ 0 & 0 & 0 \end{bmatrix}, \quad G = \begin{bmatrix} 1 & 0 & 0 \\ 0 & 0 & 0 \\ 0 & 0 & 0 \\ 0 & 0 & 0 \end{bmatrix}.$$

Here, only a single grant is made because the initialization of the output arbiters caused all three to pick the same input. In practice, such complete alignment of arbiters rarely happens. The impact of input- vs. output-first separable allocators is explored further in Exercise 19.6.

Two techniques are commonly used to improve the quality of matchings produced by separable allocators. First, the high-priority inputs of the different arbiters in a given stage can be staggered to reduce the likelihood that multiple input arbiters will all select a request destined for the same output. For example, the wavefront allocator (Section 19.4), while not a separable allocator, sets the high-priority input of each input arbiter to be offset by one position from the previous input and *rotates* this priority each cycle. The PIM (Section 19.3.1), iSLIP (Section 19.3.2), and LOA (Section 19.3.3) allocators also act to stagger the input priorities using different methods.

Second, if time permits, the matching of a separable allocator can be improved by performing multiple iterations. In each iteration, requests that conflict, in either row or column, with all grants made in previous iterations are eliminated and a separable allocation is performed on the remaining requests. Any grants made during an iteration are accumulated with the grants from previous iterations to generate the cumulative grant matrix. For example, applying a second iteration to the example shown above would result in all requests being suppressed except for r_{32}, which proceeds without interference to generate g_{32}. The second request matrix R_2, second intermediate matrix X_2, second grant matrix G_2, and cumulative grant matrix for this second iteration are

$$R_2 = X_2 = G_2 = \begin{bmatrix} 0 & 0 & 0 \\ 0 & 0 & 0 \\ 0 & 0 & 0 \\ 0 & 0 & 1 \end{bmatrix}, \quad G = \begin{bmatrix} 1 & 0 & 0 \\ 0 & 0 & 0 \\ 0 & 1 & 0 \\ 0 & 0 & 1 \end{bmatrix}.$$

19.3.1 Parallel Iterative Matching

Parallel iterative matching (PIM) performs multiple iterations of a *random* separable allocation. Each arbitration during each iteration is performed with *randomized* priority. Randomizing the arbitrations acts to probabilistically stagger the input arbiters, which makes it unlikely that they will all pick the same input. Randomizing the arbitrations also eliminates any possibility of pattern-sensitive starvation in which one input is repeatedly locked out due to deterministic priority adjustments.

A single iteration of random separable allocation was described in Section 17.2.2 and the throughput of such an allocator as a function of switch speedup was derived in Equation 17.1. The convergence of PIM is considered in Exercise 19.5.

19.3.2 iSLIP

iSLIP is a separable allocation method that uses round-robin arbiters and updates the priority of each arbiter *only when that arbiter generates a winning grant*. iSLIP can be used either in a single pass, or as an iterative matching algorithm. By rotating the winning arbiters, iSLIP acts to stagger the priority of the input arbiters, resulting in fewer conflicts at the output stage. The update of a priority only occurs when an arbitration results in a grant and, as in a round-robin arbiter, priorities are updated so that a winning request has the lowest priority in the next round.

For an example of iSLIP operation, suppose our request matrix is as shown in Equation 19.2 and that the input and output arbiters start with input 0 being high priority. That is, the input priority vector is initially $I_0 = \{0, 0, 0, 0\}$ and the output priority vector is initially $O_0 = \{0, 0, 0\}$. Further, assume that the allocator uses a single iteration and that the same request matrix is presented on each cycle.

On the first cycle, the allocation is identical to that shown in Figure 19.6. Because only input arbiters 0 and 2 generated intermediate requests that subsequently generated winning grants, only these two arbiters have their priority advanced, giving $I_1 = \{1, 0, 2, 0\}$. Similarly, the priority of output arbiters 0 and 1 are advanced giving $O_1 = \{1, 3, 0\}$. Note that even though input arbiter 1 has a winner, r_{10} wins generating x_{10}, the priority of this input arbiter is not advanced because x_{10} loses at the output stage. An input arbiter's priority is only advanced if its winner prevails at output arbitration.

With these new priority vectors, the arbitration on the second cycle results in:

$$X_1 = \begin{bmatrix} 0 & 1 & 0 \\ 1 & 0 & 0 \\ 0 & 1 & 0 \\ 0 & 1 & 0 \end{bmatrix}, \quad G_1 = \begin{bmatrix} 0 & 0 & 0 \\ 1 & 0 & 0 \\ 0 & 0 & 0 \\ 0 & 1 & 0 \end{bmatrix}.$$

After this cycle, input arbiters 1 and 2 advance, giving $I_2 = \{1, 1, 2, 2\}$. Output arbiters 1 and 2 again advance, giving $O_2 = \{2, 0, 0\}$. The process continues with:

$$X_2 = \begin{bmatrix} 0 & 1 & 0 \\ 0 & 1 & 0 \\ 0 & 1 & 0 \\ 0 & 0 & 1 \end{bmatrix}, \quad G_2 = \begin{bmatrix} 0 & 1 & 0 \\ 0 & 0 & 0 \\ 0 & 0 & 0 \\ 0 & 0 & 1 \end{bmatrix}, \quad \begin{aligned} I_3 &= \{2, 1, 2, 0\} \\ \\ O_3 &= \{2, 1, 0\} \end{aligned}$$

Note that while our iSLIP example has been shown in the context of an input-first separable allocator, iSLIP can also be implemented using an output-first separable allocator.

19.3.3 Lonely Output Allocator

Single-pass separable allocators often give poor matchings because many input arbiters all pick the same *popular* output while the few requests for *lonely* outputs are less likely to make it through the input stage. The *lonely output allocator*, shown in Figure 19.7, overcomes this problem by adding a stage before the input arbiters that counts the number of requests for each output. The input arbiters then give priority to requests for outputs that have low request counts — the *lonely outputs*. This reduces the number of conflicts at the output stage, resulting in a better matching.

For example, the operation of a lonely output allocator on our example request matrix is shown below in terms of the contents of the R, C, X, and G matrices. The count (C) matrix tags each request with the number of requests competing for the requested output. In this example, there are two requests for outputs 0 and 2 and four requests for output 1. Hence, the C matrix is identical to the R matrix, but with all non-zeros in columns 0 and 2 replaced by 2 and all non-zeros

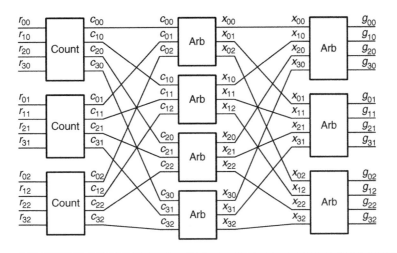

Figure 19.7   A lonely output allocator. The first stage counts the number of requests for each output. The input allocators in the second stage then give priority to requests for outputs that have the lowest counts (lonely outputs). This reduces conflicts in the output arbiters in the final stage.

in column 1 replaced by 4. The input arbiters select one request per row, giving priority to the lower-numbered requests[4] generating matrix X. Giving priority to low-count requests causes intermediate request matrix X to have entries in more columns leading to fewer conflicts in the output stage. This, for example, causes the arbiter for input 3 to assert x_{32} giving output 2 its only request. The output stage selects one request in each to generate the grant matrix G.

$$R = \begin{bmatrix} 1 & 1 & 1 \\ 1 & 1 & 0 \\ 0 & 1 & 0 \\ 0 & 1 & 1 \end{bmatrix}, \quad C = \begin{bmatrix} 2 & 4 & 2 \\ 2 & 4 & 0 \\ 0 & 4 & 0 \\ 0 & 4 & 2 \end{bmatrix}, \quad X = \begin{bmatrix} 1 & 0 & 0 \\ 1 & 0 & 0 \\ 0 & 1 & 0 \\ 0 & 0 & 1 \end{bmatrix}, \quad G = \begin{bmatrix} 1 & 0 & 0 \\ 0 & 0 & 0 \\ 0 & 1 & 0 \\ 0 & 0 & 1 \end{bmatrix}$$

19.4 **Wavefront Allocator**

The wavefront allocator, unlike the separable allocators described above, arbitrates among requests for inputs and outputs simultaneously. The structure of the wavefront allocator is shown in Figure 19.8 and the logic of each allocator cell is show in Figure 19.9. The wavefront allocator works by granting row and column tokens

---

4.   In this example, ties between two requests with the same count (for example, between c_{00} and c_{02}) are broken by giving priority to the lower-numbered request. In general, lonely output arbiters typically break ties using priority that rotates when the arbitration results in a grant — as with iSLIP.

to a *diagonal* group of cells, in effect giving this group priority. A cell with a row (column) token that is unable to use the token passes the token to the right (down), wrapping around at the end of the array. These tokens propagate in a wavefront from the priority diagonal group, hence the name of the allocator. If a cell with a request receives both a row and a column token, either because it was part of the original priority group or because the tokens were passed around the array, it grants the request and stops the propagation of tokens. To improve fairness, the diagonal group receiving tokens is rotated each cycle. However, this only ensures weak fairness. (See Exercise 19.2.)

In an $n \times n$ arbiter, diagonal group k contains cells x_{ij} such that $(i + j) \bmod n = k$. Thus, for example, in the 3×3 allocator of Figure 19.8, priority group 0, selected by signal p_0, consists of cells x_{00}, x_{21}, and x_{12}. Because each diagonal group must contain exactly one cell from each row and from each column, all wavefront allocators must be square. A non-square wavefront allocator can be realized, however, by adding dummy rows or columns to square off the array. For example, the 4×3 allocator of our examples above can be realized using a 4×4 array. The priority groups in an allocator need not be diagonals as long as they each contain one element of each row and column. (See Exercise 19.2.)

The details of the allocator cell are shown in Figure 19.9. When the cell is a member of the priority group, signal `pri` is asserted, which generates both a row token `xpri` and a column token `ypri` via a set of OR gates. If the cell is not a member of the priority group, row tokens are received via signal `xin` and column tokens via `yin`. If a cell has a row token `xpri`, a column token `ypri` and a request

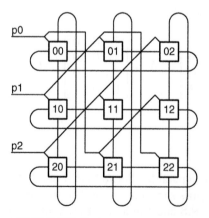

Figure 19.8   A wavefront allocator simultaneously performs input and output arbitration by passing row and column tokens in a 2-D array of allocator cells. Each cycle, row, and column tokens are granted to one of the diagonal groups of cells (p_0, p_1, or p_2). If a cell with a request has both a row token, and a column token, it grants the request. A cell with a token that is unable to make a grant passes the token along its row or column.

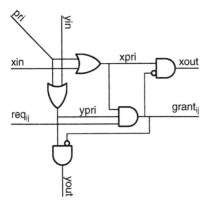

Figure 19.9   Logic diagram of a wavefront allocator cell. To grant a request, a cell needs to acquire both a row token `xpri` and a column token `ypri`. Both tokens are generated when the cell is a member of the priority diagonal group and `pri` is asserted. Otherwise, tokens are received on the `xin` and `yin` lines. A cell that receives one or more tokens and does not use them to grant a request passes the tokens along on the `xout` and `yout` lines.

req_{ij}, it generates a grant `grant`$_{ij}$ via a 3-input AND gate. If a grant is asserted, it disables further propagation of the row and column tokens via a pair of AND gates. Otherwise, if a grant is not asserted, row (column) tokens are passed to the next cell to the right (down) via `xout` (`yout`).

This cell is a 2-D generalization of the iterative arbiter cell of Figure 18.5. In two dimensions, the inhibition of the token or carry must be performed by the grant (as in Figure 19.9) rather than the request (as in Figure 18.5) to avoid blocking a token when only one of the two tokens is received. As with the iterative arbiter of Figure 18.5, the wavefront arbiter appears to contain logic with combinational cycles. However, since one of the priority inputs is asserted at all times, all cycles are broken by an OR gate with an asserted input. Thus, the logic is acyclic. However, the cyclic structure causes problems with many modern timing-analysis CAD tools.

Figure 19.10 shows the operation of a wavefront allocator on our example 4×3 request matrix. A 4×4 wavefront allocator is used to handle this non-square problem with no requests in column 3. In the figure, cells making requests are shaded and cells receiving grants are shaded darker. Asserted priority and token signals are highlighted with thick lines.

In the example, priority signal p_3 is asserted, granting tokens to cells x_{30}, x_{21}, x_{12}, and x_{03}. Of these cells, x_{21} has a request and thus uses the tokens to make a grant. The other three cells, x_{30}, x_{12}, and x_{03}, propagate their tokens to the right and down. For example, x_{30} propagates its row token to the right, to x_{31}. Even though x_{31} has a request, it cannot use this row token, since x_{21} has consumed the column token for column 1, so it passes the token on to x_{32}. Cell x_{32} receives a row token from x_{30} and a column token from x_{12}, and thus is able to make a grant. In a similar manner,

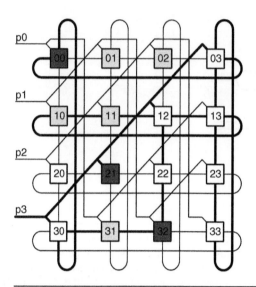

Figure 19.10 Example of wavefront allocation. Shaded cells represent requests and darkly shaded cells represent requests that are granted. Thick lines show priority and token signals that are asserted.

x_{00} makes a grant after receiving a column token from x_{30} and a row token from x_{03}. Note that the row token for row 1 and the column token for column 3 are never used and thus propagate the entire way around the cycle.

19.5 Incremental vs. Batch Allocation

In many routers, an allocation is needed only once every i cycles for some integer i. For example, a router in which each flit is 4 times the width of the internal datapaths, and hence takes 4 cycles to traverse the switch, need only perform an allocation once every $i = 4$ cycles. This reduced allocation frequency can be exploited by running $i = 4$ iterations of PIM or iSLIP, and hence improving the cardinality of the matching generated by the allocator. Alternatively, instead of performing such a *batch* allocation, the allocation can be performed *incrementally* with a portion of the allocation completed each cycle.

With incremental allocation, a simple one-cycle allocator is used to generate the best possible allocation each cycle. The flits that receive grants on a given cycle begin traversing the switch immediately and the corresponding rows and columns of the request matrix are removed from consideration until the flits complete their transmission, just as in an iterative allocator. The difference from the iterative allocator is that each flit begins transmission as soon as it receives a grant, rather than waiting for the end of the batch period. Because each flit may start transmission during a different cycle, they also may finish transmission during different cycles. This means

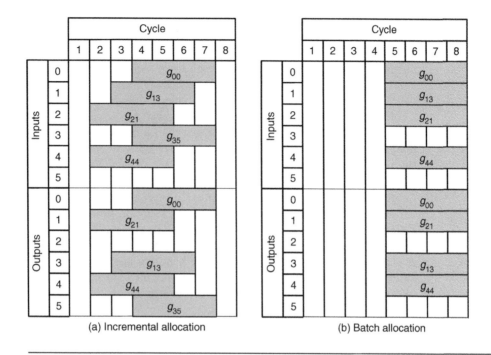

(a) Incremental allocation   (b) Batch allocation

Figure 19.11   Incremental vs. batch allocation. (a) With incremental allocation, requesters are granted resources as soon as they are available. This reduces latency but may reduce throughput and raise fairness issues. (b) With batch allocation, all requesters are granted resources at the same time, in a batch.

different ports become available during different cycles, rather than all becoming available at the same time with batch allocation. This makes it much more difficult to provide fair allocation.

Figure 19.11 shows the application of incremental and batch allocation to the request matrix R of Equation 19.1. In this example, grants g_{21} and g_{44} are asserted during cycle 1, grant g_{13} is asserted during cycle 2, and grants g_{00} and g_{35} are asserted during cycle 3. With the incremental allocator (Figure 19.11[a]) each flit begins traversing the switch on the cycle following the grant assertion. With batch allocation (Figure 19.11[b]), on the other hand, transmission of all flits is deferred until cycle 5 after all 4 iterations of the allocator have completed.

Incremental allocation reduces latency and enables allocation to handle variable-length packets that are not segmented into flits. These advantages come at the expense of slightly reduced bandwidth and reduced fairness. Latency is reduced because each flit may traverse the switch as soon as it receives a grant rather than waiting for an entire batch time. In a lightly loaded network, most flits depart after a single allocation cycle. Bandwidth is slightly reduced because switching the start time of a flit results in one or more idle cycles that would be filled with a batch scheduler.

The ability to start and end switch traversal at arbitrary times also enables the allocator to handle variable length packets. Each packet holds the switch resources (input and output) for its entire length before releasing them. To do this with a batch allocator would require rounding up each packet length to the next whole batch size. While switching packets directly, rather than switching flits, eliminates segmentation overhead, it also increases contention latency unless packet length is limited since a competing packet potentially may have to wait for a maximum-length packet. A compromise is to segment packets larger than a certain size X (for example, two flits) and switch packets smaller than this size without segmentation. This bounds the segmentation overhead to be less than $X/(X+1)$ (for example, $\frac{2}{3}$) while also bounding contention delay to two flit times.

Special care is required to guarantee any type of fairness with an incremental scheduler. Consider the situation of Figure 19.11 in which there is a continuous assertion of r_{21} and r_{00} and in which a single packet has asserted r_{20}. For r_{20} to be serviced, it must acquire both input 2 and output 0. However, when input 2 becomes available in cycle 2, output 0 is still busy handling g_{00}. Thus, r_{20} cannot be granted input 2 and a greedy scheduler will repeatedly allocate input 2 to r_{21}. Similarly, a greedy allocator will always allocate output 0 to r_{00} in cycle 4. As long as a greedy allocator is used, r_{20} will never be served as long as r_{00} and r_{21} are asserted and are being served in different cycles.

Fairness can be guaranteed in an incremental allocator by allowing a request to acquire input ports and output ports independently after they have waited for a given amount of time. In the example above, after r_{20} has waited for a period of time, say 16 cycles, it raises its priority to allow independent port acquisition. It then acquires input 2 on cycle 2 and holds this input idle until it acquires output 0 on cycle 4.

19.6 Multistage Allocation

Some applications require multiple stages of allocation. For example, in a router, we might want to grant requests to high-priority packets first and then allocate any remaining resources to lower-priority packets. In another application, we might want to grant requests for multicast traffic first (see Exercise 19.3) and then allocate remaining ports to unicast traffic.

Such prioritized multistage allocation can be performed as shown in Figure 19.12. An example is shown in Equation 19.3. The high-priority allocation is performed first — generating a grant matrix Ga from request matrix Ra. A mask matrix Ma is then constructed from grant matrix Ga that masks all rows and columns that contain a grant. A bit Ma_{ij} of the mask matrix is set to zero, masking the downstream request, if any bit in row i or column j of Ga is set. That is,

$$Ma_{ij} = \left(\neg \bigvee_{k=0}^{n-1} Ga_{kj}\right) \bigwedge \left(\neg \bigvee_{k=0}^{m-1} Ga_{jk}\right).$$

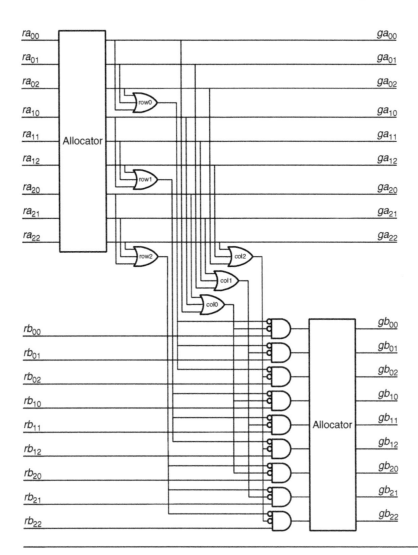

Figure 19.12 Two-stage allocation. An allocation is performed on request matrix *Ra*, generating grant matrix *Ga*. This grant matrix is also used to create a mask — eliminating rows and columns used by *Ga* from a second request matrix *Rb*. The masked version of *Rb* is then used to perform a second allocation generating grant matrix *Gb*. Because of the masking, *Gb* is guaranteed not to interfere with *Ga*.

This mask is then AND-ed with request matrix *Rb* to eliminate requests that would conflict with a grant in *Ga*. The remaining requests $Rb' = Rb \wedge Ma$ are input to a second allocator that generates grant matrix *Gb*. The process can be applied to more than two stages with the requests at each stage being masked by all of the upstream grants.

$$Ra = \begin{bmatrix} 1 & 0 & 1 & 0 \\ 0 & 0 & 0 & 0 \\ 1 & 0 & 0 & 0 \\ 0 & 0 & 0 & 0 \end{bmatrix}, \quad Ga = \begin{bmatrix} 0 & 0 & 1 & 0 \\ 0 & 0 & 0 & 0 \\ 1 & 0 & 0 & 0 \\ 0 & 0 & 0 & 0 \end{bmatrix}, \quad Ma = \begin{bmatrix} 0 & 0 & 0 & 0 \\ 0 & 1 & 0 & 1 \\ 0 & 0 & 0 & 0 \\ 0 & 1 & 0 & 1 \end{bmatrix},$$

$$Rb = \begin{bmatrix} 1 & 1 & 1 & 1 \\ 0 & 1 & 1 & 1 \\ 0 & 0 & 1 & 1 \\ 0 & 0 & 0 & 1 \end{bmatrix}, \quad Rb' = \begin{bmatrix} 0 & 0 & 0 & 0 \\ 0 & 1 & 0 & 1 \\ 0 & 0 & 0 & 0 \\ 0 & 0 & 0 & 1 \end{bmatrix}, \quad Gb = \begin{bmatrix} 0 & 0 & 0 & 0 \\ 0 & 1 & 0 & 0 \\ 0 & 0 & 0 & 0 \\ 0 & 0 & 0 & 1 \end{bmatrix}$$

$$(19.3)$$

19.7 Performance of Allocators

To measure the performance of an allocator, the allocator can be used to schedule flit transmission across a crossbar switch. In order to isolate the allocator from other factors, such as finite buffer size, some ideal assumptions are made about the switch. First, each input to the switch feeds into an infinite pool of buffers. Additionally, each input buffer pool is split into *virtual output queues*. Virtual output queues are essentially a set of virtual-channel buffers, with one virtual channel per output. When an incoming flit arrives at an input, it is immediately stored in a virtual output queue based on its desired output port. This prevents *head-of-line blocking*, where data destined for one port of a switch is blocked behind data waiting for another port of the switch. Finally, all latencies, such as routing and switch traversal, are ignored. The delay of a flit is simply the time it spends in the input buffers — if it arrives and is immediately scheduled, then this time is zero. For the following experiments, an 8 × 8 switch and uniform traffic are assumed.

Figure 19.13 shows the average delay vs. offered traffic of each allocator described in this chapter. PIM saturates first, at around 66% offered traffic, and LOA improves slightly on this, saturating at approximately 69%. However, LOA does offer good average latency until its saturation point. Both the wavefront and iSLIP allocators do eventually approach 100% throughput, which is beyond the latency scale of the graph. Also, as shown, the wavefront allocator also offers significantly lower latency compared to the simpler iSLIP technique.

While these single-iteration, separable techniques have reasonable performance given their low complexity, additional iterations can significantly improve this performance. Figure 19.14 shows the performance of PIM with multiple iterations. While the single iteration PIM1 saturates at 66%, two iterations extend the saturation point to near 90%, and with three iterations throughput approaches 100%. Little performance benefit is seen beyond three iterations for this configuration. Not shown is the benefit of additional iterations for iSLIP. Although iSLIP achieves 100% throughput

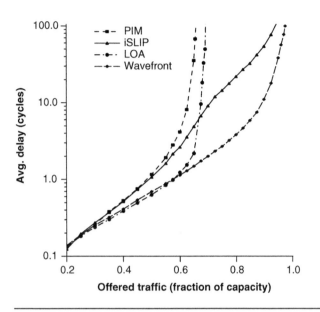

Figure 19.13 Performance of single iteration allocators. Each curve shows the average delay versus offered traffic for an *8 × 8* crossbar under uniform traffic.

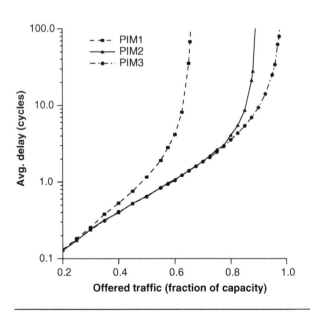

Figure 19.14 Performance of PIM with multiple iterations (e.g., PIM1 is PIM with a single iteration) on an *8 × 8* crossbar switch under uniform traffic.

on uniform traffic with a single iteration, an additional iteration greatly reduces the latency, making it competitive with wavefront. Only small improvements are realized for increasing the number of iSLIP iterations beyond two in this case.

As mentioned in Section 17.2.2, another method to improve the performance of an allocator is to provide an input speedup, an output speedup, or both to the corresponding switch. Figure 19.15 shows an LOA with different input and output speedups. Compared to the case with no speedup (IS=1,OS=1), which saturates at approximately 69%, an input speedup of two (IS=2,OS=1) improves both the saturation point, near 95%, and the latency of an LOA. Adding an output speedup with the input speedup (IS=2,OS=2) improves the switch throughput to 100%.

We can also speed up the crossbar and allocator relative to the channel rate to improve the performance of a simple allocator: for each packet arriving at the channel rate, S packets can traverse the crossbar, where S is the speedup. In Figure 19.16, the performance of LOA with speedup is shown. A 25% speedup relative to the channel rate (S=1.25) gives an approximately 25% increase in the saturation throughput to approximately 85%. This trend continues for a 50% speedup (S=1.5), where the throughput approaches 98%. Beyond this point (S=1.75), additional speedup provides almost no gain in either throughput or latency.

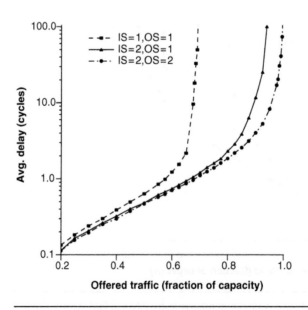

Figure 19.15 Performance of LOA with input and output speedup on an *8 × 8* crossbar switch under uniform traffic.

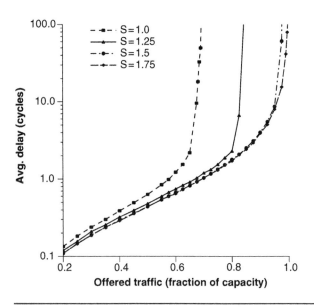

Figure 19.16 Performance of LOA with speedup relative to the channel rate on an 8 × 8 crossbar switch under uniform traffic.

19.8 **Case Study: The Tiny Tera Allocator**

As mentioned in Section 7.4, the Tiny Tera is a fast packet switch originally architected at Stanford University and later commercialized by Abrizio [125]. The core of the Tiny Tera is organized around a 32-port crossbar that must be reconfigured every 51 ns. The designers chose the iSLIP allocation algorithm with an output-first arbitration and three iterations (Section 19.3.2) for the design, which required both pipelining and aggressive arbiter designs to meet their timing goals [79].

When performing multiple iterations of the iSLIP algorithm, it may seem that both stages of arbitration in the separable allocator must be completed before beginning the next iteration. Then the result of the iteration can be used to mask (deactivate) any request corresponding to a matched input or output for the next iteration. Using this approach, each iteration would take the time required for two arbitration steps. However, a clever optimization allows the iteration time to be cut nearly in half.

The optimization used in the Tiny Tera is based on the observation that if *any* output arbiter in the first stage of the allocator selects a particular input, that input will be matched after the second stage of arbiters. Therefore, the input masks can be computed by appropriately OR-ing the results of the first stage of arbitration. The pipelined architecture along with the logic to compute the input masks im_i for a

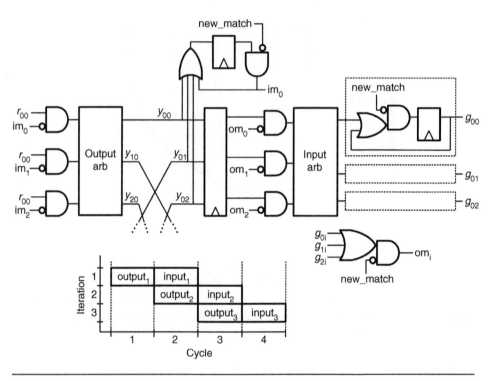

Figure 19.17 Logic for the pipelined implementation of a multiple iteration iSLIP allocator used in the Tiny Tera packet switch. For clarity, a 3 × 3 allocator is used, of which only a portion is shown. The new_match signal is asserted during the first cycle of a new allocation. The pipeline diagram for the operation of the allocator is also shown — the arbitration for outputs occurs in parallel with arbitration for inputs from the previous iteration.

3 × 3 iSLIP allocator are shown in Figure 19.17. Creating the input masks from the results of the first stage reduces the iteration time to that of a single arbitration plus the time required for the OR, which is generally much smaller than the arbitration time, especially for the large arbiters used in the Tiny Tera.

The second pipeline stage of the allocator contains the arbiters for the inputs, which produce the grant signals. These grants are OR-ed similarly to generate the output masks om_i. However, instead of using the output masks at the original requests, they mask the inputs to the second stage of arbiters. This does not prevent the propagation of spurious requests from already matched outputs through the first stage of allocators. However, the output masks are applied before the input arbiters, so these spurious requests never affect the final grants.

To illustrate the application of the input and output masks during the allocator's pipeline, consider the 2 × 2 allocation with 2 iterations shown in Figure 19.18. As shown, all possible requests are asserted. During the first iteration, both output

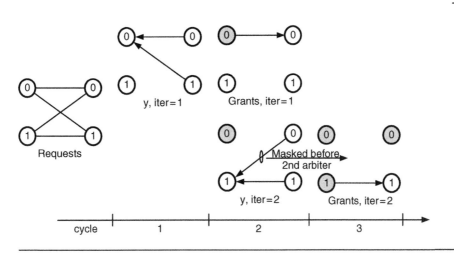

Figure 19.18 Example of the pipelined iSLIP algorithm for a 2 × 2 allocation with 2 iterations. During each step of the allocation, masked inputs (outputs) are indicated by left (right) nodes, gray.

arbiters choose input 0. This information is used to mark (gray) input 0 as matched during subsequent cycles. In the second cycle, the first iteration's input arbitration grants the request from input 0 to output 0. Simultaneously, both output arbitrations of the second iteration select input 1. By the beginning of the third cycle, the output mask correctly reflects the fact that output 0 is matched. This allows the spurious request from input 1 to output 0 to be masked before the second iteration's grants are computed.

19.9 **Bibliographic Notes**

The earliest algorithms for computing a maximum matching are due to Hall and to Ford and Fulkerson. While the augmenting path algorithm has complexity $O(|V||E|)$, Hopcroft and Karp developed an $O(|V|^{1/2}|E|)$ time algorithm [83]. As shown by McKeown et al., a maximum-size matching does not guarantee a 100% throughput on non-uniform traffic, while a more complex maximum weight matching does [126]. Several techniques have since been introduced to approximate maximum weight matchings and still achieve 100% throughput [183, 71]. Parallel iterative matching was first developed by Anderson et al. [11]. iSLIP was first studied and described by McKeown [123] and implemented in the Tiny Tera packet switch [125, 79]. Tamir developed the wavefront allocator [180]. Incremental allocation was used to advantage in the router of the Alpha 21364 [130, 131].

19.10 **Exercises**

19.1 *Performance of a separable allocator.* Find the best possible and worst possible grant matrix that could be generated by a single-pass separable allocator for the following request matrix:

$$R = \begin{bmatrix} 1 & 1 & 1 & 1 & 1 & 1 \\ 1 & 1 & 1 & 1 & 1 & 0 \\ 1 & 1 & 1 & 1 & 0 & 0 \\ 1 & 0 & 0 & 0 & 0 & 0 \\ 0 & 1 & 0 & 0 & 0 & 0 \\ 0 & 0 & 1 & 0 & 0 & 0 \end{bmatrix}$$

19.2 *Randomization and history in a wavefront allocator.* Consider a 4×4 wave-front allocator with the priority groups wired as in Figure 19.10 and the request matrix

$$R = \begin{bmatrix} 1 & 0 & 0 & 1 \\ 0 & 0 & 1 & 1 \\ 0 & 1 & 1 & 0 \\ 1 & 1 & 0 & 0 \end{bmatrix}.$$

(a) If the priority group is incremented each cycle ($p_0, \ldots, p_3, p_0, \ldots$) and the request matrix R remains fixed, what fraction of grants is given to the corresponding entries of R? How does this affect the throughput of a switch using this allocator?

(b) Does randomly choosing a priority group each cycle improve the performance for this request matrix relative to the original strategy of simply incrementing the priority group?

(c) In general, the wavefront allocator cells can be prioritized in any pattern that contains exactly one cell from each row and column (the prioritized cells must be a permutation). If the cell priorities are set each cycle using a random permutation, how does the allocator perform on R relative to the previous two approaches?

19.3 *Multicast allocation.* Explain how to extend a simple separable allocator to handle multicast allocation in addition to unicast allocation. The multicast allocator accepts two additional inputs per input port: a multicast request r_m, which is asserted when the port wishes to simultaneously allocate several outputs, and a multicast bit vector b_m with one bit for each output to indicate which outputs are to be allocated for the multicast.

19.4 *Incremental multicast allocation.* In a router, a multicast request can be handled either atomically, by insisting that the entire multicast be allocated at once, or incrementally, by dividing the multicast set into smaller groups. In the extreme case, a multicast set can be divided into a number of unicasts — one from the input to each of the multicast outputs. Explain how to build a router that performs greedy incremental multicast allocation by allocating as many outputs as are available each cycle and then deferring the remaining outputs until later cycles.

19.5 *Convergence of PIM.* What is the average number of iterations for PIM to converge to a maximal matching under uniform traffic? What is the worst-case number of iterations for convergence over all possible request patterns?

19.6 *Input- vs. output-first allocation with input speedup.* Consider the allocation of an SN input by N output switch, where S is the input speedup and N is the number of ports.

    (a) If a single-pass PIM allocator is used and requests are uniform (all inputs request outputs with equal probability), what is the throughput of input-first allocation? What if output-first allocation is used?

    (b) Now assume the number of ports N is large. What integer value of S gives the largest difference in throughput between the input-first and output-first allocators? What is that difference?

19.7 **Simulation:** Compare the performance of a four-iteration PIM allocator on a switch with a speedup of one to a single iteration PIM allocator on a switch with a speedup of two.

CHAPTER 20

Network Interfaces

The interface between an interconnection network and the network client can often be a major factor in the performance of the interconnection network. A well-designed network interface is unobtrusive — enabling the client to use the full bandwidth at the lowest latency offered by the network itself. Unfortunately, many interfaces are not so transparent. A poorly designed interface can become a throughput bottleneck and greatly increase network latency.

In this chapter, we look briefly at the issues involved in three types of network interfaces. Processor-network interfaces should be designed to provide a high-bandwidth path from the processor to the network without incurring the overhead of copying messages to memory or the bottleneck of traversing an I/O interface. A successful interface should minimize processor overhead and be *safe*, preventing an errant process from disabling the network.

Shared-memory interfaces are used to connect a processor to a memory controller via an interconnection network. They may implement a simple remote memory access or a complex cache coherence protocol. Latency is critical in a shared-memory interface because such interfaces are in the critical path for remote memory accesses.

Line-card interfaces connect an external network channel with an interconnection network that is used as a switching fabric. The primary function of the line-card interface is to provide queueing and packet scheduling. queueing is provided between the input line and the fabric and between the fabric and the output line. Input queues are typically provided for each output subport, packet class pair so that packets destined for one subport will not block packets to a different subport and so packets of a lower priority class will not block packets of a higher priority class. As with the processor-network interface, there are issues of both performance and safety in the design of such queueing systems. At the output side, the queues match the rate of the fabric (which typically has speedup) to the rate of the output line and may perform rate shaping for different classes of traffic.

20.1 Processor-Network Interface

Many applications of interconnection networks at their core involve passing messages between processors attached to the network. The networks in message-passing parallel computers obviously fall into this category, but less obviously, most I/O networks and many packet switching fabrics also involve passing messages between processors associated with the I/O devices or line interfaces.

Figure 20.1 shows a number of processor nodes, $P_1, \ldots, P_N$ that communicate by exchanging messages over an interconnection network. In most applications, message lengths are bimodal. Short messages, about 32 bytes in length, are sent to make requests and for control (read disk sector x). Long messages, 1 Kbyte or longer, are sent to transfer blocks of data (the contents of disk sector x). We are concerned with the latency and throughput achieved for both short and long messages. In practice, achieving good throughput on short messages is usually the hardest problem.

A good message-passing interface must do two things: it must provide a low-overhead path to the network, and it must prevent a misbehaving process from using the network to interfere with other processes. The access path should be low-latency so a short message can be sent in a few cycles. The path should also avoid traversing low-bandwidth choke points in the system — like the memory interface. In general, a network interface should avoid copying messages, particularly to or from memory, since this adds latency and often results in a bandwidth bottleneck.

A key aspect in the design of a network interface is where it attaches to the processor node. Figure 20.2 shows a typical processor node. A processor chip, which contains a register file and on-chip cache memory, connects to DRAM memory and I/O devices via a bridge chip.[1] A network interface can attach to any of these points.

The most efficient network interfaces attach directly to the processor's internal registers. This permits small messages to be composed directly out of processor

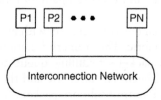

Figure 20.1   A number of processors, P_1 to P_N, pass messages to one another over an interconnection network.

---

1.   In modern PCs this is called the *north bridge*.

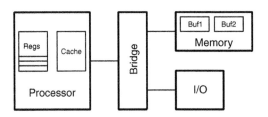

Figure 20.2   A typical processor node consists of a processor with internal registers and cache memory connected to DRAM memory and I/O devices via a bridge chip. The network interface can be connected to the I/O bus, integrated with the processor, or integrated with the bridge chip.

registers or the cache without the latency associated with traversing the off-chip memory interface. Unfortunately, most network interfaces attach to the I/O bus because this requires the least modification to existing components. These interfaces transfer messages between the interconnection network and memory. This incurs considerable latency as it can take more than 30 cycles to cause an external bus cycle on modern processors. It also causes every word of the message to traverse the memory interface twice at either end of the communication. This places a heavy load on memory bandwidth and can become a bandwidth bottleneck in some situations. This I/O-based network interface is similar in design to a standard peripheral interface and will not be discussed further here.

20.1.1 **Two-Register Interface**

A simple two-register interface to an interconnection network is illustrated in Figure 20.3.[2] Messages are sent via a single *network output register*, to which each word of outgoing messages is written and messages are received via a single *network input register*, from which each word of incoming messages is read. To send a message, the processor simply moves each word of the message to the network output register. A special move instruction is used to transfer the last word of the message and terminate the message. To receive a message, the processor reads the words of the incoming message from the network input register. Each read dequeues the next word of the message so that the next readfrom the same register returns the next word of the register. To synchronize with the arrival of messages, the processor may either test for the presence of a message before reading the input register or may block on the input register until a message arrives.

The two-register interface provides a simple interface that efficiently handles short messages without incurring any memory overhead. Short messages can be sent directly from data in the processor registers and can be received directly into

---

2. This type of register network interface was used for both sends and receives in the MARS accelerator [4] and for sends in the J-Machine [53].

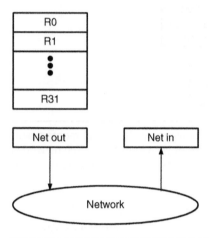

Figure 20.3   A two-register network interface. Messages are sent by moving one word at a time to a *network output register*. Messages are received by reading from a *network input register*.

the processor registers. This interface, however, has two limitations. First, for long messages, particularly those that must transfer blocks of memory-resident data, the processor is tied up serving as a DMA engine, transferring data between memory and the network interface registers. This processor overhead can be prohibitive in some applications.

A second problem with the two-register interface is that it does not protect the network from software running on the processor. A misbehaving processor can send the first part of a message and then delay indefinitely sending the end of the message. The partial message can tie up network resources such as buffers and virtual channels indefinitely, interfering with other processes' and nodes' use of the network. A process can also tie up the network, interfering with other processes, by failing to read a message from the input register.

For a network interface to be *safe* it must guarantee that a process cannot indefinitely hold *shared* network resources. Any shared resource, such as a buffer or a virtual channel used to send a message, must be released within a bounded amount of time regardless of the behavior of the sending or receiving processes. Resources that may be held indefinitely should not be shared. For example, a virtual channel on each physical channel can be dedicated to every process if the process cannot guarantee that it will be released in a bounded amount of time.

20.1.2   **Register-Mapped Interface**

One approach to solving the safety problem of the two-register interface is to send a message atomically from a subset of the processor's general purpose registers, as illustrated in Figure 20.4.[3] A processor composes a message in the processor registers and then sends the message atomically into the network interfaces with a single *send* instruction that specifies the registers that contain the first and last words of the message. This mechanism for message transmission is safe, since there is no way for a processor to leave a partial message in the network. This interface is very limiting, however, as it prevents processors from sending long messages, which forces long messages to be segmented, causes register pressure by consuming general registers,[4] and still forces the processor to act as a DMA engine.

20.1.3 **Descriptor-Based Interface**

A descriptor-based message send mechanism, shown in Figure 20.5, overcomes the limitations of the register send mechanism of Figure 20.4. With this approach, the processor composes the message in a set of dedicated message descriptor registers. This register set is large enough to hold a working set of message descriptors. Each descriptor may contain an immediate value to be inserted into the message, a reference to a processor register, or a reference to a block of memory. The example message shown in the figure contains one of each of these descriptor types. The descriptor-based message send is safe and eliminates the processor overhead associated with register interfaces. In effect, it offloads the processor overhead to a co-processor that steps through the descriptors and composes the message.

20.1.4 **Message Reception**

Performing a message receive in a safe manner and without processor overhead is most easily accomplished by dedicating a co-processor, or separate thread of a

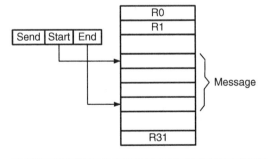

Figure 20.4 Short messages can be sent directly from the processor's general purpose registers.

---

3. This type of message send mechanism was implemented in the M-Machine [112] (Section 20.4).
4. The register pressure issue can be addressed by forming messages in a special register set rather than in the general-purpose registers.

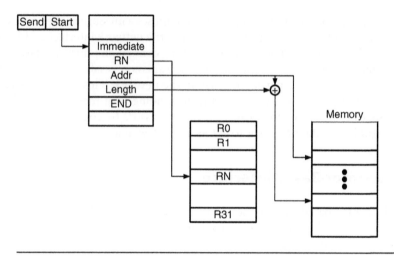

Figure 20.5   A descriptor-based register network interface. A message is composed in a set of dedicated send registers that contain descriptors. Each descriptor may contain an immediate value, a reference to a processor register, or a reference to a block of memory.

multi-threaded processor, to receiving messages.[5] The message thread handles simple messages itself and queues more complex messages for handling by the user's thread. Some common message types, such as shared memory references, may even be handled by dedicated hardware for efficiency. This interface is safe in that the receive thread can be validated to always remove the message from the network in a bounded amount of time.

20.2 Shared-Memory Interface

In a shared-memory multiprocessor (Section 1.2.1) an interconnection network is used to carry messages from processors to memories. In a system that does not permit remote caching, such as the Cray T3E, the messages are simple read and write requests and replies. In a system that supports coherent caching of remote data, such as the SGI Origin 2000, a larger vocabulary of messages is used to implement a cache coherence protocol. In either case, two network interfaces are used. A processor-network interface formats messages in response to processor load and store operations that miss the cache and cache line evictions. At the memory controller side, a memory-network interface receives requests from the network, carries out the requested action, and sends reply messages. In a typical shared-memory processing node, these two interfaces are co-located and may share network injection and extraction ports. However, their functions are logically separate. Because latency is critical

---

5.   The M-Machine [112] uses two separate receive threads to handle two classes of arriving messages.

in the interconnection networks for shared-memory multiprocessors, these interfaces are optimized to inject request messages in response to processor or memory events in just a few clock cycles.

20.2.1 **Processor-Network Interface**

A simplified block diagram of a processor-network interface is shown in Figure 20.6. Each time the processor performs a load or store operation, it places a request in the memory request register (Req Reg).[6] The request record specifies the type of the request (read or write, cacheable or uncacheable, and so on), the physical address to be accessed,[7] and for write requests the data to be written. Each request is tagged so that the processor can identify the corresponding reply when it is returned from the memory system. In many systems, the tag encodes how the reply is to be handled (for example, store the reply data into register R62). The request is first presented to the cache. If the address being accessed resides in the cache, the read or write operation is performed in the cache, and the cache places a reply, including the requested data for a read, into the memory reply register.

If the request misses in the cache, it is posted to a miss-status holding register (MSHR) [105] and the status of the MSHR is initialized. The action taken in response to an MSHR depends on the type of operation and whether or not the machine supports a cache coherence protocol. First, let us consider a read operation on a simple machine that does not permit caching of remote data. In this case, the MSHR status is initialized to *pending read*. Upon seeing this status, the message transmit block formats a read request message, addresses it to the node containing the requested address, and injects it into the network. After the message is injected, the status of the request is updated to *read requested*. A translation step is sometimes required to convert the address to a node number — that is, to convert the memory address to a network address. The read request message contains the destination node, the type of message (read), and the address to be read.

The network will ultimately return a read reply message in response to the request. The address field of the read reply message is used to identify the MSHR(s) waiting for the requested data. All matching MSHRs are updated with the data and their status is changed to *read complete*. The completed MSHRs are forwarded in turn to the processor reply register, where the tag field is used by the processor to direct the data to the proper location. As each completed operation is forwarded to the processor, its status is changed to *idle*, freeing the MSHR to handle another request.

----

6. In high performance processors, the memory system is designed to accept several (typically two to four) memory requests per cycle. For simplicity, we consider only a single request per cycle.

7. In most systems, the processor presents a virtual address, which is translated to a physical address in parallel with the cache access.

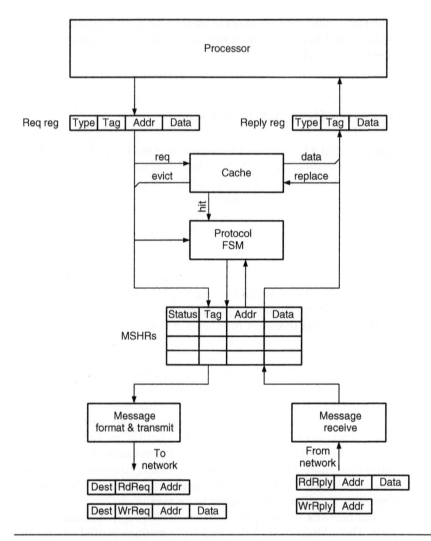

Figure 20.6   A processor-network interface for a shared-memory multiprocessor. Processor memory requests that miss the cache are posted to a miss-status holding register (MSHR) and a request message is transmitted to perform the requested operation. Reply messages received from the network are matched to the pending operation in an MSHR and the completed operation is forwarded to the processor.

Uncacheable writes are handled in a manner identical to reads except that data is included in the request message and not in the reply message. Also, a write to an address with a pending request requires a second write request message to be sent, and a mechanism is required to guarantee that the two writes to the same address are received in order.

The MSHRs act as a *scoreboard* for all outstanding requests. When a request misses in the cache, an entry is made in the MSHR and the status is initialized. Agents that handle requests (the protocol FSM and the message transmit block) monitor the status of the MSHR entries, and upon detecting an entry in a state that requires an action, initiate the appropriate action. This monitoring is often done by using a one-hot encoding of the status field and triggering an agent on the logical OR of the bits reflecting a particular state.

The MSHRs also serve to combine requests to the same location. If a second read is requested to a location that already has a pending read, the address match will be detected when the second request is posted to an MSHR and no redundant read request message will be sent. When a reply from the first request is received, it will satisfy all pending requests for that address.

The number of MSHRs determines the number of memory references that can be pending at any given point in time. When all MSHRs are full, the next memory reference that misses the cache must stall in the request register until an MSHR becomes available. Typical designs have between 4 and 32 MSHRs. Shared-memory network interfaces that handle much larger numbers of outstanding references can be built by eliminating the MSHRs and forwarding the entire state of each request with the request message. In this section, however, we restrict our discussion to MSHR-based interfaces.

20.2.2 Cache Coherence

In a machine that supports caching of remote data with a coherence protocol, operation is similar to that described above for uncacheable reads and writes, with three main differences. First, all operations are performed in units of cache lines. A read request, for example, reads an entire cache line that is then stored in the local cache. Second, the protocol requires a larger vocabulary of messages. Separate messages are used, for example, to request a cache line in a read-only state and a read-write state. Additional messages are also used to forward and invalidate cache lines. Finally, coherence protocols require the processor-network interface to send messages in response to messages received, not just in response to processor actions.

A complete discussion of cache coherence protocols is beyond the scope of this book.[8] However, a simple coherence protocol requires the processor to send: a read request message (sent on a read miss), a read exclusive message (sent on a write miss to acquire a line), a writeback message (sent to evict a dirty line to memory), a forward message (sent to forward a dirty cache line to a new owner), and an invalidation acknowledgment (sent to acknowledge that a clean line has been invalidated). The processor is required to handle receipt of a read reply message (with a read-only cache line), a forward message (with a read-write cache line), an invalidation request (asking for a read-only line to be invalidated), and a forward request (asking for an

---

8.   The interested reader is referred to [115, 116].

exclusive cache line to be forwarded). The receipt of each of these messages causes an existing MSHR entry to be updated, or (for invalidation or forward requests) a new MSHR entry to be created. The status field of the MSHR entry triggers the protocol FSM and the message transmit unit to carry out any actions needed in response to the message. For example, an invalidation request requires the protocol state machine to invalidate the specified line and to send a reply message to signal that the invalidation is complete. Similarly, a forward request message requires the protocol FSM to invalidate a line and forward its contents to a specified node in a forward message.

Coherence messages that carry data carry an entire cache line. Cache line sizes on modern machines vary from 8 bytes (one word on the Cray X-1) to 512 bytes (the L2 line size on the IBM Power4), and a line size of 128 bytes is typical. Transfer of a 128 bytes line is typically done one or two words (8 to 16 bytes) at a time and thus takes 8 to 16 cycles. To minimize latency, message injection is pipelined with transfer of the cache line from the cache. The header of the message is injected as soon as it is formatted from the data in the MSHR entry, rather than waiting for the entire cache line to be read. Each word of the cache line is then injected as it is read. Also, to reduce latency, many protocols read (and send) the *requested word* first, sending the rest of the line in a wrapped order after this critical word.

A key issue in the design of processor-network interfaces for cache-coherent shared-memory multiprocessors is *occupancy*, which is the amount of time a critical resource is busy with (occupied by) each memory access. In a well-designed interface, the resources (the cache, the MSHRs, and the message transmit and receive units) are occupied for only a single cycle (or a single cycle per word) for each memory access. In some interfaces, such as those that use software to implement the coherence protocol, a resource may be occupied for tens of cycles (or longer). In such cases, this busy resource quickly becomes a bottleneck that limits throughput.

20.2.3 Memory-Network Interface

A memory-network interface is shown in Figure 20.7. This interface receives memory request messages sent by the processor-memory interface and sends replies. Like the processor-memory interface, it is optimized for low latency.

Messages received from the network are used to initialize a transaction status holding register (TSHR). A small request queue is used to hold a few request messages when all TSHRs are busy to delay the point at which requests back up into the network. Each TSHR, analogous to the MSHR on the processor side, tracks the status of a pending memory transaction. The TSHRs are monitored by the memory bank controllers and the message transmit unit, and changes in a TSHR status field trigger the appropriate action in these units. Bank controllers for each of the N memory banks perform read and write operations as required by the pending transactions, moving data between the data fields of the TSHRs and the memory banks. The message transmit unit formats and transmits messages in response to completed transactions.

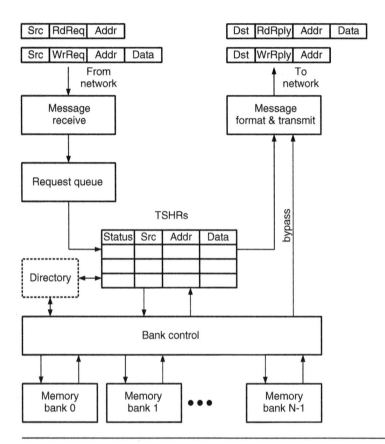

Figure 20.7 Memory-network interface. Messages requesting memory operations are received from the network and queued. Pending transactions are held in TSHRs while they access the directory and/or memory banks. Completed transactions result in messages being formatted and injected back into the network.

A non-cacheable read request, for example, initializes a TSHR with status *read pending* and sets the address and source node fields. When the memory bank matching the bank-select bits of the address is available, it starts a memory access and changes the status to *bank activated*. Two cycles before the first word is returned from the memory bank, it sets the status of the TSHR to *read complete*. This status triggers the message transmit unit to format a read reply message header addressed to the source node (Src) with the requested address. By the time the header has been injected into the network, the first word from the selected memory bank is available and words from the memory bank are injected directly into the network to complete the reply message. Finally, the TSHR entry is marked *idle*.

If the requests are simple reads and writes, and if they are guaranteed to complete in order, the design can be simplified by replacing the TSHRs with a simple queue, or a queue per memory bank. As each request gets to the head of the request queue, it waits for the memory bank it needs to become available (stalling all of the following requests). Once the memory bank is available, the request is initiated and placed into a pending request queue, which takes the place of the TSHRs. As the memory operations complete, they are matched with their requests — now at the head of the pending request queue — and the message transmit unit uses the combined information to send a replay message. Queues are simpler and less costly than TSHRs, but cannot handle the actions required by a more complex protocol.

For a cache coherent request, a TSHR serves to hold the state of a transaction between protocol messages.[9] Consider, for example, a read-exclusive request. The request creates a TSHR entry and sets the status to *read-exclusive directory pending*. This activates the directory unit, which determines the current state of the requested cache line. If the line is in a shared state, the directory unit sets the TSHR status to *read pending, invalidate pending* and updates the TSHR with a list of the nodes sharing the line, and a count of these nodes (in fields not shown). The selected (by the address) memory bank is triggered to start reading the requested line by the *read pending* portion of the status. In parallel, the message transmit unit is triggered by the *invalidate pending* portion of the status to send invalidate messages one at a time, updating the count in the TSHR as each is sent. When all the invalidate requests have been sent, the TSHR status is set to *awaiting invalidate reply*. As each invalidate reply is received, a count is updated, and when all have been received, the status becomes *invalidate complete*. If the read is also complete, this triggers the message transmit unit to send the reply message.[10]

20.3   Line-Fabric Interface

In a packet switch or router that employs an interconnection network as a switching fabric, the network interface must provide queueing both before and after the interconnection network, as shown in Figure 20.8. The input queues are required to prevent interference between packets destined for different outputs. If an output *A* becomes momentarily congested, it is unacceptable to block all packets entering the network while they wait on a packet destined to *A*. In a network switch, blocked packets are eventually dropped because there is no mechanism to provide backpressure to stop packet arrival on the incoming line.

To avoid this head-of-line blocking at the input of a packet switch, the switch provides a separate *virtual output queue* at each input for packets destined to each output. A packet to a blocked output is queued before the fabric, allowing packets destined to other outputs to proceed. In practice, queues are usually provided not just for each output port of the network, but for each class of traffic times each subport of each output port. This prevents high-priority traffic from being blocked by low-

---

9.   This transient state could be held in the directory, but it saves directory space and accesses to factor it out into the TSHRs.

10.   Many machines optimize this process by having the invalidate acknowledgments forwarded to the receiving node to shorten the critical path of a read-exclusive transaction to three hops from four.

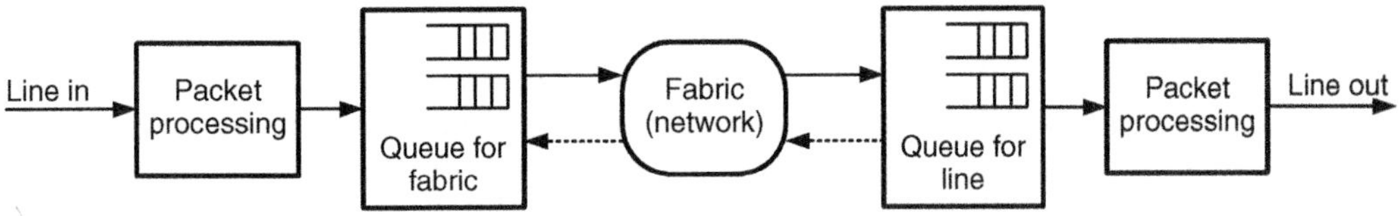

Figure 20.8   A packet router or switch requires queueing of packets before and after the interconnection network. The input queue holds packets being scheduled for transmission over the network. The output queue holds packets being scheduled for transmission over the line out.

priority traffic and prevents a subport of an output port (for example, one 1 Gbit Ethernet interface on a 10-interface line card) from blocking traffic to other subports.

At the output of the interconnection network, a second set of queues holds packets while they are scheduled for transmission on the output line. This set of queues is required because the fabric typically has speedup; the bandwidth from the interconnection network to the line card is higher than the bandwidth of the line out. Fewer buffers are required at the exit side of the line card, just one per subport × class. However, the buffers on the exit side are usually quite large, since in many applications they must buffer traffic during transient overloads of an output node that may last 10 ms or longer.

For example, consider a router that has 256 line cards, each with 20 Gbits/s of capacity divided among eight 2.5-Gbits/s subports. An interconnection network is used for the fabric that connects the line cards and provides one input port and one output port to each line card. The line cards define 4 classes of service with a strict priority required between the classes. In this case, each line card must provide a total of 8 K input queues (256 ports × 8 subports/port × 4 classes). On the output side, each line card need only provide 32 queues (8 subports × 4 classes).

A typical line card also includes packet processing logic in both the input and output paths. This logic rewrites packets and updates statistics counters. Its operation is independent of the interface to the fabric and will not be discussed further.

In some applications, end-to-end flow control independent of the interconnection network is provided from the output queue manager to the input queue manager. This is illustrated by the dotted lines in the figure. This flow control is usually implemented by sending dedicated control packets over the interconnection network to start or stop the flow of packets from a particular input queue.

To prevent low-priority packets from blocking high-priority packets, and to prevent traffic to a congested output subport from blocking traffic of the same class to a different output subport, the interconnection network must be *non-interfering* for packets from different classes and destined to different output subports. One brute force approach to providing this non-interference is to provide a virtual network (a set of virtual channels for each physical channel) for each subport × class.

In practice, when a packet arrives on the input line, it is classified and assigned an output port by the packet processor. The input queue manager[11] then enqueues

---

11.   The block that performs the queue manager and scheduler functions is often called a *traffic manager.*

it in the appropriate input queue. If the queue was empty, a request is sent to the fabric scheduler, also part of the queue block in Figure 20.8. The fabric scheduler keeps track of the status of the input queues and of the interconnection network input port. Using this information, it repeatedly selects the highest priority packet to an unblocked output from the set of waiting packets and inserts this packet into the network.[12]

Because the interconnection network has a higher bandwidth than the input line, the input queues are nearly always empty and the few that are non-empty are usually quite short. Only when an output becomes blocked does an input queue grow to any significant length.[13] Because most queues are short, the input queue manager can keep almost all of its queued packets in on-chip memory, avoiding the power dissipation required to write these packets to off-chip memory and then read them back.

A block diagram of a queue manager and scheduler is shown in Figure 20.9. The queue manager maintains a state vector S, on-chip head and tail pointers h and t, and off-chip head and tail pointers H and T. The state vector indicates whether the queue resides entirely on-chip (at addresses indicated by h and t) or whether the tail of the queue is off-chip (at the addresses indicated by H and T) and the head of the queue is on-chip (as indicated by h and t).

Packets arrive tagged with a queue number. On arrival, this number is used to look up the queue state. If the state indicates that the off-chip portion of the queue is empty and there is room to append the packet to the on-chip portion of the queue,

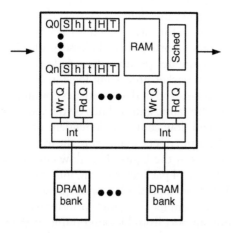

Figure 20.9 A queue manager keeps short queues in on-chip memory and overflows long queues to an off-chip, banked, DRAM memory.

---

12. In some applications, the scheduler may apply more sophisticated scheduling policies. For example, it may *rate shape* the traffic by metering out packets from certain queues at regular intervals.
13. When this occurs, some routers start applying a random early discard policy to the blocked input queue rather than waiting for these packets to get to the output node before applying this policy.

then the packet is inserted in the on-chip queue. Otherwise, it is inserted off-chip. All queue accesses to off-chip memory are striped across multiple DRAM banks to balance the load on the banks. Read and write queues associated with each bank buffer requests until the bank becomes available.

While the queue manager inserts packets into the queues, the scheduler removes them. When the scheduler selects a non-empty queue to supply the next packet, it dequeues the packet from the on-chip queue. The head of all of the queues is always on-chip, so the scheduler never reads packets from off-chip memory. If, after dequeueing a packet, the size of an on-chip queue that has a non-empty off-chip tail falls below a low watermark, a request to transfer sufficient data from the off-chip queue to fill the on-chip queue is initiated.

Storing the heads of queues on-chip and the tails of queues off-chip results in very little off-chip memory traffic because most queues never get long enough to require the off-chip queue tail.

Because memory bandwidth is costly in terms of power and pin-count, it is important that packets be queued, at most, once on their way into the line card and, at most, once on their exit from the line card. This is analogous to avoiding memory copies in a processor-memory interface. Regrettably many routers do not follow this principle, queueing the packet multiple times: once in the packet processor, once in a traffic manager that performs traffic shaping, and once in a fabric scheduler. With careful design, the same functionality can be realized with the packet being written to and read from memory a single time.

20.4 Case Study: The MIT M-Machine Network Interface

The M-Machine is an experimental multicomputer built at MIT and Stanford to demonstrate mechanisms for fine-grain communication between multiple on-chip multithreaded processors [93]. The M-Machine includes a 2-D torus interconnection network with a register-mapped network interface [112]. The interface supports both message-passing and shared-memory models of computation. It provides low-overhead communication without sacrificing safety and isolation of processes.

An M-Machine consists of a number of processing nodes connected in a 2-D torus network. Each processing node was based on a Multi-ALU Processor (MAP) chip (Figure 20.10). Each MAP chip contained three 64-bit multithreaded processors, a memory subsystem, a 2-D torus router, and a network interface. Threads running the the same thread-slot of different on-chip processors could efficiently communicate and synchronize via registers. There are many interesting aspects of the M-Machine architecture. In this section, we will focus on its network interface.

On the M-Machine, messages are directly composed in the processor registers and sent atomically via a message *SEND* instruction, as described in Section 20.1.2 and shown in Figure 20.4. Each of the threads on each of the three processors on a MAP chip has fourteen 64-bit integer registers and fifteen 64-bit floating-point

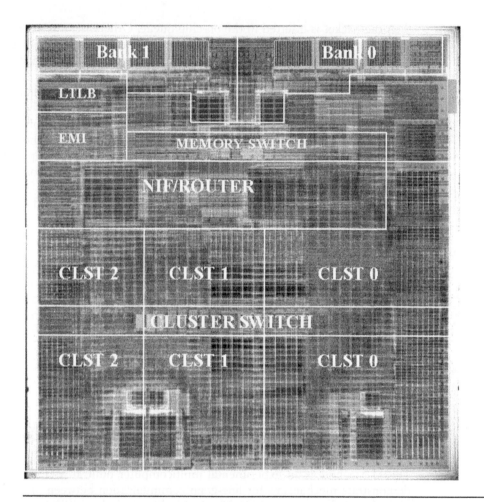

Figure 20.10   The Multi-ALU Processor MAP chip is the single-chip processing node used in the M-Machine. The chip contains three 64-bit multithreaded processors, an on-chip two-bank cache, a memory controller, a network interface and a 2-D torus router.

registers.[14] A thread composes a message in a contiguous set of these registers starting with register I4 or F4 (integer register 4 or floating-point register 4) and then sends the message by executing a send instruction.

Figure 20.11 shows the format of an M-Machine *SEND* instruction. The instruction has four fields: length, dest, handler, and CCR. *Length* specifies the number of registers to be read starting at I4 to compose the body of the message. *Dest* specifies the register containing the destination virtual address. This address is translated to

---

14.   On the prototype die only one of the three processors has a floating-point unit and floating-point registers.

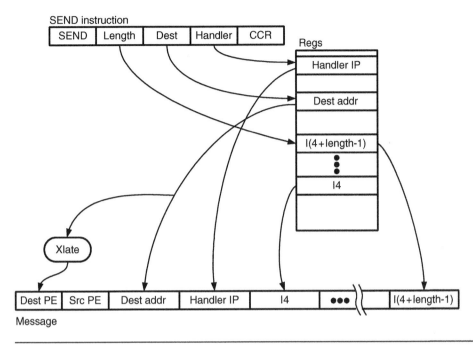

Figure 20.11 The M-Machine *SEND* instruction sends the message composed in the register file starting at register I4 (or F4 for *FSEND*). The instruction specifies the length of the message (the number of registers to read), the destination virtual address, the message handler that should be run upon message receipt, and a condition-code register to be set when the message has been accepted by the network interface.

determine the destination processing element (PE) and is also include in the message. The *handler* is the virtual address of the code that should be run to handle the message upon receipt. This supports a *message-driven* model of computation [53]. Finally, the CCR field specifies a condition-code register that is initially set false and then set true once the message is launched into the network — at which point the registers are free to be overwritten. Using the CCR field to signal completion allows completion of the SEND instruction to be overlapped with subsequent instructions.

The register-mapped M-Machine SEND mechanism was designed to retain the efficiency of the two-register SEND instruction of the preceeding J-Machine while achieving safety by making the entire message send an atomic operation. Before the send commits (as signaled by the CCR being set) no network resources are used and all state is in the processor registers, which are swapped on a process switch. Once the send commits, the message has been transferred entirely into the network input queue where, because of the deadlock and livelock freedom of the network, it will eventually be delivered to the destination node. Because the send is atomic, there is no danger of a thread starting to send a message and then faulting or swapping out and leaving the network interface blocked by a half-sent message, and hence unusable by other threads — as could occur on the J-Machine.

The major limitation of the M-Machine SEND mechanism was due to the small size of the M-Machine register files — 14 integer and 15 floating-point registers. With such a small register set, message size was limited to a maximum of 10 or 11 and composing even modest-sized messages caused considerable register pressure, resulting in spilling registers to the stack. The mechanism would be much more effective with larger register files.

To ensure system-level (as opposed to network-level) deadlock freedom, the M-Machine employed a *return-to-sender* convention. Each processing node maintained a buffer to handle returned messages. A free-space counter FS reflected the available space in this buffer. Before sending a message, a node would check that $FS > L$ (where L is the message length) and then decrement FS by L, guaranteeing that there is sufficient room in the buffer to hold the message being sent. When a message was successfully received, an acknowledgement was sent that caused FS to be incremented by L — returning the reserved space to the pool. If a receiving node was unable to accept a message, it returned it to the sending node. Returns were made on a separate set of virtual channels and injection/extraction buffers to avoid request-reply deadlock. The sending node would buffer the returned message in the space reserved for it and retry the send.

M-Machine network reception was via a pair of registers, as illustrated in Figure 20.12. Arriving messages are enqueued into one of two receive queues — one for each of two logical networks. Only one is shown in the figure. For each queue, a receive thread runs in a dedicated thread slot on the multithreaded processing node. The receive thread can read the next word in the queue (the head of the queue) by reading register 115 (the MsgBody register). To skip the remaining words of the current message and advance to the head of the next message, the receive thread reads register 116 (the MsgHead register). If the requested word, body or head, has not yet arrived in the queue, the receive thread blocks until it arrives. After reading either 115 or 116, the receive pointer is advanced to move the head of the queue to just after the word read.

Because the receive thread must always be available to remove messages from the network, it is permitted to perform only short, bounded computations in response to arriving messages. This regulated by requiring the handler instruction pointer (IP) word of each message to be an unforgeable pointer to a *message handler*, which has been verified to have the required behavior. The system message handler reads the handler IP and jumps to it. For messages that can be handled quickly with no chance of a fault or delay (such as acknowledge, physical memory write, physical memory read, and so on), the message handler performs the work directly. For other message types, the message handler enqueues the message in an appropriate system queue and returns to handle the next message.

The M-Machine provided special support for implementing shared memory on top of the register-based messaging system described above. Memory requests queried the on-chip cache and local translation lookaside buffer (LTLB). If the location requested was in the cache or mapped to local memory, the access completed in hardware. If the access missed in the LTLB, or if the LTLB indicated that a remote

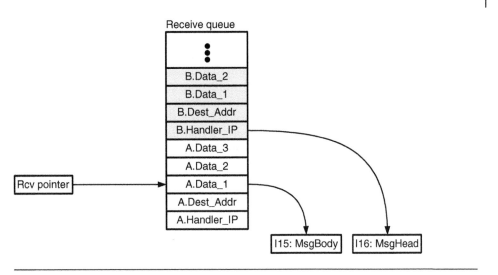

Figure 20.12   M-Machine message reception. Arriving messages are stored in one of two receive queues. Reading from I15 (MsgBody register) reads the next word of the message. Reading from I16 (MsgHead register) reads the header of the next message. In either case, the receive pointer is advanced to point to the word after the word just read.

access was required, the required information — the reason for the event, the address in question, the data to be written (if any), and the continuation (thread ID and register number) to return data to — was enqueued in the node's event queue. A thread running in a dedicated event-handler thread slot processed events in this queue in exactly the same manner that the message handler threads processed messages from the two message queues.

For example, the event handler thread would handle a remote read by sending a message to the address in question with the handler IP of the remote read handler and containing the continuation information as payload. On the remote node, the message handler thread would jump to the remote read handler IP, perform the read, and send a reply message. The resulting remote memory access times were just slightly longer than for machines that performed remote memory accesses with dedicated hardware [32]. Implementing these handlers in software, however, enabled experimentation with different coherence protocols and memory policies.

20.5 **Bibliographic Notes**

Most early processor network interfaces were attached to the I/O bus and used either program transfers or DMA to transfer messages. Joerg and Henry study alternative network interface architectures and locations [82]. The MARS accelerator used a two-register interface for message send and receive [4]. The J-Machine used a send instruction to compose messages out of the register file, but received messages to local memory [53]. The AP1000 improved the speed of memory-based network interfaces by associating the interface with the cache memory [80]. The M-machine implemented a send instruction that transmitted a contiguous group of registers as a message [112]. The SHRIMP multicomputer [22, 23] uses an I/O attached processor-memory interface connected to standard workstation nodes. It reduces overhead by mapping windows of memory between the address spaces of nodes. Myrinet [24] also uses an I/O attached interface that includes a local processor for protocol processing. The Berkeley NOW project [10] is an example of a multicomputer built using such an I/O attached interface. Such I/O attached message interfaces are often used with message-passing libraries such as Berkeley's Active Messages [189] or Illinois Fast Messages [139]. The use of MSHRs to allow multiple cache misses to be overlapped was introduced by Kroft [105]. DASH, one of the first network-based shared-memory multiprocessors is described by Lenoski et al. [115]. The SGI Origin 2000 is one of the first commercial machines of this type [108]. Shared memory multiprocessors are covered by Lenoski and Weber [116]. Split on-chip off-chip queueing is described in U.S. Patent 6,078,565 [16] and by Iyer et al. [87].

20.6 **Exercises**

20.1 *Comparing message interface overhead.* Consider sending a short message (128 bits) and a long message (32 Kb) using (a) a two-register interface, (b) a register-mapped interface, and (c) a descriptor-based interface. Assume that the short message resides initially in processor registers and the long message resides initially in memory. Compare the processor overhead (both the time required to send the message and the processor's occupancy) of sending each length of message on each interface.

20.2 *Cache coherence protocol.* Consider a cache coherence protocol in which each cache line can be in one of three states on each processor: invalid (this processor doesn't have a copy), shared (this processor has a read-only copy), and exclusive (this processor has an exclusive copy that may be *dirty*). Describe the sequence of messages that must be sent to handle a read or a write to a cache line in each of the possible states.

20.3 *Protecting a two-register interface.* Explain how a malicious thread can tie up a network using a two-register interface indefinitely. Write a small code fragment illustrating the problem. Suggest a method to prevent this problem from occurring.

20.4 *Long messages with register-mapped interface.* Suppose you have a processor that has 64 general purpose registers and a register-mapped interface as described in Section 20.1.2 and you need to send a 1,024-word message from a buffer in memory. Write a short code fragment that performs this message send. Suggest a way to reduce the overhead of sending long messages.

20.5 *Format a descriptor-based message.* Write down the register contents for a descriptor-based message that sends a 1,024-word memory buffer to another node. The message should include the destination address, a header identifying the type and length of the message, and the data itself.

20.6 *Single-memory line-network interface.* Consider a line-network interface in which packets are buffered in memory only on the output side of the fabric. A small on-chip queue (100 packets) is all that is provided on the input side of the fabric. Suppose that your router must support 128 line cards, each of which handles 4 classes of traffic. Also, assume (unrealistically) that input traffic is uniformly distributed over the output nodes. How can you guarantee that no packets will be dropped from the small input queue? Sketch a solution.

CHAPTER 21

Error Control

Many applications of interconnection networks require high reliability and availability. A large parallel computer requires that its interconnection network operate without packet loss for ten thousands of hours. An Internet router can accept a small amount of packet loss, but the router itself must remain *up* with an availability of 99.999% — five minutes of downtime allowed per year. I/O systems have similar availability requirements.

Interconnection networks are often composed of hundreds (or thousands) of components — routers, channels, and connectors — that collectively have failure rates higher than is acceptable for the application. Thus, these networks must employ *error control* to continue operation without interruption, and possibly without packet loss, despite the transient or permanent failure of a component.

21.1 Know Thy Enemy: Failure Modes and Fault Models

The first step in dealing with errors is to understand the nature of component failures and then to develop simple models that allow us to reason about the failure and the methods for handling it. We classify failures that may occur in our system as *failure modes*. A failure mode is a physical cause of an error. Gaussian noise on a channel, corrosion on a connector, cold solder joint failure, an open output driver in a power supply, alpha-particle strikes, electromigration of a conductor on a chip, threshold voltage shift in a device, operator removing the wrong module, and software failure are examples of failure modes.

Because failure modes are often complex and arcane, we develop simple *fault models* that describe the relevant behavior of the failure mode while hiding most of

Table 21.1 Failure modes and fault models for a typical interconnection network.

Failure Mode	Fault Model	Typical Value	Units
Gaussian noise on a channel	Transient bit error	10^{-20}	BER (errors/bit)
Alpha-particle strikes on memory (per chip)	Soft error	10^{-9}	SER (s^{-1})
Alpha-particle strikes on logic (per chip)	Transient bit error	10^{-10}	BER (s^{-1})
Electromigration of a conductor	Stuck-at fault	1	MTBF (FITs)
Threshold shift of a device	Stuck-at fault	1	MTBF (FITs)
Connector corrosion open	Stuck-at fault	10	MTBF (FITs)
Cold solder joint	Stuck-at fault	10	MTBF (FITs)
Power supply failure	Fail-stop	10^4	MTBF (FITs)
Operator removes good module	Fail-stop	10^5	MTBF (FITs)
Software failure	Fail-stop or Byzantine	10^4	MTBF (FITs)

the unneeded complexity. Table 21.1 gives a partial list of common failure modes in interconnection networks. The table also shows typical values for the failure rates of these different modes.[1] Some failures, such as Gaussian noise and alpha-particle strikes, cause transient faults that result in one or more bits being in error but do not permanently impair machine operation. Others, such as a connector failure or electromigration of a line, cause a permanent failure of some module.

Transient failures are usually modeled with a bit-error rate (BER) or soft-error rate (SER). These rates have dimensions of s^{-1} and the inverse of these rates is the time between errors.[2] At one level, we model the permanent failures with a stuck-at fault model in which we assume that some logical node is stuck at logic one or zero. Other permanent failures we model as fail-stop faults in which we assume some component (link or router) stops functioning and informs adjacent modules that it is out of service. These failures are usually described in terms of their mean-time between failures (MTBF) also in units of time (often expressed in hours). Sometimes such failure rates are expressed in failures in 10^9 hours (FITs). Often, we design systems to reduce stuck-at faults, or even excessively frequent transient faults, to a fail-stop fault. A component, such as a channel, will monitor its own execution and shut itself down (fail-stop) when it detects an error.

---

1. The values in this table are only an indication of the general magnitude of these error rates for typical 2003 systems. Error rates are very sensitive to a number of technology factors and can vary greatly from system to system. In doing any error analysis, make sure to get the correct value for the technologies you are using.
2. Link BER in errors/bit can be converted to an error rate in errors/s by multiplying by the link bandwidth in bits/s.

While at first glance the error rates in Table 21.1 appear quite small, in a large system, they add up to significant totals. Consider, for example, a 1,024-node 3-D torus network with 10 Gbits/s channels, each with a BER of 10^{-15}. At first, 10^{-15} seems like a very small number. However, multiplying by the channel rate of 10^{10} bits/s gives a failure rate of 10^{-5} errors/s per channel. Summing over the 6,144 channels in the system gives an error rate of 6×10^{-2} errors/s — an error every 16 seconds. An even more reliable link with a BER of 10^{-20} gives an aggregate error rate in this system of 6×10^{-7} — about 2 errors per month. In practice, we can build very reliable systems from such links by controlling the effects of errors.

Some types of failures are Byzantine in that rather than stopping operation, the system continues to operate, but in a malicious manner, purposely violating protocols in an attempt to cause adjacent modules to fail. Byzantine failures are extremely difficult to deal with, and fortunately are quite rare in practice. To avoid Byzantine failures, we design systems with sufficient self-checking to shut down failing modules before they can run amok. Software failures can sometimes exhibit Byzantine behavior. However, they can be dealt with just like hardware failures, by monitoring their execution and stopping execution if they violate an invariant.

The failures of components such as power supplies, cooling fans, and clock generators are handled by redundancy complemented by field replacement. Redundant power supplies are provided so that the failure of any single supply is *masked*, the failure is easily detected by supply monitoring hardware, and the supply is replaced before a second supply is likely to fail. Cooling fans use a similar $N + 1$ form of redundancy. On systems that require a global clock, multiple clocks are distributed, with each module switching over when a clock fault is detected and using a local phase-locked loop (PLL) to provide a reliable clock during the transient. For the remainder of this chapter, we will assume that good engineering practice has been applied to critical infrastructure such as power, cooling, and clocks, and will confine our attention to link and router failures.

The failure rates shown in Table 21.1 refer to the failure rates of these components during the bulk of their lifetimes. Many types of components have much higher failure rates at the beginning and end of their lifetimes as shown in Figure 21.1. Marginal components tend to fail after just a few hours of operation, a phenomena called *infant mortality*, leading to a high failure rate early in life. After components have survived a few hundred hours, the marginal components have been weeded out and failure rate is relatively constant until the component begins to wear out. Once wearout starts to occur, failure rate again rises. To eliminate infant mortality, we typically *burn-in* components, operating them for a period of time (and under stressful conditions) to weed out marginal components before installing them in a system. Similarly, we eliminate wearout by replacing life-limited components before they begin to fail. For example, we may burn-in a router chip for 100 hours at elevated temperature, voltage, and frequency before installing it in a system to prevent failures due to infant mortality. We will replace a disk drive that has an expected lifetime of 10^5 hours (about 10 years) after 5×10^4 hours (about 5 years) to prevent failures due to wearout.

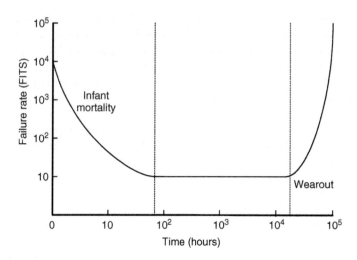

Figure 21.1   The *bathtub* curve shows how failure rate varies over the lifetime of a component. Failure rate is higher at the start of life as marginal components fail — a phenomenon named *infant mortality*. During the middle of a component's life, failure rate is relatively constant. As a component nears the end of its life, failure rate again increases due to *wearout*.

For the rest of this section, we will assume that our system and components have been designed so that all failure modes have been reduced to either transient errors or fail-stop of a router, channel, or network interface. Further, we assume that a combination of burn-in and planned replacement have reduced all error rates to constants. Transient errors may occur as either errors in bits transmitted over a channel or as state bits spontaneously flipping state. We will discuss some error detection and monitoring methods below that are used to reduce other failures to the fault models we will consider.

21.2 The Error Control Process: Detection, Containment, and Recovery

All error control involves three basic steps: detection, containment, and recovery. As in dealing with any problem, the first step is to recognize that a problem exists. In an interconnection network, this is the step of error *detection*. For example, we may detect a bit error on a channel by checking the parity or check character for a flit. Once we have detected a fault or error, we must *contain* the error to prevent its propagation. Continuing our example, if the error has corrupted the virtual channel identifier portion of the flit, we must prevent the flit from erroneously updating the virtual channel state of the wrong virtual channel. Finally, the third step of the error control process is to *recover* from the error and resume normal operation. In our example, we might recover from the bit error by requesting a retransmission of the flit — either at the link level or from the original source.

21.3 **Link Level Error Control**

Most interconnection networks use a hierarchy of error control that starts at the physical link level and then continues to the router level, network level, and system or end-to-end level. At the link level, link logic acts to mask link errors (possibly adding delay) and shuts the link down when errors cannot be masked — reducing the fault to a fail-stop fault model. Logic modules at the two ends of the link work together to detect, contain, and recover from bit errors on the link. In the event of a hard error, the link logic either reconfigures the link around the error or shuts the link down.

21.3.1 **Link Monitoring**

Error detection at the link level is performed by encoding redundant information on the link, using an *error control code* (ECC).[3] Simple parity is sufficient to detect any single bit error. However, most links use a cyclic-redundancy check (CRC) of sufficient length that the probability of a multibit error going undetected becomes vanishingly small. An n-bit CRC field will detect all but 1 in 2^n multibit errors and all that involve fewer than n bits.

The error check can be made at different levels of granularity: flit, packet, or multipacket frame. Although performing checks over large-sized units (such as frames) is more efficient, it delays detection of an error until the entire unit is received making containment difficult. By the time a frame-level CRC detects an error, the erroneous flit or packet may already have propagated over several additional channels and corrupted the internal state of several routers. To avoid such error propagation, many routers perform checks on every flit and may additionally separately protect critical header information so it can be safely acted on before the entire flit is received and validated.

As an example of link monitoring, Figure 21.2 shows a flit format that includes two CRC fields to check for bit errors on the link. Field CRC1 checks the header fields (virtual-channel ID, flit type, and credit) that are received during the first cycle. The longer CRC2 field is a CRC over the entire flit (including CRC1). Providing a separate CRC for the header fields that arrive during the first cycle of the flit allows these fields to be used immediately without waiting for validation by the whole-flit CRC. Including the header fields and their CRC in the calculation of the longer CRC applies the greater error detection capability of the longer CRC to the header

---

3.   A detailed discussion of ECC is beyond the scope of this book. For a good treatment of the subject, see [21].

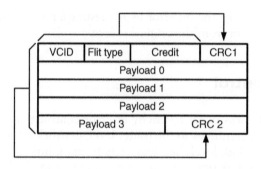

Figure 21.2   A typical flit format including error checking fields. Four-bit CRC1 is a CRC over the header information that arrives during the first cycle of the flit. Eight-bit CRC2 is a longer CRC that covers the entire flit but does not arrive until the fifth cycle. Providing a separate CRC for the header information allows this information to be used safely before the final CRC is received and validated.

information as well — although the error will have already propagated if it was missed by the short CRC and caught by the long.

Link monitoring should be continuous. If an idle channel goes unchecked, an error can remain latent for a long period until the next flit arrives. If the link is idle, special idle flits should be sent over the link with random contents to exercise and continuously test the link. Alternatively, a continuous frame-based monitoring scheme can be used in addition to flit-based monitoring.

21.3.2   Link-Level Retransmission

Once an error has been detected, the link logic must proceed to the step of containment. Most link errors are contained by *masking* the error so that it is never visible to the router logic. Masking both contains the error and recovers from the error in a single step. Retransmitting the faulty flit is the simplest method to mask the error. Masking can also be performed by using a forward error-correcting (FEC) code. With a FEC code, sufficient information is sent with the unit of protection (usually a flit) to not only detect the error, but also to correct it. However, in an interconnection network with a relatively short round-trip latency, retransmission is usually preferred because it is simpler and can correct more errors with lower overhead.

A simple retransmission system is shown in Figure 21.3 and its timing is illustrated in Figure 21.4. As the transmitter transmits each flit, it retains a copy in a transmit flit buffer until correct Receipt is acknowledged.[4] Each transmitted flit is

---

4.   In an implementation that employs an output buffer to handle switch output speedup, the transmit flit buffer is usually combined with the output buffer.

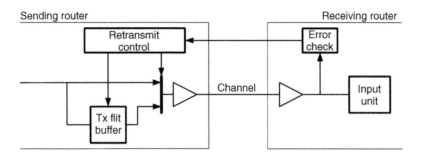

Figure 21.3 Link errors can be masked, with a small delay, by retransmitting faulty flits. As each flit is transmitted, it is also stored in a small flit buffer. If an error is detected, the flit is retransmitted from the buffer.

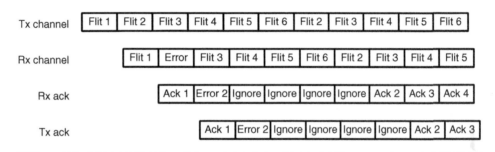

Figure 21.4 Timing diagram for link-level retransmission. Flit 2 is received in error. The receiver detects the error, signals the transmitter to retransmit, and begins ignoring flits. The transmitter sends four additional flits (flits 3 through 6) before it receives notification of the error from the receiver. Upon receiving notification, the transmitter retransmits flits 2 through 6.

tagged (typically with its address in the transmit flit buffer) to facilitate identification of flits in acknowledgments. As each flit is received, the receiver checks the flit for errors. If the flit is received correctly, an acknowledgment is sent to the transmitter. The acknowledgment identifies the flit being acknowledged (using the tag transmitted with the flit) and indicates the reception status — received correctly, received in error, or ignored.[5] Upon receiving this acknowledgment, the transmitter discards its copy of the flit. If a flit is received in error, the transmitter switches the multiplexer and retransmits the flit in question. In most cases, the error is transient and the flit is received correctly on the second attempt.

5.   Of course, the acknowledgment itself must be checked for errors and retransmitted if incorrect. This is usually accomplished by piggybacking the acknowledgment on a flit traveling in the reverse direction and retransmitting the whole flit if any part (including the acknowledgment) is in error.

While it is only necessary to retransmit the one faulty flit, this would reorder the flits on the channel and considerably complicate the router input logic. For example, a body flit might be received before the head flit of its packet. To avoid such complications, it is easier to simply roll back transmission to the faulty flit and retransmit all flits in the transmit flit buffer starting at that point. Any new flits arriving while these are being transmitted are added to the tail of the flit buffer. When the transmit pointer into the flit buffer reaches the end, the multiplexer switches back to transmitting flits directly from the switch.

An example of retransmission is shown in the timing diagram of Figure 21.4. Each time the transmitter sends a flit, the receiver sends an acknowledgment or error indication. Flit 2 is corrupted during transmission. Upon receipt, the receiver checks the CRC of flit 2, detects an error, and sends an error indication to the transmitter. The transmitter sends four more flits (3 through 6) before it receives notice of the error. These flits are ignored by the receiver and retained in the transmit flit buffer. When the transmitter receives notification of the error, it rolls back transmission to flit 2 and resends all flits starting at that point. The receiver stops ignoring flits and resumes normal operation upon receipt of the retransmitted flit 2. From the receiver's point of view, the error is masked, the only difference from operation without the error is a delay of five flit times — as required for the round-trip error notification.

The transmit buffer is managed using three pointers, as shown in Figure 21.5. In the absence of errors, the transmit pointer, and tail pointer act as a head and tail pointer for a FIFO transmit queue. New flits are added to the buffer at the tail pointer and flits are transmitted starting at the transmit pointer. If the router is transmitting directly from the switch (no delay in the transmit flit buffer) the transmit and tail pointers point to the same location. Unlike a FIFO queue, however, a location in the flit buffer cannot be reused once the flit it contains has been transmitted. The flit must be retained until it is acknowledged. The acknowledge pointer identifies the oldest flit that has not yet been acknowledged. Each time an

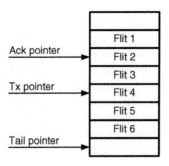

Figure 21.5   Retransmission from the transmit flit buffer is managed by three pointers: the ack pointer identifies the next flit to be acknowledged, the transmit pointer identifies the next flit to be transmitted, and the tail pointer indicates the next free location in the buffer. The figure shows the pointers just before flit 4 is retransmitted in Figure 21.4.

acknowledge is received the acknowledge pointer advances, freeing one flit buffer. If an error indication is received, the transmit pointer is reset to the acknowledge pointer.

21.3.3 Channel Reconfiguration, Degradation, and Shutdown

One aspect of the containment process is to prevent a known bad component from continuing to disrupt traffic. If a channel has repeated errors, or even a BER that is much higher than expected, it is likely that some portion of the channel has suffered a hard error. To detect such persistent errors, each channel is provided with an error counter that keeps track of the error rate on that channel. If the error rate exceeds a threshold, the channel is determined to be faulty. For example, if the expected error rate on one channel is 10^{-5} errors per second (about 1 error per day) and a channel logs more than 10 errors in a 10^4 second interval (3 hours), the channel is declared faulty and taken out of service.[6] This process of BER monitoring and faulting of channels is typically performed by supervisor software that continuously polls the status of network components.

Some multibit channels provide one or more spare bits, allowing the channel to be reconfigured around errors affecting a single bit. When a hard error is detected on such a channel, a diagnostic procedure is run to determine which bit(s) is (are) in error and, if possible, the channel is reconfigured around the bad bits.

Figure 21.6 shows an 8-bit channel with 1 spare bit. A set of eight 2:1 multiplexers at the transmitter and receiver allow the 8 bits of the channel $d_0, \ldots, d_7$ to be steered over any working 8 of the 9 signal lines $s_0, \ldots, s_8$. For example, when line s_3 fails (perhaps due to a bad connector), the channel is reconfigured by shifting bits $d_3 - d_7$ to the left at the transmitter so they traverse lines $s_4 - s_8$. The bits are shifted back at the receiver.

Some channels do not provide spare bits, but can be reconfigured to transmit at a reduced rate, using only the good signal lines of a channel. For example, when 1 line of an 8-bit channel fails, transmission will continue on the remaining 7 lines at $\frac{7}{8}$ of the original rate. This type of reconfiguration is an example of *graceful degradation*. Instead of completely failing the link when the first bit fails, it gracefully degrades, losing bandwidth a little at a time. We can apply graceful degradation at the network level even if our links fail-stop on the first hard bit error.

If a channel cannot be reconfigured to mask the hard error, the channel remains shut down (reducing the channel error to a fail stop) and error control is performed at the next level of the hierarchy — at the network level by routing around the failed channel. Flits that are in the transmit flit buffer when a channel is shut down are either returned to an input controller for rerouting, or dropped.

---

6. A channel that fails on every flit will be taken out of service after the first 10 failures. There is no need to wait the whole 10^4-second interval.

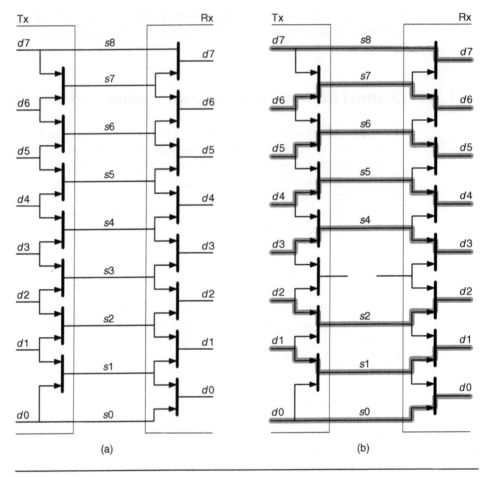

Figure 21.6   (a) An 8-bit channel with 1 spare bit. (b) Bit 3 of the channel fails and the channel is reconfigured by shifting bits 3 through 7 to the left.

A complication arises when a packet is split across a failed channel. The head flit may be several hops downstream and the tail flit upstream when the channel goes down. The headless tail of the packet, upstream of the failed link, must be dropped, since it has no access to the routing information in the head flit and, hence, does not know where it is going.[7] More problematic is the tailless head. The head flit will continue propagating on to the destination, allocating virtual channels along the way. Unless these resources are freed, other packets needing to use these virtual channels will be unable to make progress. To free these resources, an input controller

---

7.   The Reliable Router [52] was able to reroute such *severed tails* by retaining a copy of the head flit at each router that contained any flits of the packet.

synthesizes a tail flit for each virtual channel that is mid-packet when the channel it connects to fails. Packets dropped when split by a failing channel may be recovered by using end-to-end error control (Section 21.6).

21.4 **Router Error Control**

While some failure modes cause link errors, others cause transient or hard router errors. Although less frequent than bit errors on channels, soft errors in router memories and logic do occur and must be controlled to build large systems that have high levels of reliability. As with link errors, the principles of detection, containment, and recovery also apply to router errors.

Router errors are most easily detected by duplicating the router logic and comparing an exclusive-OR of representative signals on a cycle-by-cycle basis. One copy of the router logic is the *master* and generates all primary outputs. The second, *shadow* copy of the logic receives the same inputs as the master copy, but its outputs are unused except to compare against the master. A *checker* compares the outputs and internal signals of the two copies. Comparing just the outputs of the two modules is usually not adequate to catch errors in a timely manner. Many errors will corrupt the internal router state without being observable at router outputs for thousands of cycles. To simplify checking, the master and shadow typically generate a small number (e.g., 32) of *compressed state* lines by exclusive-ORing a large number (thousands) of internal signals and module outputs together. The checker then compares the corresponding compressed state lines from the master and shadow. If a single one of any of the thousands of internal signals differs, the compressed state lines will differ and an error will be detected.

Errors in router memories and internal buses may be detected using an ECC. Single error correcting, double error detecting (SECDED) codes are commonly used on memories and buses to mask bit errors. An n bit memory or bus can be protected by a SECDED code by adding $\log_2(n) + 1$ check bits to the n-bit word being stored or transmitted. To mask memory errors, it is important to periodically *sweep* the memory by reading each location and correcting any single-bit errors. Detecting and correcting errors soon after they occur — by sweeping — reduces the probability of having an uncorrectable second error occur before the first is corrected.

Router errors are also detected via consistency checks. Protocol consistency is checked to ensure, for example, that there is at least one tail flit between one head flit and the next head flit. Credit consistency is periodically checked to ensure that the input controller at the receiving end of a channel and the output controller at the sending end of a channel agree on the number of free buffers available in the input controller. State consistency checks ensure that the inputs received are appropriate for the state of the input controller or virtual channel. If any of these consistency checks are violated, an error is detected and must then be contained.

Once a router error is detected, the error must be contained and recovered from. For errors that cannot be masked, the simplest method of containment is to stop the router or the portion of the router (such as the input controller) having the error — taking it out of service. Stopping the component involved with the error reduces the error to a fail-stop and prevents the fault from corrupting the state of adjacent routers. For purposes of containment, we divide the router into *fault-containment regions*. Typically, each input controller is a separate fault-containment region, which can be taken out of service separately and the entire router is an enclosing fault-containment region, which is taken out of service when an error in the allocator or other common logic is detected.

If the error is transient, the router or input controller can be restarted after resetting all state to its initial condition. Packets en route through the faulty component are dropped — just as packets enroute through a faulty link are dropped. When a failed component is restarting, it must synchronize its state with that of adjacent modules. For example, when a router is restarting, it must send a control flit to each adjacent router to restart the channels between the two routers and initialize the credit count.

On a hard failure, the failed component cannot be restarted and must be replaced. Part of a reliable design is a provision to replace failed modules while the system is operating — often called *hot swapping*. Typically, all active components, such as routers, are replaceable in this manner, although the granularity of replacement (often called a field replaceable unit, or FRU) may be larger than a single router. For example, four routers may be packaged on a single printed-circuit card. The entire card must be removed, taking all four routers out of service, to replace one bad router.

An alternative to field replacement is to overprovision the system with sufficient extra units (channels and routers) so that it is highly probable that the system will complete its planned lifetime with an adequate level of performance. This alternative is attractive if the system is deployed in a location (e.g., orbit) where field servicing is difficult.

21.5 Network-Level Error Control

The network level is the next level of the hierarchy of error control that started at the link-level. At the network level, we model link and router failures as fail-stop links and routers must route packets around these failed components. This network-level error control is most easily realized using adaptive routing. The out-of-service links are simply made unavailable to the adaptive routing function and all packets are routed using one of the remaining available links.

Network level error control can also be realized with table-based oblivious routing. Immediately after failure, packets continue to be routed via the table and any packets selecting a path involving the failed link are either dropped or locally rerouted. A packet is locally rerouted by replacing a single hop over a failed link (east) with a series of hops that reaches the same intermediate point (north, east, south). Eventually, the routing tables are recomputed to avoid the failed links.

21.6 **End-to-end Error Control**

If a packet is dropped during the containment of a link or router failure, it may be recovered by using end-to-end packet retransmission. This operates in a manner similar to that of link-level flit retransmission, except that packets rather than flits are retransmitted and the retransmission is done over the entire route — from source to destination — not over a single link.

End-to-end recovery starts by retaining a copy of each packet sent on the sending node. This copy is held until an acknowledgment is received. If a timeout expires before an acknowledgment is received, or if a negative acknowledgment is received (as may occur for some types of errors), the packet is retransmitted. The process repeats as necessary until an acknowledgment is received, at which time the packet is discarded, freeing space in the transmit packet buffer.

It is possible with end-to-end retransmission for the destination of a packet to receive two copies of the packet. This may happen, for example, if retransmission occurred just before receiving a delayed acknowledgment. Duplicate packets are dropped by giving each packet a serial number and retaining at each node a list of the normal packets received and the retransmitted packets received in the last T flit intervals. As each packet arrives, if it is a retransmission (indicated by a bit in the header), its serial number is compared against the list of normal packets. If it is a normal packet, its serial number is compared against the list of retransmitted packets. In either case, the received packet is dropped if it is a duplicate. The check against normal packets is costly, but performed rarely. The more common check against retransmitted packets is inexpensive, as there are usually no retransmitted packets and at most one or two.[8]

With end-to-end packet retransmission layered on top of network-level, router-level, and link-level error control, an interconnection network can be made arbitrarily reliable. However, the clients of the interconnection network, such as processing nodes, network line cards, or I/O devices, are still subject to failure. Fortunately, the same principles of reliable design can be applied to these devices (often by using redundant clients attached to different network terminals), resulting in a complete system of high reliability.

21.7 **Bibliographic Notes**

Siewiorek [167] gives a good overview of reliable system design, including a discussion of failure mechanisms, fault models, and error control techniques. Error control codes are described in [21]. A more mathematically rigorous treatment is given in [19]. The MIT Reliable Router [52] incorporates many of the technologies described in this chapter, including link monitoring, link-level retry, link shutdown, and the network-level fault masking technique. Adaptive routing algorithms [36, 118]

---

8.   The Reliable Router used a *unique token protocol* to avoid the need for a receiver duplicate check.

are useful in routing around faulty links after they have been shut down. End-to-end error control is described in [157], where it is argued that it is both necessary and sufficient.

21.8 **Exercises**

21.1 *System MTBF.* Consider a 1,024-node, 3-D torus system with 2 logic chips and 8 memory chips per node. Each node is connected to its 6 neighbors via links with 16 Gbits/s of bandwidth each. Compute the MTBF for this system using the numbers in Table 21.1.

21.2 *MTBF with retry.* For the machine of Exercise 21.1, recompute the MTBF assuming that link-level retry masks all link errors, all routers are self-checked, and end-to-end packet retransmission is employed.

21.3 *Link-level retransmission.* Consider a link-level retransmission system as described in Section 21.3.2. Sketch a diagram showing what happens when a flit is received correctly but the acknowledgment in the reverse direction is received in error.

21.4 *Router restart.* Suppose a router X detects an internal fault, shuts down, and restarts. Describe how the input links and output links of the router must be sequenced to bring the router back on-line while causing a minimum of disturbance to an upstream router W and a downstream router Y. In particular, describe (a) what happends to a packet with tail and body flits in W and head flit in X, (b) what happens to a packet with body and tail flits in X and head flit in Y, and (c) what happens to a new packet arriving from W to X following the packet with head flits in X when X restarts.

21.5 *Relationship of bandwidth and error rate.* Suppose a router uses an I/O link technology that operates with an error rate of $\exp(-(t_{bit} - 200\,\text{ps}))$. The bit-time of the link is one over its frequency, $t_{bit} = \frac{1}{f}$. Further, suppose the router has a requirement to operate with a flit error rate of less than 10^{-15}. At what *effective* frequency (accounting for ECC overhead) can the link operate with (a) no retry, (b) retry using simple parity (that will detect any single-bit error), and (c) retry using an ECC that will detect any double-bit error. Use a flit payload of 64 bits, 1 bit of overhead for parity, and 6 bits of overhead for the double-error detecting code.

21.6 *Link-level vs end-to-end retransmission.* Saltzer et al. [157] argue that a system with end-to-end error control needs no further error control. In this question we will investigate this argument.

(a) Consider an interconnection network that sends 1,024-bit packets divided into 64-bit flits. Each packet travels an average of 10 hops, and each link has a BER of 10^{-5}. Link bandwidth is 1 Gbit/s and router latency is 20 ns per hop. The system checks each packet at the destination and requests retransmission if the packet is in error. To simplify analysis, assume that the error control code detects all possible errors and that delivery of the ack or nack message is error free. What is the probability of a packet being received in error? What is the resulting throughput and average latency for a packet?

(b) Now consider the same system, but with link-level retry in addition to end-to-end retry. Each link checks each flit of each packet as it is received and requests a retry if the flit is received in error. How does link-level retry affect the throughput and average latency?

CHAPTER 22

Buses

Buses are the simplest and most widely used of interconnection networks. A bus connects a number of modules with a single, shared channel that serves as a broadcast medium. A bus is usually implemented as a set of signal lines (or a single line) that is connected to all of the modules. One module transmits messages over the bus that are received by all of the other modules. In many situations, the message is addressed to one specific module and is ignored by the others. Often, the recipient of a message will respond by sending a reply message back to the sender of the original message. To read from memory, for example, a processor sends a message addressed to a memory module specifying the word to be read. The memory module responds by sending a message back to the processor with the requested data completing the transaction. A bus protocol determines which module has permission to transmit at any given time and defines the messages and transactions between modules.

Buses have two key properties that are often exploited by higher-level communication protocols implemented on top of the bus: *broadcast* and *serialization*. Sending a multicast or broadcast message over a bus is no more expensive than sending a point-to-point message because every message transmitted on a bus is physically broadcast to all of the modules. Thus, it is easy to distribute global information over buses. Because only one module can transmit a message over the bus at any given time, messages are serialized — they occur in a fixed, unambiguous order. Snooping cache-coherence protocols exploit both of these properties. The address of a cache line being written is broadcast to all modules so they can invalidate (or update) their local copies, and writes are serialized so that it is clear what the last value written to a particular address is. Such protocols become considerably more complex on a general interconnection network where broadcast is not free and where serialization requires explicit synchronization.

Because they are simple and inexpensive, buses have been widely used in app-lications including data transfer within datapaths, processor-memory interconnect, connecting the line cards of a router or switch, and connecting I/O devices to a processor. Buses, however have limited performance for two reasons. First, it is elec-trically difficult to make a multi-drop bus operate at high speeds [55]. Also, buses are inherently serial — only one message can be sent over the bus at a time. In applications that demand more performance than a bus can provide, point-to-point interconnection networks are used.

22.1 Bus Basics

Figure 22.1 shows the datapath of a typical bus that connects four modules A through D. Each module is connected to the bus through a bidirectional interface that enables it to drive a signal T onto the bus when the transmit enable ET is asserted and to sample a signal off the bus onto an internal signal R when a receive enable ER is asserted. For module A to send a message to module C, module A asserts its transmit enable signal ET_A to drive its transmit signal T_A onto the bus. During the same *cycle*, module C asserts its receive enable signal ER_C to sample the message off the bus onto its internal receive signal R_C.

Physically, the bus may be a single conductor, a serial bus, or a set of conductors that carries an entire message broadside (a parallel bus). At intermediate points between these two extremes, a message may be sequenced over a smaller set of parallel conductors, taking several cycles to transmit a message.

Electrically, buses are very difficult to operate at high speeds because of the stubs and impedance discontinuities caused by each connection to a module [55]. The electrical issues of this interface are beyond the scope of this book. Logically, the transmit interface must permit each module to drive a signal onto the bus when that module's transmit enable is asserted. The transmit interface may be a tri-state driver (Figure 22.2[a]), an open-drain driver (Figure 22.2[b]), or a dotted-emitter driver (Figure 22.2[c]). The latter two interfaces have the advantage that overlap in the

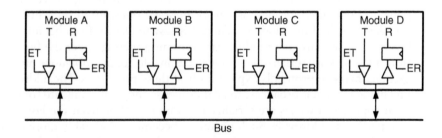

Figure 22.1   The datapath of a typical bus. Four modules A through D communicate over a shared bus. At a given time, any one module may drive its transmit signal T onto the bus when the transmit enable ET is asserted. The signal is broadcast on the bus and can be received by any or all of the modules by asserting their respective receive enable ER signals.

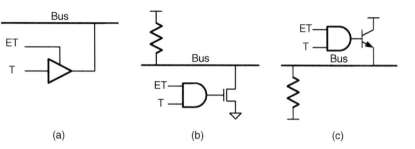

Figure 22.2 Typical bus transmitter interfaces: (a) a tri-state driver, (b) an open-drain driver, and (c) a *dotted emitter driver*.

transmit enable signals does not result in a power-to-ground short. The receive interface consists of a receiver appropriate for the signal levels on the bus and a register that captures the message on the bus when the receive enable is asserted. For a serial (or multicycle) bus, this register may assemble a message over several bus cycles.

Buses operate in units of *cycles, messages*, and *transactions*. As in any interconnection network, a *message* is a logical unit of information transferred from a transmitter to a set of receivers. For example, to read a memory location, a processor module sends a message containing an address and control information to one or more memory modules. In a serial bus (or a bus with fewer parallel lines than the message length), each message requires a number of *cycles* during which one phit of information is transferred across the bus line(s). Finally, a *transaction* consists of a sequence of messages that are causally related. All transactions are initiated by one message and consist of the chain or tree of messages generated in response to the initiating message. For example, a memory read transaction includes a request message containing an address from the processor to the memory modules, and a reply message from the selected memory module to the processor containing the requested data.

Buses may be *externally sequenced* or *internally sequenced*. In an externally sequenced bus, all transmit and receive enable signals are controlled by a central, external sequencer. In an internally sequenced bus, each module generates its own enable signals according to a bus protocol. For example, a microcoded processor often uses an externally sequenced bus: centralized control logic generates the enable signals to control data transfers between registers and function units. Most processor-memory buses, on the other hand, are internally sequenced. The processor generates its own transmit enable when it gains control of the bus, and the memory modules monitor the messages on the bus to decided when to receive request messages and transmit reply messages.

However they are sequenced, a bus cycle may be synchronous or asynchronous. (See Figure 22.3.) With a synchronous bus, a bus cycle is a cycle of the bus clock. The transmitter drives the bus starting at the beginning of the clock cycle and the receiver samples the data off the bus at the end of the clock cycle.[1]

---

1. To tolerate clock skew between modules, some buses have the receiver sample data off the bus before the end of the clock cycle. For example, the NuBus used on many Apple Macintosh computers sampled data off the bus $\frac{3}{4}$ of the way through the clock cycle.

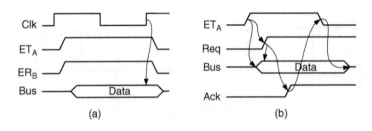

(a)   (b)

Figure 22.3   Bus synchronization: (a) In a synchronous bus, each cycle is synchronized with a clock signal. (b) In an asynchronous bus, each cycle is sequenced by the edges of request and acknowledge signals.

In an asynchronous bus, each bus cycle is sequenced by request and acknowledgment signals. To perform a bus cycle, the transmitter drives the data onto the bus and then asserts the request signal *Req*. Upon seeing the request signal, the receiver samples the data off the bus and asserts an acknowledge signal *Ack* to indicate that the data transfer is complete. Upon receiving the acknowledge signal, the transmitter turns its driver off, freeing the bus for the next transfer.

Sending a message on an internally sequenced bus involves the steps of arbitration, addressing, transfer, and acknowledgment. Arbitration determines which module gets to initiate a transaction. The winner of the arbitration is often called the *bus master*. In the simplest buses, one module (e.g., the processor) is always the bus master and no arbitration is needed. In more complex systems, a simple protocol is used to elect a master.

The addressing phase selects the module (or modules) to receive a message. Some buses perform addressing as a separate step, sending a broadcast address message before each directed message, while others send the address as a part of each message during the transfer. An example of such an internally addressed message is shown in Figure 22.4. In this case, all of the modules on the bus receive the message and then examine the control and address fields to determine if the message applies to them. In serial buses, control and addressing information is sent first so a module can examine these phits of the message and stop receiving as soon as it determines that it is not a target of the message. Note that the address field here is the address of a module on the bus and *not* an address at a higher level of protocol (such as a memory address). A memory address, for example in a memory read request message, is considered data by the bus protocol.

Control	Address	Data

Figure 22.4   A bus message with internal addressing. The message consists of a control field that specifies the type of message (such as memory read), a bus address, and data to be transferred (such as the memory address).

The transfer phase actually moves the data from the transmitting module to the receiving module. On simple buses, this is the only phase. Once the transfer is complete, an acknowledge phase may follow in which the receiving module acknowledges error-free receipt. If no acknowledge is received or if the receiver flags an error during transmission, the transmitter may attempt error recovery, for example by retrying the message.

Figure 22.5 shows a simple parallel bus that connects a single processor P to sixteen memory modules M_0 through M_{15}. The bus consists of parallel signal lines that carry 4 bits of control, 4 bits of module address, and 32 bits of data. The control field identifies the type of bus cycle. The bus is configured so an idle cycle is indicated if no module is driving the bus. The address field selects a memory module (one of 16) for a read or a write transaction. The data field carries the payload of all messages: memory addresses and data. The processor is the only module that initiates transactions on the bus, so no arbitration is required. Addressing is performed in parallel with transfer using the four address lines to select the memory module for each transaction.

The timing diagram in the figure illustrates how read and write transactions are performed on this bus. Both transactions involve two messages, one from the processor to a particular memory module, M_4 for the read transaction and M_3 for the write, and a second message from the selected memory module back to the

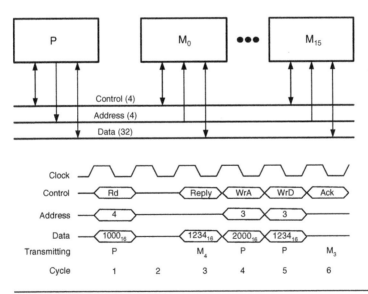

Figure 22.5   A simple bus connects a processor P to 16 memory modules $M_0, \ldots, M_{15}$ via parallel control, module address, and data signals. A memory read transaction consists of a single *Rd* cycle that sends an address message to a memory module and a single *Reply* cycle in which the memory module responds with the requested data. A memory write transaction takes two cycles *WrA* and *WrD* to send a write message to the memory module. The module responds with a single-cycle *Ack* message.

processor. For the read transaction, each of these messages takes a single cycle. For the write transaction, the write request requires two cycles to transport both the address and data over the data lines.

The read transaction takes place in cycles 1 through 3. During cycle 1, the processor initiates the transaction by driving the control lines with *Rd* to send a read request message to a memory module. The address lines select the target memory module, in this case, M_4, and the data lines carry the memory address within M_4 to be read (in this case, 1000_{16}). The bus is idle during cycle 2 as the memory module accesses the requested data. The memory module drives the bus with a reply message, control lines contain *Reply*, and the data lines contain the data in cycle 3. Even though module M_4 drives the bus in cycle 3, there is no need for arbitration because it is simply responding to the processor's request and not initiating a transaction. No other module can respond at this time, so no arbitration is needed.

The write transaction follows the read in cycles 4 through 6. The processor initiates the transaction by sending a two-cycle write request message to memory module M_3. Cycle 4 is a write address cycle in which the control carries *WrA* and the data carries the address to be written, 2000_{16}. A write data cycle follows in cycle 5 with the control indicating the cycle type with *WrD* and the data lines carrying the data to be written, 1234_{16}.

In either the read or the write transaction, if the addressed memory module did not reply within some time limit, the transaction would *timeout*. After the timeout, control of the bus returns to the processor module, which may then initiate a new transaction. A bus timeout prevents a transaction to an unimplemented address or an unresponsive addressed module from hanging up the bus indefinitely.

22.2 Bus Arbitration

In buses where more than one module may initiate transactions, modules must arbitrate for access to the bus before initiating a transaction. Any of the arbiters described in Chapter 18 can be used for this purpose.

Radial arbitration, so named because the request and acknowledge lines run *radially* like the spokes on a wheel to and from a central arbiter, is illustrated in Figure 22.6. The request and grant signals between the modules and the arbiter typically share a connector with the bus lines and are bused together, but rather are routed individually to the arbiter, as shown. When a module wishes to initiate a transaction, it asserts its request line. When it receives a grant it becomes bus master and may initiate a transaction. Fairness rules may require the module to drop its request during the last cycle of its transaction to allow another module access to the bus. Alternatively, the arbiter may monitor the bus transaction, deassert the grant signal during the last cycle of the transaction, and grant the bus to another module during the same cycle.

Figure 22.7 illustrates the timing of bus arbitration for a synchronous bus. Module 1 requests the bus during cycle 1, receives a grant during cycle 2, and starts its transaction during cycle 3. Compared to a case in which module 1 is the only bus master, or is already bus master, arbitration adds two cycles to the latency of

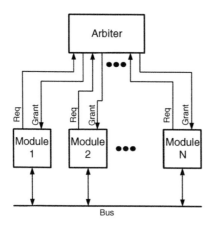

Figure 22.6 A bus with *radial* arbitration. Each module that wishes to initiate a transaction asserts its request signal to a central arbiter. When the arbiter responds with a grant, the module becomes bus master and may initiate a transaction.

a transaction even when uncontested. Module 2 requests the bus after it has been granted to module 1 and must wait until after module 1 releases the bus before it can initiate its transaction. There is a two-cycle idle period between the end of the first transaction and the start of the second due to arbitration. This idle period could be eliminated by pipelining the arbitration, overlapping arbitration for the next transaction with execution of the current transaction.

To eliminate the need for a central arbiter, some older buses used *daisy-chain* arbitration, as shown in Figure 22.8. A fixed-priority arbiter (Section 18.3) is distributed across the modules. Module 1 has its carry in, carry0, tied high and, hence, will

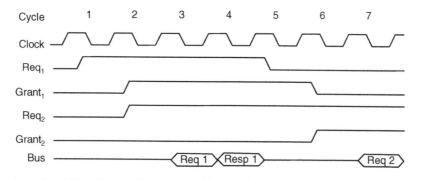

Figure 22.7 Timing of bus arbitration for a synchronous bus. Module 1 requests the bus in cycle 1, is granted the bus in cycle 2, performs a transaction in cycles 3 and 4, and relinquishes the bus in cycle 5 by deasserting its request. Module 2 requests the bus in cycle 2 but must wait until cycle 6 to receive its grant.

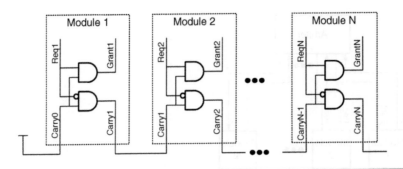

Figure 22.8   *Daisy-chain* arbitration. A fixed-priority arbiter is distributed across the modules, with each module passing a grant signal to the next module unless it requires the bus.

always receive a grant when it makes a request. If it does not make a request, the carry is passed down the chain to module 2, and so on. Daisy-chain arbiters are of mostly historical interest. Because of their fixed priority and the high delay of running the carry chain across multiple modules, they are rarely used in modern systems.

A sequential distributed arbiter uses the bus lines of a *wire-or* bus to perform the actual arbitration. If a bus is implemented with open drain or emitter-follower transmitters (Figure 22.2[b] or [c]) the signal on each bus line is the logical OR (possibly with negative logic) of the signals driven by all enabled transmitters. The logic performed by this ORing of signals can be exploited to perform arbitration one bit of module address at a time by having each module perform the algorithm of Figure 22.9. The algorithm starts with the most-significant bit of the module priority. During each cycle of the algorithm, all competing modules drive their priority onto the bus. The bus ORs these priorities together. The modules then check to see if the current bit of the bus signal, the OR of all priorities, is higher than the current bit of their priority. If it is, they drop out of competition. After scanning over all of the bits, only the highest priority competing module is left participating and becomes bus master.

```
For module i
    bit = n-1 ;                          bit of priority currently being tested
    participating = 1                    this module still a contender
    priority = pri(i)                    priority of module i
    while (bit>=0 && participating)      for each bit
        drive priority onto bus          one bus cycle here
        if (bus[bit] > priority[bit])    someone has higher priority
            participating = 0            drop out of competition
        bit = bit-1                      next bit
    master = participating               only winning module still participating
```

Figure 22.9   Algorithm for distributed arbitration using a wire-or bus.

An example of distributed arbitration is shown in Table 22.1. Three modules with priorities 9, 10, and 7 (1001, 1010, and 0111) compete to become bus master. In the first cycle, all three priorities are ORed together driving the bus to a value of 15 (1111). Comparing to the most significant bit, the module with priority 7 drops out of the competition. The other two modules then OR their priorities on the bus, giving a result of 1011. Neither module drops out when they are comparing against bit 2, since the value on the bus is zero. The module with priority 9 drops out when modules are comparing against bit 1, leaving the module with priority 10 to become bus master.

The priorities used in the distributed arbitration scheme can be allocated in any manner as long as they are unique. A fixed-priority arbitration scheme can be realized by having each module use its address as its priority. A least recently served scheme can be implemented by having the bus master take the lowest priority, zero, and incrementing the priority of all other modules before each arbitration cycle.

Distributed arbitration has the advantage that it can be implemented using the regular bus lines and without central arbitration logic. However, it has the disadvantage of being slow and of tying up the bus lines during arbitration, which prevents arbitration from being overlapped with bus transactions. Distributed arbitration is slow because each iteration of the algorithm requires a complete bus traversal to allow the wire-or signals to settle. In contrast, with a fast radial arbitration scheme, once the request signals are gathered at the central arbiter, each iteration requires only a few gate delays to complete — hundreds of picoseconds versus many nanoseconds.

Another alternative, which has been implemented on a number of shared-memory multiprocessors, is to use a replicated arbiter. A copy of the arbiter is placed on each bus module and all requests are distributed to all copies of the arbiter. At reset time, all copies of the arbiter are initialized to the same state. During each arbitration cycle, all of the arbiter copies start in the same state, receive the same set of requests, and generate the same grant. The module that receives the grant uses this information locally. All other modules ignore the grant.

This replicated-arbiter approach eliminates the need for a central arbiter, but has a number of significant disadvantages. First, the number of bus lines is increased, and the loading of request lines is increased by the need to distribute all requests to

Table 22.1 Example of distributed arbitration. Three modules with priorities 1001, 1010, and 0111 compete to become bus master.

		Participating		
Bit	Bus	1001	1010	0111
3	1111	1	1	0
2	1011	1	1	0
1	1011	0	1	0
0	1010	0	1	0

all modules. More importantly, this approach commits the cardinal sin of replicating state. The arbiter state is replicated in N copies of the arbiter for N modules. If the state of any of these arbiter copies diverges, for example, due to a soft error or synchronization error, two modules may be given a grant in the same cycle. To prevent this from happening, additional complexity must be introduced to constantly compare the state of the replicated arbiters and keep them synchronized.

22.3 High Performance Bus Protocol

22.3.1 Bus Pipelining

In Figure 22.7 the bus is idle for much of the time, waiting on arbitration. This results in a significant loss of performance in buses used in applications — such as shared memory multiprocessors, in which arbitration is performed for almost every bus transaction. The duty factor of a bus can be improved in these cases by pipelining the phases of a bus transaction.

Figure 22.10 shows the pipeline diagram and reservation table for a memory write transaction (a) and for a memory read transaction with a single-cycle memory latency (b). Each diagram shows the bus cycles needed to complete the transaction along the top and the resources involved down the left side. The read transaction starts with a three-cycle arbitration sequence: an *AR* cycle that asserts an arbiter request, followed by an *ARB* cycle during which the arbiter makes a decision, and an

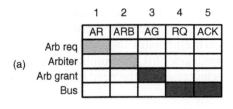

Figure 22.10 Pipeline sequences and reservation tables for (a) a memory write transaction and (b) a memory read transaction. The figure shows the sequence of bus cycles along the top and a list of resources down the right side. A darkly shaded box indicates that a resource is exclusively used by a transaction during a particular cycle. A lightly shaded box indicates that a resource may be shared.

AG cycle in which the grant from the arbiter is signaled back to the requester. After the arbitration, the read transaction consists of a request *RQ* in which the address is sent to the memory module, a processing cycle *P* in which the memory access is performed, and a reply cycle *RPLY* when the data is returned to the requester.

The reservation tables show which resources are busy during each cycle. The darkly shaded boxes show the resources that are used exclusively by the transaction during a given cycle. Only one transaction can receive a grant during a given cycle and only one transaction can use the bus during a given cycle. The lightly shaded boxes show resources that may be shared during a given cycle. Any number of transactions may make requests and arbitrate during each cycle.

Reservation tables give us a simple rule for initiating transactions. A transaction with fixed delays can be initiated in any cycle when its reservation table does not request any exclusive resources that have already been reserved by earlier transactions. Using this rule, we see that a read transaction can be initiated one cycle after another read, but a third read would need to wait two cycles after the first two. We also see that a write transaction must wait three cycles after a read to issue.

Figure 22.11 shows the timing of a sequence of six transactions on a pipelined bus. On an unpipelined bus, in which arbitration does not start until the previous transaction is complete, this sequence would take 34 cycles to complete two 5-cycle writes and four 6-cycle reads with no overlap. The figure shows that the pipelined bus completely hides the arbitration latency through overlap. Also, in some cases, the bus is able to overlap transactions with non-conflicting reservations. For example, in cycle 12 transaction *Read 5* issues its request during the *P* cycle of transaction *Read 4*.

However, the bus still idles during the *P* cycle of a read transaction followed by a write transaction because of mismatch between the read and write pipelines. For example, transaction *Write 2* cannot issue its request until cycle 7 to avoid a collision between its *ACK* and the *RPLY* from transaction *Read 1*. This situation is worse when dealing with transactions that have variable delay. For example, if a read transaction

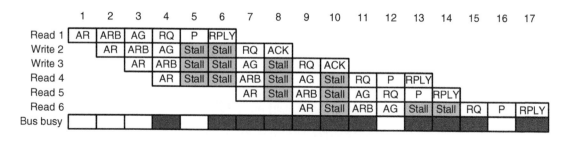

Figure 22.11 Execution of four reads and two writes on a pipelined bus. The reads and writes have timing, as shown in Figure 22.10. The sequence completes in 17 cycles as compared to 34 cycles for an unpipelined bus.

took between zero and twenty *P* cycles, no other transaction could initiate until each read was complete, idling the bus during all of the *P* cycles.

22.3.2 Split-Transaction Buses

Bus idling between the messages of a transaction can be eliminated by splitting the transaction into two transactions and releasing the bus for arbitration between the two messages. Such *split-transaction* buses are typically employed to free the bus during long, variable-length waiting periods between request and reply messages. By treating the reply messages as separate transactions that must arbitrate to become bus master before sending its message, other transactions can make use of the bus during the waiting period.

Figure 22.12 shows the sequence of Figure 22.11 executed on a split transaction bus. Each transaction in Figure 22.11 is expanded into two transactions: one that arbitrates for the bus and sends the request *RQ* message, and one that arbitrates for the bus and sends the reply *RPLY* or acknowledge *ACK* message. For example, the *Read 1* transaction ends after sending its *RQ* message in cycle 4. The reply is sent by the *Rply 1* transaction in cycle 8.

Because arbitration is assumed to take three cycles on this bus, the minimum latency between a request and a reply or acknowledge is four cycles, substantially increasing the latency of individual operations. In exchange for this increase in latency, bus utilization is maximized. As long as there are transactions waiting to initiate, the bus is used every cycle to send an *RQ*, *RPLY*, or *ACK*.

Figure 22.13 shows that the throughput gains of split transactions can be considerable when transactions have long and variable delays between request and response. The figure shows a sequence of three transactions. Each transaction may take anywhere from one cycle (no wait cycles) to six cycles to complete. A pipelined bus without split transactions must serialize these transactions as shown in Figure 22.13(a)

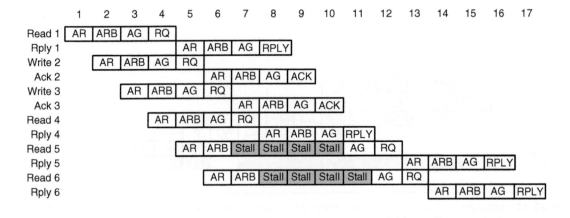

Figure 22.12 Execution of the sequence of Figure 22.11 on a split-transaction bus. Latency for individual operations is increased, but bus utilization is also increased by eliminating idle *P* cycles.

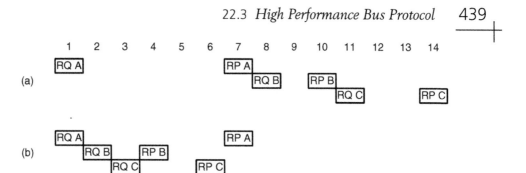

Figure 22.13 A sequence of variable-delay transactions: (a) on a pipelined bus, (b) on a split-transaction bus. For clarity, arbitration, which takes one cycle, is not shown. The split transaction bus allows the later transactions to run during the waiting cycles of the first transaction.

because each transaction must reserve the bus for its reply starting with the cycle after the request. The pipelining hides the arbitration (not shown), but does not allow overlap of transactions.

With a split-transaction bus, the three requests can be overlapped completely, as shown in Figure 22.13(b). Because the bus need not be reserved for the reply from *RQ A*, request messages *RQ B* and *RQ C* and their respective replies can all be sent during the waiting time of the first request. If reply *RP C* were delayed one cycle, it would be ready to use the bus during the same cycle as *RP A*. The arbitration would grant the bus to one of these two replies and the other would be delayed a cycle.

A split-transaction bus, like a general interconnection network, requires a means to identify the reply associated with each request. In a bus without split transactions, the reply is identified by the time it appears on the bus. With a split-transaction bus, depending on arbitration, replies may appear in an arbitrary order. For example, replies *RP C* and *RP A* in Figure 22.13 could easily appear in the opposite order.

To identify replies, each request includes a *tag* that uniquely identifies this request from all other pending requests — those requests that have not yet received a reply. The requester transmits the tag in the request message. The responder remembers the tag and includes it in the reply message. By matching tags, the requester is able to determine the reply that corresponds to a particular request.

In some buses, such as the bus of the original Sequent Symmetry multiprocessor, the tag is explicitly sent as a separate field of the request message. In the case of the Symmetry, tags are assigned to the requester from a central pool when the requester wins bus arbitration. In other systems, such as the SCSI peripheral bus, the tag is formed implicitly from the source address, destination address, and sequence number fields of the request message.

22.3.3 **Burst Messages**

The overhead of a bus transaction is considerable. Arbitration, addressing, and some form of acknowledgment must be performed for each transaction. If transactions transfer only a single word, the overhead can easily be larger than the payload — an overhead greater than 100%. The overhead of a transaction, expressed as a percentage, can be reduced by increasing the amount of data transferred in each message. When multiple words are being transferred, as, for example, when loading or storing a cache line or transferring a block of data to a peripheral, sending a block of many words in each message is much more efficient than sending a single word. Choosing the amount of data transferred during each message is equivalent to choosing the flit size for the bus.

Figure 22.14 illustrates the reduced overhead of burst mode messages. Figure 22.14(a) shows a bus with two cycles of overhead per message that transfers just a single word during a three-cycle message for an efficiency of 1/3. In Figure 22.14(b) four words are sent in each message, increasing the efficiency to 2/3 or four words transferred in six cycles. With an eight-word message efficiency would be further increased to 4/5. In general, with x cycles of overhead and n words of data transferred, the efficiency will be $n/(n + x)$.

One might think that when transferring a large block of data the largest possible burst size should be employed to bring the efficiency as close as possible to one. Increasing burst size, however, increases the maximum amount of time that a high-priority requester may need to wait to gain access to the bus. If a request is made during the first cycle of the message of Figure 22.14(b), the requester will need to wait six cycles to receive a grant. If the burst size were 256 words, the requester would need to wait 258 cycles. Such long arbitration latencies are unacceptable in many applications.

To allow the efficiency of long burst messages and at the same time provide low latency to higher priority requesters, some buses allow burst messages to be interrupted and either resumed or restarted. Figure 22.15 shows an eight-data-word message A being interrupted by a single data word message B and then resumed. The first busy cycle of the resumed message is used to transmit a tag to identify the message being resumed. This is needed in systems where the interrupting message may itself be interrupted, resulting in multiple messages waiting to be resumed. Alternatively, an interrupted message can be aborted and restarted from scratch.

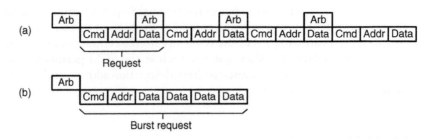

Figure 22.14 Burst message: (a) Each bus message transfers a single word of data for an efficiency of 1/3. (b) A four-word burst message amortizes the two cycles of overhead over four words to bring the efficiency up to 2/3.

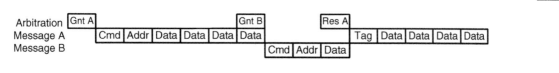

Figure 22.15 Burst interrupt and resume. Message A is interrupted by message B and then resumed.

However, abort and restart schemes can adversely affect performance because of the redundant data transfers required.

Interrupting long messages is analogous to dividing long packets into multiple flits, so higher-priority packets need not wait for the whole packet to be transmitted before being allocated channel bandwidth. One key difference is that each flit of a packet must carry per-flit overhead, whereas the overhead of resuming a bus message is only incurred if the message is actually interrupted.

22.4 **From Buses to Networks**

Many communication tasks that were formally performed using buses are now performed with networks. As these tasks are migrated, many designers attempt to replicate the characteristics of a bus on a point-to-point network. Some characteristics, like having a common interface for all modules, are easily replicated on a network. Others, such as serialization of all transactions and broadcast to all modules, are more difficult to emulate.

One of the major advantages of a bus is that it provides a common interface for all communicating modules. This facilitates modularity and interoperability between modules. For example, most small computers today use a Peripheral Component Interconnect (PCI) bus to connect to peripherals. Many computers and many peripherals are designed to interface to this common bus and any PCI peripheral can be used with any PCI computer.

The common interface of the bus is shared by most networks. The network provides an interface (see Chapter 20) that is common to all modules connected to the network. To date, such interfaces have not been standardized. However, this situation is changing with the development of network interface standards such as PCI-Express, Infiniband, and Fibre Channel Switched Fabric. Already, the modularity facilitated by standard bus interfaces is being realized by networks.

Many systems would like to maintain the serialization and broadcast properties of the bus, but cannot use a shared medium because of electrical constraints. For example, the parasitic resistance and capacitance of on-chip interconnects requires repeaters every few millimeters to maintain signaling speed. Thus, a large on-chip bus cannot be realized as a single electrical node driven and sensed by many modules.

A communication system that is logically equivalent of a bus can be realized entirely with short point-to-point links, as shown in Figure 22.16. Each module drives its signal into an OR-network when it is enabled and drives zeros into the OR-network otherwise. Thus, the output of the OR-network always reflects the

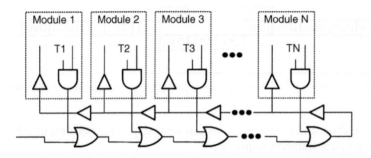

Figure 22.16 A bus can be implemented logically with no shared media by ORing the transmit outputs of all of the modules together. Only the currently transmitting module drives non-zero data into the OR network. A repeater chain distributes the result of the OR network to the receive port of each module.

value being driven by the currently enabled module. This value is then distributed to all of the modules by a repeater chain. The high delay of the linear OR chain and linear repeater chain can be somewhat reduced by building tree networks in place of both chains. (See Exercise 22.5.)

Some systems would like to maintain the semantics of a serial broadcast bus, for example, to support a snooping cache-coherence protocol, but cannot accept the performance limitations of sending only one message at a time. Such systems usually start out by making the bus very wide — for example, 256 bits wide to transfer a 32-byte cache line in parallel. However, while this can speed large data transfers, it does not address the problem of needing to run multiple address cycles at once.

The next step in the search for performance is to run multiple parallel buses, or at least multiple parallel address buses. For example, with four parallel address buses, four address cycles can be run simultaneously. Serialization is still maintained by assigning an ordering to the multiple buses. An address on bus i arriving during cycle t is considered to have arrived before an address on bus $j > i$ during the same cycle. The expense of the bus interfaces increases more than linearly with the number of buses since logic and connections are required to check each address against the addresses on all *earlier* buses for collisions. While this is an expensive approach to scaling bandwidth, it has been used successfully. The Sun Ultra Enterprise 10,000, which was the leading SMP server for many years, employed a four-way duplicated bus system [35].

The ultimate step in the search for performance is to give up on the semantics of the bus and move to a network. This has advantages of performance, cost, and scalability. Moreover, the cost of a full broadcast is avoided for the majority of messages that do not require broadcast. When broadcast is required, it can be provided either via a hardware multicast facility or by sending multiple messages, possibly via a distribution tree. Serialization can also be provided when required by sending all messages that must be serialized to a common node. The order of receipt serializes the messages. This approach serializes only related messages, whereas a bus serializes all messages, needlessly slowing communication.

Ordering is also required by some protocols that have traditionally been implemented on the bus. In a shared memory machine, for example, two write messages A and B from a processor P to a memory M need to be delivered in the same order they are sent or the memory may be left in an incorrect state. Buses, since they carry out only one communication at a time, inherently keep messages ordered. In a network, however, it is possible for message B to pass message A somewhere in the network. Ordering can be enforced in a network by requiring ordered messages to always follow the same path (and use the same virtual channels), or by tagging them with sequence numbers and reordering them at the destination.

Of course, running bus-based protocols on a network is a bit like trying to attach a propeller to a jet engine. If one is building a machine around an interconnection network, it is better to design protocols that are optimized for the characteristics of the network, rather than to force-fit protocols intended for a serial, broadcast medium. Today, for example, the largest shared memory machines use general interconnection networks running directory-based protocols that do not require broadcast.

22.5 **Case Study: The PCI Bus**

Perhaps the most widely used bus today is the PCI bus [143], which is used to connect peripherals to many different types of computer systems, ranging from embedded systems to large servers. The PCI bus also brings together many of the properties of buses described in this chapter.

The PCI bus is a synchronous bus with multiplexed address/data lines and pipelined arbitration. It is available in both 32-bit and 64-bit versions and operates at speeds of 33 MHz, 66 MHz, and 133 MHz. In the widest and highest speed configuration, 64-bit datapath at 133 MHz, the bus is able to sustain burst rates of 1 Gbyte/s.

Table 22.2 shows the major signals of the PCI bus. These 42 signal lines are the core of a PCI bus. Other signals are included in the full bus specification for configuration, error reporting, and other functions. These will not be discussed here.

A PCI-bus transaction consists of an address cycle followed by one or more data cycles. Wait cycles may be interspersed between these address and data cycles as needed to synchronize the initiator and the target. The PCI-bus supports both single-word and burst transactions. For example, Figure 22.17 shows a two-word PCI read transaction. The transaction is initiated by the master (initiator) asserting FRAME#[2] in cycle 1. In this cycle, the master also drives the address on the AD lines and the bus command (memory read) on the C/BE lines. The master then asserts IRDY# in cycle 2 to indicate that it is ready to receive the first word of data. However, since TRDY# is not asserted, no data transfer takes place in this cycle. This idle cycle is required in all read transactions to *turn the AD bus around* — from the master

---

2. The pound sign "#" is used in the PCI bus specification (and elsewhere) to denote a low-true signal — that is, a signal that is true (asserted) when it is in the logic 0 state.

Table 22.2 Major PCI bus signals.

Name	Description
CLK	Bus clock: All bus signals are sampled on the rising edge of CLK.
AD[31:0]	32-bit address/data bus: During an address cycle, these lines are driven by the bus master (initiator) with the address of the target device. During data cycles, these lines are driven with data by the data provider, the initiator for writes, and the target for reads.
C/BE[3:0]	Command/byte-enable: During an address cycle, these lines encode the bus *command* (such as memory read). During data cycles, these lines are used as byte enables, specifying which bytes of a word are to be written.
FRAME#	Transaction frame: The initiator starts a bus transaction by asserting FRAME#. The first clock cycle in which FRAME# is asserted is the address cycle. The last data cycle is started when FRAME# is deasserted.
IRDY#	Initiator ready: Indicates when the initiator is ready to transfer data. Data is transferred when both the initiator and target are ready (that is, when IRDY# and TRDY# are both asserted).
TRDY#	Target ready: Indicates when the target is ready to transfer data.
REQ#	Bus request: A module asserts REQ# to request access to the bus as an initiator or bus master.
GNT#	Bus grant: A central (radial) arbiter examines the REQ# signals from each module and asserts at most one GNT# signal. The GNT# signal indicates which module will become the initiator when the current transaction completes.

driving to the target driving without overlap of drive current. The target provides the data in cycle 3 and asserts TRDY#. Since IRDY# and TRDY# are both asserted, the first word of data is transferred in this cycle. In cycle 4, the target provides the second word of data and TRDY# remains asserted. However, this time IRDY# is not asserted, so the second data transfer must wait on the master until cycle 5. During cycle 5, the master asserts IRDY#, allowing the second data transfer to complete. Also during this cycle, the master deasserts FRAME#, indicating that this will be the last data transfer of this transaction. In cycle 6, both FRAME# and IRDY# are deasserted, indicating the end of the transaction. The next bus master detects this state and is free to initiate the next transaction in cycle 7.

A write transaction is nearly identical to the read transaction shown in Figure 22.17, except that the initiator rather than the target drives the AD lines during the data cycles. During a write, no idle cycle is required for bus turnaround after the address cycle. This is because the same module, the initiator, is driving both the address and the data.

Figure 22.18 shows how PCI bus arbitration is pipelined. During cycle 1 modules 1 and 2, both request to become bus masters by asserting their respective REQ# lines. The global arbiter decides to grant bus mastership to module 1 and asserts GNT1# in cycle 2. Module 1 receives this grant and starts a transaction by asserting FRAME# in cycle 3. As soon as module 1 starts its transaction, the arbiter negates GNT1# and asserts GNT2# in cycle 4. This signals to module 2 that it will become bus

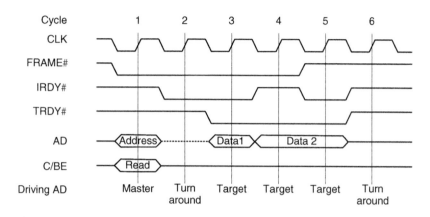

Figure 22.17 A PCI bus read transaction.

master as soon as the current transaction completes. Module 1 continues to be bus master, completing its transaction. In cycle 5, completion is signaled by FRAME# and IRDY# both being deasserted. Module 2 becomes bus master when it recognizes this completion and starts a transaction in cycle 6 by asserting FRAME#.

PCI transactions are not split. A read transaction includes both the request message from the initiator to the target and the reply message from the target back to the initiator. PCI, however, includes a mechanism for aborting and retrying a transaction to avoid tying up the bus during a long latency operation. A target device can latch the request information and then abort a transaction by asserting STOP# rather than TRDY#. The master then retries the transaction at a later time, hopefully after the target has completed the long latency operation and can

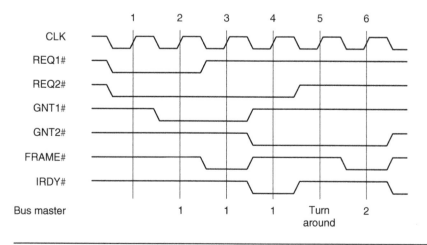

Figure 22.18 PCI bus arbitration.

provide the requested data. The PCI specification calls such an exchange a *delayed transaction*. Delayed transactions share with split transactions the ability to allow other bus transactions to proceed during long latency operations. The delayed transaction mechanism, however, is considerably less efficient than a split transaction mechanism and is only intended to handle exceptional cases. A split transaction mechanism is intended to handle most or all bus requests, and hence must be more efficient.

The PCI specification includes a set of protocols for module initialization. In addition to addressing modules using the AD bus during an address cycle, an alternate addressing mechanism using a radial select signal, IDSEL, is used to select a module by its slot number. Configuration read and write cycles can be run using IDSEL addressing to read and write the configuration registers of a module. In a typical initialization process, some configuration registers are read to identify the module type. The controller then writes configuration registers to set the address of the module, and to configure module options.

22.6   Bibliographic Notes

Buses have been around long enough that it is hard to identify when they first appeared. They certainly existed in the early computers of the 1940s and 1950s. Starting with Digital's Unibus, buses became a standard peripheral interface. Both the PCI bus [143] and the Small Computer System Interconnect (SCSI, pronounced *scuzzy*) bus [173] are modern examples of peripheral interface buses. For a long period of time (1960s to 1980s), a typical minicomputer or microcomputer used a bus to connect memory modules to the CPU. In modern PCs this has been replaced by point-to-point connection through a north-bridge chip. Perhaps one of the most interesting and high-performance buses today is Rambus's DRDRAM bus used to connect to high-bandwidth DRAM chips [41]. The on-chip network offered by Sonics uses the OR-tree approach to realizing a bus with point-to-point interconnects [193]. The Sun Ultra-Enterprise 10,000 [35] also realizes a bus with point-to-point interconnects. The UE 10,000 also provides multiple address buses to increase throughput. It is perhaps a good advertisement for why one should use a network rather than a bus. *The Digital Bus Handbook* [72] gives a more complete overview of bus technology than there is room for in this chapter.

22.7   Exercises

22.1   *Multiplexed versus non-multiplexed bus.* Consider a bus on which all transactions require sending an address message with a 32-bit address field and a 4-bit control field from the bus master to a bus slave. Read transactions, that comprise 70% of all transactions, are then completed by having a 32-bit data message sent from the slave to the master. Write transactions, that comprise the remaining 30% of all transactions, are completed by having a 32-bit data message sent from the master to the slave. This write data message can be sent at the same time as the address message, if resources permit. You have a total of 40 bus lines to use for the design of your bus. Suggest a bus design that maximizes bus throughput. Consider both multiplexed (shared data and address lines) and non-multiplexed buses and assume that buses are maximally pipelined.

22.2 *Early win distributed arbitration.* In the distributed arbitration method shown in Figure 22.9, a full arbitration takes b cycles for b module address bits. Explain how this arbitration could be optimized to be faster in some cases. Simulate the arbitration for i modules randomly requesting the bus for $0 < i \leq b$ and $b = 8$. Plot the average time required for arbitration as a function of i.

22.3 *Pipelined bus with separate reply.* Draw a timing diagram for the transaction sequence of Figure 22.11 on a system with a separate reply bus, assuming that the reply bus is separately arbitrated. How much faster is this sequence of transactions completed with the separate reply bus?

22.4 *Virtual channels on buses.* Suppose a bus is built employing messages that are subdivided into flits, each of which is tagged with a virtual-channel number to enable large messages to be interrupted by short, high-priority messages. Plot the amount of overhead vs. maximum wait time as a function of flit size. Assume that the maximum message size is 64 Kbytes and 8 virtual channels are supported.

22.5 *Tree-based buses.* Show how the logical bus using point-to-point links of Figure 22.16 can be realized with lower delay by arranging both the OR network and the repeater chain into trees. Assume that this bus is being placed on a 12mm square chip, each module is 2mm square, and a repeater or logic gate must be placed each 2mm along a line to maintain signaling speed. Further assume that a signal takes one unit of time to traverse 2mm of wire and one gate. Compare the maximum latency of this 36-module system with linear OR and repeater networks and tree-structured OR and repeater networks.

22.6 **Simulation.** As an alternative to tagging requests and replies on a split-transaction bus, a bus can be designed so that the replies are forced to occur in the same order as the requests. Thus, each reply can be identified by its position in the stream of replies. Suppose the delay of a transaction is uniformly distributed between 1 and T_{max} cycles. Use simulation to compare the average latency of a tagged bus and an ordered-reply bus.

CHAPTER 23

Performance Analysis

As we have seen throughout the book, the design of an interconnection network begins with a specification of performance requirements combined with some packaging and cost constraints. These criteria then drive the choice of topology, routing, and flow-control for the particular network. To guide our initial decisions for each of these aspects of network design, we also introduced some simple metrics to estimate performance, such a zero-load latency and throughput. Of course, these metrics involved some assumptions about the network. For example, in most of our analysis of routing algorithms and topology, we assumed ideal flow control. Although it is reasonable to work with simple models in the early stages of network design, more detailed models become necessary to accurately characterize the exact performance of the network.

In this chapter, we first define the basic performance measures of any network and discuss how to estimate and interpret these measures. We then introduce a basic set of analytical tools that are useful for modeling the behavior of networks at a high level, including both queuing models and probabilistic methods. In Chapter 24, we discuss the use of detailed network simulators. Finally, simulation results from several network configurations are presented in Chapter 25.

23.1 Measures of Interconnection Network Performance

While there are many different ways to measure and present the performance of a particular network, we will focus on three basic quantities: throughput, latency, and fault tolerance. While these names sound generic, their exact definition depends strongly on how we measure them or, more specifically, the measurement setup.

449

The standard setup for interconnection networks is shown in Figure 23.1. To measure the performance of an interconnection network, we attach *terminal instrumentation* to each terminal or port of the network. The instrumentation includes a packet source that generates packets according to a specified traffic pattern, packet length distribution, and interarrival time distribution. Separating the packet source from the network at each terminal is an infinite depth *source queue*. The source queues are not part of the network being simulated, but serve to isolate the traffic processes from the network itself. Between the packet source and the source queue, an input packet measurement process counts injected packets and measures the start time of each packet injected. It is important that this measurement process be placed before the source queue rather than after the queue so that packets that have been generated by the source but not yet injected into the network are considered and so that packet latency includes time spent in the source queue. A complementary measurement process at each output terminal counts packets and records each packet's finish time. Throughput is measured by counting the packets arriving at each output and latency is measured by subtracting the start time from the finish time for each packet.

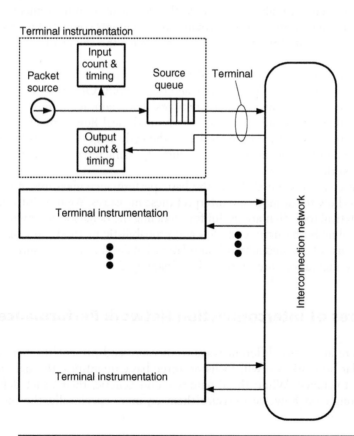

Figure 23.1   Standard interconnection network measurement setup.

This *open-loop* measurement configuration enables the traffic parameters to be controlled independently of the network itself. Without the source queues, a packet source may attempt to inject a packet at a time when the network terminal is unable to accept traffic — for example, when the network's input buffers are full. In such a case, the traffic produced by the source is influenced by the network and is not the traffic pattern originally specified. Because our goal is generally to evelute the network on a *specific* traffic pattern, we will used open-loop measurements for throughput, latency, and fault tolerance in the following sections.

Closed-loop measurement systems, in which the network influences the traffic, are useful for measuring overall system performance. For example, the performance of a multicomputer may be estimated by running a simulation in which the terminal instrumentation is replaced by simulations of the multicomputer's processors. The traffic generated consists of the inter-processor or processor-memory messages generated by running an application. Since the processors interact with the network, their injection rate is not readily controlled. For example, processors waiting for memory responses will make fewer requests due to limits on the number of outstanding requests. Rather, a more typical application of this simulation setup would be to test the sensitivity of the application run time to network parameters such as bandwidths, routing algorithms, and flow control.

Most often, we are interested in measuring *steady-state* performance of a network: the performance of a network with a stationary traffic source after it has reached equilibrium. A network has reached equilibrium when its average queue lengths have reached their steady-state values. Steady-state measurements are meaningful only when the underlying process being measured is *stationary* — that is, the stastics of the process do not change over time. Transient performance measures — the response of a network to a change in traffic or configuration — are sometimes of interest as well, but the bulk of our measurements will be steady state.

To measure steady-state performance, we run an experiment (or simulation) in three phases: warm-up, measurement, and drain. First, we run N_1 *warm-up* cycles to bring the network to equilibrium. During the warm-up phase, packets are not timed or counted. Once warm-up is complete, we run N_2 *measurement* cycles. During this phase, every packet entering the source queue is tagged with its start time. We refer to these tagged packets as *measurement* packets. Finally, during the *drain* phase, we run the network long enough for *all* of the measurement packets to reach their destination. As these packets arrive at the destination, their finish time is measured and they are logged. In addition, all packets arriving at the destination during the measurement interval are counted. Throughput measures are computed from the packet counts collected at the destination during the measurement phase. Latency measures are computed from the start and finish times of *all* measurement packets. It is important that the simulation be run long enough to measure the finish time of every measurement packet or the tail of the latency distribution will be truncated.[1]

---

1. If a network is not fair, it may take a large number of cycles for the last of the measurement packets to be delivered. If a network is subject to starvation, the drain phase may never complete.

During the warm-up and drain phases, the packet source continues to generate packets; however, these packets are not measured. They do, however, affect the measurement by providing background traffic that interacts with the measurement packets. The determination of the cycle counts N_1 and N_2 is discussed in Section 24.3.

23.1.1 Throughput

Throughput is the rate at which packets are delivered by the network for a particular traffic pattern. It is measured by counting the packets that arrive at destinations over a time interval for each flow (source-destination pair) in the traffic pattern and computing from these flow rates the fraction of the traffic pattern delivered. Throughput, or *accepted traffic*, is to be contrasted with demand, or *offered traffic*, which is the rate at which packets are generated by the packet source.

To understand how network throughput is related to demand, we typically plot throughput (accepted traffic) as a function of demand (offered traffic), as shown in Figure 23.2. At traffic levels less than *saturation*, the throughput equals the demand and the curve is a straight line. Continuing to increase the offered traffic, we eventually reach *saturation*, the highest level of demand for which thoughput equals demand. As demand is increased beyond saturation, the network is not able to deliver

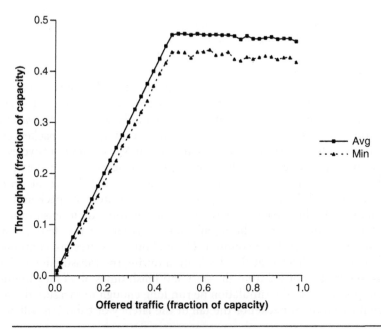

Figure 23.2   Throughput vs. offered traffic for an 8-ary 2-mesh under bit-complement traffic using dimension-order routing. Both the average throughput and the minimum throughput over all source-destination pairs are shown. Because the throughput does not degrade beyond the saturation point, the network is stable.

packets as fast as they are being created — or at least not for the traffic pattern we require. Above saturation, a *stable* network continues to deliver the peak throughput on the specified traffic pattern. Many networks, however, are *unstable* and their throughput drops beyond saturation. The network of Figure 23.2, for example, is stable with a saturation throughput of 43% of capacity. Figure 23.3 shows the performance of a similar network that also saturates at 43% of capacity. However, this network is unstable as indicated by the sharp drop in throughput just beyond the saturation point caused by unfairness in the flow control mechanism. We investigate these two networks further as part of the simulation examples in Section 25.2.5.

We typically present throughput as a fraction of network capacity. This gives a more intuitive understanding of the performance of the network and allows direct comparison between networks of differing sizes and topologies. For example, in Figure 23.2, the total capacity of this 8-ary 2-mesh is $\Theta_{U,M} = 4b/k = b/2$ where b is the channel bandwidth. Since each source is contributing an equal amount to the offered traffic, each source generates packets at a rate equal to $b\lambda/2$ where λ is the offered traffic as a fraction of capacity. Saturation occurs at 43% of this capacity.

For some unstable networks, the total number of packets delivered remains constant beyond saturation, but the distribution of these packets does not match the specified traffic pattern. To correctly measure throughput of such networks, throughput must be measured on the specific traffic pattern. To do this, we apply the specified traffic at all network inputs and measure the accepted traffic for each flow separately. The throughput on the specified traffic pattern is the *minimum* ratio of accepted traffic to offered traffic over all of the flows.

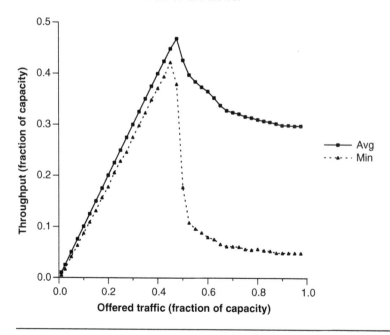

Figure 23.3 Throughput vs. offered traffic for an 8-ary 2-mesh under bit-complement traffic using dimension-order routing. Throughput drops drastically after saturation, revealing an unfair network.

Mathematically, we start with a traffic matrix Λ in which each row and each column of Λ sums to unity. For offered traffic of $\alpha\Lambda$, each source i generates packets destined for each destination j with a rate of $\alpha\lambda_{ij}$. At the destinations we record the arriving traffic from each source and compute a rate λ'_{ij} for each flow. Throughput is then defined as the minimum ratio of λ' to λ over all flows:

$$\Theta = \min_{ij}\left(\frac{\lambda'_{ij}}{\lambda_{ij}}\right). \qquad (23.1)$$

When demand is increased beyond saturation, the traffic matrix delivered to the destinations can be thought of as taking the form:

$$\Lambda' = \Theta\Lambda + X \qquad (23.2)$$

That is, the traffic delivered to the destination consists of Θ units of the desired traffic pattern plus extra traffic, X, where X is a strictly positive matrix. In measuring throughput, we give no credit for this extra traffic since it does not conform to our specified traffic pattern.

Network instability is often due to fairness problems in the network. Historically, only the average accepted traffic across all flows has been reported beyond saturation — in effect, giving credit for X. However, by reporting the minimum accepted traffic over all of the flows, we can see if any flows are becoming starved as offered traffic increases. Starvation is generally a result of unfair flow control. In Figure 23.2, the accepted traffic remains approximately constant beyond the saturation point, so we can be confident that this network is allocating an even amount of bandwidth to each flow even beyond saturation. In Figure 23.3, however, throughput falls dramatically after saturation, suggesting that the network may have unfair flow control.

It is also possible for the accepted traffic to decrease as the offered traffic increases, even if the network is fairly allocating bandwidth between the flows. This situation can be caused by non-minimal routing algorithms. Because the routing decisions are made with imperfect information, misrouting often increases beyond the saturation point due to the increasing contention. This increases the average number of hops a packet travels, increasing channel load in turn. Without careful design of a routing algorithm to avoid this, throughput can drop off quickly beyond the saturation point.

A similar situtation can occur with adaptive routing algorithms that use deadlock-free escape paths. As offered traffic increases, more packets are forced onto the escape paths and the accepted throughput eventually degrades to the throughput of the escape routing algorithm, not the adaptive algorithm. These are discussed further in Section 25.2.

While the accepted vs. offered traffic curve shows the detailed throughput performance of the network under a specific traffic pattern, the performance of many traffic patterns can be quickly presented by plotting a histogram of saturation throughputs over those patterns. A typical experiment of this form might be to generate a large number of random permutations, say $10^3 - 10^6$, for the traffic patterns. Examples of this plot are included in Section 25.1.

23.1.2 **Latency**

Latency is the time required for a packet to traverse the network from source to destination. Our evaluations of latency until this point have mainly focused on the zero-load latency of the network. This ignores latency due to contention with other packets over shared resources. Once we include contention latency, through modeling or simulation, latency becomes a function of offered traffic and it becomes instructive to plot this function. Figure 23.4 shows an example of latency vs. offered traffic graph.

In measuring latency, we apply traffic using the measurement setup of Figure 23.1. We typically sweep offered traffic $\alpha\Lambda$ from $\alpha = 0$ to saturation throughput, $\alpha = \Theta$. Latency is infinite and cannot be measured at traffic levels $\alpha > \Theta$. For each packet, latency is measured from the time the first bit of the message is generated by the source to the time the last bit of the message leaves the network. Overall latency is reported as the average latency over all packets. In some cases, it is also useful to report histograms of packet latency, worst-case packet latency, and packet latency statistics for individual flows. As with throughput, we present offered traffic as a fraction of capacity.

Latency vs. offered traffic graphs share a distinctive shape that starts at the horizontal asymptote of zero-load latency and slopes upward to the vertical asymptote of saturation throughput. At low offered traffic, latency approaches zero-load latency.

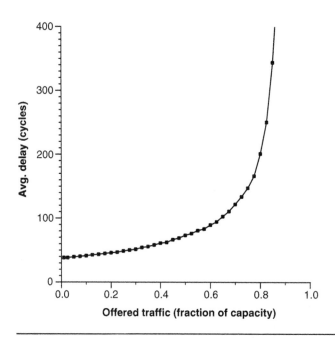

Figure 23.4 Average latency vs. offered traffic for an 8-ary 2-mesh under uniform traffic with dimension-order routing.

As traffic increases, increased contention causes latency to increase as packets must wait for buffers and channels. Eventually, the latency curve approaches a vertical asymptote as the offered traffic approaches the saturation throughput of the network. The shape of this curve is largely explained by basic queuing theory (Section 23.2.1).

Although is it generally useful to look at the average latency of packets, it can also be informative to study the latency distribution over all packets or just a subset of packets. For example, with virtual-channel flow control, some channels may be reserved for high-priority traffic. In this case, separate plots of high - and normal priority traffic can reveal the effectiveness of the prioritorization scheme. Another interesting subset of traffic can be all the packets that travel between a particular source-destination pair. This can be especially useful in non-minimal routing algorithms and gives a general indication the length of paths between the source-destination pair. Several latency distribution graphs are presented in Section 25.2.

23.1.3 Fault Tolerance

Fault tolerance is the ability of the network to perform in the presence of one or more faults. The most significant bit of information about a network's fault tolerance is whether it can function at all in the presence of faults. We say that a network is *single-point fault tolerant* if it can continue to operate (deliver packets between all non-faulty terminals) in the presence of any one node or channel fault. For many systems, being single-point fault tolerance is sufficient, because if the probably of one fault is low, the chance of having multiple faults simultaneously is extremely low. Moreover, most systems with high availability are actively maintained, so faulty components can be replaced before another fault occurs.

The requirement for fault tolerance drives a number of design decisions. For example, if a network must be single-point fault tolerant, it cannot use deterministic routing lest a single channel failure disconnect all node pairs communicating over that channel. We say that a network *degrades gracefully* in the presence of faults if performance is reduced gradually with the number of faults. The performance of a fault-tolerant network is explored in Section 25.3.

23.1.4 Common Measurement Pitfalls

In measuring interconnection network performance a number of common errors have been made on numerous occasions. The authors confess to committing all of these sins at one time or another and hope that by describing them here, you can learn from our mistakes and not be doomed to repeat them.

No source queue: All to often, network performance measurements are taken without the proper use of the source queue in the measurement setup. Figure 23.5 shows two ways in which the effects of the source queue can be omitted from network measurements. In Figure 23.5(a), there is simply no source queue. If the packet source

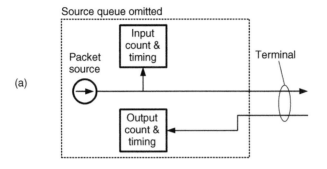

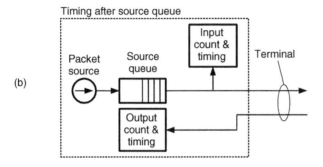

Figure 23.5   Attempting to measure network performance without an infinite source queue. The resulting system is closed loop. The traffic pattern applied is not the intended pattern and the latency computed does not account for time waiting to enter the network. (a) The source drops or delays packets when the input channel is busy. (b) Packet counting and timing is started at the output of the source queue rather than the input.

generates a packet at a time when the input port is busy, the source must either drop the packet, or delay injection until the input is ready. In the latter case, the generation of subsequent packets must be delayed as well, otherwise a queue would be required. In either case, the traffic pattern being generated is affected by contention in the network. In Figure 23.5(b), a source queue is used but packets are counted and packet timing is started at the output of the source queue rather than at the input.

Measurements taken without the source queue are invalid for two reasons. First, the traffic pattern being applied at the measurement points[2] is not the intended pattern. For example, consider the simple two-node network of Figure 23.6. We set up the packet sources to inject one unit of traffic from node 1 to 0 and the same amount from 0 to 1. However, by measuring the packets after the source queue, we count and time only the 0.1 unit of traffic that is actually accepted for routing from

---

2.   The points at which packet counting and timing takes place.

Figure 23.6   A two-node network with one slow link is to route one unit of traffic from 0 to 1 and one unit of traffic from 1 to 0. Measuring at the output, rather than the input of the source queue the actual traffic pattern measured is 0.1 unit of traffic from 0 to 1 and one unit of traffic from 1 to 0.

0 to 1, but we do count the full unit of traffic in the opposite direction. This is not the traffic pattern that we are interested in measuring.

The second problem with omitting the source queue is that packet latency is understated. In networks with shallow queues and blocking flow control, a large fraction of the contention latency at high loads is in the source queue. Once the packet actually gets into the network, it is delivered fairly quickly. Omitting the time spent waiting for an input channel can greatly underestimate the *real* latency, which is the time from when the packet was created to when it was delivered.

Average instead of minimum delivered traffic: While omitting the source queue distorts the measurement at the input side, reporting the *average* rather than *minimum* delivered traffic across all flows distorts the measurement at the output side. The problem here, again, is that the traffic pattern being measured is not the one specified.

Consider the network of Figure 23.6. If we average the traffic received across the two destinations, we would conclude that the network saturated at 0.55 units of traffic — the average of 1 and 0.1. However, this is not the case. The network is not delivering

$$\Lambda = 0.55 \begin{bmatrix} 0 & 1 \\ 1 & 0 \end{bmatrix}, \tag{23.3}$$

but rather is delivering

$$\Lambda = 0.1 \begin{bmatrix} 0 & 1 \\ 1 & 0 \end{bmatrix} + \begin{bmatrix} 0 & 0 \\ 0.9 & 0 \end{bmatrix}. \tag{23.4}$$

That is, the network is delivering 0.1 units of the specified traffic pattern plus some extra traffic. Taking the minimum traffic received across the flows gives the true picture.

In networks where some fraction of traffic is starved at high loads, reporting average delivered traffic can completely hide the problem. For example, chaotic routing [26] gives packets in the network absolute priority over packets entering the network. For some traffic patterns, this causes flows from some inputs to be completely starved at high loads because they are unable to merge into busy channels. Reporting average received traffic makes it appear as if the network is *stable*, continuing to deliver peak througput at loads beyond saturation, while in fact the network is *unstable* because some flows are completely throttled at these high loads.

Combined latency and accepted traffic graphs: Some researchers attempt to show both a latency vs. offered traffic curve and an accepted vs. offered traffic curve in a single plot, as shown in Figure 23.7.[3] The plot shows latency as a function of *accepted*, not *offered*, traffic. Points along the curve indicate different amounts of offered traffic. In an unstable network, where accepted traffic declines beyond saturation, the curve is often shown to wrap back to the left, as shown in the figure.

The biggest problem with this format is that it clearly shows that a source queue is not being included in the measurement. If the source queue were considered, then latency would be infinite at all offered traffic levels above saturation and the curve would not wrap back, but rather would reach an asymptote at the peak accepted traffic.

BNF charts also do not usually reflect *steady-state* performance. In most unstable networks, steady-state-accepted traffic drops abruptly to a post-saturation value at offered traffic levels even slightly greater than saturation and then stays at this lower level as offered traffic is increased further. This is because source queues start growing without bound as soon as saturation is reached. In the steady state, these queues are infinite — increasing offered traffic beyond saturation does not change their steady-state size, only how fast they get there. Of course, running simulations to measure the steady-state performance of networks beyond saturation takes a long time. Most BNF charts that show a gradual fold back in accepted traffic as offered traffic is

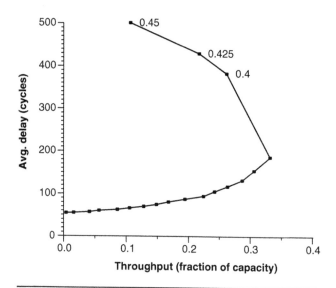

Figure 23.7   An incorrect method of presenting latency and accepted traffic data. This graph plots latency (vertical axis) as a function of accepted traffic (horizontal axis). Points beyond saturation are labeled with their corresponding offered traffic as a fraction of capacity.

3.   This form of plot is sometimes called "Burton Normal Form" after Burton Smith or BNF for short — not to be confused with Backus Naur Form.

increased indicate that the simulation has not been run long enough to reach the steady-state value.

Not measuring *all* the packets *generated* during the test interval: Often, latency measurements will be taken by averaging the latency of all packets *received* at destinations during the measurement interval rather than all packets *generated* during this interval. This is incorrect because it uses a biased sample of packets to compute average latency. This sample excludes long-latency packets that do not arrive until long after the measurement interval and hence truncates the latency distribution and underestimates average latency. Any unbiased sample of packets generated after warm-up could be used for latency measurement. However, selecting packets based on when they arrive is not unbiased.

Reporting results *only* on random traffic: Uniform random traffic is quite benign because it naturally balances load across network channels. Although not quite as nice as nearest-neighbor traffic, it is very close to being a best-case workload. It can hide many network sins — in particular, poor routing algorithms. In most cases much more insight can be gained by complementing random traffic with a number of adversarial traffic patterns (Section 3.2) and a large sample of randomly generated permutations.

23.2 **Analysis**

A number of tools are available to measure the performance of an interconnection network. Analysis, simulation, and experiment all play roles in network evaluation. We typically start using analysis to estimate network performance using a mathematical model of the network. Analysis provides approximate performance numbers with a minimum amount of effort and gives insight into how different factors affect performance. Analysis also allows an entire family of networks with varying parameters and configuration to be evaluated at once by deriving a set of equations that predicts the performance of the entire family. However, analysis usually involves making a number of approximations that may affect the accuracy of results.

After approximate results are derived using analysis, simulations are usually performed to validate the analysis and give a more accurate estimate of performance. With good models, simulations provide accurate performance estimates but require more time to generate estimates. Each simulation run can take considerable time and evaluates only a single network configuration, traffic pattern, and load point. Also, simulation gives summary results without directly providing insight into what factors lead to the results.[4] A simulation is as accurate as its models. A simulator that closely matches most router timing, arbitration, and channel delays will give very accurate

---

4.   However, with appropriate instrumentation, simulations can be used to diagnose performance issues.

results. A simulator that ignores router timing and arbitration may give results of very poor accuracy.

Once a network is constructed, experiments are run to measure the actual network performance and to validate the simulation models. Of course, at this point it is difficult, time consuming, and expensive to make changes if performance problems are encountered.

In this section we deal with analysis, the first step in network performance evaluation. In particular, we introduce two basic tools for analytic performance analysis: queuing theory and probability theory. queuing theory is useful for analyzing networks in which packets spend much of their time waiting in queues. Probability theory is more useful in analyzing networks in which most contention time is due to blocking rather than queuing.

23.2.1 **Queuing Theory**

Packets in an interconnection network spend a lot of time waiting in queues. Packets wait in the source queue that is a part of our measurement setup and the input buffers at each step of the route are also queues. We can derive approximations of some components of packet latency by using queuing theory to predict the average amount of time that a packet spends waiting in each queue in the system.

A complete treatment of queuing theory is beyond the scope of this book. However, in this section, we introduce the basics of queuing theory and show how they can be applied to analyzing simple interconnection networks. The interested reader should consult a book on queuing theory [99] for more details.

Figure 23.8 shows a basic queuing system. A *source* generates packets with a rate of λ packets per second. The packets are placed in a *queue* while waiting for service. A *server* removes the packets from the queue in turn and processes them with an average service time of T seconds, giving an average service rate of $\mu = 1/T$ packets per second.

In an interconnection network, the packet source is either a terminal injecting packets into the interconnection network or a channel delivering packets from another node. The packet arrival rate λ is either the traffic rate from an input terminal, or the superposition of traffic on a network channel. Similarly, the server is either a terminal removing packets from the network or a channel carrying packets to another node. In both cases, the service time T is the time for the packet to traverse the channel or terminal and hence is proportional to packet length $T = L/b$.

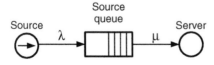

Figure 23.8 A basic queuing system. A packet source injects packets with rate λ into a queue. A server removes packets from the queue with rate $\mu = 1/T$.

To simplify our analysis, we often make a number of assumptions that we know are not completely accurate. To start, we will assume that queues have infinite length and that both input inter-arrival times and output service times are exponentially distributed.[5] An exponential process is also called *memoryless* because the probability that a new event occurs, such as a packet arrival, is independent of previous events. While this is a simple model for arrivals and certainly a simplification for service times, making these assumptions allows us to model our simple queuing system with a Markov chain, as shown in Figure 23.9.

The Markov chain of Figure 23.9 represents the state transitions of a queue. Each circle in the figure represents a state of the queue and is labeled with the number of entries that are in the queue in that state. New packets arrive in the queue with a rate of λ, and thus states transfer to the next state to the right at a rate of λ. Similarly, states transition to the state to the left at a rate of μ corresponding to the rate of packets leaving the queue. In the steady state, the rate into a state must equal the rate out of a state. For state p_0, we can express this equilibrium condition as

$$\lambda p_0 = \mu p_1.$$

Here, λp_0 is the rate out of state p_0 and μp_1 is the state into p_0. We can write this as

$$p_1 = \frac{\lambda}{\mu} p_0.$$

Or, if we define the duty factor of the system as $\rho = \frac{\lambda}{\mu}$, we can rewrite this as

$$p_1 = \rho p_0.$$

Writing the equilibrium equation for p_1, we can derive that $p_2 = \rho p_1$ and repeating this process for each state gives us

$$p_i = \rho p_{i-1} = \rho^i p_0.$$

Then, using the constraint that $\sum_i p_i = 1$, we see that

$$p_0 = 1 - \rho,$$

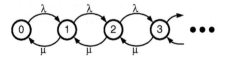

Figure 23.9   A Markov chain model of a queue. Each state reflects a different number of packets in the queue. A transition to a state that has greater occupancy occurs with rate λ. A transition to a state with lower occupancy occurs with a rate of μ.

5.   A variable x is exponentially distributed if $P(x < y) = 1 - \exp(-y/\mu)$.

and

$$p_i = (1 - \rho)\rho^i.$$

Now that we have an expression for the probability of being in each state, we can compute the expected number of entries in the queue as

$$E(N_Q) = \sum_i i p_i = \sum_i i(1 - \rho)\rho^i = \frac{\rho}{1 - \rho}. \qquad (23.5)$$

Because an arriving packet must wait for each of these queued packets to be served, the expected waiting time is

$$E(T_W) = E(T)E(N_Q) = \frac{T\rho}{(1 - \rho)} = \frac{\rho}{\mu(1 - \rho)}. \qquad (23.6)$$

This relationship between waiting time T_W, service time T, and the number of packets in the queue N_Q is a useful relationship, often referred to as Little's law.

We can calculate the variance of the number of queue entries in a similar manner as

$$E(\sigma_{N_Q}^2) = \sum_i i^2 p_i = \sum_i i^2(1 - \rho)\rho^i = \frac{\rho}{(1 - \rho)^2}. \qquad (23.7)$$

We have now calculated the properties of the famous M/M/1 queuing system. That is, a system with exponentially distributed inter-arrival times, exponentially distributed service times, and a single server. To highlight the effect of variance in service time on queuing behavior, we note (without derivation) that the expected occupancy of an M/D/1 queue, with deterministic service time is half as much as an M/M/1 queue:

$$E(N_Q) = \frac{\rho}{2(1 - \rho)} \quad (\text{M/D/1}). \qquad (23.8)$$

In general an M/G/1 queue with an arbitrary service time distribution has an occupancy of

$$E(N_Q) = \frac{\lambda^2 \overline{X^2}}{2(1 - \rho)} \quad (\text{M/G/1}) \qquad (23.9)$$

where $\overline{X^2}$ is the second moment of the service time [20]. Because, in many networks packet length and hence service time is constant, or at least closer to constant than exponential, we typically use Equations 23.8 or 23.9 rather than using Equation 23.5.

At the risk of stating the obvious, it is worth pointing out that Equations 23.5 through 23.9 are *invalid* for $\lambda \geq \mu$ or $\rho \geq 1$. These equations evaluate to a negative value in this invalid region when in fact, latency is infinite, or undefined, when $\lambda \geq \mu$. All too often, these expressions are combined into a larger equation that is then evaluated in this invalid region.

Figure 23.10 illustrates how queuing theory is used to model a simple four-terminal butterfly network. We assume for now that uniform traffic is applied to the

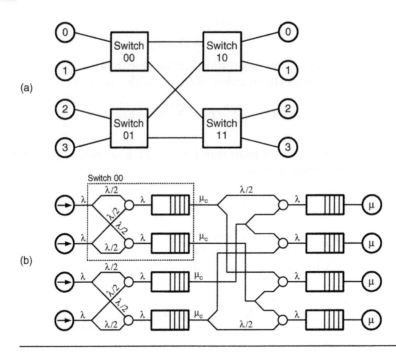

Figure 23.10 Modeling a network as a queuing system. (a) A 2-ary 2-fly network delivers uniform traffic. (b) A queuing model of the network of (a). Sources have rate λ, each switch output is modeled as a queue, internal channels are servers with rate μ_c, and destinations are servers with rate μ.

network.[6] Each source becomes a packet source that generates a stream of packets with rate λ. Each destination becomes a server with rate μ or, equivalently, service time $T = 1/\mu$.

Each 2×2 switch is modeled as a splitter on each input, a combiner on each output, and a pair of output queues. Each splitter takes the λ rate input stream and splits it into two $\lambda/2$ streams, one destined for each of the switch outputs. Note that if the traffic were non-uniform, the splitter could split the stream unevenly with different amounts of traffic to each switch output. At each switch output, a combiner takes the $\lambda/2$ rate streams from each input and combines them into a λ rate stream into the output queue. The output queue of each switch accepts the traffic stream summed by the combiner — this defines the arrival rate of the queue. The queues are served by the internal channel with rate μ_c for the first stage and by the terminal with rate μ for the second stage.

The splitter and combiner in the switch model take advantage of the properties of the Poisson arrival process. Splitting a stream of packets with an exponential inter-arrival time gives multiple streams of packets, each with exponential inter-arrival

---

6.　In Exercise 23.1, we will consider a more challenging traffic pattern.

times that have the sum of the rates of the individual streams equal to the rate of the input stream. Similarly, combining multiple streams with exponential inter-arrival times gives a single stream with exponential inter-arrival times that have a rate given by the sum of the rates of the input streams.

We model the switch as an output queued switch (whether it is or not) because there is no easy way to model the switch contention of an input queued switch, or a switch with virtual output queues, in a queuing model. If in fact the output channel is idled some fraction of time due to switch contention, this can be modeled by reducing the service rate out of the queue to account for this idle time.

If we assume that the packets have equal length, then the servers, both the internal channel and the destination terminal, have deterministic service time. Hence, we model the queues using the M/D/1 model of Equation 23.8, giving a contention latency for this network of

$$T_c = \frac{\lambda}{2\mu_c(\mu_c - \lambda)} + \frac{\lambda}{2\mu(\mu - \lambda)}. \tag{23.10}$$

The first term is the time spent waiting in the queue of the first stage switch and the second term is the time spend waiting in the second stage queue. In general, an expression for queuing model latency includes a term for each queue in the system between source and destination. If our network had non-uniform traffic, the *average* latency expression would be written as the weighted sum of the latency of each distinct path through the network (Exercise 23.1).

Figure 23.11 plots contention latency T_C as a function of offered traffic λ for the case in which internal channel bandwidth is 1.5 times the terminal bandwidth, $\mu_c = 1.5\mu$. The curve is normalized by setting $\mu = 1$. The figure shows three curves: the average waiting time in each queue and the sum of these times, the total contention latency. We see that the single-queue curves have the characteristic $\frac{\rho}{1-\rho}$ shape of queue delay. The overall curve is dominated by the delay of the more heavily loaded of the two queues — in this case, the second stage queue. In general, the *bottleneck* queue always dominates the latency-throughput curve, both when queues are in a series, as in this example, and when queues are in parallel, as occurs with unbalanced load (Exercise 23.1).

23.2.2 **Probabilistic Analysis**

To complement queuing theory, we use probabilistic analysis to study networks and network components that are unbuffered. We have already seen an example of using probabilistic analysis to estimate the performance of a network with dropping flow control in Section 2.6. In this section, we show how to estimate the performance of networks with blocking flow control.

Consider, for example, a 2×2 switch that has blocking flow control and no queuing, as shown in Figure 23.12. Suppose the service time on each output port is identical, deterministic, and equal to T_o; the input traffic rates are equal, $\lambda_1 = \lambda_2 = \lambda$; and traffic from both inputs is uniformly distributed across the two outputs,

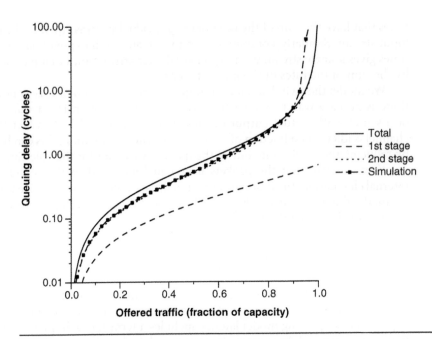

Figure 23.11 Contention latency vs. offered traffic (λ) curve for queuing model of Figure 23.10 for $\mu = 1$ and $\mu_c = 1.5$. Zero-load contention latency is zero. The network saturates at $\lambda = 1$. This graph shows the time spent waiting in each queue as well as the overall contention latency.

$\lambda_{ij} = \lambda/2 \; \forall i, j$. To calculate T_i the amount of time an input channel — say, i_1 — is busy, we first compute the probability P_w that a packet on i_1 will need to wait.

$$P_w = \frac{\lambda T_o}{2} = \frac{\rho_o}{2}. \tag{23.11}$$

This is just the fraction of time that the desired output is busy handling traffic from the other input.

In the event that a packet arriving on i_1 does have to wait, the average waiting time is

$$T_{wc} = \frac{T_o}{2}. \tag{23.12}$$

Combining Equations 23.11 and 23.12, we see that over all packets, the average waiting time is

$$T_w = P_w T_{wc} = \frac{\lambda T_o^2}{4} = \frac{\rho_o T_o}{4}.$$

Thus, the input busy time is

$$T_i = T_o + T_w = T_o \left(1 + \frac{\lambda T_o}{4} \right) = T_o \left(1 + \frac{\rho_o}{4} \right). \tag{23.13}$$

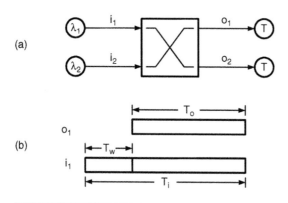

Figure 23.12 Calculating latency and throughput for a switch without buffering. (a) A 2 × 2 switch with no buffering connects two packet sources to two servers. (b) The average input channel busy time T_i is the sum of the output service time T_o and the average time spent blocked waiting for the output T_w.

From Equation 23.13 we see that the busy time at the input of the switch is increased by an amount proportional to the duty factor of the output link $\rho_o = \lambda T_o$.

The average waiting time given a collision, T_{wc}, depends strongly on the variance of the service time. For a deterministic service time, Equation 23.12, an average packet reaches the busy switch when it is half done with a packet from the other input and hence waits $T_o/2$. If output service times were exponentially distributed, the expected waiting time given a collision is twice this amount:

$$T_{wc} = T_o, \quad \text{(Exponentially distributed T)}$$

$$T_w = \frac{\lambda T_o^2}{2} = \frac{\rho_o T_o}{2},$$

$$T_i = T_o \left(1 + \frac{\lambda T_o}{2}\right) = T_o \left(1 + \frac{\rho_o}{2}\right).$$

This increase in waiting time with variance of service time occurs because packets are more likely to wait on packets with long service times than on packets with short service times. This is because the packets with long service times account for more of the total busy time. When we analyze multistage networks, the variance may be different at each stage. For example, with deterministic output service time, the switch of Figure 23.12 has non-zero variance.

23.3 **Validation**

The ability of queuing theory, probabilistic analysis, or simulation to predict the performance of an interconnection network is only as good as the model used in the analysis. To the extent that the model captures all of the relevant behavior of the interconnection network, analysis can be quite accurate and can provide a powerful tool to evaluate large portions of the design space quickly and to shed insight on factors affecting performance. However, incomplete or inaccurate modeling of a key aspect of the network or making an improper approximation can lead to analyses that give results that differ significantly from actual network performance. The same is true of simulation models. An inaccurate model leads to inaccurate simulation results.

Validation is the process of checking the accuracy of an analytical model or of a simulation against known good data. For example, if we have latency vs. offered traffic data collected from an actual network, we can validate a simulation model or analytical model using this data by generating the same numbers via analysis and simulation. If the numbers match, we have confidence that the model accurately accounts for the factors that are important for *that* particular network under the particular operating conditions tested. It is likely that the models will also accurately handle similar networks and similar operating conditions.

If we have high confidence in a simulation model, we will sometimes validate analytical models against simulation. This can be useful in cases where experimental data is not available. Figure 23.11 is an example of this type of validation.

To validate a model, we need to compare its results against known data at a representative set of operating points and network configurations. Validating a model at one operating point does not imply that the model will be accurate at vastly different operating points, where different factors affect performance. For example, validating that a model gives the same zero-load latency as the actual network does not imply that it will accurately predict the latency near saturation.

23.4 Case Study: Efficiency and Loss in the BBN Monarch Network

BBN Advanced Computer Systems designed the BBN Monarch [156] as a follow-on to the Butterfly TC-2000 (Section 4.5). The Monarch incorporated many interesting architectural features. Unfortunately, the machine was never completed. The Monarch was a uniform-memory-access shared memory machine. N processors (up to 64 Kbits) made accesses to N memory modules over a butterfly-like network. The network used dropping flow control, hashed memory addresses to randomize the destination memory module, and access combining to handle hot-spot traffic. Accesses were made synchronously to simplify combining. Time was divided into *frames*, and each processor could initiate one memory access at the start of each frame. Accesses to the same location during the same frame were combined at an internal node of the network and a single combined access packet was sent on to the memory module.

The Monarch network consisted of alternating stages of switches and concentrators, as shown in Figure 23.13. Each switch had up to 12 inputs and up to 32 outputs and could resolve up to 4 address bits. The output ports of a switch were configured in groups to allow multiple simultaneous packets to proceed in the same direction. For example, a switch could be configured with 16 output groups of 2 ports each or 8 output groups of 4 ports each. A packet arriving at any input of the switch

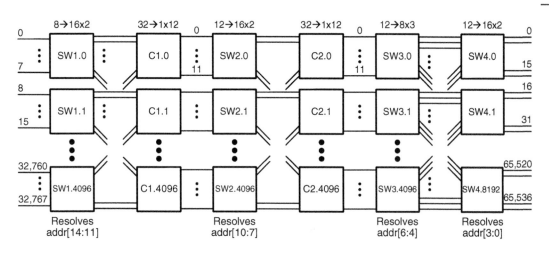

Figure 23.13 The BBN Monarch network (1990) was composed of alternating stages of switches and concentrators. The concentration ratio at each stage was adjusted to maximize efficiency and minimize loss due to the dropping flow control. Each channel operated serially with a bandwidth of 350 Mbits/s.

would arbitrate for one of the output channel groups. If it succeeded in acquiring any channel of the requested group, it would proceed to the next stage of the network. Otherwise, it would be dropped.

The concentrator components acted to multiplex a large number of lightly loaded channels onto a smaller number of more heavily loaded channels. This might be done, for example, just before an expensive channel that must traverse several packaging levels. Each concentrator chip had up to 32 inputs and up to 12 outputs. All inputs and outputs were equivalent. Each frame, up to 12 packets, arriving on any 12 of the 32 inputs would be routed to the 12 outputs. Any additional packets arriving on other inputs during that frame would be dropped.

We can use probabalistic analysis to compute the probability of packet loss at each stage of the Monarch network. We assume that all input channels are loaded uniformly and that traffic is distributed uniformly over the outputs. The address hashing and access combining help make this assumption realistic. Let the probability that an input is busy (the input duty factor) be P_i on a switch with I inputs, G output groups, and C channels per output group. Then the probability that a particular input i has a packet for a particular output group j is $P_{ij} = P_i/G$. Then, over all of the inputs, the probability that there are k requests for output j is given by the binomial distribution

$$P_k = \binom{I}{k} \left(\frac{P_i}{G} \right)^k \left(1 - \frac{P_i}{G} \right)^{I-k}.$$

The probability that an output channel is busy, P_o, can be computed by summing over the cases of output channel occupancy k. For $k \leq C$, the probability that output channel j is not busy is $C - k/C$. For $k \geq C$ the probability is zero. This gives us the summation:

$$P_o = 1 - \sum_{k=0}^{C-1} \left(\frac{C-k}{C} \right) P_k$$

$$= 1 - \sum_{k=0}^{C-1} \left(\frac{C-k}{C} \right) \binom{I}{k} \left(\frac{P_i}{G} \right)^k \left(1 - \frac{P_i}{G} \right)^{I-k}. \qquad (23.14)$$

Now that we have the output duty factor, the efficiency of the stage, the fraction of packets that arrive at an input that successfully acquire an output, can be calculated as the total output traffic divided by the total input traffic:

$$P(\text{transmit}) = \frac{GCP_o}{IP_i}, \qquad (23.15)$$

and the loss is just

$$P(\text{drop}) = 1 - P(\text{transmit}) = 1 - \frac{GCP_o}{IP_i}. \qquad (23.16)$$

Table 23.1 lists the connection pattern of a 64-K-port Monarch network and presents an analysis of packet loss in that network. For each stage, we apply Equation 23.14 to compute P_o for that stage and hence P_i for the next stage. Given P_i and P_o for each stage, we then compute the efficiency and loss by using Equations 23.15 and 23.16.

Table 23.1 shows that the 64-K Monarch network configured as shown has a loss of less than 10% with 100% input load. This is achieved at significant cost by overprovisioning the links so that the inputs to the switches (except the first) are less than 65% busy and the inputs to the concentrators are less than 25% busy. We explore the relationship between link duty factor and loss in Exercise 23.7.

Adding buffering to a network can reduce loss by delaying a packet until an output is idle rather than dropping it. With buffering (but no backpressure) a packet is dropped only when all outputs are busy and all buffers are full. We explore the analysis of a Monarch-like network with buffering in Exercise 23.9.

23.5 **Bibliographic Notes**

A good general reference to queuing theory is Kleinrock [99], while Bertekas [20] presents modeling techniques more specialized to networks. Examples of probabilistic analyses of networks are given by Dally [46] and Agarwal [2]. An analysis modeling partially full finite queues is also given by Dally [47]. Pinkston and Warnakulasuriya develop models for deadlocks in k-ary n-cube networks [152].

Table 23.1 Specification and analysis of a 65,536-port Monarch network (from [156]). For each of the six stages (four switch stages and two concentrator stages) the columns I, G, and C specify the connectivity of the network by describing the number of inputs I, separately addressed output groups G, and channels per output group C. The remaining columns show the results of analyzing this network configuration assuming a 100% duty factor on the input channels of switch 1. P_i gives the input load to each stage (equal to P_o of the previous stage). The efficiency and loss columns give the probability of a packet being transmitted or dropped, respectively, at each stage.

Stage	I	G	C	P_i	Efficiency	Loss
Switch 1	8	16	2	1.00	0.977	0.023
Concentrator 1	32	1	12	0.24	0.993	0.007
Switch 2	12	16	2	0.65	0.975	0.025
Concentrator 2	32	1	12	0.24	0.995	0.005
Switch 3	12	8	3	0.63	0.986	0.014
Switch 4	12	16	2	0.62	0.977	0.023
TOTAL					0.907	0.093

23.6 **Exercises**

23.1 *Non-uniform traffic — a queuing model.* In the example of Figure 23.10, consider the case in which sources 0 and 1 distribute their traffic uniformly over destinations 2 and 3 and vice versa. Compute the latency vs. offered traffic curve for the queuing model under this traffic pattern. What is the average occupancy of the first stage and second stage switch queues?

23.2 *Queuing delay in a concentrator.* A 4:1 concentrator is fed by four terminals that each transmit 16-byte packets at a rate of 1 Gbit/s with an average rate of one packet per microsecond (an average bandwidth of 128 Mbits/s). The network channel of the concentrator has a bandwidth of 2 Gbits/s. Assuming that the inter-arrival interval is exponentially distributed, compute the following: (a) the probability that the 2 Gbits/s network channel is over-subscribed (more than two inputs transmitting packets simultaneously), and (b) the average waiting time (queuing delay) seen by a packet at the input of the concentrator.

23.3 *Queuing model of dropping flow control.* Derive the queuing model given in Exercise 2.9. Assume packets are dropped halfway through the network on average and are immediately reinjected into an input queue.

23.4 *Throughput of a circuit-switched butterfly.* Use probabilistic analysis to calculate expected throughput of a 2-ary 3-fly with circuit switching. Assume uniform traffic and a service time of $T_0 = 4$ cycles.

23.5 *Probabilistic modeling of a k input switch.* Extend Equation 23.13 for a k input switch with uniform traffic.

23.6 *Probabilistic modeling of arbitrary traffic patterns.* Extend the equation you derived in Exercise 23.5 to handle an arbitrary traffic matrix in which λ_{ij} is the traffic from input i to output j.

23.7 *Tradeoff of duty factor and loss in the Monarch network.* Plot a graph of network loss as a function of (a) switch input duty factor and (b) concentrator input duty factor for Monarch networks similar to the one analyzed in Table 23.4. You will need to vary the parameter C for each stage to adjust the duty factor.

23.8 *Tradeoff of cost and loss in the Monarch network.* Assume that the cost of the Monarch network is proportional to switch and concentrator pin count. Plot a graph of network loss as a function of network cost by varying G and C. At each point on the graph, experimentally determine the network configuration that requires minimum cost (pin-count) to achieve a given loss.

23.9 *Buffering in the Monarch network.* The probability of dropping a packet, Equation 23.16, in a network with dropping flow control can be greatly reduced by adding some buffering and dropping a packet only when all buffers are full. Redo the analysis of Table 23.4 assuming that each input has buffering for F packets. With each frame, all new input packets and all buffered packets compete for the outputs. For packets requesting a particular output group j, the first C packets acquire output channels. The first F packets at each input that are not granted an output channel are buffered. Any additional packets at each input are dropped. What is the dropping probability for the configuration of Table 23.4 for the case in which $F = 8$?

23.10 **Simulation.** For the model and traffic pattern of Exercise 23.1, plot the saturation throughput as a function of internal channel bandwidth μ_c. Compare this model to a simulation of the network.

23.11 **Simulation.** Compare the throughput result from Exercise 23.4 to a simulation of the network. Comment on any differences.

23.12 **Simulation.** For the buffered Monarch network of Exercise 23.9, compare the model with simulation. Comment on any differences.

CHAPTER 24

Simulation

The queueing theory and probabilistic analysis techniques discussed in Chapter 23 can model many aspects of a network, but there are some situations that are simply too complex to express under these models. In these cases, simulation is an invaluable tool. However, simulation is a double-edged sword — while it can provide excellent models of complex network designs, simulators and simulations are equally complex. To that end, the quality of simulation results is only as good as the methodology used to generate and measure these results.

In this chapter, we address the basics of simulation input, measurement, and design. Not all network simulators need to model the intricate details of router microarchitecture, and we begin with a discussion of the different levels of simulation detail available to a network designer. As important as choosing the appropriate modeling accuracy is selecting the workload or input for a network simulation. Both application-driven and synthetic techniques for generating workloads are covered. Once the input and model for a simulation are defined, the performance of the network must be measured. Several statistical approaches for both measuring networks and assessing the accuracy of those measurements are presented. Finally, the basics of simulator design are introduced along with several issues specific to network simulators.

24.1 **Levels of Detail**

Before simulation of a network begins, the designer must choose an appropriate level of detail for the simulation. Using an extremely accurate simulator is always safe in the sense that modeling inaccuracies will be avoided, but it is also expensive in terms of the time required to both write and run the simulator. A more reasonable and efficient approach is to choose a level of simulation detail that will capture the aspects of the network's behavior important to the designer and avoid simulating unnecessary details of the network.

A typical range of levels of detail for network simulations is shown in Figure 24.1. The least detailed *interface level* provides only the functionality of the network interface combined with simple packet delivery. This level is also often referred to a *behavioral* simulation. Behavioral simulations are most useful in early design stages, where aspects of a design which use the network, such as coherence protocols, need to be tested. Although the details of the final network implementation may not be known, the interface level can still provide the general behaviors of the network.

The *capacity level* is introduced as an intermediate level between purely behavioral and very detailed network models. At this level, basic constraints are placed on the capabilities of the network, such as channel bandwidths, buffering amounts, and injection/ejection rates. With these capacities in place, initial performance of the network can be assessed and the simulator is still simple enough to be quickly changed based on preliminary results.

Finally, the microarchitectural details of the network are incorporated at the *flit level*. Individual flits are modeled in this level, requiring the introduction of structures

Interface level

- Models network interfaces and provides packet delivery.
- Generates simple approximations for packet latency based purely on distance.

Capacity level

- Adds simple constraints on resource capacities, such as channel bandwidths or bounds on the total number of packets in flight.
- Resource contention can affect packet latency.

Flit level

- Resource usage is tracked on a flit-by-flit basis, requiring a detailed modeling of most router components, including buffers, switches, and allocators.
- Packet latency is accurately modeled in terms of flit times or router cycles.

Hardware level

- Adds implementation details of the hardware, yielding area and timing information.
- Latencies can be expressed in terms of absolute time (that is, seconds).

Figure 24.1 A hierarchy of simulation levels of detail. The topmost levels provide the least accuracy, but simulate quickly. Moving down the list gives increasing simulation detail at the cost of longer simulation times.

for buffering, switching, and arbitration. Assuming a correct modeling of these structures, the performance information at this level of detail can be very accurate, but is generally expressed in terms of flit times (cycles). By moving to the *hardware level*, flit times are replaced with the timing information from a physical design and detailed information about the implementation cost/area of the router microarchitecture is determined.

24.2 Network Workloads

The network *workload* refers to the pattern of traffic (such as packets) that is applied at the network terminals over time. Understanding and modeling load enables us to design and evaluate topologies and routing functions. The most realistic traffic patterns are *application-driven workloads* generated directly by the clients using the network. For example, in a shared-memory interconnect, the network traffic consists of coherence messages send between the various processing nodes. If we model not only the network, but also the application running on the processors, our workload is exactly that required by the application. While application-driven workloads give us the most accurate modeling of the network, it is often difficult achieve a thorough coverage of expected traffic with these methods exclusively. Following our shared-memory example, workloads are tied directly to applications, so expanding the set of workloads involves creating new applications, which is generally quite expensive. Alternatively, a carefully designed *synthetic workload* can capture the demands expected for the interconnection network, while also remaining flexible.

24.2.1 Application-Driven Workloads

Ideally, the performance of an interconnection network is measured under application-driven workloads. That is, the sequences of messages applied to the network are generated directly from the intende d application(s) of the network. One approach for generating these workloads is to simulate the network clients in addition to the network itself. For example, in a processor-memory interconnect, models for both the processors and memory modules would be required. Then, performance of the interconnection network is evaluated by running relevant applications on the processing elements. The memory and inter-processor messages generated by the application are the network traffic. This "full-system" simulation approach is often called an *execution-driven workload*. An execution-driven workload is certainly accurate, but one drawback is that feedback from the network influences the workload. This can make it difficult for a designer to isolate bottlenecks in the interconnection network — any change in the network design not only affects the network, but it can also affect the workload.

An alternative to simultaneously modeling the network and its clients is to instead capture a sequence of messages from an application of the network and then "replay" this sequence for the network simulation. These *trace-driven workloads* can either be captured from a working system or from an execution-driven simulation, as discussed above. For example, in an IP router application, a sequence or *trace* of packet arrivals could be recorded for a period of time. For each arrival, the time, length, and destination of the packet would be stored. Then, this sequence is simply recreated for the network simulation. In a trace captured from an execution-driven simulation, the network being simulated might offer only a low level of detail so the simulation can be performed quickly. This trace could then be reused for a detailed simulation of the network only. Since the traces are captured in advance, feedback from the network does not affect the workload. Although this can reduce accuracy, for some applications this may not be a significant issue. For example, in an IP router, feedback only affects the workload on long time scales.

24.2.2 Synthetic Workloads

As mentioned previously, application-driven workloads can be too cumbersome to develop and control. This motivates the inclusion of synthetic workloads, which capture the salient aspects of the application-driven workloads, but can also be more easily designed and manipulated. We divide our discussion of synthetic workloads into three independent axes: injection processes, traffic patterns, and packet lengths. In some situations, these axes may not be truly independent and we highlight this issue in Exercise 24.3.

The first parameter of interest for any injection process is the average number of packets it injects per cycle, or its *injection rate*. Perhaps the simplest means of injecting packets at a rate r is with a *periodic process*. Figure 24.2(a) shows the behavior of a periodic injection process with rate r. The binary process A indicates a packet injection when its value is 1 and is 0 otherwise. As shown, the period T between injections is always $1/r$, so the average injection rate is obviously r.

From inspection, the periodic injection process may seem overly simplistic, as it does not incorporate fluctuations that might be expected from a realistic injection

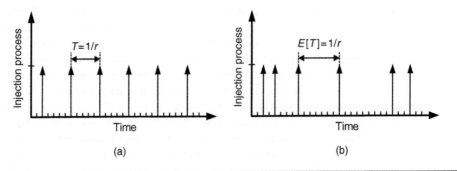

Figure 24.2 The behavior of a periodic (a) and Bernoulli (b) injection process with rate r over time. As shown, the periodic source has a constant period $T = 1/r$ and the Bernoulli process has an expected period of $E[T] = 1/r$.

process. This leads to perhaps the most common injection processes used in network simulations, the *Bernoulli process*. For a Bernoulli process with rate r, the injection process A is a random variable with the probability of injection a packet equal to the process rate, $P(A = 1) = r$. This is equivalent to flipping a weighted coin with a probability r of heads each cycle — if the flip results in heads, a packet is injected. As shown in Figure 24.2(b), this gives geometrically spaced packet injections.

Although the Bernoulli process does incorporate randomness into the injection process, it is still quite simple as it lacks any state. This prevents it from modeling time-varying or correlated traffic processes. In practice, many traffic sources are in fact time-varying. For example, Internet traffic is diurnal as more people are on-line during the day than at night. The Bernoulli process may still be a good estimate for short periods of time in a slowly varying process, but many injection processes also vary rapidly on small time scales.

A rapidly changing injection process is often called *bursty*. Many models have been proposed to capture particular burstiness aspects of different traffic sources such as voice connections, video streams, and HTTP traffic. One popular model for modeling burstiness is the Markov modulated process (MMP). In an MMP, the rate of a Bernoulli injection process is modulated by the current state of a Markov chain.

An example of a two-state MMP is shown in Figure 24.3. As labeled, the two-states represent an "on" and "off" mode for the injection process, with the respective injection rates being r_1 and 0. Each cycle, the probability of transitioning from the off state to the on state is α and from on to off is β. This MMP describes a bursty injection process, where during the bursts injections occur with rate r_1 and outside a burst, the injection process is quiet. Also, the average length of a burst is given by $1/\beta$ and the average time between bursts is $1/\alpha$. To determine the injection rate of this system, the steady-state distribution between the on and off states is determined. First, let x_0 represent the probability of being in the off state and x_1 the probability of the on state. Then, in steady-state

$$\alpha x_0 = \beta x_1.$$

Since $x_0 + x_1 = 1$, the steady-state probability of being in the on state is

$$x_1 = \frac{\alpha}{\alpha + \beta}.$$

Figure 24.3 A two-state Markov-modulated process (MMP). In the off state, no injections occur and in the on state injections are Bernoulli with rate r_1.

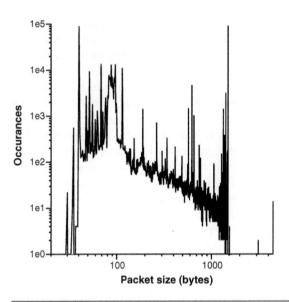

Figure 24.4   A distribution of Internet (IP) packet lengths captured at the University of Memphis on October 4, 2002.

Therefore, the injection rate of this MMP is

$$r = r_1 x_1 = \frac{\alpha r_1}{\alpha + \beta}.$$

A more complex MMP is studied in Exercise 24.1.

The basics of traffic patterns were presented in Section 3.2, which leaves the determination of packet lengths. A straightforward approach for choosing packet lengths is to use a packet length distribution captured from an application-driven workload, as described in Section 24.2.1. Figure 24.4 shows an example of such a trace gathered from an IP router. Then, for each injected packet, the corresponding length is chosen with a probability equal to the fraction of occurrences of that packet length in the captured distribution.

24.3 **Simulation Measurements**

When estimating network performance, there are two main sources of error: *systematic error* and *sampling error* [66]. Systematic errors are errors introduced by bias in the measurements or simulation itself and for network simulation are generally a result of the initialization of the simulator. By choosing an appropriate *warm-up period* for a simulation, as discussed in Section 24.3.1, the impact of systematic errors can be minimized.

Once the network is warmed up, it has necessarily reached a *steady state*. That is, the statistics of the network are stationary and no longer change with time. At this point, the focus shifts to sampling the network so that an accurate estimate of a particular network parameter can be determined. Two common sampling approaches, the *batch means* and *replication* methods, are presented in Section 24.3.2. Using one of these methods combined with an confidence interval (Section 24.3.3) for the measurement provides a rigorous, statistical approach to both measure a network parameter and also to assess the accuracy of that measurement.

24.3.1 Simulator Warm-Up

For simplicity, most simulators are initialized with empty buffers and idle resources before any packets are injected. While easy to implement, this introduces a systematic error into any measurement of the network. For example, packets that are injected early in the simulation see a relatively empty network. These packets have less contention and therefore traverse the network more quickly. However, as buffers begin to fill up, later packets see more contention, increasing their latencies. Over time the influence of the initialization becomes minimal, and at this point the simulation is said to be *warmed up*. By ignoring all the events that happen before the warm-up point, the impact of systematic error on measurements can be minimized. Unfortunately, there is no universal method for determining the length of the warm-up period, but most approaches follow the same basic procedure:[1]

1. Set the initial warm-up period T_{wu} based on a heuristic.
2. Collect statistics, ignoring samples during the estimated warm-up period.
3. Test the remaining samples to determine if they are stationary. If stationary, use T_{wu} as the warm-up period. Otherwise, increase T_{wu} and repeat steps 2 and 3.

Although many complex approaches have been proposed for the details of the above procedure, we describe a relatively simple approach that tends to work well in practice. First, the initial estimate of the warm-up period is picked to be a number of events that can be quickly simulated (on the order of 100 to 1,000 events). This small initial guess limits the overhead of estimating the warm-up period for simulations that quickly reach steady state.

Statistics are collected from a simulation run, ignoring events during the warm-up period. For the example we discuss here, we are estimating average packet latency, so batch averages of the events are computed. That is, a single sample point is the

---

1. This procedure is adapted from [142].

average of many individual packet latencies. The number of samples per average is called the batch size and should be statistically significant: at least 30 to 50 samples. For our example, the batch size is 100, so the first sample represents the first 100 packet arrivals, the second the second 100 arrivals, and so on.

Using the samples after the warm-up period, a linear fit of the data is performed. If the line is approximately flat within some predetermined precision, the network is considered in steady state. Otherwise, the warm-up is lengthened and the procedure repeated.

Figure 24.5 shows an example time-evolution of the packet latency in a mesh network. Even with a batch size of 100 samples, a single simulation does not seem steady after the first 6,000 packet arrivals. This is simply because we are sampling a random process, which naturally introduces some sampling error. By averaging several independent simulator runs into an *ensemble average*, this sampling error can be greatly reduced. Figure 24.5 shows that with an ensemble of 10 independent simulation runs, the system appears to stabilize after around 4,000 sample points.

While the warm-up period can be estimated from a single simulator run, an ensemble of several runs will greatly reduce the chance of underestimating the warm-up period.[2] Of course, this comes at the expense of additional simulation time.

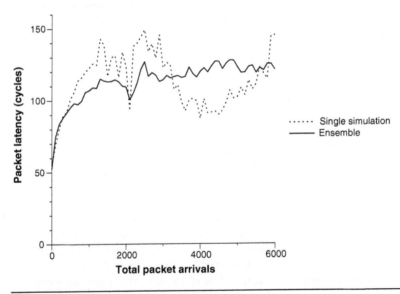

Figure 24.5 Average packet latency in an 8-ary 2-mesh at 70% capacity. Each sample is a time average of 100 individual packet latencies and the ensemble average is composed from 10 independent simulation runs.

---

2. It is generally not useful to create an ensemble of more than 20 to 30 independent runs.

24.3.2 **Steady-State Sampling**

Given a simulation that has reached steady state after an appropriate warm-up period, the next step is to measure the underlying process. We present two basic approaches for making these measurements, the *batch means method* and the *replication method*, both of which are reasonable techniques to use in a network simulator.

In the batch means method, measurements are taken from a long, single simulation run. Like the batching used to smooth the samples when determining the warm-up period, the samples are split into many batches and statistics are first accumulated for these batches. Unlike our warm-up batches though, the batch size used for this approach is selected based on the entire length of the simulation so that there are 20 to 30 batches total. Given a set of observations $\{X_0, X_1, \ldots, X_{n-1}\}$ from a single simulation, individual batches B are created based on the number of batches k and batch size s. For simplicity, we let $n = sk$ and then batch means are computed as

$$\bar{B}_i = \frac{1}{s}\sum_{j=0}^{s-1} X_{si+j}, \quad 0 \le i < k.$$

Then the sample mean is simply the mean of the batch means:

$$\bar{B} = \frac{1}{k}\sum_{i=0}^{k-1} \bar{B}_i.$$

The usefulness of batching might be unclear at this point because the overall mean we have computed $\bar{B}$ is equal to the mean of original samples. However, as we will see in the next section, an estimate of the standard deviation of the underlying process is useful for accessing the quality of our sample mean. For batch means, an estimate of the standard deviation σ_S is

$$\sigma_S^2 = \frac{1}{k-1}\sum_{i=0}^{k-1}(\bar{B} - \bar{B}_i)^2.$$

Since each sample in the batch means method is an average over many of the original samples, the variance between batch means is greatly reduced, which in turn reduces the standard deviation of the measurements. This leads to better confidence in our estimates of the mean.

The main drawback of the batch means approach comes from an analysis of its statistical implications. Ideally, one would want each batch collected to be independent from all the other batches. Otherwise, measures such as the overall mean $\bar{B}$ could be biased by correlation between the batches. In reality, batches from a single simulation run are not independent because the packets queued in the network at the end of one batch remain in the network at the beginning of the next batch. This effect can be reduced by making the batches as large as possible, which explains the choice of using only 20 to 30 total batches.

An alternative sampling approach, the replication method, solves the problem of correlation between batches by taking samples from many independent runs of the network simulator. Instead of a single long run, as in batch means, several smaller runs are collected. Generally, using more than 30 total runs yields little improvement in accuracy; rather, the length of the individual runs should be increased. Also, it is important that each individual run contain at least hundreds of samples to ensure statistical normality. For the kth simulation run and its observations $\{X_0^{(k)}, X_1^{(k)}, \ldots, X_{n-1}^{(k)}\}$ a single mean is computed:

$$\bar{R}_k = \sum_{i=1}^{n-1} X_i^{(k)}, \quad 0 \le k < r.$$

The sample mean $\bar{R}$ is then the mean of the means from the individual runs

$$\bar{R} = \sum_{k=0}^{r-1} R_k$$

and the standard deviation can be estimated as

$$\sigma_S^2 = \frac{1}{r-1} \sum_{k=0}^{r-1} (\bar{R} - \bar{R}_k)^2.$$

While replication eliminates correlation between the mean samples R_i, it is more susceptible to systematic errors introduced by initialization. Unlike the batch means method that has only a single initialization to affect the samples, many initializations are performed in the replication method. To minimize the impact of this on the replication means, it is suggested that 5 to 10 times more samples be collected in each run as were discarded during the warm-up period.

24.3.3 Confidence Intervals

Once a designer has collected samples of a simulation, either through the batch means or replication method, the next logical question is: How well does this data represent the underlying process? Confidence intervals are a statistical tool that let us quantitatively answer this question. Given a particular group of samples, a confidence interval is a range of values that contains the true mean of the process with a given level of confidence.

To compute a confidence interval, it is assumed that the underlying process being measured is stationary and normally distributed. For network simulations, the warm-up procedure described in the previous section ensures that the process is stationary. While we cannot assume the process being measured is normally distributed, the central limit theorem tells us that the cumulative distribution of many observations of an arbitrary random process approaches the normal distribution. Because most simulations capture hundreds to thousands of samples, this is an issue that can generally be ignored.

For a set of n observations of the stationary process $\{X_0, X_1, \ldots, X_{n-1}\}$, the sample mean $\bar{X}$ is computed as

$$\bar{X} = \frac{1}{n} \sum_{i=0}^{n-1} X_i$$

and is our best guess at the actual mean of the process $\tilde{X}$. A $100(1 - \delta)$ percent confidence interval then bounds the range in which the actual mean falls in terms of the sample mean:

$$\bar{X} - \frac{\sigma_S t_{n-1,\delta/2}}{\sqrt{n}} \leq \tilde{X} \leq \bar{X} + \frac{\sigma_S t_{n-1,\delta/2}}{\sqrt{n}}.$$

This equation tells us that there is a $100(1 - \delta)$ percent chance that the actual mean of the parameter being measured $\tilde{X}$ is within the interval defined by the sampled mean $\bar{X}$ plus or minus an error term. As shown, the error term falls off as $n^{-1/2}$ and is also proportional to both the sample's standard deviation σ_S and the parameter $t_{n-1,\delta/2}$. The standard deviation can be estimated directly from the samples as

$$\sigma_S^2 = \frac{1}{n-1} \sum_{i=0}^{n-1} (\bar{X} - X_i)^2$$

or from the standard deviations corresponding to the sampling methods discussed in Section 24.3.2.

The last parameter of the error term $t_{n-1,\delta/2}$ is *Student's t-distribution* and roughly accounts for the quality of our estimate of the standard deviation based on the number of samples used to compute σ_S minus one $n - 1$, often referred to at the degrees of freedom, and the half confidence interval $\delta/2$. Several values of t are given in Table 24.1 for 95% and 99% confidence intervals. More values of t can be found in standard mathematical tables, such as those in [201].

Table 24.1 Student's *t*-distribution for 95% and 99% confidence levels.

$n - 1$	95%	99%
1	6.3137	31.821
5	2.0150	3.3649
10	1.8125	2.7638
50	1.6759	2.4033
100	1.6602	2.3642
∞	1.6449	2.3263

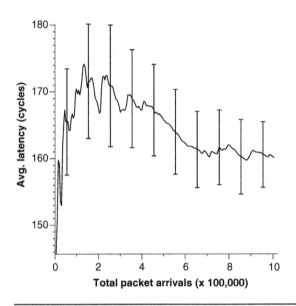

Figure 24.6   Average packet latency and 95% confidence intervals in an 8-ary 2-mesh near saturation sampled using the batch means method (30 batches total) vs. the number of arrivals sampled. The upper and lower limits of the confidence intervals are indicated by the level of the horizontal stops at the ends of the vertical lines. Simulation starts after a 5,000 arrival warm-up.

Figure 24.6 shows the confidence intervals given by the batch means method on average latency versus the number of packet arrivals sampled. At each intermediate confidence interval shown, all samples up to that point are split into 30 equal-sized batches. While the convergence of the confidence intervals may appear slow, the interval size follows the expected $n^{-1/2}$ shape. The slow convergence is also explained by the fact that the network under test is operating near saturation. In terms of relative confidence, the intervals close to 5% of the average latency in approximately 300,000 arrivals while taking nearly 1,000,000 arrivals to close to roughly 2.5%. Also note that the final average latency is not within the confidence range of the of first several confidence intervals, indicating the negative affects of correlation between batches on the accuracy of the intervals.

24.4 Simulator Design

Although a general discussion of discrete-event simulator design is beyond the scope of this book, there are several basic areas that any network simulator designer or user should understand. We begin with a simple explanation of two basic approaches to simulator design, *cycle-based* and *event-driven* simulators. We then address an issue specific to network simulation — modeling the infinite source queues used to decouple the injection process from the network. Next, almost any simulator needs a source of random numbers to model the injection processes and we discuss the key issues associated with random number generation. Finally, we provide some practical advice for designers encountering unexpected behavior from their simulator.

24.4.1 Simulation Approaches

There are two common approaches for designing a network simulator: *cycle-based* and *event-driven* simulation. To explain both design approaches and highlight their differences, we will use both to model a simple output-queued router node. In this node, arriving packets are immediately forwarded to a single output queue corresponding to their output port. The output queues have enough write bandwidth to simultaneously accept packets from all the input ports. For simplicity, the total delay of the node is assumed to be one simulator cycle, and an infinite buffer will also be used.

In cycle-based simulation, time proceeds in two phases, generally, one phase is loosely associated with reading global state and the other with writing that state. By separating the simulation into reading and writing phases, any procedure that needs to read the global state can be invoked before the procedures that can update this state.

To illustrate this concept, consider our example of the shared memory switch modeled using cycle-based simulation (Figure 24.7). At the beginning of a simulation cycle, packets are arriving at each node and the first phase of the simulation occurs. Any function that reads the global state, such as `ReadInputs` in our case, must be registered with the simulator and is invoked during this first phase. For this example, `ReadInputs` is called once per node simply to read any packets arriving at the inputs of the node and store them in the appropriate queue.

In the second phase of the cycle, the functions that write global state are invoked — `WriteOutputs` for this example. `WriteOutputs` selects an appropriate packet for each output and writes it to the inputs of the next node to be read by `ReadInputs` in the subsequent cycle. This simple two-phase procedure is repeated for the duration of the simulation. While the definition of the functionality of either stage is loose, the critical invariant is that all the functions within a phase can be evaluated in any order without changing the outcome of the simulation. This is explored further in Exercise 24.6.

An alternative to cycle-based simulation is event-driven simulation. Unlike cycle-based simulations, event-driven simulations are not tied to a global clock, allowing for significantly more flexibility in modeling. This is especially useful when the underlying design is asynchronous. Event-driven simulations are built on a very simple framework of individual *events*. Each event is a data structure with three fields: an invocation time, an action (function call), and data (function arguments). Simulation proceeds by creating a *event queue* of all pending events, sorted by their execution

```
void ReadInputs ( int node ) {
  Packet *p;
  int input, output;

  // Visit each input port at the node and read
  // the arriving packets

  for ( input = 0; input < NumInputs ( ); inputs++ ) {
    p = ReadArrival ( node, input );

    if ( Valid ( p ) ) {
      output = OutputPort ( node, p );
      AddToQueue ( node, output, p );
    }
  }
}

void WriteOutputs ( int node ) {
  Packet *p;
  int output;

  // Visit each output queue at the node and select
  // a packet to be forwarded to the next node

  for ( output = 0; output < NumOutputs ( ); outputs++ ) {
    if ( ! OutputQueueEmpty ( node, output ) ) {
      p = SelectFromQueue ( node, output );
      WriteOutgoing ( node, output, p );
    }
  }
}
```

Figure 24.7   A C-like code snippet for a simple cycle-based simulation of a network of output queued switches.

times. The pending event with the lowest time is removed from the list and its corresponding action is invoked. Event actions can update state as well as issue future events that occur as a result of the given action.

To better understand the nature of event-driven simulation, the same output queued switch design considered in the cycled-based case is shown in Figure 24.8 written for an event-driven simulator. Simulation begins with the creation of an `Arrival` event for a packet at a given node. As before, the arriving packet is queued corresponding to its output port, but in the event-driven simulator, we also need an event to trigger the scheduling of output. Our implementation gives this responsibility to the first packet that arrives in the queue to ensure that a non-empty queue

```
void Arrival( int node, Packet *p ) {
  int output;

  // Add the arriving packet to the queue
  // corresponding to its desired output

  output = OutputPort( node, p );

  // If the queue is empty, add an
  // output scheduling event

  if ( OutputQueueEmpty( node, output ) ) {
    AddEvent( 1, ScheduleOutput, node, output );
  }

  AddToQueue( node, output, p );
}

void ScheduleOutput( int node, int output ) {
  Packet *p;
  int    next;

  // Select a packet from the output queue
  // and forward it to the next node

  p    = SelectFromQueue( node, output );
  next = DownStreamNode( node, output );

  AddEvent( 1, Arrival, next, p );

  // If the output queue still contains
  // packets, another ScheduleOutput event
  // should occur in the next cycle

  if ( ! OutputQueueEmpty( node, output ) ) {
    AddEvent( 2, ScheduleOutput, node, output );
  }
}
```

Figure 24.8 A C-like code snippet for a simple event-driven simulation of a network of output queued switches.

always gets a scheduling event and no duplicate events are created. The new event is placed on the event queue by the AddEvent call. The first argument gives the time the event should be invoked relative to the current time which for this example is one cycle from the current time. The `AddEvent` call takes the action name, `ScheduleOutput`, followed by its arguments.

The `ScheduleOutput` action is responsible for selecting a packet from the output queue and forwarding it to the next node in the network. The forwarding is performed by creating another `Arrival` event for the packet at the next node. One more step is necessary to ensure that the output is scheduled any time the output queue is empty. If any packets remain in the queue, another `ScheduleOutput` event is scheduled two time units in the future. The two-time-unit space is required because we have synchronized all arrivals on every other time step, leaving the remaining time steps for output scheduling, much like the two-phase approach used in the cycle-based simulator. Again, any events assigned to the same time step must be able to be performed in any order without affecting the correctness of the simulation.

Unlike the simple two-phase simulation loop of a cycle-based simulator, an event-driven simulation must maintain a sorted list of events in order to determine the next event to be executed. Additionally, insertion into this list should be efficient. This naturally leads to a heap-based implementation. A simple binary heap gives $O(\log n)$ insertion and removal of the next event, but specialized data structures designed specifically for event-driven simulators yield even more efficient implementations. For example, the calendar queue [31] gives $O(1)$ insertion and removal of the next event.

24.4.2 **Modeling Source Queues**

As discussed in Section 23.1, the standard open-loop measurement setup for an interconnection network decouples each injection process from the network with an infinite queue. Although the presence of this queue does allow independent control of the injection process, it can cause some practical problems for networks that are being simulated beyond saturation. In this case, the source queue's size becomes unbounded and stores a number of packets roughly proportional to the length of the simulation. If the simulator naively allocated memory for every packet in this queue, the memory footprint of the simulation would also be unbounded. To further complicate any attempt to compactly represent the input queues, it could be necessary to track the exact age of the packets in the queue, which would be needed, for example, by a flow control technique that uses age-based priorities.

In a straightforward implementation, the injection processes are operated in lock-step with the rest of the network simulation. During each simulator cycle, the injection process has an opportunity to add a new packet to its corresponding injection queue. However, for any given cycle, the network sees only the first packet in any source queue and additional queued packets do not immediately affect the simulation. This observation leads to a simple and efficient solution for modeling the source queues without explicitly storing all the packets. Instead of forcing each injection process to be synchronized with the network, they are allowed to lag behind, operating in the past. Then the time of the injection process is advanced only when its corresponding source queue becomes empty.

An example of the operation of a source injection queue is shown in Figure 24.9. A portion of the injection process history is shown as a segment of tape, with slots corresponding to each simulation cycle. If the slot is empty (grayed) no packet is injected during that cycle; otherwise, the slot is filled with a packet to be injected. Similarly, a source queue is shown as grayed when empty and labeled with the corresponding packet name when occupied. Because of the lagging injection process, the source queue needs enough storage for only a single packet.

The example begins with both the network time and injection process time synchronized at cycle 0 and no packet in the source queue (Figure 24.9[a]). As the network time is advanced, the state of the source queue is checked. Since it is empty in this case, the injection process time is also advanced until a packet is injected or

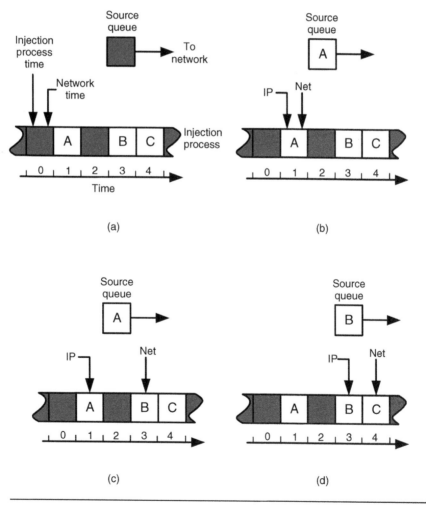

Figure 24.9 Example operation of a lagging injection process.

it matches the network time. This causes packet A to be injected and the injection process time to advance one cycle (Figure 24.9[b]). This procedure is repeated for cycles 2 and 3, but since the source queue is still occupied by packet A, the network time advances two cycles and the injection process time remains fixed at cycle 1 (Figure 24.9[c]). Packet A leaves the source queue during cycle 3, leaving the queue empty at the beginning of cycle 4. The injection process time is first advanced to cycle 2, but since no packet is injected, it continues to cycle 3, causing packet B to be injected (Figure 24.9[d]).

24.4.3 Random Number Generation

Because many of the traffic processes used in network simulations are stochastic, a network simulator needs to generate random numbers in order to properly implement these sources. For example, the following C-like pseudo-code models a typical Bernoulli injection process with rate r:

```
if ( random( ) < r ) {
    inject_packet( );
}
```

The function `random` generates a floating-point value uniformly distributed between 0 and 1. This code might look deceivingly simple; however, we have hidden the sticky issue of actually generating the random number.

First, there are several methods for generating *truly random* numbers that are employed in many digital systems, especially those involved in security and encryption. These approaches generally sample a natural random noise source, such as thermal noise (Johnson noise) in a resistor or a reverse-biased diode. However, for practical reasons, most software opts instead for *pseudo-random* number generators (PRNGs).

Most software PRNGs have the same basic structure. The state of the PRNG is stored using a *seed* value. Then, when the user requests a new random value, the PRNG algorithm computes both the "random" value and a next state. The value is returned to the user and the seed is updated with the next state. Of course, there is nothing actually random about this process, but as long as the PRNG algorithms are chosen appropriately, the outputs from the function will meet many statistical tests of randomness.

Unfortunately, for some of the most common random number generators available to programmers, the PRNG algorithms are sub-optimal. Notable examples are the `rand` and `drand48`, which are part of many C libraries. The low bits generated by `rand` tend to have small periods and are not very random. Although `drand48` is a slight improvement, it still suffers from some significant problems. Instead, designers should adopt random number generators that have a rigorous theoretical justification in addition to empirical verification. Good examples include Knuth's [101], Matsumoto and Kurita's [121], and Park and Miller's [141].

Finally, a good PRNG can have an important practical advantage over a truly random number source. Because PRNGs are deterministic, as long as their initial seed is fixed, the sequence of random numbers, and therefore the behavior of the program, is also deterministic. This allows randomization between runs by changing the seed value, but also allows repeatability of runs for recreation of specific results and for debugging.

24.4.4 **Troubleshooting**

As with most simulators, the best methodology for using them is to *first* develop an intuition and back-of-the-envelop or modeling calculation for the results that you *expect*. Then use the simulator to verify this intuition. If you follow this procedure, most of the time your results will be close to that of the simulation. However, there will be disagreements from time to time and there are some general techniques for tracking down these problems in a network simulator:

- Verify that you are simulating your network with enough detail.
- Check for unfairness or load imbalance between resources. This is an especially common problem and is often overlooked by simple models.
- Narrow down the problem by identifying the bottleneck resource. For example, if you increase the flow control resources (such as the number of virtual channels and buffer sizes), but a network still under-performs a throughput estimate, then the flow control is probably not the culprit. Now continue with the routing function, and so on.
- Gather statistics. Is all the traffic going where it is supposed to? Are a few packets staying in the network much, much longer than the majority? Anomalies in statistics are often a good clue.
- Configure the simulator to a simple case, such as uniform traffic, Bernoulli arrivals, low offered load, and large buffers, so it can be compared to an analytical solution.
- Simulators can have bugs, too. Compare against another simulator or delve into your current simulator's code.

24.5 **Bibliographic Notes**

For more information on the basics of simulation, including simulator design, input processes, measurements, and PRNGs, both [29] and [109] are excellent references.

There are many sources for application-driven workloads. Parallel computing benchmarks such as SPLASH [169, 194] or database benchmarks [77] are possible tests for processor interconnection networks. Internet packet traces are widely available on-line from a varity of sources.

Using MMPs to model voice and data traffic is discussed by Heffes and Lucantoni [81], while Jain and Routhier introduce packet trains as another technique for modeling burstiness [90]. However, it is important to note that even models that capture burstiness can be quite simple compared to the statistics of real traffic flows. For example, Leland et al. observed that Ethernet traffic can have a self-similar nature, which cannot modeled at all in many simple models [202]. The impact of packet size on interconnection network performance is addressed by Kim and Chien [96], who found that the interaction between long and short messages can have a significant effect.

Knuth [101] was an early pioneer in the rigorous analysis of PRNGs. The survey paper by Park and Miller [141] introduced a *minimal standard* PRNG, which still stands as a well-tested and adequate technique. More recent advances in PRNG are surveyed in [111]. Many implementations of truly random number generators exist — for example, [149] uses a large on-chip resistor as a noise source.

24.6 **Exercises**

24.1 *Four-state MMP.* Compute the average injection rate r for the four-state MMP shown in Figure 24.10. Explain how this process is related to the simple on-off process shown in Figure 24.3.

24.2 *Performance of a (σ, ρ) regulator.* Consider the on-off MMP in Figure 24.3 as the unregulated source to a (σ, ρ) regulator (Section 15.2.1). Let $\sigma = 1$ and $\rho = 1/2$, with one token created every other time slot. What conditions on the "on" injection rate r_1 of the injection process are necessary so that this system is stable (that is, the packet queue in the regulator has a finite expected size)? Create a Markov chain model of this system to determine the average size of the regulators packet queue for $\alpha = \frac{1}{4}$ and $\beta = \frac{1}{8}$. Simulate this model for different values of r_1, record the average queue size and use Little's law, Equation 23.6, to also find the expected delay of the regulator.

24.3 *Correlated workloads.* Although the three aspects of network workloads were presented as independent in this chapter, many network applications could have strong correlations between several of these aspects. For example, longer packets may be more infrequent than shorter packets when communicating between a particular

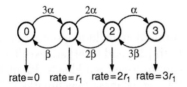

Figure 24.10 A four-state MMP.

source-destination pair. Describe an application in which this situation may arise and propose an approach for incoporating this correlation into the injection process.

24.4 *Fairness of a coin.* You are given a coin and asked whether the coin is *fair*. That is, does the coin indicate heads 50% of time and tails the other 50%? The first 11 flips of the coin give the sequence

$$\{H, T, T, T, H, T, H, H, T, H, T\},$$

which contains 5 heads (H) and 6 tails (T).

(a) Can you say, with 95% confidence, that this coin is nearly fair (49% to 51% chance of heads)? If not, is there any property of the coin you can say with the same confidence? Since the number of degrees of freedom in this example is $11 - 1 = 10$, the Student's t-distribution in Table 24.1 will be helpful. It may also be useful to assign a value of 1 to a flip of heads and 0 to tails.

(b) If you flip the coin 90 more times (101 total samples) and see 36 more heads (41 total) can you make a stronger statement about the fairness of this coin?

24.5 *Fairness of a die.* Perform a similar experiment to the one described in Exercise 24.4, but replace the coin with a fair die. Roll the die until you are 95% confident that the mean value is $(1+2+3+4+5+6)/6 = 3.5$ within 0.05. How many rolls did it take? (It will be faster to simulate your die with a PNRG.) Use Student's t-distribution for an infinite number of degrees of freedom when computing the confidence intervals. Are you also confident that the die is fair — that is, are all sides equally likely? If not, explain how you could be.

24.6 *A single-phase simulation.* Consider a "single-phase" simulation approach for the output queued switch example in Figure 24.7 where the `ReadInputs` and `WriteOutputs` functions are merged. Explain how problems in determining the order in which to evaluate nodes could arise. Are there situations in which this single-phase approach could work? Hint: Consider the topology of the network under simulation.

24.7 *A lagging injection process.* Write pseudo-code for a lagging injection process described in Section 24.4.2 to be called once per network cycle. Use the following functions to access the state of the network: `sourceq_empty()` returns true if the source queue does not contain a packet and false otherwise, `get_net_time()` returns the current network time, and `inject_packet()` runs the injection process for a single cycle and returns true if a packet was injected, false otherwise. Assume the lagging source queue's time is stored in the variable `q_time`.

24.8 *Quality of a PRNG.* The following code implements a common type of PRNG known as a multiplicative linear congruential generator.[3]

---

3. This particular generator was used in a FORTAN library as mentioned in [141].

```
int seed;

int random( ) {
    seed = ( 65539*seed ) % 0x8000000L;
    return seed;
}
```

The seed value is set before first calling the random routine. Also note that % is C's modulo operator and that the hexadecimal value 0x8000000 is 2^{31}.

A typical use of this code in a network simulator might be to select a destination node in a 64-node network under uniform traffic:

```
dest = random( ) % 64;
```

Implement this random number generator and its corresponding call to generate random destination nodes. Comment on the "randomness" of the destinations created. Hint: This is *not* a good random number generator.

CHAPTER 25

Simulation Examples

Now that we have discussed aspects of network design along with the tools of network simulation, several simulation examples are presented. These examples are not meant as a detailed study of any particular aspect of network or router design. Rather, they are designed to both introduce several useful experiments that can be performed on a typical interconnection network and emphasize some interesting and perhaps counter-intuitive results.

All simulations in this chapter were performed with the detailed, flit-level simulator described in Appendix C. Unless otherwise stated, the routers are input-queued with an input speedup of 2 and virtual-channel flow control. There are 8 virtual channels per input port and each virtual channel contains 8 flit buffers, for a total of 64 flits of buffering per input port. All packets are 20 flits in length. Both virtual-channel and switch allocation is performed using the iSLIP algorithm. Realistic pipelining is assumed and the per-hop latency of the routers is 3 cycles.

25.1 Routing

As we have seen in previous chapters, routing is a delicate balance between low latencies at low offered traffic and a high saturation throughput as traffic increases. We first focus on the latency of routing algorithms and examine the connection between our simple metrics of zero-load latency and ideal throughput and the actual performance of the network. Interestingly, different algorithms achieve different fractions of their ideals. In addition to typical aggregate latency measures, we also examine the distribution of message latencies induced by different routing algorithms. The second set of experiments focuses completely on the throughput of routing algorithms and compares the performance of two algorithms on a random sampling of traffic patterns.

495

25.1.1 **Latency**

In this set of experiments, the impact of routing from a latency perspective is explored on an 8-ary 2-mesh. We begin with perhaps the most common graph in interconnection networks research — latency versus offered traffic under uniform traffic, shown in Figure 25.1. The graph compares the performance of four routing algorithms: dimension-order routing (DOR), the randomized, minimal algorithm described in Section 9.2.2 and [135] (ROMM), Valiant's randomized algorithm (VAL), and a minimal-adaptive routing algorithm (MAD). The MAD implementation is created with Duato's algorithm using dimension-order routing as the deadlock-free sub-function.

At low traffic, zero-load latency gives an accurate estimate of the simulated latencies. Using the time it takes a flit to traverse a channel as our definition of cycles, the router model has a delay of $t_r = 3$ cycles and the serialization latency is 20 cycles because the packet length is 20 flits. So, for example, the minimal algorithms have a zero-load latency of

$$T_0 = t_r H_{avg} + T_s = 3 \left(\frac{16}{3} \right) + 20 = 36 \text{ cycles,}$$

as shown in the figure. Similarly, the zero-load latency of VAL is computed as 52 cycles. Of course, as traffic increases, the contention latency begins to dominate and the vertical asymptotes of the latency curves are determined by the saturation throughputs of the different routing algorithms.

Since we are accurately modeling flow control, the routing algorithms saturate at a fraction of their ideal throughputs. For the minimal algorithms (DOR, ROMM, and MAD), the ideals are 100% of the network's capacity, while Valiant would ideally

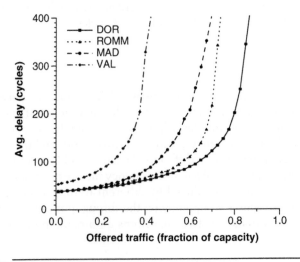

Figure 25.1   Performance of several routing algorithms on an 8-ary 2-mesh under uniform traffic.

achieve only 50% of capacity. Perhaps surprisingly, the simulation reveals that the algorithms achieve different fractions of their capacity. For example, DOR nears 90% of its ideal, with ROMM and MAD only reaching around 75% of theirs. The reason for this difference is because we have partitioned the virtual channels to avoid deadlock for the ROMM and MAD algorithms. DOR is naturally deadlock-free in the mesh; therefore, each route can freely use any of the virtual channels. As with most partitionings, this creates the opportunity for load imbalance, which in turn reduces the achieved throughput of the ROMM and MAD algorithms. Although VAL also requires partitioned resources to avoid deadlock, it still achieves about 85% of its ideal do to its natural load-balancing properties.

Figure 25.2 shows another latency vs. offered traffic curve for the same topology and same four routing algorithms, but with the transpose traffic pattern. This asymmetric pattern is more difficult to load-balance as demonstrated by the poor performance of DOR, which saturates at about 35% of capacity. ROMM fares better, improving the throughput to roughly 62%. However, MAD outperforms all the algorithms, more than doubling the throughput of DOR and saturating past 75% of capacity. VAL beats DOR in this case, again reaching about 43%. Performance on difficult traffic patterns such as transpose is often important for networks, but so is the performance on easy, local patterns such as neighbor traffic, as shown in Figure 25.3. Again, while the minimal algorithms all have the same ideal throughput on the neighbor pattern, DOR's simplicity and natural deadlock freedom give it an advantage over the more complex ROMM and MAD algorithms. As expected, VAL performs the same as in the previous two traffic patterns.

In addition to using aggregate latencies, as in the previous three simulations, individual packet latencies can give insight into the range and distribution of latencies observed in a particular simulation. For example, Figure 25.4 shows a distribution

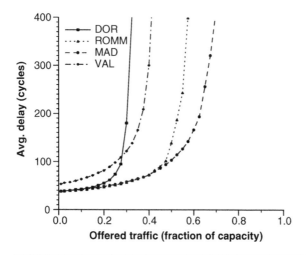

Figure 25.2   Performance of several routing algorithms on an 8-ary 2-mesh under transpose traffic.

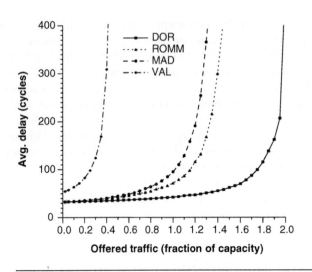

Figure 25.3   Performance of several routing algorithms on an 8-ary 2-mesh under neighbor traffic.

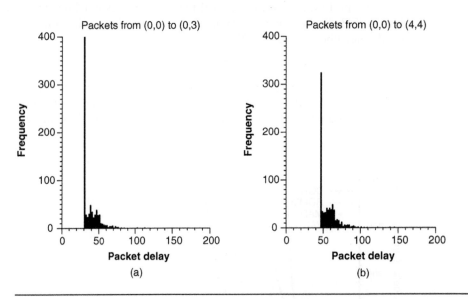

Figure 25.4   Latency distributions for (a) packets traveling between (0,0) and (0,3) and (b) between (0,0) and (4,4) in an 8-ary 2-mesh under uniform traffic with dimension-order routing and offered traffic at 20% of network capacity.

of packet latencies between two source-destination pairs taken from the previous simulation of an 8-ary 2-mesh under uniform traffic with dimension-order routing. The offered traffic is held at 20% of the network's capacity. At this low load, little contention is observed by the packets and most are delivered in the minimum number of cycles, denoted by the large spikes in the distributions at their left edges. The increased latency of the packets from (0,0) to (4,4) is simply due to their larger number hops.

The same simulation performed with Valiant's routing algorithm reveals a more interesting distribution (Figure 25.5). Because packets first travel to a random intermediate node, there is a wide range of path lengths. For each particular path length, a distribution similar to ones observed in dimension-order routing is created and the net distribution for Valiant's algorithm is simply a weighted superposition of many of these distributions. As seen in the figure, when routing from (0,0) to (0,3), most packets travel a non-minimal distance, giving the distribution a bell shape. The source and destination are further apart for the packets traveling from (0,0) to (4,4), which increases the chance the intermediate is in the minimal quadrant. When the intermediate falls within this quadrant, the overall path is minimal, explaining the shift of the distribution toward the left in this case.

25.1.2 **Throughput Distributions**

We now shift our focus exclusively to the throughput performance of particular routing algorithms. Although the standard traffic patterns used to test an interconnection network reveal the performance of the network at its extremes, these patterns do not necessarily give a good indication of the average behavior of a network. To remedy the limitations of using a few traffic patterns, the throughput of the network can be tested over a sampling of many random permutation traffic patterns.

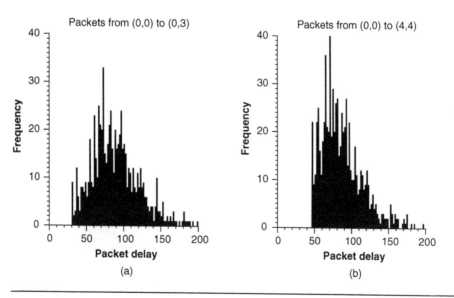

Figure 25.5   Latency distributions for (a) packets traveling between (0,0) and (0,3) and (b) between (0,0) and (4,4) in an 8-ary 2-mesh under uniform traffic with Valiant's routing algorithm and offered traffic at 20% of network capacity.

For this experiment, the throughput of an 8-ary 2-cube network is tested for both dimension-order and minimal adaptive routing over a sample of 500 random permutations. The resulting distribution of saturation throughputs for both routing algorithms is shown in Figure 25.6. Dimension-order routing's distribution shows two distinct peaks, centered near 27% and 31% of capacity, and has an average throughput of approximately 29.4% of capacity over the 500 samples. Minimal adaptive routing has a more even distribution with a single wide peak near 33% of capacity and an average throughput of about 33.3%. This throughput is approximately 13.3% greater than dimension-order routing, illustrating the potential benefits of adaptive routing.

25.2 Flow Control Performance

The choice of routing algorithm used in a network sets an upper-bound on the achievable throughput (ideal throughput) and a lower-bound on the packet latency (zero-load latency). How closely a network operates to these bounds is determined by the flow control mechanism used. The experiments presented in this section focus on the relationship between aspects of the network and its workload that can significantly affect the performance of a flow control mechanism.

25.2.1 Virtual Channels

In a typical design of a virtual channel router, a fixed amount of hardware resources is set aside to implement the virtual-channel buffers. The decision that needs to

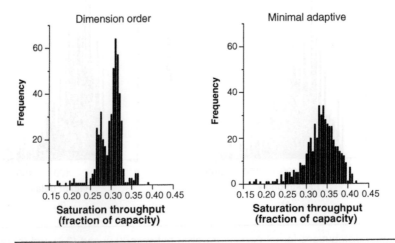

Figure 25.6 Distributions of saturation throughputs for dimension-order and minimal adaptive routing over a sampling of 500 random permutation traffic patterns on an 8-ary 2-cube.

be made is how to partition these resources to maximize the performance of the network. For example, do a few virtual channels, each with deep buffers, perform better than many virtual channels with shallower buffers?

The performance of several different virtual-channel partitionings is shown in Figure 25.7 for an 8-ary 2-mesh network. The total amount of buffering (the product of the number of virtual channels times the individual virtual-channel depth) is held constant across each of the configurations. Several trends can be observed in the experiment. First, the throughput of the network tends to increase as the number of virtual channels is increased. Although beyond the bounds of the graph, the saturation throughput of the 8-virtual-channel case is slightly higher than that of the 4-virtual-channel case. This fact is hidden by the second trend, which is that increasing the number of virtual channels tends to increase latency below saturation. The larger latency is simply a result of the increased interleaving of packets that occurs with more virtual channels, which tends to "stretch" the packets across the network. This interleaving effect can be reduced by giving priority in the switch allocator to packets that are not blocked and also won the allocation in the previous round. However, the designer must be careful to avoid both starvation issues and fairness problems that could arise with variable size packets, for example.

One exception to the throughput trend occurs for the 16-virtual-channel case. The fact that this configuration has a different zero-load latency is a key indicator of the underlying problem. Because our router model includes pipelining latencies, the buffer's credit loop latency is greater than one cycle. Furthermore, once the virtual-channel buffer depth is too small to cover this latency, the virtual channel can no longer sustain 100% utilization and stalls waiting for credits. This is the same effect

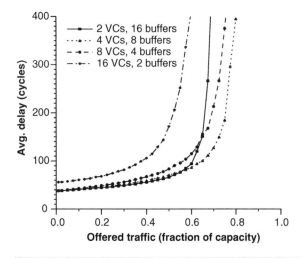

Figure 25.7 Latency vs. offered traffic for an 8-ary 2-mesh with various virtual channel partitionings under uniform traffic. The data labels indicate the number of virtual channels and the depth of and individual VC buffer.

described in Section 16.3. These credit stalls affect both the zero-load latency and saturation throughput, as shown.

Finally, while it is reasonable to approximate the hardware cost of virtual channels strictly by the total amount of buffering, it is not always the case that increasing the number of virtual channels is free in terms of latency. Generally, as the number of virtual channels increases, virtual-channel allocation time increases, which can affect the pipelining of the router. This, in turn, tends to increase the zero-load latency and may slightly decrease the saturation throughput because of the additional time required to reallocate a virtual channel. Pipelining issues, along with the depth required to cover the credit loop as mentioned above, often limit performance in partitionings that create close to the maximum number of virtual channels with very shallow buffers. However, there are other important issues, such as non-interference, that may still make these extreme partitionings attractive design points.

25.2.2 Network Size

The size of a network can have a significant effect on the fraction of its ideal throughput it can achieve, as shown in Figure 25.8. The figure shows latency vs. offered traffic curves for four different mesh networks under uniform traffic. It is important to note that the channel sizes and routers for each of the networks are exactly the same. This implies that the capacities of the networks are related by their radices. For example, the 4-ary 3-mesh and 4-ary 4-mesh have a capacity of $4b/k = b$ and the 8-ary 2-mesh has a capacity of $4b/k = b/2$, or half that of the radix-4 networks.

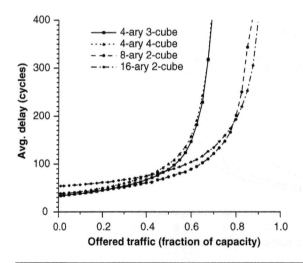

Figure 25.8   Latency vs. offered traffic for several mesh networks under uniform traffic with dimension-ordered routing. The injection processes of all networks are Bernoulli with a packet size of 20 flits. Individual channel widths and routers for each network are the same.

Although it might be natural to expect that different networks built from identical routers would achieve roughly the same fraction of capacity, the figure reveals a different trend. Instead of the achieved fraction of capacity being constant, it seems to be a function of the radix of the network. The radix-4 networks both begin to saturate at approximately 65%, while the radix-8 and radix-16 networks saturate near 80% and 83%, respectively. Further simulation confirms the trend of achieved capacity being determined by the network's radix.

One can explain these results by considering how the different network sizes interact with the flow control mechanisms. Using the fraction of capacity allows us to compare the curves of networks with different capacities, but it hides the absolute throughput injected by each node. From our previous remarks on the capacities of the networks, we know that the nodes in the radix-4 networks can ideally inject twice as much traffic before saturating than the nodes in the 8-ary 2-mesh and four times as much as the nodes in the 16-ary 2-mesh. Although the individual injection processes used for each node in the simulations are the same, the aggregate traffic is significantly different. For the radix-4 network, fewer nodes are injecting more traffic, while for the larger radix networks, more nodes inject less traffic. This difference in traffic greatly affects the flow control: a small number of intense sources generates more instantaneous load (burstiness) than a larger number of less intense sources. Mitigating this burstiness is the task of the flow control, and how effectively this can be done is largely a function of the amount of buffering at each node. However, because the networks all use identical routers, the smaller-radix networks are able to mitigate less burstiness and therefore achieve a lower fraction of their capacity.

25.2.3 Injection Processes

As mentioned in the previous experiment on the affects of network size, the burstiness of the underlying traffic can affect the efficiency of a network's flow control. We explore these ideas further in this section by explicitly varying the burstiness of the injection processes used in the network simulation.

Perhaps the simplest source of burstiness in networks is the size of the packets themselves. Even if a source is injecting at a very low average rate, the minimum unit of injection is a packet, which may contain many flits. This can be thought of as a burst of incoming flits all destined to the same node. The affects of packet size on network performance are shown in Figure 25.9.

The dominant trend in the data is both the increasing latency and decreasing throughput that comes from larger packet size. Larger packets are already handicapped in terms of latency because of their longer serialization overhead, as reflected in the zero-load latencies shown in the figure. Additionally, the flow control is not perfect and has more difficultly utilizing the resources as the packets get long. For example, when the packet size is 40 flits, it is spread across at least 5 routers because the buffer depth at each router is only 8 flits. If this packet becomes momentarily blocked, the resources of at least 5 routers also become blocked, accounting for the

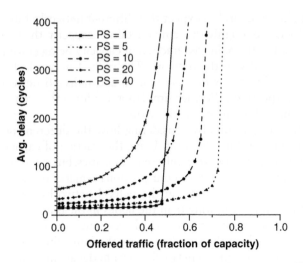

Figure 25.9　Latency vs. offered traffic in an 8-ary 2-cube under uniform traffic with dimension-ordered routing using different packet sizes. Each curve's corresponding injection process uses fixed-size packets and this packet size (PS) is indicated by its label in flits.

reduction in saturation throughput. This is in contrast to the the smaller packets, which are not spread across as many resources.

The outlier from the overall trend is the case in which the packet size is one (PS = 1). Because our router model adopts the conservative approach for reallocating virtual channels, shown in Figure 16.7(a) of Section 16.4, several cycles are necessary before a virtual channel can be reused. As the packet size gets smaller, more and more of the virtual-channel time is spent in this reallocation period. Because of this, the one flit packet greatly reduces the *effective* number of virtual channels in the routers. Further simulations confirm this observation and the anomaly for one-flit packets disappears when either the reallocation time is reduced or the number of virtual channels is increased.

The affect of the injection process is further explored by considering a mesh network with a two-state MMP, as discussed in Section 24.2.2. Packet lengths are again fixed at 20 flits. Figure 25.10 shows the performance of a network under several parameter values for the MMP. Each MMP has two parameters, α and β, that control both the spacing and duration of bursts. $\frac{1}{\alpha}$ sets the average spacing between "on" bursts. This gives an average spacing of 1, 200, and 400 cycles for the three curves. Similarly, the β parameter can be interpreted as one over the average duration of a burst period, so the first curve has an infinite burst period while the second and third curves have average burst periods of 100 and 50 cycles, respectively.

The infinite burst length of the first MMP means that it is always in the on state and therefore reduces to a Bernoulli injection process. Because of this, the α parameter of this MMP is arbitrary and does not affect the steady state. From the analysis in Section 24.2.2, the injection rate during the burst period is $1 + \beta/\alpha$ times the average injection rate. The larger the ratio of β to α, the more intense the injection

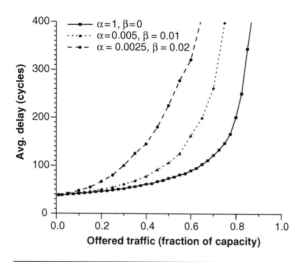

Figure 25.10 Latency vs. offered traffic in an 8-ary 2-mesh under uniform traffic with dimension-ordered routing and several MMPs. Each MMP is the simple two-state, on-off process shown in Figure 24.3 and the different α and β parameters used to generate each curve are shown in the labels.

rate is during the burst period. In the second MMP, $\beta/\alpha = 0.01/0.005 = 2$, so at an offered load of 40% of capacity, the process alternates between periods of no packet injections and Bernoulli injections at a load of 40(1+2) = 120% of capacity. As expected, this bursty behavior increases average latency and reduces the saturation throughput (Figure 25.10). The third MMP has shorter, more intense bursts, with $\beta/\alpha = 0.02/0.0025 = 8$. Consequently, the average latency and saturation throughput show further degradation vs. the second curve.

25.2.4 **Prioritization**

While most metrics of network latency focus on the aggregate or average latency of messages through the network, different flow control approaches can greatly affect the distribution of individual message latencies. Controlling these distributions is important for applications sensitive to worst-cast delay and jitter, fairness, or when different message priorities exist in the same network.

Consider a network that contains two message classes. One class might support real-time video traffic that requires both low delay and low jitter. Another class may contain latency-tolerant data transfers. To ensure that the high-priority video traffic maintains its low delay requirements, it is given absolute priority over any data transfer traffic.

Figure 25.11 shows a latency distribution from creating two priority classes in a 2-ary 6-fly network. For this experiment, 10% of the traffic is treated as high-priority, the remaining 90% is low-priority, and data is taken near saturation. To incorporate priority into the router model, separable allocators (Section 19.3) constructed from

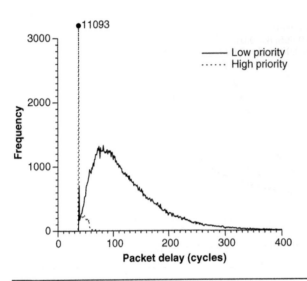

Figure 25.11   A latency distribution of packets in a 2-ary 6-fly with a two-level prioritization scheme. High priority packets (dotted line) make up approximately 10% of the traffic and have absolute priority over low-priority packets (solid line). Measurements are taken near saturation.

prioritized arbiters are used. The prioritized arbiters simply select the requester that has the highest priority and break ties using a round-robin policy.

The resulting distribution shown in the figure plots the number of occurrences for each particular latency. Despite the fact that the network is near saturation, 11,093 of the 15,612, or approximately 71%, of the high-priority messages sampled are delivered at the network's minimum latency of 37 cycles and almost 99% of the high-priority messages are delivered within 70 cycles. The distribution of the low-priority messages reflects the high average latency expected near saturation. Although over 98% of the low-priority traffic is delivered in 300 cycles, the tail of the distribution continues to over 700 cycles.

The level of differentiation between the two traffic classes is possible because the high-priority traffic compromises only a small fraction of the overall network traffic. Because this priority is absolute, the high-priority traffic sees a very lightly loaded network — less than 10% of capacity. Therefore, it experiences very little contention delay. However, as the fraction of high-priority traffic becomes larger, the differentiation continues to diminish. In an extreme where the high-priority traffic represents nearly all of the total traffic, there is little benefit to prioritization.

Another important message prioritization scheme approximates age-based fairness, which is described in Section 15.4.1. As in the previous example of a two priority levels, age-based fairness is implemented by using prioritized allocators in the routers. However, in this case contention between requesters is decided by granting access to the oldest of the requesting packets. A packet's age is measured as the number of network cycles that have elapsed since its initial injection into the network.

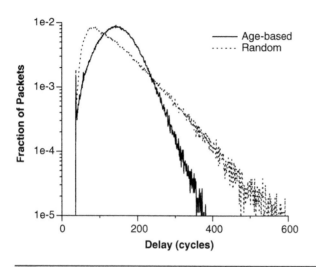

Figure 25.12   A latency distribution of packets in a 2-ary 6-fly with and without age-based arbitration. Measurements are taken near saturation.

The distribution of packet latencies in a 2-ary 6-fly with and without age-based arbitration is shown in Figure 25.12. As before, the network is operated near saturation as the measurements are taken. While more packets are delivered quickly (in less than about 50 cycles) when no age-based arbitration is used, the tail of the distribution is long. A significant number of packets take more than 600 cycles to deliver without age-based arbitration. In contrast, fewer packets are delivered quickly with age-based arbitration, but the tail of the distribution is kept tighter and most packets arrive within 400 cycles.

25.2.5 **Stability**

As a network approaches and exceeds its saturation throughput, the focus of the designer generally shifts from latency to the underlying fairness of the flow control technique. If a saturated channel is not fairly allocated between flows, an unstable network can result — some flows become starved and their throughput can drop dramatically as the load increases beyond saturation. Figure 25.13 shows the throughput of an unstable network along with two flow control mechanisms that implement fairness and ensure stability. To prevent greedy flows from masking starved flows, the figure shows minimum accepted throughput, as explained in Section 23.1.1.

The performance of the three flow control techniques is similar below saturation and all three reach saturation at approximately 43% of the network's capacity. Beyond saturation, the throughputs begin to diverge. When no fairness mechanisms are used, the throughput plummets to less than 5% of capacity as the offered traffic continues to increase. The unfairness is produced by an effect analogous to the parking lot example of Section 15.4.1 — packets that require fewer hops and

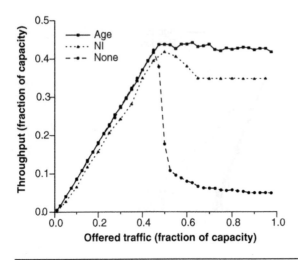

Figure 25.13   Throughput of the minimum flow vs. offered traffic for an 8-ary 2-mesh under bit-complement traffic using dimension-order routing. Several flow control techniques are shown: age-based arbitration (Age), a non-interfering network with a separate virtual channel for each destination (NI), and a network with no priority or isolation (None).

therefore fewer resource arbitrations get a higher proportion of the bandwidth of these resources. In contrast, the addition of age-based arbitration results in very stable throughputs beyond saturation. A non-interfering network, with one virtual channel per destination, is also stable beyond saturation, but does suffer some degradation in throughput before stabilizing at about 35% of capacity.

25.3 Fault Tolerance

For many interconnection networks, operation in the presence of one or more faults is an important attribute. Additionally, it is desirable for these networks to degrade gracefully in the presence of faults. An example of graceful degradation is shown in Figure 25.14.

In this experiment, an 8-ary 2-mesh network is simulated with a variable number of failed links (horizontal axis) using the fault-tolerant variant of planar-adaptive routing described in Exercise 14.8. For each number of failures, the saturation throughput of the network (vertical axis) under uniform traffic is measured. Since different arrangements of failed links may affect the saturation throughput more or less severely, each throughput point is an average of 30 different arrangements of failed links.[1] Along with the average, the sample's standard deviation is also plotted

---

1.   For simplicity of presentation, we generate faults in the network so that the resulting fault regions are convex. This ensures that all nodes in the network are still connected using planar-adaptive routing. See [36] for further information.

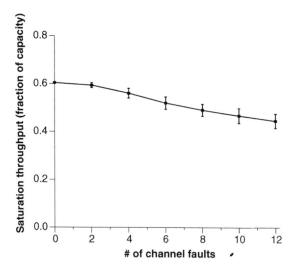

Figure 25.14 Saturation throughput of an 8-ary 2-mesh under uniform traffic vs. the number of failed links. Fault-tolerant planar-adaptive routing is used. Each point shows the average throughput over a sampling of random failures and the error bars indicate one standard deviation from the average.

using vertical error bars. The top and bottom of each error bar represents the average throughput plus and minus the standard deviation, respectively.

The throughput of the non-faulty network is just above 60% of capacity and the network's grace in the presence of a small number of faults is illustrated by the corresponding small drop in throughput. The network continues to remain resilient even as the number of faults grows to 12 with only a slight increase in the rate of throughput degradation. Also, the standard deviation slowly increases with the number of faults, indicating the potentially greater impact of a collection of faults vs. a single fault in isolation. For example, if many faults are clustered in a small area of the network, the number of channels that can access nodes in that area may be reduced, which, in turn, can significantly increase load on the remaining channels. As the number of faults increases, the chance of these clusters of faults nearly isolating a node also increase and, in extreme cases, the network may become partitioned such that there is no available path between some particular nodes.

APPENDIX A

Nomenclature

Symbol	Description	Unit of Measurement
b	Channel bandwidth, the product of signal frequency and channel width, $b = fw$.	bits/s
B_B	Bisection bandwidth, the sum of the bandwidth of the channels that cross a minimum bisection of a network.	bits/s
B_C	Channel bisection, the number of channels that cross a minimum bisection of a system.	channels
B_n	Node bandwidth, the sum of the bandwidth of the channels entering and leaving a node. This is the product of node pinout (wire capacity) and signal frequency, $B_n = fW_n$.	bits/s
B_s	System bandwidth, the bandwidth of the maximum number of wires that can cross the midpoint of the system. This is the product of system wire capacity and signal frequency, $B_s = fW_s$.	bits/s
C	The set of channels in a network.	
d	Duty factor.	
D	Distance, the physical length of the channels traversed along a path from source to destination.	m
δ	Node degree, the number of channels that terminate on a node.	channels
f	Frequency, the bandwidth per signal of a channel.	1/s

Symbol	Description	Unit of Measurement
F	Number of flit buffers per virtual channel.	
γ	Channel load, ratio of the bandwidth on a channel to the bandwidth on the network input, or, stated differently, the load placed on a channel by a unit load at the network input.	
H	Hop count, the number of channels traversed along a path from source to destination.	channels
l	Channel length, the physical length of a channel.	m
λ_{xy}	Routing probability, the fraction of all traffic that routes from node x to node y.	
L	Packet length.	bits
N	The set of nodes in a network.	
N	The number of terminal nodes in a network.[1]	
P	the number of ports on a router.	
R_{xy}	Routes, the set of minimal routes from node x to node y.	
R'_{xy}	Routes, the set of all routes minimal and non-minimal from node x to node y.	
t	Channel delay, the time required for a signal to traverse a channel.	s
t_r	Router delay, the delay through a single router.	s
t_c	Credit latency, the delay in clock cycles seen by a credit passing through a router, on both ends.	cycles
t_{ck}	Clock period.	s
t_{crt}	Credit round-trip latency, the round trip delay from a flit leaving the SA stage of a router to a credit returning, enabling the next flit to leave the SA stage using the same buffer.	cycles
t_f	Flit latency, the delay in clock cycles seen by a flit passing through a router.	cycles
T	Latency, the time from when the head flit of a packet enters the network to when the tail flit of a packet departs the network.	s
T_h	Head latency, the time for the head of a packet to reach the destination.	s
T_r	Routing latency, the fraction of T_h due to delay through routers along the path, $T_r = Ht_r$.	s

---

1.   We are overloading the symbol N. Wherever the meaning is not clear from the context, it will be explicitly stated.

Symbol	Description	Unit of Measurement
T_s	Serialization latency, the time required to send a packet over a link, $T_s = b/L$.	s
T_w	Time of flight, the fraction of T_h due to propagation delay over the wires of the channels.	s
Θ	Throughput, the maximum traffic that the network can accept from each network input.	bits/s
v	Signal velocity, the rate at which signals travel over channels, $t = \frac{l}{v}$.	m/s
w	Channel width.	bits
W_n	Node wire capacity, the number of signals that can enter and leave a node. For node-limited topologies, W_n limits network bandwidth, while for bisection-limited topologies, W_s limits network bandwidth.	signals
W_s	System wire capacity, the number of signals that can pass across the midpoint of a system. This is the fundamental limit on bisection bandwidth, B_B, along with wire frequency f.	signals

APPENDIX B

Glossary

accepted traffic — See **throughput**.

adaptive routing — With adaptive routing, the path taken by a packet is determined dynamically based on the state of the network. For example, a path may be chosen to avoid congested or faulty channels.

ASIC — Application-specific integrated circuit.

availability — The fraction of time a system is operating properly, which is commonly expressed by a number of "nines." For example, a system with five nines of availability is up 99.999% of the time.

backpressure — Information about the utilization of *downstream* resources. Backpressure information is used by flow control to prevent overflow of buffers and can be used by an adaptive routing algorithm to avoid congested resources, for example.

blocking — A network is blocking if it cannot handle all circuit requests that are a *permutation* of the inputs and outputs.

CAM — Content-addressable memory.

capacity — The *throughput* of a network on uniform, random traffic. Normalizing the throughput of a network to its capacity (expressed as a fraction of capacity) allows a meaningful comparison of throughputs of different networks.

dateline — A conceptual line across a channel of a ring network (or within a single dimension of a torus). When a packet starts in a lower dateline class, then switches to the upper dateline class as the packet crosses the dateline, the cyclic dependencies inherent in the ring are removed.

deadlock —A deadlock occurs when a set of agents holding resources are waiting on another set of resources such that a cycle of waiting agents is formed. Most networks are designed to avoid deadlock, but it is also possible to recover from deadlock by detecting and breaking cyclic wait-for relationships.

deterministic routing — With deterministic routing, the path a packet takes is only a function of its source and destination. Moreover, packets between a particular source-destination pair all follow the same path. Thus, deterministic routing does not take advantage of any *path diversity* in the topology and is subject to poor worst-case performance.

downstream — For a particular channel, packets traveling through that channel travel downstream. Also, the router at the destination of the channel is the downstream router. Relative to a particular packet, the downstream resources are those that will be encountered as the packet's route continues.

ECL — Emitter coupled logic.

escape channels — Escape channels provide a deadlock-free set of paths available to every packet. Then, additional channels can be used to provide routing flexibility without the constraint of an acyclic channel dependency graph — any packet trapped in a wait-for cycle in these additional channels can always be "drained" via the escape channels.

fault tolerance — The ability of a network to detect, contain, and recover from faulty resources.

FPGA — Field-programmable gate array.

flit — A flow control dig**it**, or flit, is the smallest unit of resource allocation in a router. Variable length *packets* are divided into one or more fixed length flits to simplify the management and allocation of resources. Flits may be divided further into *phits* for handling by the router datapath.

flow — A *flow* is a sequence packets traveling between a single source-destination pair and is the unit at which *quality of service* is provided. It is possible for a source or destination to support multiple flows concurrently.

flow control — Flow control is the scheduling and allocation of a network's resources, such as channel bandwidth, buffer space, and control state.

folding — Folding a topology combines nodes by taking advantage of a reflective symmetry. For example, folding a Clos network combines the first and third stage switching elements to form a fat-tree network. Similarly, a butterfly network with *n* extra stages can be folded. Folding can also be used to refer to the interleaving of nodes along the dimensions of a torus network. This type of folding eliminates the need for long, wrap-around connections in the packaging of a torus.

hot-spot — A hot-spot resource is one whose demand is significantly greater than other, similar resources. For example, a particular destination terminal becomes a hot-spot in a shared memory multicomputer when many processors are simultaneously reading from the same memory location (for example, a shared lock or data structure).

jitter — The maximum difference in the *latency* between two packets within a *flow*. Low jitter is often a requirement for video streams or other real time data for which the regularity of data arrival is important. The jitter times the bandwidth of a flow gives a lower bound on the size of buffer required.

latency — The time required to deliver a unit of data (usually a packet or message) through the network, measured as the elapsed time between the injection of the first bit at the source to the ejection of the last bit at the destination.

livelock — Livelock occurs when a packet is not able to make progress in the network and is never delivered to its destination. Unlike *deadlock*, though, a livelocked packet continues to move through the network.

load balance — The measure of how uniformly resources are being utilized in a network. A network is load-balanced if all the (expensive) resources tend to *saturate* at the same *offered traffic*.

loss — The fraction of messages that is dropped by the network. For some applications, such as a shared-memory multicomputer interconnect, no packet loss is allowed because a lost or malformed message will generally result in a system crash. However, for other applications, such as a packet switch fabric, a small fraction of messages can be lost without adversely affecting performance.

LSB — Least significant bit.

message — Messages are the logical unit of data transfer provided by the network interfaces. Because messages do not always have a bounded length, they are often broken into smaller *packets* for handling within the network.

MIMD — Multiple-instruction-multiple-data parallel computer.

minimal — A route between a source-destination pair is minimal if it contains the smallest possible number of hops between that pair. In torus and mesh networks, the set of nodes along the union of all minimal routes between a source-destination pair forms the minimal quadrant.

MSB — Most significant bit.

multicast — A multicast packet can be sent to multiple destinations. A broadcast is a multicast in which a packet is sent to all destinations.

non-blocking —A network is non-blocking if it can simultaneously handle all circuit requests that are a *permutation* of the inputs and outputs. A non-blocking network can always handle a request for a circuit from any idle input to any idle output.

non-interfering — As in a *non-blocking* network, a non-interfering network provides full bandwidth between inputs and outputs for all request patterns. However, a non-interfering network does not provide strict latency isolation between *flows*. Rather, no flow is allowed to deny service to another flow for more than a predetermined amount of time.

oblivious routing — With oblivious routing, the set of paths a packet can take are only a function of its source and destination. Randomization is then used to select a path for a particular packet from among the set of possible paths. This randomization allows oblivious routing to both take advantage of *path diversity* and achieve better *load-balance* than *deterministic routing*.

offered traffic —The amount of traffic (in bits/s) generated by the source terminals of the network. If the network is below *saturation*, all the offered traffic is accepted by the network and thus the offered traffic equals the *throughput* of the network.

packet — Packets are the unit of routing within an interconnection network. *Messages* are broken into one or more variable, but bounded, length packets for processing by the network. All data contained within a packet follow the same *route* through the network and packets are reassembled into messages at the destination node.

path diversity —The path diversity of a network is the number of distinct paths between each source-destination pair. Higher path diversity enables both better *fault tolerance* and *load balance* within the network, but at the cost of longer paths through the network.

PCI — Peripheral component interconnect.

permutation — A *traffic pattern* in which each input sends traffic to exactly one output and each output receives traffic from exactly one input. Thus, the entries of the corresponding traffic matrix Λ are either zero or one.

phit —A physical digit, or phit, is the smallest unit of data processed by a router. One or more phits are combined to form a *flit*.

quality of service (QoS) — The bandwidth, latency, and/or jitter received by a particular flow or class of traffic. A QoS policy differentiates between *flows* and provides services to those flows based on a contract that guarantees the QoS provided to each flow, provided that the flow complies with restrictions on volume and burstiness of traffic.

RAM — Random-access memory.

reliability — The probability that a network is *working* at a given point in time.

routing — The process of choosing a path for a packet through the network. Ideally, the path should *load-balance* the channels while maintaining a short path length.

saturation — A resource is in saturation when the demands being placed on it are beyond its capacity for servicing those demands. For example, a channel becomes saturated when the amount of data that wants to be routed over the channel (in bits/s) exceeds the bandwidth of the channel. The *saturation throughput* of a network is the smallest rate of *offered traffic* for which some resource in the network is saturated.

self-throttling — A network is self-throttling if its *offered load* naturally decreases as the network approaches *saturation*. A shared-memory multiprocessor, for example, is self-throttling because each processor can support only a small number of outstanding requests. If the network approaches saturation, the message latency increases and the outstanding request queues of the processors fill up, preventing any new requests from being issued.

serialization — Serialization occurs when a large piece of data, such as a packet or flit, is squeezed over a narrow resource, such as a channel. The data must be transferred across the narrow resource over a period of several cycles, thus incurring a serialization latency.

SIMD — Single-instruction-multiple-data parallel computer.

SONET — Synchronous optical network.

speedup — Provisioning resource(s) with a capacity greater than that required in the ideal case to compensate for other compromises or imperfections in a network. For example, a network might support only two-thirds of its ideal throughput due to load imbalance. By increasing the bandwidth of all the network's resources by 1.5 times (providing a speedup of 1.5), the original ideal throughput can be realized.

stiff backpressure — Analogous to the stiffness of a spring in a mechanical system, *backpressure* is stiff if *downstream* congestion is quickly relayed to upstream routers. Stiff backpressure allows rapid adaptation to congestion or *hot-spot* resources, but can also result in an overcorrection for a momentary load imbalance.

stable — A network is stable if its *throughput* remains constant (does not drop) as *offered traffic* continues to increase beyond the *saturation* throughput.

STS — Synchronous transport signal. (STS-N refers to STS level-N, an N x 51.84Mbits/s signal.)

throughput — The amount of traffic (in bits/s) delivered to the destination terminals of the network. If the network is below *saturation*, all the offered traffic is accepted by the network and thus the *offered traffic* equals the throughput of the network.

topology — The static arrangement of router nodes, channels, and terminals in a network.

traffic — The sequence of injection times and destinations for the packets being offered to the network. This sequence is often modeled by a static traffic pattern that defines the probability a packet travels between a particular source-destination pair and an arrival process.

TTL — Transistor-transistor logic.

unicast — A unicast packet has a single destination terminal (as opposed to *multicast*).

upstream — For a particular channel, credits or other flow-control information along the back channel travel upstream. Also, the router at the source of the channel is the upstream router. Relative to a particular packet, the upstream resources are those that have already been visited along the packet's route.

APPENDIX C

Network Simulator

Although the designer of an interconnection network should have strong intuition in regard of the performance of that network, an accurate simulator is still an important tool for verifying this intuition and analyzing specific design tradeoffs. To aid in this process, the simulator written for this book is freely available at `http://cva.stanford.edu/`. All the results from Chapters 19 and 25 were created using this simulator. The simulator models the network at the flit level and includes support for multiple topologies and routing algorithms. Buffering, speedup, and pipeline timing of the routers are fully configurable. Also, several simple traffic models and patterns are available. Internally, the simulator uses the two-phase cycle-based simulation approach described in Section 24.4.1.

Bibliography

[1] Bülent Abali and Cevdet Aykanat. "Routing algorithms for IBM SP1." In *Proc. of the First International Parallel Computer Routing and Communication Workshop (PCRCW)*, pages 161–175, Seattle, 1994.

[2] Anant Agarwal. "Limits on interconnection network performance." *IEEE Transactions on Parallel and Distributed Systems*, 2(4):398–412, 1991.

[3] P. Agrawal, W.J. Dally, W.C. Fischer, H.V. Jagadish, A.S. Krishnakumar, and R. Tutundjain. "MARS: A multiprocessor-based programmable accelerator." *IEEE Design and Test of Computers*, 4(5):28–37, Feb. 1987.

[4] Prathima Agrawal and William J. Dally. "A hardware logic simulation system." *IEEE Transactions on Computer-Aided Design of Integrated Circuits and Systems*, 9(1):19–29, Jan. 1990.

[5] Hamid Ahmadi and Wolfgang E. Denzel. "Survey of modern high performance switching." *IEEE Journal on Selected Areas in Communications*, 7(7):1091–1103, Sept. 1989.

[6] Ravindra K. Ahuja, Thomas L. Magnanti, and James B. Orlin. *Network flows: Theory, algorithms, and applications.* Upper Saddle River, New Jersey Prentice-Hall, Inc. 1993.

[7] Sheldon B. Akers and Balakrishnan Krishnamurthy. "A group-theoritic model for symmetric interconnection networks." *IEEE Transactions on Computers*, 38(4):555–566, April 1989.

[8] James D. Allen, Patrick T. Gaughan, David E. Schimmel, and Sudhakar Yalamanchili. "Ariadne — an adaptive router for fault-tolerant multicomputers." In *Proc. of the International Symposium on Computer Architecture (ISCA)*, pages 278–288, Chicago, 1994.

[9] Robert Alverson, David Callahan, Daniel Cummings, Brian Koblenz, Allan Porterfield, and Burton Smith. "The Tera computer system." In *Proc. of the International Conference on Supercomputing*, pages 1–6, 1990.

[10] T. Anderson, D. Culler, and D. Patterson. "A case for NOW (networks of workstations)" IEEE Micro, 15(1): 54–64, Feb. 1995.

[11] Thomas E. Anderson, Susan S. Owicki, James B. Saxe, and Charles P. Thacker. "High speed switch scheduling for local area networks." *ACM Transactions on Computer Systems*, 11(4):319–352, Nov. 1993.

[12] P. Baran. "On distributed communication networks." *IEEE Transactions on Communications Systems*, 12(1):1–9, March 1964.

[13] George H. Barnes, Richard M. Brown, Maso Kato, David J. Kuck, Daniel L. Slotnick, and Richard A. Stokes. "The ILLIAC IV computer." *IEEE Transactions on Computers*, 17(8):746–757, Aug. 1968.

[14] K. E. Batcher. "Sorting networks and their applications." In *AFIPS Conference Proceedings 32*, page 307. Montvale, N.J.: AFIPS Press. 1968.

[15] K.E. Batcher. "The flip network in STARAN." In *Proc. of the International Conference on Parallel Processing*, pages 65–71, 1976.

[16] Simoni Ben-Michael, Michael Ben-Nun, and Yifat Ben-Shahar. "Method and apparatus to expand an on chip fifo into local memory." United States Patent 6,078,565. June 2000.

[17] V.E. Beneš. Rearrangeable three stage connecting networks. *Bell System Technical Journal*, 41:1481–1492, 1962.

[18] V.E. Beneš. *Mathematical theory of connecting networks and telephone traffic*. New York: Academic Press. 1965.

[19] Elwyn R. Berlekamp. *Algebraic Coding Theory, Revised Edition*. Berkeley, Aegean Press, June 1984.

[20] D. Bertsekas and R. Gallager. *Data Networks*. Upper Saddle River, N.J.: Prentice-Hall, Inc. 2nd edition, 1992.

[21] Richard E. Blahut. *Algebraic Codes for Data Transmission*. Cambridge University Press, 2002.

[22] Matthias A. Blumrich, Richard D. Alpert, Yuqun Chen, Douglas W. Clark, Stefanos N. Damianakis, Cezary Dubnicki, Edward W. Felten, Liviu Iftode, Kai Li, Margaret Martonosi, and Robert A. Shillner. "Design choices in the SHRIMP system: An empirical study." In *Proc. of the International Symposium on Computer Architecture (ISCA)*, pages 330–341, 1998.

[23] Matthias A. Blumrich, Kai Li, Richard Alpert, Cezary Dubnicki, Edward W. Felten, and Jonathan Sandberg. "Virtual memory mapped network interface for the SHRIMP multicomputer." In *Proc. of the International Symposium on Computer Architecture (ISCA)*, pages 142–153, April 1994.

[24] Nannette J. Boden, Danny Cohen, Robert E. Felderman, Alan E. Kulawik, Charles E. Seitz, Jakov N. Seizovic, and Wen-King Su. "Myrinet: a gigabit-per-second local area network." *IEEE Micro*, pages 29–36, Feb. 1995.

[25] Kevin Bolding. "Non-uniformities introduced by virtual channel deadlock prevention." *Technical Report UW-CSE-92-07-07*, University of Washington, July 1992.

[26] Kevin Bolding, Melanie Fulgham, and Lawrence Snyder. "The case for chaotic adaptive routing." *IEEE Transactions on Computers*, 12(46):1281–1292, Dec. 1997.

[27] Rajendra V. Boppana and Suresh Chalasani. "A comparison of adaptive wormhole routing algorithms." In *Proc. of the International Symposium on Computer Architecture (ISCA)*, pages 351–360, 1993.

[28] Allan Borodin and John E. Hopcroft. "Routing, merging, and sorting on parallel models of computation." *Journal of Computer and System Sciences*, 30:130–145, 1985.

[29] Paul Bratley, Bennett L. Fox, and Linus E. Schrage. *A guide to simulation.* New York: Springer-Verlag, 2nd edition, 1986.

[30] Mark S. Brirrittella, Richard E. Kessler, Steven M. Oberlin, Randal S. Passint, and Greg Thorson. "System for allocating messages between virtual channels to avoid deadlock and to optimize the amount of message traffic on each type of virtual channel." United States Patent 5,583,990. Dec. 1996.

[31] Randy Brown. "Calendar queues: A fast $O(1)$ priority queue implementation of the simulation event set problem." *Communications of the ACM*, 31(10):1220–1227, Oct. 1988.

[32] Nicholas P. Carter. *Processor Mechanisms for Software Shared Memory.* Ph.D. thesis, Massachusetts Institute of Technology, Feb. 1999.

[33] Philip P. Carvey, William J. Dally, and Larry R. Dennison. "Apparatus and methods for connecting modules using remote switching." United States Patent 6,205,532. March 2001.

[34] Philip P. Carvey, William J. Dally, and Larry R. Dennison. "Composite trunking." United States Patent 6,359,879. March 2002.

[35] Alan Charlesworth. "Starfire — extending the SMP envelope." *IEEE Micro*, 18(1):39–49, Jan./Feb. 1998.

[36] Andrew A. Chien and Jae H. Kim. "Planar-adaptive routing: Low-cost adaptive networks for multiprocessors." In *Proc. of the International Symposium on Computer Architecture (ISCA)*, pages 268–277, 1992.

[37] Charles Clos. "A study of non-blocking switching networks." *Bell System Technical Journal*, 32:406–424, 1953.

[38] P. Close. "The iPSC/2 node architecture." In *Proc. of the Conference on Hypercube Concurrent Computers and Applications*, pages 43–55, Jan. 1988.

[39] R. Cole and J. Hopcroft. "On edge coloring bipartite graphs." *SIAM Journal on Computing*, 11:540–546, 1982.

[40] Richard Cole, Kirstin Ost, and Stefan Schirra. "Edge-coloring bipartite multigraphs in $O(E \log D)$ time." *Combinatorica*, 21(1):5–12, 2001.

[41] Richard Crisp. "Direct Rambus technology: The new main memory standard." *IEEE Micro*, 17(6):18–28, Nov./Dec. 1997.

[42] Rene L. Cruz. "A calculus for network delay, part I: Network elements in isolation." *IEEE Transactions on Information Theory*, 37(1):114–131, Jan. 1991.

[43] Rene L. Cruz. "A calculus for network delay, part II: Network analysis." *IEEE Transactions on Information Theory*, 37(1):132–141, Jan. 1991.

[44] William J. Dally. "Virtual-channel flow control." In *Proc. of the International Symposium on Computer Architecture (ISCA)*, pages 60–68, May 1990.

[45] William J. Dally. "Express cube: Improving the performance of k-ary n-cube interconnection networks." *IEEE Transactions on Computers*, 40(9):1016–1023, Sept. 1991.

[46] William J. Dally. "Performance analysis of k-ary n-cube interconnection networks." *IEEE Transactions on Computers*, 39(6):775–785, June 1991.

[47] William J. Dally. "Virtual-channel flow control." *IEEE Transactions on Parallel and Distributed Systems*, 3(2):194–205, March 1992.

[48] William J. Dally and Hiromichi Aoki. "Deadlock-free adaptive routing in multicomputer networks using virtual channels." *IEEE Transactions on Parallel and Distributed Systems*, 4(4):466–475, April 1993.

[49] William J. Dally, P. P. Carvey, and L. R. Dennison. "The Avici terabit switch/router." In *Proc. of the Symposium on Hot Interconnects*, pages 41–50, Aug. 1998.

[50] William J. Dally, Philip P. Carvey, Larry R. Dennison, and Allen P. King. "Internet switch router." United States Patent 6,370,145. April 2002.

[51] William J. Dally, Andrew Chang, Andrew Chien, Stuart Fiske, Waldemar Horwat, John Keen, Richard Lethin, Michael Noakes, Peter Nuth, Ellen Spertus, Deborah Wallach, and Scott D. Wills. "The J-machine." In *Retrospective in 25 Years of the International Symposia on Computer Architecture*, pages 54–58, 1998.

[52] William J. Dally, Larry R. Dennison, David Harris, Kinhong Kan, and Thucydides Xanthopoulos. "The Reliable Router: A reliable and high-performance communication substrate for parallel computers." In *Proc. of the First International Parallel Computer Routing and Communication Workshop (PCRCW)*, Seattle, May 1994.

[53] William J. Dally, J. A. Stuart Fiske, John S. Keen, Richard A. Lethin, Michael D. Noakes, Peter R. Nuth, Roy E. Davison, and Gregory A. Fyler. "The message-driven processor — a multicomputer processing node with efficient mechanisms." *IEEE Micro*, 12(2):23–39, April 1992.

[54] William J. Dally and John Poulton. "Transmitter equalization for 4-Gbps signaling." *IEEE Micro*, 17(1):48–56, Jan./Feb. 1997.

[55] William J. Dally and John W. Poulton. *Digital Systems Engineering*. Cambridge University Press, 1998.

[56] William J. Dally and Charles L. Seitz. "The torus routing chip." *Journal of Parallel and Distributed Computing*, 1(3):187–196, 1986.

[57] William J. Dally and Charles L. Seitz. "Deadlock free message routing in multiprocessor interconnection networks." *IEEE Transactions on Computers*, 36(5):547–553, May 1987.

[58] Alan Demers, Srinivasan Keshav, and Scott Shenker. "Analysis and simulation of a fair queueing algorithm." *Proc. of ACM SIGCOMM*, 19(4):1–12, Sept. 1989.

[59] Willibald Doeringer, Günter Karjoth, and Mehdi Nassehi. "Routing on longest-matching prefixes." *IEEE/ACM Transactions on Networking*, 4(1):86–97, Feb. 1996.

[60] José Duato. "A new theory of deadlock-free adaptive routing in wormhole networks." *IEEE Transactions on Parallel and Distributed Systems*, 4(12):1320–1331, Dec. 1993.

[61] José Duato. "A necessary and sufficient condition for deadlock-free adaptive routing in wormhole networks." *IEEE Transactions on Parallel and Distributed Systems*, 6(10):1055–1067, Oct. 1995.

[62] José Duato. "A necessary and sufficient condition for deadlock-free routing in cut-through and store-and-forward networks." *IEEE Transactions on Parallel and Distributed Systems*, 7(6):841–854, Aug. 1996.

[63] Jose Duato, Sudhakar Yalamanchili, Blanca Caminero, Damon S. Love, and Francisco J. Quiles. "MMR: A high-performance multimedia router — architecture and design trade-offs." In *Proc. of the International Symposium on High-Performance Computer Architecture (HPCA)*, pages 300–309, 1999.

[64] A.M. Duguid. "Structural properties of switching networks." *Technical Report BTL-7*, Brown University, 1959.

[65] T-Y. Feng. "Data manipulating functions in parallel processors and their implementations." *IEEE Transactions on Computers*, 23(3):309–318, March 1974.

[66] George S. Fishman. *Discrete-Event Simulation: Modeling, Programming, and Analysis*. New York: Springer-Verlag, 2001.

[67] L.R. Ford and D.R. Fulkerson. "Maximal flow through a network." *Canadian Journal of Mathematics*, pages 399–404, 1956.

[68] Edward Fredkin. "Trie memory." *Communications of the ACM*, 3(9):490–499, August 1960.

[69] Mike Galles. "Scalable pipelined interconnect for distributed endpoint routing: The SGI SPIDER chip." In *Proc. of the Symposium on Hot Interconnects*, pages 141–146, Aug. 1996.

[70] D. Gelernter. "A DAG-based algorithm for prevention of store-and-forward deadlock in packet networks." *IEEE Transactions on Computers*, 30(10):709–715, Oct. 1981.

[71] Paolo Giaccone, Balaji Prabhakar, and Devavrat Shah. "Towards simple, high-performance schedulers for high-aggregate bandwidth switches." In *Proc. of IEEE INFOCOM*, pages 1160–1169, New York, June 2002.

[72] Joseph Di Giacomo. *Digital Bus Handbook*. McGraw-Hill Professional, Jan. 1990.

[73] Christopher J. Glass and Lionel M. Ni. "The turn model for adaptive routing." In *Proc. of the International Symposium on Computer Architecture (ISCA)*, pages 278–287, 1992.

[74] S. Jamaloddin Golestani. "A stop-and-go queueing framework for congestion management." *Proc. of ACM SIGCOMM*, 20(4):8–18, Aug. 1990.

[75] Allan Gottlieb, Ralph Grishman, Clyde P. Kruskal, Kevin P. McAuliffe, Larry Rudolph, and Marc Snir. "The NYU Ultracomputer — Designining a MIMD shared memory parallel computer." *IEEE Transactions on Computers*, 32(2):175–189, Feb. 1983.

[76] Luis Gravano, Gustavo D. Pifarré, Pablo E. Berman, and Jorge L.C. Sanz. "Adaptive deadlock- and livelock-free routing with all minimal paths in torus networks." *IEEE Transactions on Parallel and Distributed Systems*, 5(12):1233–1251, Dec. 1994.

[77] Jim Gray, editor. *The Benchmark Handbook.* San Mateo, CA: Morgan Kaufmann, 2nd edition, 1993.

[78] K.D. Gunther. "Prevention of deadlocks in packet-switched data transport systems." *IEEE Transactions on Communications,* 29(4), 1981.

[79] Pankaj Gupta and Nick McKeown. "Designing and implementing a fast crossbar scheduler." *IEEE Micro,* 19(1):20–28, Jan./Feb. 1999.

[80] Kenichi Hayashi, Tunehisa Doi, Takeshi Horie, Yoichi Koyanagi, Osamu Shiraki, Nobutaka Imamura, Toshiyuki Shimizu, Hiroaki Ishihata, and Tatsuya Shindo. "AP1000+: Architectural support of PUT/GET interface for parallelizing compiler." In *Proc. of. Architectural Support for Programming Languages and Operating Systems (ASPLOS),* pages 196–207, San Jose, CA, 1994.

[81] H. Heffes and D.M. Lucantoni. "A Markov modulated characterization of packetized voice and data traffic and related statistical multiplexer performance." *IEEE Journal on Selected Areas in Communications,* 4(6):856–868, Sept. 1986.

[82] Dana S. Henry and Christopher F. Joerg. "A tightly-coupled processor-network interface." In *Proc. of. Architectural Support for Programming Languages and Operating Systems (ASPLOS),* pages 111–121, 1992.

[83] J. E. Hopcroft and R. M. Karp. "An $n^{5/2}$ algorithm for maximum matching in bipartite graphs." *SIAM Journal on Computing,* 2:225–231, 1973.

[84] Robert W. Horst. "TNet: a reliable system area network." *IEEE Micro,* 15(1):37–45, Feb. 1995.

[85] BBN Advanced Computers Incorporated. "Butterfly parallel processor overview." BBN Report No. 6148, March 1986.

[86] INMOS. *The T9000 Transputer Products Overview Manual,* 1991.

[87] Sundar Iyer, Ramana Rao Kompella, and Nick McKeown. "Analysis of a memory architecture for fast packet buffers." In *Proc. of the IEEE Workshop on High Performance Switching and Routing (HPSR),* pages 368–373, Dallas, TX, May 2001.

[88] Jeffrey M. Jaffe. "Bottleneck flow control." *IEEE Transactions on Communications,* 29(7):954–962, July 1981.

[89] A. Jain, W. Anderson, T. Benninghoff, D. Berucci, M. Braganza, J. Burnetie, T. Chang, J. Eble, R. Faber, O. Gowda, J. Grodstein, G. Hess, J. Kowaleski, A. Kumar, B. Miller, R. Mueller, P. Paul, J. Pickholtz, S. Russell, M. Shen, T. Truex, A. Vardharajan, D. Xanthopoulus, and T. Zou. "A 1.2GHz Alpha microprocessor with 44.8Gb/s chip pin bandwidth." In *Proc. of the IEEE International Solid-State Circuits Conference (ISSCC),* pages 240–241, San Francisco, Feb. 2001.

[90] Raj Jain and Shawn A. Routhier. "Packet trains – measurements and a new model for computer network traffic." *IEEE Journal on Selected Areas in Communications*, 4(6):986–995, Sept. 1986.

[91] Christos Kaklamanis, Danny Krizanc, and Thanasis Tsantilas. "Tight bounds for oblivious routing in the hypercube." In *Proc. of the Symposium on Parallel Algorithms and Architectures (SPAA)*, pages 31–36, 1990.

[92] Manolis Katevenis, Panagiota Vatsolaki, and Aristides Efthymiou. "Pipelined memory shared buffer for VLSI switches." *Proc. of ACM SIGCOMM*, 25(4):39–48, Oct. 1995.

[93] Stephen W. Keckler, William J. Dally, Daniel Maskit, Nicholas P. Carter, Andrew Chang, and Whay Sing Lee. "Exploiting fine-grain thread level parallelism on the MIT multi-ALU processor." In *Proc. of the International Symposium on Computer Architecture (ISCA)*, pages 306–317, Barcelona, Spain, July 1998.

[94] P. Kermani and L. Kleinrock. "Virtual-cut through: a new computer communications switching technique." *Computer Networks*, 3(4):267–286, 1979.

[95] R.E. Kessler and J.L. Schwarzmeier. "Cray T3D: a new dimension for Cray Research." In *Proc. of the IEEE Computer Society International Conferrence (COMPCON)*, pages 176–182, Feb. 1993.

[96] Jae H. Kim and Andrew A. Chien. "Network performance under bimodal traffic loads." *Journal of Parallel and Distributed Computing*, 28(1):43–64, Jul. 1995.

[97] Jae H. Kim and Andrew A. Chien. "Rotating combined queueing (RCQ): bandwidth and latency guarantees in low-cost, high-performance networks." In *Proc. of the International Symposium on Computer Architecture (ISCA)*, pages 226–236, May 1996.

[98] Jae H. Kim, Ziqiang Liu, and Andrew A. Chien. "Compressionless routing: A framework for adaptive and fault-tolerant routing." *IEEE Transactions on Parallel and Distributed Systems*, 8(3):229–244, March 1997.

[99] Leonard Kleinrock. *Queuing Systems*, Volume 1. New York: John Wiley & Sons, Inc., 1975.

[100] Leonard Kleinrock and Farouk Kamoun. "Hierarchical routing for large networks: Performance evaluation and optimization." *Computer Networks*, 1:154–174, 1977.

[101] Donald E. Knuth. *Seminumerical Algorithms*. Reading, Mass.: Addison-Wesley, 3rd edition, 1997.

[102] Donald E. Knuth. *Sorting and Searching*. Reading, Mass.: Addison-Wesley, 2nd edition, 1998.

[103] D.König. "Graphok és alkalmazásuk a determinánsok és a halmazok elméletére [Hungarian]." *Mathematikai és Természettudományi Értesito*, 34:104–119, 1916.

[104] S. Konstantinidou and L. Snyder. "The Chaos Router: A practical application of randomization in network routing." *Proc. of the Symposium on Parallel Algorithms and Architectures (SPAA)*, pages 21–30, 1990.

[105] David Kroft. "Lockup-free instruction fetch/prefetch cache organization." In *Proc. of the International Symposium on Computer Architecture (ISCA)*, pages 81–88, 1981.

[106] Clyde P. Kruskal and Marc Snir. "A unified theory of interconnection network structure." *Theoretical Computer Science*, 48(3):75–94, 1986.

[107] Vijay P. Kumar, T. V. Lashman, and Dimitrios Stiliadis. "Beyond best effort: Router architectures for the differentiated services of tomorrow's internet." *IEEE Communications Magazine*, pages 152–164, May 1998.

[108] James Laudon and Daniel Lenoski. "The SGI Origin: a ccNUMA highly scalable server." In *Proc. of the International Symposium on Computer Architecture (ISCA)*, pages 241–251, June 1997.

[109] Averill M. Law and David W. Kelton. *Simulation Modeling and Analysis*. New York: McGraw Hill, 3rd edition, 2000.

[110] D. H. Lawrie. "Access and alignment of data in an array processor." *IEEE Transactions on Computers*, 24:1145–1155, Dec. 1975.

[111] Pierre L'Ecuyer. "Random numbers for simulation." *Communications of the ACM*, 33(10):85–97, Oct. 1990.

[112] Whay Sing Lee, William J. Dally, Stephen W. Keckler, Nicholas P. Carter, and Andrew Chang. "Efficient, protected message interface in the MIT M-Machine." *IEEE Computer*, 31(11):69–75, Nov. 1998.

[113] Charles E. Leiserson. "Fat-trees: Universal networks for hardware efficient supercomputing." *IEEE Transactions on Computers*, 34(10):892–901, October 1985.

[114] Charles E. Leiserson, Zahi S. Abuhamdeh, David C. Douglas, Carl R. Feynman, Mahesh N. Ganmukhi, Jeffrey V. Hill, W. Daniel Hillis, Bradley C. Kuszmaul, Margaret A. St Pierre, David S. Wells, Monica C. Wong-Chan, Shaw-Wen Yang, and Robert Zak. "The network architecture of the Connection Machine CM-5." *Journal of Parallel and Distributed Computing*, 33(2):145–158, 1996.

[115] Daniel Lenoski, James Laudon, Kourosh Gharachorloo, Wolf-Dietrich Weber, Anoop Gupta, John Henessy, Mark Horowitz, and Monica S. Lam. "The Stanford Dash multiprocessor." *IEEE Computer*, 25(3):63–79, March 1992.

[116] Daniel E. Lenoski and Wolf-Dietrich Weber. *Scalable Shared-Memory Multi-processing.* Morgan Kaufmann, 1995.

[117] Sigurd L. Lillevik. "The Touchstone 30 gigaflop DELTA prototype." In *DMCC*, pages 671–677, April 1991.

[118] Daniel H. Linder and Jim C. Harden. "An adaptive and fault tolerant worm-hole routing strategy for k-ary n-cubes." *IEEE Transactions on Computers*, 40(1):2–12, Jan. 1991.

[119] Alain. J. Martin. "The TORUS: An exercise in constructing a processing surface." In *Proc. of the* 2nd *Caltech Conference on VLSI*, pages 527–537, Jan. 1981.

[120] G. M. Masson and B. W. Jordan Jr. "Generalized multi-stage connection networks." *Networks*, 2:191–209, 1972.

[121] Makoto Matsumoto and Yoshiharu Kurita. "Twisted GFSR generators." *ACM Transactions on Modeling and Computer Simulation*, 2(3):179–194, July 1992.

[122] Anthony J. McAuley and Paul Francis. "Fast routing table lookup using CAMs." In *Proc. of IEEE INFOCOM*, pages 1382–1391, San Francisco, March 1993.

[123] Nick McKeown. "The iSLIP scheduling algorithm for input-queued switches." *IEEE/ACM Transactions on Networking*, 7(2):188–201, April 1999.

[124] Nick McKeown, Venkat Anatharam, and Jean Warland. "Achieving 100% throughput in an input-queued switch." In *Proc. of IEEE INFOCOM*, pages 296–302, San Franciso, 1996.

[125] Nick McKeown, Martin Izzard, Adisak Mekkittikul, Bill Ellersick, and Mark Horowitz. "The Tiny Tera: A packet switch core." *IEEE Micro*, 17(1):26–33, Jan./Feb. 1997.

[126] Nick McKeown, Adisak Mekkittikul, Venkat Anantharam, and Jean Walrand. "Achieving 100% throughput in an input-queued switch." In *Proc. of IEEE INFOCOM*, pages 296–302, San Francisco, March 1996.

[127] J. McQuillan. "Adaptive routing algorithms for distributed computer networks." *Technical Report BBN Tech. Rep. 2831*, Cambridge, Mass.: Bolt Beranek and Newman Inc., May 1974.

[128] Dikran S. Meliksetian and C.Y. Roger Chen. "Optimal routing algorithm and the diameter of the cube-connected cycles." *IEEE Transactions on Parallel and Distributed Systems*, 4(10):1172–1178, Oct. 1993.

[129] P.M. Merlin and P.J. Schweitzer. "Deadlock avoidance in store-and-forward networks — I: Store and forward deadlock." *IEEE Transactions on Communications*, 28(3):345–352, March 1980.

[130] Shubhendu S. Mukherjee, Federico Silla, Peter Bannon, Joel Emer, Steve Lang, and David Webb. "A comparative study of arbitration algorithms for the Alpha 21364 pipelined router." In *Proc of. Architectural Support for Programming Languages and Operating Systems (ASPLOS)*, pages 223–234, San Jose, CA, Oct. 2002.

[131] Shuhendu S. Mukherjee, Peter Bannno, Steven Lang, Aaron Spink, and David Webb. "The Alpha 21364 network architecture." In *Proc. of the Symposium on Hot Interconnects*, pages 113–117, Aug. 2001.

[132] J. Nagle. "On packet switches with infinite storage." *IEEE Transactions on Communications*, 35(4):435–438, April 1987.

[133] T. Nakata, Y. Kanoh, K. Tatsukawa, S. Yanagida, N. Nishi, and H. Takayama. "Architecture and the software environment of parallel computer Cenju-4." *NEC Research and Development Journal*, 39:385–390, Oct. 1998.

[134] nCUBE Corporation. *nCUBE Processor Manual*, 1990.

[135] Ted Nesson and S. Lennart Johnsson. "ROMM routing on mesh and torus networks." In *Proc. of the Symposium on Parallel Algorithms and Architectures (SPAA)*, pages 275–287, Santa Barbara, CA, 1995.

[136] Michael D. Noakes, Deborah A. Wallach, and William J. Dally. "The J-machine multicomputer: An architectural evaluation." In *Proc. of the International Symposium on Computer Architecture (ISCA)*, pages 224–235, May 1993.

[137] Satoshi Nojima, Eiichi Tsutsui, Haruki Fukuda, and Masamichi Hashimoto. "Integrated services packet network using bus matrix switch." *IEEE Journal on Selected Areas in Communications*, 5(8):1284–1292, Oct. 1987.

[138] Peter R. Nuth and William J. Dally. "The J-machine network." In *Proc. of the International Conference on Computer Design*, pages 420–423, Oct. 1992.

[139] Scott Pakin, Vijay Karamcheti, and Andrew A. Chien. "Fast Messages: Efficient, portable communication for workstation clusters and MPPs." *IEEE Concurrency*, 5(2):60–73, April/June 1997.

[140] J.F. Palmer. "The NCUBE family of parallel supercomputers." In *Proc. of the International Conference on Computer Design*, 1986.

[141] Stephen K. Park and Keith W. Miller. "Random number generators: good ones are hard to find." *Communications of the ACM*, 31(10):1192–1201, Oct. 1988.

[142] Krzysztof Pawlikowski. "Steady-state simulation of queueing processes: A survey of problems and solutions." *ACM Computing Surveys*, 22(2):123–170, June 1990.

[143] PCI Special Interests Group, Portland, OR. *PCI Local Bus Specification*, 2001. Revision 2.3.

[144] M.C. Pease. "The indirect binary *n*-cube microprocessor array." *IEEE Transactions on Computers*, 26(5):458–473, May 1977.

[145] Li-Shiuan Peh and William J. Dally. "Flit-reservation flow control." In *Proc. of the International Symposium on High-Performance Computer Architecture (HPCA)*, pages 73–84, Toulouse, France, Jan. 1999.

[146] Li-Shiuan Peh and William J. Dally. "A delay model and speculative architecture for pipelined routers." In *Proc. of the International Symposium on High-Performance Computer Architecture (HPCA)*, pages 255–266, Monterrey, Mexico, Jan. 2001.

[147] Tong-Bi Pei and Charles Zukowski. "VLSI implementation of routing tables: tries and CAMs." In *Proc. of IEEE INFOCOM*, pages 512–524, April 1991.

[148] Larry R. Peterson and Bruce S. Davie. *Computer Networks: a systems approach*. San Francisco: Morgan Kaufmann, 2nd edition, 2000.

[149] Craig S. Petrie and J. Alvin Connelly. "A noise-based IC random number generator for applications in cryptography." *IEEE Transactions on Circuits and Systems I: Fundamental Theory and Applications*, 47(5):615–621, May 2000.

[150] Greg Pfister. *High Performance Mass Storage and Parallel I/O*, "An Introduction to the InfiniBand Architecture," Chapter 42, pages 617–632. IEEE Press and Wiley Press, 2001.

[151] Gregory F. Pfister and V. Alan Norton. "Hot spot contention and combining in multistage interconnection networks." *IEEE Transactions on Computers*, 34(10):943–948, Oct. 1985.

[152] Timothy Mark Pinkston and Sugath Warnakulasuriya. "Characterization of deadlocks in *k*-ary *n*-cube networks." *IEEE Transactions on Parallel and Distributed Systems*, 10(9):904–921, Sept. 1999.

[153] Franco P. Preparata and Jean Vuillemin. "The cube-connected cycles: a versatile network for parallel computation." *Communications of the ACM*, 24(5):300–309, May 1981.

[154] Martin De Prycker. *Asynchronous Transfer Mode: Solution for Broadband ISDN*. Prentice Hall, 3rd edition, 1995.

[155] J. Rattner. "Concurrent processing: a new direction in scientific computing." In *AFIPS Conference Proceedings, National Computer Conference*, volume 54, pages 157–166, 1985.

[156] Randall D. Rettberg, William R. Crowther, Philip P. Carvey, and Raymond S. Tomlinson. "The monarch parallel processor hardware design." *IEEE Computer*, 23(4):18–28, April 1990.

[157] Jerome H. Saltzer, David P. Reed, and David D. Clark. "End-to-end arguments in system design." *ACM Transactions on Computer Systems*, 2(4):277–288, Nov. 1984.

[158] Nicola Santoro and Ramez Khatib. "Labeling and implicit routing in networks." *The Computer Journal*, 28(1):5–8, 1985.

[159] Michael D. Schroeder, Andrew D. Birrell, Michael Burrows, Hal Murray, Roger M. Needham, Thomas L. Rodeheffer, Edwin H. Satterthwaite, and Charles P. Thacker. "Autonet: a high-speed, self-configuring local area network using point-to-point links." *IEEE Journal on Selected Areas in Communications*, 9(8):1318–1335, Oct. 1991.

[160] Loren Schwiebert. "Deadlock-free oblivious wormhole routing with cyclic dependencies." *IEEE Transactions on Computers*, 50(9):865–876, Sept. 2001.

[161] Steven L. Scott and Greg Thorson. "Optimized routing in the Cray T3D". In *Proc. of the First International Parallel Computer Routing and Communication Workshop (PCRCW)*, pages 281–294, Seattle, May 1994.

[162] Steven L. Scott and Gregory M. Thorson. "The Cray T3E network: Adaptive routing in a high performance 3D torus." In *Proc. of the Symposium on Hot Interconnects*, pages 147–156, Aug. 1996.

[163] Charles L. Seitz. "The Cosmic Cube." *Communications of the ACM*, 28(1):22–33, Jan. 1985.

[164] Charles L. Seitz, W. C. Athas, C. M. Flaig, A. J. Martin, J. Seizovic, C. S. Steele, and W.-K. Su. "The architecture and programming of the Ametek series 2010 multicomputer." In *Proc. of the Conference on Hypercube Concurrent Computers and Applications*, pages 33–36, 1988.

[165] Carlo H. Séquin. "Doubly twisted torus networks for VLSI processing arrays." In *Proc. of the International Symposium on Computer Architecture (ISCA)*, pages 471–480, 1981.

[166] H. J. Siegel. "Interconnection networks for SIMD machines." *IEEE Transactions on Computers*, 12(6):57–65, June 1979.

[167] Daniel Siewiorek and Robert Swarz. *Theory and Practice of Reliable System Design*. Digital Press, Dec. 1983.

[168] Arjun Singh, William J. Dally, Brian Towles, and Amit K. Gupta. "Locality-preserving randomized oblivious routing on torus networks." In *Proc. of the Symposium on Parallel Algorithms and Architectures (SPAA)*, pages 9–19, Winnipeg, Manitoba, Canada, Aug. 2002.

[169] Jaswinder Pal Singh, Wolf-Dietrich Weber, and Anoop Gupta. "SPLASH: Stanford parallel applications for shared-memory." *ACM SIGARCH Computer Architecture News*, 1(20):5–44, March 1992.

[170] Rajeev Sivaram, Craig B. Stunkel, and Dhabaleswar K. Panda. "HIPIQS: A high-performance switch architecture using input queuing." *IEEE Transactions on Parallel and Distributed Systems*, 13(3):275–289, May 1998.

[171] David Slepian. "Two theorems on a particular crossbar switching network." 1952.

[172] Daniel L. Slotnick, W. Carl Borck, and Robert C. McReynolds. "The Soloman computer." In *Proc. of the AFIPS Spring Joint Computer Conference*, volume 22, pages 97–107, New York: Spartan Books, 1967.

[173] Small computer systems interface — 2 (SCSI-2).   ANSI X3.131-1994 (R1999), 1999.

[174] Burton J. Smith. "Architecture and applications of the HEP multiprocessor computer system." In *Proc. of SPIE: Real-Time Signal Processing IV*, volume 298, pages 241–248, 1981.

[175] Harold S. Stone. "Parallel processing with the perfect shuffle." *IEEE Transactions on Computers*, 20(2):153–161, Feb. 1971.

[176] C. B. Stunkel and P. H. Hochschild. "SP2 high-performance switch architecture." In *Proc. of the Symposium on Hot Interconnects*, pages 115–121, Stanford, Ca., Aug. 1994.

[177] Craig B. Stunkel, Jay Herring, Bulent Abali, and Rajeev Sivaram. "A new switch chip for IBM RS/6000 SP systems." In *Proc. of the ACM/IEEE Conference on Supercomputing*, Portland, Or., Nov. 1999.

[178] Craig B. Stunkel, Dennis G. Shea, Don G. Grice, Peter H. Hochschild, and Michael Tsao. "The SP1 high-performance switch." In *Proc. of the Scalable High Performance Computing Conference*, pages 150–157, May 1994.

[179] H. Sullivan and T. R. Bashkow. "A large scale, homogeneous, fully distributed parallel machine." In *Proc. of the International Symposium on Computer Architecture (ISCA)*, pages 105–124, March 1977.

[180] Yuval Tamir and Hsin-Chou Chi. "Symmetric crossbar arbiters for VLSI communication switches." *IEEE Transactions on Parallel and Distributed Systems*, 4(1):13–27, Jan. 1993.

[181] Yuval Tamir and Gregory L. Frazier. "High performance multi-queue buffers for VLSI communication switches." In *Proc. of the International Symposium on Computer Architecture (ISCA)*, pages 343–354, June 1988.

[182] Yuval Tamir and Gregory L. Frazier. "Dynamically-allocated multi-queue buffers for VLSI communication switches." *IEEE Transactions on Computers*, 41(6):725–737, June 1992.

[183] Leandros Tassiulas. "Linear complexity algorithms for maximum throughput in radio networks and input queued switches." In *Proc. of IEEE INFOCOM*, pages 533–539, New York, April 1998.

[184] Fouad A. Tobagi. "Fast packet switch architectures for broadband integrated services digital networks." *Proc. of the IEEE*, 78(1):137–167, Jan. 1990.

[185] Brian Towles and William J. Dally. "Worst-case traffic for oblivious routing functions." In *Proc. of the Symposium on Parallel Algorithms and Architectures (SPAA)*, pages 1–8, Winnipeg, Manitoba, Canada, Aug. 2002.

[186] Anjan K. V. and Timothy Mark Pinkston. "An efficient, fully adaptive deadlock recovery scheme: DISHA." In *Proc. of the International Symposium on Computer Architecture (ISCA)*, pages 201–210, 1997.

[187] L. G. Valiant and G. J. Brebner. "Universal schemes for parallel communication." In *Proc. of the ACM Symposium of the Theory of Computing*, pages 263–277, Milwaukee, Minn., 1981.

[188] J. van Leeuwen and R. B. Tan. "Interval routing." *The Computer Journal*, 30(4): 298–307, 1987.

[189] Thorsten von Eicken, David E. Culler, Seth Copen Goldstein, and Klaus Erik Schauser. "Active messages: A mechanism for integrated communication and computation." In *Proc. of the International Symposium on Computer Architecture (ISCA)*, pages 256–266, Gold Coast, Australia, 1992.

[190] Abraham Waksman. "A permutation network." *Journal of the ACM*, 15(1): 159–163, Jan. 1968.

[191] Duncan J. Watts and Steven H. Strogatz. "Collective dyanmics of 'small-world' networks." *Nature*, 393:440–442, June 1998.

[192] C. Whitby-Strevens. "The transputer." In *Proc. of the International Symposium on Computer Architecture (ISCA)*, pages 292–300, June 1985.

[193] Drew Wingard. "MicroNetwork-based integration for SOCs." In *Proc. of the Design Automation Conference*, pages 673–677, Las Vegas, June 2001.

[194] Steven Cameron Woo, Moriyoshi Ohara, Evan Torrie, Jaswinder Pal Singh, and Anoop Gupta. "The SPLASH-2 programs: Characterization and methodological considerations." In *Proc. of the International Symposium on Computer Architecture (ISCA)*, pages 24–36, June 1995.

[195] C.W. Wu and T. Feng. "On a class of multistage interconnection networks." *IEEE Transactions on Computers*, 29(8):694–702, August 1980.

[196] Yuanyuan Yang and Gerald M. Masson. "Nonblocking broadcast switching networks." *IEEE Transactions on Computers*, 40(9):1005–1015, Sept. 1991.

[197] Kenneth Yeager. The MIPS R10000 superscalar microprocessor. *IEEE Micro*, 16(2):28–40, April 1996.

[198] Ki Hwan Yum, Eun Jung Kim, Chinta R. Das, and Aniruddha S. Vaidya. "MediaWorm: a QoS capable router architecture for clusters." *IEEE Transactions on Parallel and Distributed Systems*, 13(12):1261–1274, Dec. 2002.

[199] Robert C. Zak, Charles E. Leiserson, Bradley C. Kuzmaul, Shaw-Wen Yang, W. Daniel Hillis, David C. Douglas, and David Potter. "Parallel computer system including arrangement for transferring messages from a source processor to selected ones of a plurality of destination processors and combining responses." United States Patent 5,265,207. Nov. 1993.

[200] L. Zhang. "VirtualClock: A new traffic control algorithm for packet-switched networks." *ACM Transactions on Computer Systems*, 9(2):101–124, May 1991.

[201] Daniel Zwillinger, editor. *CRC Standard Mathematical Tables and Formulae*. Chemical Rubber Company Press, 30th edition, 1995.

[202] Will E. Leland, Murad S. Taqque, Walter Willinger, and Daniel V. Wilson. "On the self-similar nature of Ethernet traffic." *IEEE/ACM Transactions on Networking*, 2(1): 1-15, Feb. 1994.

Index

Topology

$$\Theta_{\text{ideal}} = \frac{b}{\gamma_{\text{max}}}$$

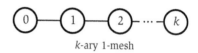

$$T_0 = \underbrace{H_{\text{min}}t_r}_{\text{router}} + \underbrace{\frac{L}{b}}_{\text{serialization}} + \underbrace{\frac{D_{\text{min}}}{v}}_{\text{wire}}$$

$$\gamma_{\text{max}} \geq \frac{NH_{\text{min}}}{C}$$

$$\gamma_{\text{max}} \geq \frac{N}{2B_C}$$

hop count bound   bisection bound

---

k-ary n-cube (Torus)

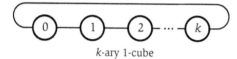

k-ary 1-cube

$$H_{\text{min}} = \begin{cases} \frac{nk}{4} & k \text{ even} \\ n\left(\frac{k}{4} - \frac{1}{4k}\right) & k \text{ odd} \end{cases}$$

$$\gamma_{\text{max}} = \begin{cases} \frac{k}{8} & k \text{ even} \\ \frac{k}{8} - \frac{1}{8k} & k \text{ odd} \end{cases}$$

uniform traffic

$$B_C = 4k^{n-1} = \frac{4N}{k} \qquad \delta = 4n$$

Two-level packaging (k even)

$$w \leq \min\left(\frac{W_n}{4n}, \frac{kW_s}{4N}\right) \qquad b = wf$$

$$T_s = \frac{L}{b} \qquad \Theta_{\text{ideal}} = \frac{8b}{k}$$

k-ary n-mesh (Mesh)

k-ary 1-mesh

$$H_{\text{min}} = \begin{cases} \frac{nk}{4} & k \text{ even} \\ n\left(\frac{k}{3} - \frac{1}{3k}\right) & k \text{ odd} \end{cases}$$

$$\gamma_{\text{max}} = \begin{cases} \frac{k}{4} & k \text{ even} \\ \frac{k}{4} - \frac{1}{4k} & k \text{ odd} \end{cases}$$

uniform traffic

$$B_C = 2k^{n-1} = \frac{2N}{k} \qquad \delta = 4n$$

Two-level packaging (k even)

$$w \leq \min\left(\frac{W_n}{4n}, \frac{kW_s}{2N}\right) \qquad b = wf$$

$$T_s = \frac{L}{b} \qquad \Theta_{\text{ideal}} = \frac{4b}{k}$$

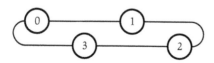

Folding to eliminate long channels
from a torus layout.

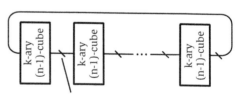

k^{n-1} channels

Recursive construction of higher
dimensional k-ary n-cubes.

k-ary n-fly (Butterfly)

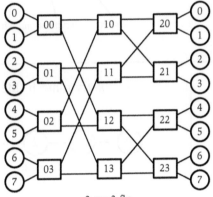

2-ary 3-fly

Clos

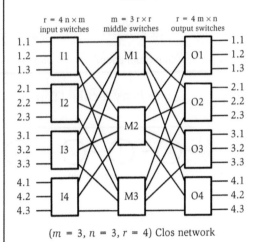

$(m = 3, n = 3, r = 4)$ Clos network

$$\underbrace{\gamma_{\max} = 1}_{\text{uniform traffic}} \qquad \underbrace{\gamma_{\max,\text{wc}} = \sqrt{N}}_{\text{worst-case } (n \text{ even})}$$

$$H_{\min} = n+1 \qquad B_C = \frac{N}{2} \qquad \delta = 2k$$

Two-level packaging:

$$w \le \min\left(\frac{W_n}{2k}, \frac{2W_s}{N}\right) \qquad b = wf$$

$$T_s = \frac{L}{b} \qquad \Theta_{\text{ideal}} = b$$

Strictly non-blocking if $m \ge 2n - 1$

Rearrangeably non-blocking if $m \ge n$

Non-blocking for fanout f multicast if

$$f \le \frac{m(m-n)}{m(n-1)}$$

Channel Slicing

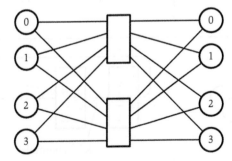

Channel sliced 4-ary 1-fly

Channel slicing a butterfly with slicing factor x:

$$n' = \frac{n}{1 + \log_k x}$$

$$T_s = \frac{xL}{b}$$

$$T_h = t_r \left(\frac{n}{1 + \log_k x}\right)$$

Units of Resource Allocation

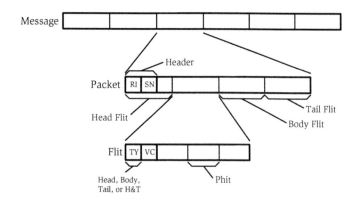

Message

Packet — Header — RI | SN | | | |
Head Flit
Tail Flit
Body Flit

Flit | TY | VC | | |
Head, Body, Tail, or H&T
Phit

Store-and-forward Flow Control

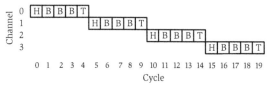

Channel
0 H B B B T
1   H B B B T
2   H B B B T
3   H B B B T

0 1 2 3 4 5 6 7 8 9 10 11 12 13 14 15 16 17 18 19
Cycle

$$T_0 = H\left(t_r + \frac{L}{b}\right)$$

Cut-through / Wormhole FC

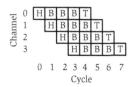

Channel
0 H B B B T
1   H B B B T
2   H B B B T
3   H B B B T

0 1 2 3 4 5 6 7
Cycle

$$T_0 = Ht_r + \frac{L}{b}$$

Credit-based Flow Control

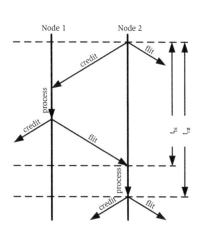

$$F \geq \frac{t_{\mathrm{crt}}b}{L_f}$$

On/Off Flow Control

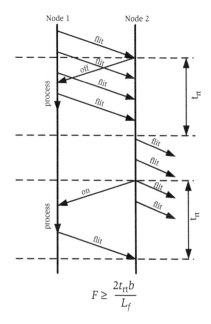

$$F \geq \frac{2t_{\mathrm{rt}}b}{L_f}$$

An Input Queued, Virtual Channel Router

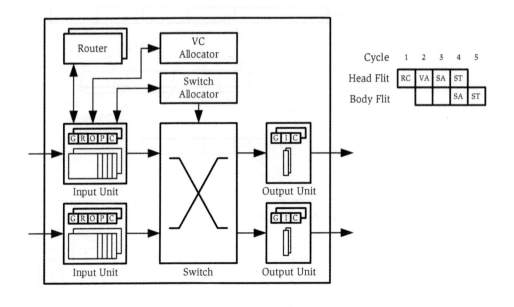

Latency Versus Offered Traffic

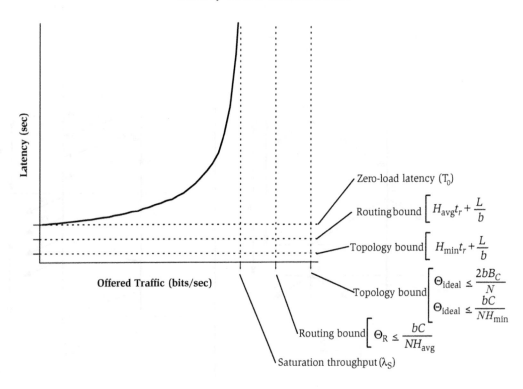

Printed and bound by CPI Group (UK) Ltd, Croydon, CR0 4YY

03/10/2024

01040320-0007